(*cardplate continues in back of book*)

Categorical Data Analysis

Categorical Data Analysis

ALAN AGRESTI

University of Florida
Gainesville, Florida

WILEY

A Wiley-Interscience Publication

JOHN WILEY & SONS

New York • Chichester • Brisbane • Toronto • Singapore

Library of Congress Cataloging-in-Publication Data:

Agresti, Alan.
  Categorical data analysis / Alan Agresti.

    p. cm. -- (Wiley series in probability and mathematical
statistics. Applied probability and statistics)
    "A Wiley-Interscience publication."
    Includes bibliographical references.
    1. Multivariate analysis. I. Title. II. Series.

  QA278.A353 1990
  519.5'35--dc20   89-22645
  ISBN 0-471-85301-1   CIP

Printed and bound in the United States of America

10 9 8 7

To Jacki

Preface

The past quarter century has seen an explosion in the development of methods for analyzing categorical data. This book summarizes those methods. It gives special emphasis to loglinear and logit modeling techniques, which share many features with linear model methods for continuous variables. It also examines contributions of noted statisticians, such as Pearson, Yule, Fisher, Neyman, and Cochran, whose pioneering efforts—and sometimes vocal debates—broke the ground for this evolution.

In response to this rapid development, most statistics and biostatistics departments have introduced courses in categorical data analysis. This book can be used as a text for such courses. The material in Chapters 1–7 forms the heart of most courses. Chapters 1–3 cover traditional methods for two-way contingency tables, and Chapters 4–7 introduce basic loglinear and logit models for two-way and multiway tables. Courses having applied emphasis can also cover material in Chapters 8–11, which present applications and generalizations of these models. Courses that emphasize theory can also cover Chapters 12 and 13, which present the theoretical foundations for modeling categorical data.

I intend this book to be accessible to the diverse mix of students who commonly take courses in categorical data analysis. But I have also written this book with practicing statisticians and biostatisticians in mind. Many statisticians (myself included) were unable to study this subject during their educational training. I hope this book enables them to "catch up" with recent advances.

The development of new methods has influenced—and been influenced by—the increasing availability of multivariate data sets with categorical responses in the social, behavioral, and biomedical sciences, as well as in public health, ecology, education, marketing, food science, and industrial quality control. And so, although this book is directed mainly to statisticians and biostatisticians, I also intend it to be helpful to users of statistical methods in these fields.

Readers should possess a background that includes regression and analysis of variance models, as well as maximum likelihood methods of statistical theory. Readers not having the theory background should be able to follow most methodological discussions, since theoretical derivations are relegated primarily to the final two chapters.

This book is much more comprehensive than my book *Analysis of Ordinal Categorical Data*, which focused on methods for variables having ordered categories. For this text, I have selected a wide range of topics for a reader's first exposure to categorical data analysis. Special features include:

1. A unified presentation of models for categorical (or continuous) responses as special cases of generalized linear models.
2. Two chapters on methods for repeated measurement data, increasingly important in biomedical applications.
3. Two theoretical chapters outlining derivations of basic asymptotic and fixed-sample-size inferences.
4. Discussion of exact small-sample procedures.
5. Historical perspective on the evolution of methods.
6. Prescriptions for how ordinal variables should be treated differently from nominal variables.
7. More than 40 examples of analyses of "real" data sets.
8. More than 400 exercises at the end of the chapters, some directed towards theory and methods and some towards applications and data analysis.
9. An appendix that describes, by chapter, computer software currently available for performing analyses presented in this book.
10. Notes at the end of each chapter that provide references for many topics not covered in the text.

Use of sophisticated methods for categorical data analysis usually requires computers. The appendix that describes computer software also gives several examples of the use of SAS, GLIM, SPSS^X, and BMDP. Readers may wish to refer to this appendix while reading the text, as an aid to implementing the methods in practice.

I thank those individuals who commented on parts of the manuscript. These include Mark Becker, James Booth, Kathryn Chaloner, Clifford Clogg, Bent Jørgensen, Joe Lang, Stuart Lipsitz, William Louv, William Stanish, and two anonymous reviewers. I thank Clifford Clogg, Alfred

DeMaris, Lisa Doyle, Stuart Lipsitz, Clint Moore, Mary Moore, Jane Pendergast, Aparna Raychaudhuri, Louise Ryan, Divakar Sharma, Laurel Smith, Stephen Stigler, James Ware, and Marvin Zelen for suggesting examples or volunteering data sets for the text. I also appreciate specific helpful comments from R. L. Anderson, Frederick Mosteller, C. R. Rao, and students at the University of Florida and Boston University.

Special thanks to Ralph D'Agostino and Marvin Zelen for invitations to visit Boston University and Harvard University during a sabbatical period in which much of this work was done. I also owe thanks to authors of other books on categorical data analysis from which I have learned much; in particular, books by Bishop et al. (1975) and Plackett (1981) have strongly influenced me over the years. Finally, thanks to Jacki Levine for encouragement and help with editing.

ALAN AGRESTI

Gainesville, Florida
February 1990

Contents

CHAPTER 1

Introduction

Statistical methodology for categorical data has only recently reached the level of sophistication achieved early in this century by methodology for continuous data. Regression methods for continuous variables developed rapidly after Francis Galton's breakthroughs in the 1880s. The strong influence of R. A. Fisher, G. Udny Yule and other statisticians on experimentation in agriculture and biological sciences ensured widespread adoption of regression and ANOVA models by the mid-twentieth century. On the other hand, despite influential articles at the turn of the century by Karl Pearson and Yule on association between categorical variables, there was little subsequent work on discrete-response models. Many important figures in the history of statistics, including R. A. Fisher, Jerzy Neyman, William Cochran, and Maurice Bartlett, made important contributions to the categorical-data literature, but analogs of regression models for categorical responses received scant attention until the past quarter century.

Most methodology discussed in this text has been developed since 1963. The recent development of methods for categorical data was stimulated by the increasing methodological sophistication of the social and biomedical sciences. Categorical scales are pervasive in the social sciences for measuring attitudes and opinions on various issues and demographic characteristics such as gender, race, and social class. Categorical scales are used in biomedical sciences to measure such factors as severity of an injury, degree of recovery from surgery, and stage of a disease. Statisticians developed regression-type models for categorical responses to meet the need for analyses of multivariate discrete data sets. Not surprisingly, this progress was fostered by statisticians having ties to the social sciences (such as Leo Goodman, Shelby Haberman, Frederick Mosteller, and Stephen Fienberg) or to the biomedical sciences (such as Joseph Berkson, Jerome Cornfield, and Gary Koch).

1

In describing methods for analyzing categorical data, this book summarizes the early bivariate-association work and the more recent work on multivariate model-building. The primary emphasis is on the latter topic.

Before outlining the topics covered, we describe the major types of categorical data.

1.1 CATEGORICAL RESPONSE DATA

A *categorical* variable is one for which the measurement scale consists of a set of categories. For instance, political philosophy may be measured as "liberal," "moderate," or "conservative"; smoking status might be measured using categories "never smoked," "former smoker," and "current smoker"; and recovery from an operation might be rated as "completely recovered," "nearly recovered," "only somewhat recovered," and "not at all recovered."

Though categorical scales are common in the social and biomedical sciences, they are by no means restricted to those areas. They occur frequently in the behavioral sciences, public health, ecology, education, and marketing. They even occur in highly quantitative fields such as engineering sciences and industrial quality control. Such applications often involve subjective evaluation of some characteristic—how soft to the touch a certain fabric is, how good a particular food product tastes, or how easy a worker finds a certain task to be.

There are many types of categorical variables. This section describes several ways of classifying them.

1.1.1 Response/Explanatory Distinction

Most statistical analyses distinguish between *response* (or "dependent") variables and *explanatory* (or "independent") variables. For instance, regression models describe how the distribution of a continuous response, such as survival time following a heart-transplant operation, changes according to levels of explanatory variables, such as age and cholesterol level.

This text is concerned with the analysis of categorical response variables. For some methods, the explanatory variables must also be categorical; for others, they can be continuous or categorical.

1.1.2 Measurement Scale Distinction

Categorical variables for which levels do not have a natural ordering are called *nominal*. Examples of nominal variables are religious affiliation (categories Catholic, Jewish, Protestant, other), mode of transportation (automobile, bus, subway, bicycle, other), choice of residence (house, apartment, condominium, other), race, gender, and marital status. For nominal variables, the order of listing of the categories is irrelevant to the statistical analysis.

Many categorical variables *do* have ordered levels. Such variables are called *ordinal*. Examples of ordinal variables are size of automobile (subcompact, compact, mid-size, large), social class (upper, middle, lower), attitude toward legalization of abortion (strongly disapprove, disapprove, approve, strongly approve), appraisal of company's inventory level (too low, about right, too high), and diagnosis of whether patient has multiple sclerosis (certain, probable, unlikely, definitely not). Ordinal variables clearly order the categories, but absolute distances between categories are unknown. While we can conclude that a person categorized as "moderate" is more liberal than a person categorized as "conservative," we cannot give a numerical value for *how much more* liberal that person is.

An *interval* variable is one that *does* have numerical distances between any two levels of the scale. For example, blood pressure level, functional life length of television set, length of prison term, income, and age are interval variables.

In the measurement hierarchy, interval variables are highest, ordinal variables are next, and nominal variables are lowest. Statistical methods designed for variables of one type can also be used with variables at higher levels, but not at lower levels. For instance, statistical methods for ordinal variables can also be used with interval variables (by using only the ordering of levels and not their distances); they cannot be used with nominal variables, since categories of such variables have no meaningful ordering. Normally, it is best to apply methods appropriate for the actual scale.

The way a characteristic is measured determines how it can be used in statistical methods. For example, the variable "education" is only nominal when measured by types of education, such as public school or private school; it is ordinal when measured by levels of education, such as grammar school, high school, college, and post-graduate; and it is interval when measured by number of years of education, using the integer values 0, 1, 2, ...

Since this book deals with categorical variables, it is primarily useful for the analysis of nominal and ordinal responses. The methods also apply to interval variables having a small number of distinct values (e.g., number of times married) or for which the values are grouped into ordered categories (e.g., education, measured as < 10 years, 10-12 years, > 12 years).

1.1.3   Continuous/Discrete Distinction

Variables are classified as *continuous* or *discrete*, according to the number of values they can attain. Actual measurement of all variables occurs in a discrete manner, due to limitations in measuring instruments. The continuous/discrete distinction is, in practice, a distinction between variables that can take on lots of values and variables that take on relatively few values. For instance, statisticians often treat discrete interval variables that can assume a large number of values (such as test scores) as if they were continuous, using them in regression models and other methods that assume continuous responses.

This book deals with certain types of discretely-measured responses—namely, (1) nominal variables, (2) ordinal variables, (3) discrete interval variables having relatively few values, and (4) continuous interval variables that are grouped into a small number of categories.

1.1.4   Quantitative/Qualitative Distinction

Nominal variables are *qualitative*—distinct levels differ in quality, not in quantity. Interval variables are *quantitative*—distinct levels have differing amounts of the characteristic of interest.

The position of ordinal variables on the quantitative/qualitative classification is fuzzy. They are often treated as qualitative, being analyzed using methods for nominal variables. But in many respects, ordinal variables more closely resemble interval variables than nominal variables. They possess important quantitative features: Each level has a *greater* or *smaller* magnitude of the characteristic than another level; and, though not often possible to measure, there is usually an underlying continuous variable present. The racial prejudice classification (none, low, high) is a crude measurement of an inherently continuous characteristic.

Though not strictly permissible, statisticians often take advantage of the quantitative nature of ordinal variables by assigning numerical scores to categories. The purpose of the scoring process might be to approximate relative distances for an underlying continuous scale. This requires good judgment and guidance from researchers who use the scale, but can lead to benefits in the variety of methods available for data analysis.

To denote categorical variables in model formulas, we will generally use letters at the end of the alphabet (e.g., X, Y, Z) for response variables. When there is a single response variable, it is denoted by Y, and the explanatory variables are then denoted by letters at the beginning of the alphabet (A, B, C) if they are qualitative and by letters at the end of the alphabet (e.g., X) if they are quantitative.

1.2 ORGANIZATION OF THIS BOOK

This book discusses models that provide a framework for descriptive and inferential analyses of categorical response variables. These models closely resemble regression models for continuous response variables, but they assume binomial, multinomial, or Poisson response distributions, rather than normal. We present two types of models in detail, *logit* and *loglinear* models. Logit models are used with binomial or multinomial responses, whereas loglinear models are used with Poisson responses. Many equivalences exist between these two types of models.

The book has four main units. The first, consisting of Chapters 2 and 3, gives descriptive and inferential methods for bivariate categorical data. These chapters introduce basic measures of association as well as classic chi-squared tests. This first unit summarizes the non-model-based methods developed prior to about 1960.

The second unit, Chapters 4–7, develops the basics of model building. Chapter 4 describes a class of generalized linear models that has loglinear and logit models as special cases. That chapter focuses on models for binary response variables, with primary emphasis on logit models. Chapter 5 introduces concepts used in analyzing multivariate categorical data, and shows how to represent association patterns by loglinear models. Chapter 6 gives the mechanics for fitting loglinear and logit models to categorical data, using the maximum likelihood approach. Chapter 7 discusses topics related to model-building, such as strategies for model selection, model diagnostics, and sample size and power considerations.

The third unit, Chapters 8–11, discusses applications and generalizations of these models. Chapter 8 shows how loglinear and logit models can efficiently utilize ordinal information. Chapter 9 discusses alternative models for response variables having more than two response categories. Generalized forms of logits receive special attention. Chapter 10 presents models for dependent samples, which occur when we measure a categorical response for matched pairs or for the same subjects at two separate occasions. Chapter 11 gives models applicable to more general forms of repeated categorical data, such as longitudinal data from several occasions.

The fourth and final unit is more theoretical. Chapter 12 develops asymptotic theory for categorical data models. This theory is the basis for gauging large-sample behavior of model parameter estimators and goodness-of-fit statistics. Chapter 13 discusses methods of estimation for categorical data analysis. Maximum likelihood estimation receives primary attention.

For categorical data analyses, it is important to link statistical theory and methodology to computational capabilities. Many methods require extensive computations, and familiarity with a statistical computer package is important for anyone planning to use them effectively. The Appendix summarizes programs that can perform the analyses described in this book, and gives examples of their use.

CHAPTER NOTES

Section 1.1: Categorical Response Data

1.1 There was *some* early work on models for categorical response data. For instance, Stigler (1986, p. 246) described work by Fechner in 1860 that influenced Bliss, Finney, and others in the development of the probit model.

1.2 Identification of scales of measurement originated in the psychology literature (Stevens 1951). Alternative scales can result from mixtures of these types. For instance, *partially ordered* scales occur when subjects respond to questions having categories ordered except for a "don't know" or "undecided" category.

PROBLEMS

1.1 In the following examples, distinguish between response and explanatory variables.

    a. Attitude toward abortion on demand (favor, oppose); gender (male, female).

    b. Heart disease (yes, no); cholesterol level.

    c. Race (white, nonwhite); gender (male, female); vote for President (Republican, Democrat, Other); income.

    d. Hospital (A, B); treatment ($T1$, $T2$); patient outcome (survive, die).

1.2 What scale of measurement is most appropriate for the following variables?

    a. Political party affiliation (Democrat, Republican, other)

    b. Highest degree obtained (none, high school, bachelor's, master's, doctorate)

    c. Patient condition (good, fair, serious, critical)

    d. Patient survival (in number of months)

    e. Location of hospital in which data collected (London, Boston, Rochester)

1.3 Describe a potential study that would have a categorical response variable. List variables that would be important in the study, distinguishing between response and explanatory variables. For each variable, identify the measurement scale, and indicate whether it would be treated as continuous or discrete.

1.4 This book discusses statistical models for *categorical* response variables. By contrast, what type of statistical model is appropriate when the response variable is *continuous*

    a. and the explanatory variables are continuous?

    b. and the explanatory variables are categorical?

    c. and the explanatory variables are a mixture of categorical and continuous?

CHAPTER 2

Describing Two-Way Contingency Tables

Chapters 2 and 3 deal with bivariate categorical relationships. This chapter introduces parameters that describe the *population* association between two categorical variables. Chapter 3 shows how to use *sample* data to make statistical inferences about these parameters.

Categorical data consist of frequency counts of observations occurring in the response categories. For categorical variables having only two levels, Section 2.2 introduces odds ratios, differences of proportions, and ratios of proportions. The odds ratio has special importance, serving as a building block for models discussed in this text. Section 2.3 presents summary measures of nominal and ordinal association for variables having more than two categories. The final section presents a historical overview. First, Section 2.1 introduces basic terminology and notation.

2.1 TABLE STRUCTURE FOR TWO DIMENSIONS

A bivariate relationship is defined by the joint distribution of the two associated random variables. The joint distribution determines the marginal and conditional distributions. Simplification occurs in a joint distribution when the component random variables are statistically independent. This section discusses these topics for categorical variables.

2.1.1 Contingency Tables

Let X and Y denote two categorical response variables, X having I levels and Y having J levels. When we classify subjects on both variables, there are IJ possible combinations of classifications. The responses (X, Y) of a

8

subject randomly chosen from some population have a probability distribution. We display this distribution in a rectangular table having I rows for the categories of X and J columns for the categories of Y. The *cells* of the table represent the IJ possible outcomes. Their probabilities are $\{\pi_{ij}\}$, where π_{ij} denotes the probability that (X, Y) falls in the cell in row i and column j. When the cells contain frequency counts of outcomes, the table is called a *contingency* table, a term introduced by Karl Pearson (1904). Another name is *cross-classification* table. A contingency table having I rows and J columns is referred to as an *I-by-J* (or $I \times J$) table.

The probability distribution $\{\pi_{ij}\}$ is the *joint distribution* of X and Y. The *marginal distributions* are the row and column totals obtained by summing the joint probabilities. These are denoted by $\{\pi_{i+}\}$ for the row variable and $\{\pi_{+j}\}$ for the column variable, where the subscript "$+$" denotes the sum over the index it replaces; that is,

$$\pi_{i+} = \sum_{j} \pi_{ij} \quad \text{and} \quad \pi_{+j} = \sum_{i} \pi_{ij} \, ,$$

which satisfy $\Sigma_i \pi_{i+} = \Sigma_j \pi_{+j} = \Sigma_i \Sigma_j \pi_{ij} = 1.0$. The marginal distributions are single-variable information, and do not pertain to association linkages between the variables.

In many contingency tables, one variable (say, Y) is a response variable and the other (X) is an explanatory variable. When X is fixed rather than random, the notion of a joint distribution for X and Y is no longer meaningful. However, for a fixed level of X, Y has a probability distribution. It is germane to study how this probability distribution of Y changes as the level of X changes. Given that a subject is classified in row i of X, let $\pi_{j|i}$ denote the probability of classification in column j of Y, $j = 1, \ldots, J$, where $\Sigma_j \pi_{j|i} = 1$. The probabilities $\{\pi_{1|i}, \ldots, \pi_{J|i}\}$ form the *conditional distribution* of Y at level i of X.

A principal aim of many studies is to compare the conditional distribution of Y at various levels of explanatory variables. For ordinal response variables, it is informative to do this using the cumulative distribution function (cdf). The conditional cdf

$$F_{j|i} = \sum_{b \leq j} \pi_{b|i}, \quad j = 1, \ldots, J \tag{2.1}$$

equals the probability of classification in one of the first j columns, given classification in row i. Suppose that for two rows h and i,

$$F_{j|h} \leq F_{j|i} \quad \text{for } j = 1, \ldots, J \, .$$

The conditional distribution in row h is then *stochastically higher* than the one in row i. This means that row h is more likely than row i to have observations at the high end of the ordinal scale. Figure 2.1 illustrates stochastic orderings for underlying continuous probability density functions corresponding to $\{\pi_{j|i}\}$ and for underlying continuous cdfs corresponding to $\{F_{j|i}\}$.

2.1.2 Independence

When both variables are response variables, we can describe the association using their joint distribution, the conditional distribution of Y given X, or the conditional distribution of X given Y. The conditional distribution of Y given X is related to the joint distribution by

$$\pi_{j|i} = \pi_{ij}/\pi_{i+} \quad \text{for all } i \text{ and } j .$$

The variables are statistically *independent* if all joint probabilities equal the product of their marginal probabilities; that is, if

$$\pi_{ij} = \pi_{i+} \pi_{+j} \quad \text{for } i = 1, \ldots, I \quad \text{and} \quad j = 1, \ldots, J . \qquad (2.2)$$

When X and Y are independent,

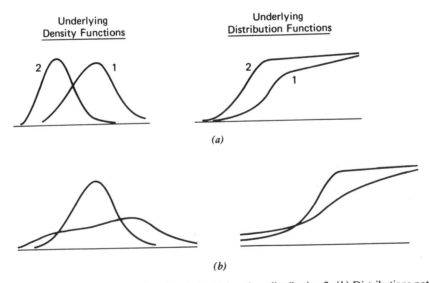

(a)

(b)

Figure 2.1 (a) Distribution 1 stochastically higher than distribution 2. (b) Distributions not stochastically ordered.

$$\pi_{j \mid i} = \pi_{ij} / \pi_{i+} = (\pi_{i+} \pi_{+j}) / \pi_{i+} = \pi_{+j} \quad \text{for } i = 1, \dots, I.$$

Each conditional distribution of Y is identical to the marginal distribution of Y. Thus, two variables are independent when the probability of column response j is the same in each row, for $j = 1, \dots, J$. When Y is a response and X is an explanatory variable, the condition $\{\pi_{j \mid 1} = \cdots = \pi_{j \mid I}$ for all $j\}$ provides a more natural definition of independence than (2.2).

Table 2.1 displays notation for joint, marginal, and conditional distributions for the 2×2 case. We use similar notation for sample distributions, with the letter p in place of π. For instance, $\{p_{ij}\}$ denotes the sample joint distribution in a contingency table. The cell frequencies are denoted by $\{n_{ij}\}$, with $n = \Sigma_i \Sigma_j n_{ij}$ being the total sample size, so

$$p_{ij} = n_{ij} / n .$$

The proportion of times that subjects in row i made response j is

$$p_{j \mid i} = p_{ij} / p_{i+} = n_{ij} / n_{i+}$$

where $n_{i+} = np_{i+} = \Sigma_j n_{ij}$.

2.1.3 Oral Contraceptive Example

Table 2.2 is an example of a 2×2 contingency table. These data are taken from a study by Mann et al. (1975) that investigated the effect of oral contraceptive use on the likelihood of heart attacks. The 58 subjects in the first column represent married women under 45 years of age

Table 2.1 Notation for Joint, Conditional, and Marginal Probabilities

Row	Column 1	Column 2	Total
1	π_{11} $(\pi_{1 \mid 1})$	π_{12} $(\pi_{2 \mid 1})$	π_{1+} (1.0)
2	π_{21} $(\pi_{1 \mid 2})$	π_{22} $(\pi_{2 \mid 2})$	π_{2+} (1.0)
Total	π_{+1}	π_{+2}	1.0

Table 2.2   Cross-Classification of Oral Contraceptive Practice by Myocardial Infarction

	Myocardial Infarction	
Oral Contraceptive Practice	Yes	No
Used	23	34
Never used	35	132
Total	58	166

Reprinted with permission from Mann et al. (1975).

treated for myocardial infarction in two hospital regions in England and Wales during 1968–1972. Each case was matched with three control patients in the same hospitals who were not being treated for myocardial infarction. All subjects were then asked whether they had ever used oral contraceptives.

We would normally regard myocardial infarction as a response variable and oral contraceptive practice as an explanatory variable. In this study, however, the marginal distribution of myocardial infarction is fixed by the sampling design, and the outcome measured for each subject is whether she ever used oral contraceptives. The study, which uses a *retrospective* design in which we "look into the past," is called a *case-control* study. Such studies are common in health-related applications, for instance to ensure a sufficiently large sample of subjects having the disease studied.

We might wish to compare women who used oral contraceptives with those who never used them in terms of the proportion who suffered myocardial infarction. These proportions refer to the conditional distribution of myocardial infarction, given oral contraceptive practice. We can compute proportions in the reverse direction, for the conditional distribution of oral contraceptive practice, given myocardial infarction status. For women suffering myocardial infarction, the proportion who had ever practiced oral contraception was $23/58 = 0.397$, while it was $34/166 = 0.205$ for women who had not suffered myocardial infarction.

When we know the proportion of the population who suffered myocardial infarction, then we can use Bayes Theorem to compute sample conditional distributions in the direction of main interest (see Problem 2.6). Otherwise we cannot estimate proportions suffering myocardial infarction at each level of oral contraceptive practice, using a retrospective sample. For Table 2.2, we do not know the population prevalence of myocardial infarction, and patients suffering myocardial infarction were probably sampled at a rate far in excess of their occurrence in the general population.

By contrast, imagine a study that samples subjects from the population of newly married women under 45. After 20 years, we observe the rates of myocardial infarction for the two groups. Such a sampling design is *prospective*. There are two types of prospective studies. In *clinical trials*, we randomly allocate subjects to the groups who will use or not use oral contraceptives. In *cohort* studies, the women make their own choice about whether to use contraceptives, and we simply observe in future time who suffers myocardial infarction. Yet another approach, a *cross-sectional* design, samples women and classifies them simultaneously on both variables. Case-control, cohort, and cross-sectional studies are called *observational* studies. We simply observe who chooses each group and who has the outcome of interest. By contrast, a clinical trial is *experimental*, the investigator having control over which subjects receive each treatment. Fleiss (1981) and Hennekens and During (1987) discussed advantages and disadvantages of the various designs.

2.2 WAYS OF COMPARING PROPORTIONS

Response variables having two categories are called *binary*. For instance, outcome of a medical treatment (cure, no cure) is binary. Studies often compare several groups on a binary response, Y. When there are I groups, we can display results in an $I \times 2$ contingency table, in which columns are the levels of Y. This section presents descriptive measures for comparing groups on binary responses, and shows extensions when Y has $J > 2$ responses.

2.2.1 Difference of Proportions

For subjects in row i, $i = 1, \ldots, I$, $\pi_{1|i}$ is the probability of response 1, and $(\pi_{1|i}, \pi_{2|i}) = (\pi_{1|i}, 1 - \pi_{1|i})$ is the conditional distribution of the binary response. We can compare two rows, say h and i, using the *difference of proportions*, $\pi_{1|h} - \pi_{1|i}$. Comparison on response 2 is equivalent to comparison on response 1, since

$$\pi_{2|h} - \pi_{2|i} = (1 - \pi_{1|h}) - (1 - \pi_{1|i}) = \pi_{1|i} - \pi_{1|h} .$$

The difference of proportions falls between -1.0 and $+1.0$. It equals zero when rows h and i have identical conditional distributions. The response Y is statistically independent of the row classification when $\pi_{1|h} - \pi_{1|i} = 0$ for all pairs of rows h and i.

For $I \times J$ contingency tables, we can compare the conditional probabilities of response j for rows h and i using the difference $\pi_{j|h} - \pi_{j|i}$.

The variables are independent when this difference equals zero for all pairs of rows h and i and all possible responses j; equivalently, when the $(I-1)(J-1)$ differences $\pi_{j|i} - \pi_{j|I} = 0$, $i = 1, \ldots, I-1$, $j = 1, \ldots, J-1$.

When both variables are responses and there is a joint distribution $\{\pi_{ij}\}$, the comparison of proportions within rows h and i satisfies

$$\pi_{1|h} - \pi_{1|i} = \pi_{h1}/\pi_{h+} - \pi_{i1}/\pi_{i+} \, . \tag{2.3}$$

For the 2×2 case,

$$P(\text{col } 1 \,|\, \text{row } 1) - P(\text{col } 1 \,|\, \text{row } 2) = \pi_{11}/\pi_{1+} - \pi_{21}/\pi_{2+} \, .$$

We can also compare columns in terms of the proportion of row-1 responses, using the difference of within-column proportions

$$P(\text{row } 1 \,|\, \text{col } 1) - P(\text{row } 1 \,|\, \text{col } 2) = \pi_{11}/\pi_{+1} - \pi_{12}/\pi_{+2} \, .$$

This does not usually give the same value as the difference of within-row proportions.

2.2.2 Relative Risk

A difference in proportions of fixed size may have greater importance when both proportions are close to 0 or 1 than when they are near the middle of the range. For instance, suppose we compare two drugs in terms of the proportion of subjects who suffer bad side effects. The difference between 0.010 and 0.001 may be more noteworthy than the difference between 0.410 and 0.401. In such cases, the ratio of proportions is also a useful descriptive measure.

For 2×2 tables, the *relative risk* is the ratio

$$\pi_{1|1}/\pi_{1|2} \, . \tag{2.4}$$

This ratio can be any nonnegative real number. A relative risk of 1.0 corresponds to independence. Comparison on the second response gives a different relative risk, $\pi_{2|1}/\pi_{2|2} = (1 - \pi_{1|1})/(1 - \pi_{1|2})$.

2.2.3 Odds Ratio

Refer again to the 2×2 table. Within row 1, the *odds* that the response is in column 1 instead of column 2 is defined to be

$$\Omega_1 = \pi_{1|1}/\pi_{2|1} \, .$$

Within row 2, the corresponding odds equals

$$\Omega_2 = \pi_{1|2}/\pi_{2|2} \, .$$

For joint distributions, the equivalent definition is

$$\Omega_i = \pi_{i1}/\pi_{i2} \, , \quad i = 1, 2 \, .$$

Each Ω_i is nonnegative, with value greater than 1.0 when response 1 is more likely than response 2. When $\Omega_1 = 4.0$, in the first row response 1 is four times as likely as response 2. The within-row conditional distributions are identical, and thus the variables are independent, if and only if $\Omega_1 = \Omega_2$.

The ratio of the odds Ω_1 and Ω_2,

$$\theta = \Omega_1/\Omega_2 \tag{2.5}$$

is called the *odds ratio*. From the definition of odds using joint probabilities,

$$\theta = \frac{\pi_{11}/\pi_{12}}{\pi_{21}/\pi_{22}} = \frac{\pi_{11}\pi_{22}}{\pi_{12}\pi_{21}} \, . \tag{2.6}$$

An alternative name for θ is the *cross-product ratio*, since it equals the ratio of the products $\pi_{11}\pi_{22}$ and $\pi_{12}\pi_{21}$ of probabilities from diagonally-opposite cells.

The odds ratio can equal any non negative number. When all cell probabilities are positive, independence of X and Y is equivalent to $\theta = 1$. When $1 < \theta < \infty$, subjects in row 1 are more likely to make the first response than are subjects in row 2; that is, $\pi_{1|1} > \pi_{1|2}$. For instance, when $\theta = 4$, the odds of the first response are four times higher in row 1 than in row 2. (This does not mean that the *probability* $\pi_{1|1}$ is four times higher than $\pi_{1|2}$; that is the interpretation of a *relative risk* of 4.0.) When $0 < \theta < 1$, the first response is less likely in row 1 than in row 2; that is, $\pi_{1|1} < \pi_{1|2}$. When one cell has zero probability, θ equals 0 or ∞.

The odds ratio does not change value when the orientation of the table is reversed so that the rows become the columns and the columns become the rows. Therefore, it is unnecessary to identify one classification as the response variable in order to calculate θ.

Values of θ farther from 1.0 in a given direction represent stronger levels of association. Two values for θ represent the same level of association, but in opposite directions, when one value is the inverse of the other. For instance, when $\theta = 0.25$, the odds of the first response are 0.25 times as high in row 1 as in row 2, or equivalently $1/0.25 = 4.0$ times as high in row 2 as in row 1. When the order of the rows is reversed or the order of the columns is reversed, the new value of θ is the inverse of the original value.

It is sometimes more convenient to use $\log(\theta)$, the natural logarithm of θ. Independence corresponds to $\log(\theta) = 0$. The log odds ratio is symmetric about this value—reversal of rows or of columns results in a change in its sign. Two values for $\log(\theta)$ that are the same except for sign, such as $\log(4) = 1.39$ and $\log(0.25) = -1.39$, represent the same level of association.

For sample cell frequencies $\{n_{ij}\}$, a sample version of θ is $n_{11}n_{22}/n_{12}n_{21}$. The sample odds ratio does not change value when both cell frequencies within any row are multiplied by a nonzero constant, or when both cell frequencies within any column are multiplied by a nonzero constant. The sample versions of the difference of proportions (2.3) and relative risk (2.4) are invariant to multiplication within rows by a constant, but they are not invariant to multiplication within columns or to row/column interchange.

An implication of the multiplicative invariance property is that the sample odds ratio estimates the same characteristic (θ) even when we select disproportionately large or small samples from marginal categories of a variable. For instance, suppose a study investigates the association between vaccination and catching a certain strain of flu. For a retrospective design, the sample odds ratio estimates the same characteristic whether we randomly sample (1) 100 people who got the flu and 100 people who did not, or (2) 150 people who got the flu and 50 people who did not, in each case classifying subjects on whether they took the vaccine. In fact, the odds ratio is equally valid for retrospective, prospective, or cross-sectional sampling designs. We would estimate the same characteristic if (3) we randomly sample 100 people who took the vaccine and 100 people who did not, and then classify them on whether they got the flu, or (4) we randomly sample 200 people and classify them on whether they took the vaccine and whether they got the flu.

2.2.4 Aspirin and Heart Attacks Example

Table 2.3 is taken from a report on the relationship between aspirin use and heart attacks by the Physicians' Health Study Research Group at Harvard Medical School. The Physicians' Health Study is a randomized

Table 2.3 Cross-Classification of Aspirin Use and Myocardial Infarction

	Myocardial Infarction		
	Fatal Attack	Non-Fatal Attack	No Attack
Placebo	18	171	10,845
Aspirin	5	99	10,933

Source: Preliminary Report: Findings from the Aspirin Component of the Ongoing Physicians' Health Study. *N. Engl. J. Med.* **318**: 262–264, 1988.

clinical trial testing whether aspirin taken regularly reduces mortality from cardiovascular disease. Every other day, physicians participating in the study took either one aspirin tablet or a placebo. The study was blind—those in the study did not know which they were taking.

Table 2.3 differentiates between fatal and nonfatal heart attacks, but we combine these outcomes for now. Of the 11,034 physicians taking placebo, there were 189 heart attacks over the course of the study, a proportion of 0.0171. Of the 11,037 taking aspirin, there were 104 heart attacks, a proportion of 0.0094. The sample difference of proportions is 0.0077. The relative risk is $0.0171/0.0094 = 1.82$. The proportion suffering heart attacks was 1.82 times higher for those taking placebo than for those taking aspirin. The sample odds ratio is $(189 \times 10,933)/(10,845 \times 104) = 1.83$.

2.2.5 Relationship Between Odds Ratio and Relative Risk

From definitions (2.4) and (2.5),

$$\text{Odds ratio} = \text{Relative risk}\left(\frac{1 - \pi_{1|2}}{1 - \pi_{1|1}}\right).$$

Their magnitudes are similar whenever the probability of response 1 is close to zero for both groups. We observed this similarity in the previous example. The relative risk must be closer than the odds ratio to the independence value of 1.0.

When the sampling design is retrospective, it is possible to construct conditional distributions within levels of the fixed response. It is usually not possible to estimate the probability of the outcome of interest, or to compute the difference of proportions or relative risk for that outcome.

We can compute the odds ratio, however, since it is determined by the conditional distributions in *either* direction. When the probability of the outcome of interest is very small, the population odds ratio and relative risk take similar values. Thus, we can use the sample odds ratio to provide a rough indication of the relative risk.

For instance, for Table 2.2, the difference of proportions and the relative risk are limited to comparisons of the proportions who used oral contraceptives. Though we cannot compute the relative risk of myocardial infarction, we can compute the odds ratio, which is 2.55. Since the probability of suffering myocardial infarction seems small regardless of oral contraceptive practice, 2.55 is also a rough estimate of the relative risk. We estimate that women who used oral contraceptives were about two and a half times as likely to suffer myocardial infarction as women who did not use them.

2.2.6   Odds Ratios for $I \times J$ Tables

Odds ratios are also useful for describing contingency tables larger than 2×2. Odds ratios for $I \times J$ tables can use each of the $\binom{I}{2} = I(I-1)/2$ pairs of rows in combination with each of the $\binom{J}{2} = J(J-1)/2$ pairs of columns. For rows a and b, and columns c and d, the odds ratio $(\pi_{ac}\pi_{bd})/(\pi_{bc}\pi_{ad})$ uses four cells in a rectangular pattern. There are $\binom{I}{2}\binom{J}{2}$ odds ratios of this type.

This set of odds ratios contains much redundant information. Consider the subset of $(I-1)(J-1)$ local odds ratios

$$\theta_{ij} = \frac{\pi_{i,j}\,\pi_{i+1,\,j+1}}{\pi_{i,j+1}\,\pi_{i+1,j}}\,, \quad i=1,\ldots,I-1\,, j=1,\ldots,J-1\,. \quad (2.7)$$

As shown in Figure 2.2, local odds ratios use cells in adjacent rows and adjacent columns. These $(I-1)(J-1)$ odds ratios determine all $\binom{I}{2}\binom{J}{2}$ odds ratios formed from pairs of rows and pairs of columns.

To illustrate, in Table 2.3, the sample local odds ratio is 2.08 for the first two columns and 1.74 for the second and third columns. In each case, the more serious outcome was more prevalent for the placebo group. The product of these two odds ratios is 3.63, which is the odds ratio for the first and third columns.

The construction (2.7) for a minimal set of odds ratios is not unique. Another basic set is

$$\alpha_{ij} = \frac{\pi_{ij}\,\pi_{IJ}}{\pi_{Ij}\,\pi_{iJ}}\,, \quad i=1,\ldots,I-1\,, j=1,\ldots,J-1\,. \quad (2.8)$$

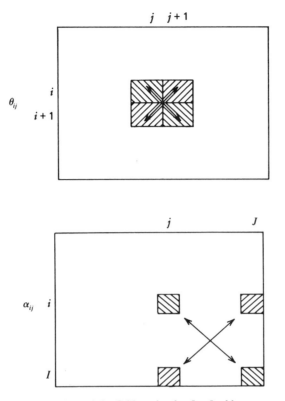

Figure 2.2 Odds ratios for $I \times J$ tables.

Each odds ratio uses the rectangular pattern of cells determined by rows i and I and columns j and J, as indicated in Figure 2.2.

When all cell probabilities are positive, conversion of the probabilities into the set of odds ratios (2.7) or (2.8) does not result in a loss of information. Given the marginal distributions $\{\pi_{i+}\}$ and $\{\pi_{+j}\}$, the cell probabilities are determined by either set of odds ratios (A way of showing this will be outlined in Section 6.7 and Problem 6.39). In this sense, $(I-1)(J-1)$ separate pieces of information can describe the association in an $I \times J$ table.

2.3 SUMMARY MEASURES OF ASSOCIATION

For 2×2 tables, a single number such as the odds ratio can summarize the association. For $I \times J$ tables, it is rarely possible to summarize association by a single number without some loss of information. However, summary indices can describe certain features of the association.

It is sometimes meaningful to reduce a set of odds ratios to a single number. For instance, if sample values of the odds ratios (2.7) are similar in value, we might summarize them by some weighted average. This is related to a model-building approach presented in Section 8.1. For now, we consider a simpler approach which achieved popularity, particularly in the social sciences, before the advent of model building for categorical data.

2.3.1 Measures of Ordinal Association

A basic question researchers usually pose when analyzing ordinal data is "Does Y tend to increase as X increases?" Bivariate analyses of interval-scale variables often summarize covariation by the Pearson correlation, which describes the degree to which Y has a linear relationship with X. Ordinal variables do not have a defined metric, so the notion of linearity is not meaningful. However, the inherent ordering of categories allows consideration of *monotonicity*—for instance, whether Y tends to increase as X does. Measures for ordinal variables that are analogous to the Pearson correlation describe the degree to which the relationship is monotone.

In a strict sense, comparisons of two subjects on an ordinal scale can answer "Which subject makes the higher response?" but not "What is the numerical difference between their responses?" When we observe the ordering of two subjects on each of two variables, we can classify the pair of subjects as *concordant* or *discordant*. The pair is *concordant* if the subject ranking higher on variable X also ranks higher on variable Y. The pair is *discordant* if the subject ranking higher on X ranks lower on Y. The pair is *tied* if the subjects have the same classification on X and/or Y.

2.3.2 Job Satisfaction Example

We illustrate concordance and discordance using Table 2.4, taken from the 1984 General Social Survey of the National Data Program in the United States as quoted by Norušis (1988). The variables are income and job satisfaction. Income has levels less than $6000 (denoted <6), between $6000 and $15,000 (6–15), between $15,000 and $25,000 (15–25), and over $25,000 ($>25$). Job satisfaction has levels very dissatisfied (VD), little dissatisfied (LD), moderately satisfied (MS), and very satisfied (VS). We treat VS as the high end of the job satisfaction scale.

Table 2.4   Cross-Classification of Job Satisfaction by Income

Income (US$)	Job Satisfaction			
	Very Dissatisfied	Little Dissatisfied	Moderately Satisfied	Very Satisfied
<6000	20	24	80	82
6000–15,000	22	38	104	125
15,000–25,000	13	28	81	113
>25,000	7	18	54	92

Source: 1984 General Social Survey; see Norušis (1988).

Consider a pair of subjects, one of whom is classified in the cell (<6, VD) and the other in the cell (6–15, LD). This pair is concordant, since the second subject is ranked higher than the first both on income and on job satisfaction. Each of the 20 subjects in cell (<6, VD) form concordant pairs when matched with each of the 38 subjects classified (6–15, LD), so there are $20 \times 38 = 760$ concordant pairs from these two cells. The 20 subjects in the cell (<6, VD) are also part of a concordant pair when matched with each of the other $(104 + 125 + 28 + 81 + 113 + 18 + 54 + 92)$ subjects ranked higher on both variables. Similarly, the 24 subjects in the cell (<6, LD) cell are part of concordant pairs when matched with the $(104 + 125 + 81 + 113 + 54 + 92)$ subjects ranked higher on both variables.

The total number of concordant pairs, denoted by C, equals

$$
\begin{aligned}
C = {}& 20(38 + 104 + 125 + 28 + 81 + 113 + 18 + 54 + 92) \\
& + 24(104 + 125 + 81 + 113 + 54 + 92) + 80(125 + 113 + 92) \\
& + 22(28 + 81 + 113 + 18 + 54 + 92) + 38(81 + 113 + 54 + 92) \\
& + 104(113 + 92) + 13(18 + 54 + 92) + 28(54 + 92) + 81(92) \\
= {}& 109{,}520 \, .
\end{aligned}
$$

The number of discordant pairs of observations is

$$
\begin{aligned}
D = {}& 24(22 + 13 + 7) + 80(22 + 38 + 13 + 28 + 7 + 18) \\
& + \cdots + 113(7 + 18 + 54) = 84{,}915 \, .
\end{aligned}
$$

In this example, $C > D$ suggests a tendency for low income to occur with low job satisfaction and high income with high job satisfaction.

Consider two independent observations from a joint probability distribution $\{\pi_{ij}\}$ for two ordinal variables. For that pair of observations

$$\Pi_c = 2 \sum_i \sum_j \pi_{ij} \left(\sum_{h>i} \sum_{k>j} \pi_{hk} \right) \text{ and } \Pi_d = 2 \sum_i \sum_j \pi_{ij} \left(\sum_{h>i} \sum_{k<j} \pi_{hk} \right)$$

$$(2.9)$$

are the probabilities of concordance and discordance. In (2.9), i and j are fixed in the inner summations, and the factor of 2 occurs because the first observation could be in cell (i, j) and the second in cell (h, k), or vice versa. Several measures of association for ordinal variables utilize the difference $\Pi_c - \Pi_d$ between these probabilities. For these measures, the association is said to be positive if $\Pi_c - \Pi_d > 0$ and negative if $\Pi_c - \Pi_d < 0$.

2.3.3 Gamma

Given that the pair is untied on both variables, $\Pi_c/(\Pi_c + \Pi_d)$ is the probability of concordance and $\Pi_d/(\Pi_c + \Pi_d)$ is the probability of discordance. The difference between these probabilities is

$$\gamma = \frac{\Pi_c - \Pi_d}{\Pi_c + \Pi_d} \tag{2.10}$$

called *gamma*. The sample version of gamma is $\hat{\gamma} = (C - D)/(C + D)$.

The properties of gamma (and $\hat{\gamma}$) follow directly from its definition. Like the correlation, its range is $-1 \leq \gamma \leq 1$. Whereas the absolute value of the correlation is 1 when the relationship between X and Y is perfectly linear, only monotonicity is required for $|\gamma| = 1$, with $\gamma = 1$ if $\Pi_d = 0$ and $\gamma = -1$ if $\Pi_c = 0$. The perfect association value $|\gamma| = 1$ occurs even when the relationship is not *strictly* monotone. If $\gamma = 1$, for instance, then for observations (X_a, Y_a) and (X_b, Y_b) on a pair of subjects a and b having $X_a < X_b$, it follows that $Y_a \leq Y_b$ but not necessarily that $Y_a < Y_b$.

Table 2.5 illustrates cross-classifications having various values of γ. Independence implies $\gamma = 0$, but the converse is not true. For instance, Table 2.5c shows a U-shaped joint distribution for which $\Pi_c = \Pi_d$ and hence $\gamma = 0$. Like the correlation, gamma treats the variables symmetrically—it is unnecessary to identify one classification as a response variable. A reversal in the category orderings of one variable simply causes a change in the sign of γ.

For the data in Table 2.4 on income and job satisfaction, $C = 109,520$ and $D = 84,915$. Of the concordant and discordant pairs, 56.33% are concordant and 43.67% are discordant. The difference of the corresponding proportions gives $\hat{\gamma} = 0.127$. There is a weak tendency in this sample for job satisfaction to be higher at higher income levels.

Table 2.5 Values of Gamma for Various Cross-Classifications

a. $\gamma = 1$	$\frac{1}{3}$	0	0
	0	$\frac{1}{3}$	0
	0	0	$\frac{1}{3}$
b. $\gamma = 1$	0.2	0	0
	0.2	0.2	0
	0	0.2	0.2
c. $\gamma = 0$	0.2	0	0.2
	0.2	0	0.2
	0	0.2	0
d. $\gamma = -1$	0	0.30	
	0.03	0.67	

For 2×2 tables, γ simplifies to

$$Q = \frac{\pi_{11}\pi_{22} - \pi_{12}\pi_{21}}{\pi_{11}\pi_{22} + \pi_{12}\pi_{21}}. \qquad (2.11)$$

This measure, which Yule (1900, 1912) introduced and called Q in honor of the Belgian statistician Quetelet, is now referred to as *Yule's Q*. It is related to the odds ratio $\theta = (\pi_{11}\pi_{22})/(\pi_{12}\pi_{21})$ by $Q = (\theta - 1)/(\theta + 1)$. Gamma is then a strictly monotone transformation of θ from the $[0, \infty]$ scale onto the $[-1, +1]$ scale.

Gamma was proposed by Goodman and Kruskal (1954). Although it is simple to interpret, other measures based on concordance and discordance can be more useful for certain purposes. These include *Kendall's tau-b* and *Somers' d*. See Note 2.2 and Problem 2.20.

2.3.4 Measures of Nominal Association

When variables in a two-way table are nominal, notions such as positive/negative association and monotonicity are no longer meaningful. It is then more difficult to describe association by a single number, and summary measures are less useful than for ordinal or interval-scale variables.

The most interpretable indices for nominal variables have the same structure as R-squared (the coefficient of determination) for interval variables. R-squared and the more general intraclass correlation coefficient and correlation ratio (Kendall and Stuart, 1979) describe the proportional reduction in variance from the marginal distribution to the conditional distributions of the response.

Let $V(Y)$ denote a measure of variation for the marginal distribution $\{\pi_{+1}, \ldots, \pi_{+J}\}$ of the response Y, and let $V(Y|i)$ denote this measure computed for the conditional distribution $\{\pi_{1|i}, \ldots, \pi_{J|i}\}$ of Y at the ith setting of an explanatory variable X. A proportional reduction in variation measure has form

$$\frac{V(Y) - E[V(Y|X)]}{V(Y)}$$

where $E[(V(Y|X)]$ is the expectation of the conditional variation taken with respect to the distribution of X. When X is a categorical variable having marginal distribution $\{\pi_{1+}, \ldots, \pi_{I+}\}$, $E[V(Y|X)] = \Sigma_i \pi_{i+} V(Y|i)$.

2.3.5 Concentration and Uncertainty Measures

One variation measure for a nominal response is

$$V(Y) = \Sigma \pi_{+j}(1 - \pi_{+j}) = 1 - \Sigma \pi_{+j}^2 .$$

This is the probability that two independent observations from the marginal distribution of Y fall in different categories. The variation takes its minimum value of zero when $\pi_{+j} = 1$ for some j, and its maximum value of $(J - 1)/J$ when $\pi_{+j} = 1/J$ all j. The conditional variation in row i is then

$$V(Y|i) = 1 - \sum_j \pi_{j|i}^2 .$$

For an $I \times J$ contingency table with joint probabilities $\{\pi_{ij}\}$, the average conditional variation is

$$E[V(Y|X)] = 1 - \sum_i \pi_{i+} \sum_j \pi_{j|i}^2 = 1 - \sum\sum \pi_{ij}^2/\pi_{i+} .$$

The proportional reduction in variation is *Goodman and Kruskal's tau*

$$\tau = \frac{\sum_i \sum_j \pi_{ij}^2/\pi_{i+} - \sum_j \pi_{+j}^2}{1 - \sum_j \pi_{+j}^2}$$

also called the *concentration coefficient*.

Goodman and Kruskal (1954) gave τ the following interpretation: Suppose we guess subjects' responses, making guesses randomly according to the marginal response distribution. That is, we guess with probability π_{+1} that the response is category 1, with probability π_{+2} that it is category 2, and so on. This is called a *proportional prediction* rule. Then

$$V(Y) = \sum \pi_{+j}(1 - \pi_{+j})$$

$$= P(\text{guess category } j) \times P(\text{incorrect guess} \,|\, \text{guess category } j)$$

is the probability of an incorrect guess. If we knew a subject were in category i of X, we would use the conditional distribution on Y at that level for the proportional predictions. The probability of an incorrect guess is then $V(Y \,|\, i)$. Averaged over the distribution of X, the conditional probability of an incorrect guess is $E[V(Y \,|\, X)]$. The measure τ is the proportional reduction in the probability of an incorrect guess obtained by making predictions on Y using the classification on X. A large value for τ represents a strong association, in the sense that we can guess Y much better when we know X than when we do not.

Theil (1970) proposed an alternative variation measure $V(Y) = \sum \pi_{+j} \log \pi_{+j}$. For contingency tables, this results in the proportional reduction in variation index

$$U = - \frac{\displaystyle\sum_i \sum_j \pi_{ij} \log(\pi_{ij} / \pi_{i+} \pi_{+j})}{\displaystyle\sum_j \pi_{+j} \log \pi_{+j}} \tag{2.12}$$

called the *uncertainty coefficient*.

The measures τ and U are well defined when more than one $\pi_{+j} > 0$. They take values between 0 and 1; $\tau = U = 0$ is equivalent to independence of X and Y; $\tau = U = 1$ is equivalent to no conditional variation, in the sense that for each i, $\pi_{j \,|\, i} = 1$ for some j. The variation measure used in τ is called the *Gini concentration*, and the variation measure used in U is the *entropy*.

A difficulty with these measures is in determining how large a value constitutes a "strong" association. When the response variable has several possible categorizations, these measures tend to take smaller values as the number of categories increases. For instance, for τ the variation measure is the probability that two independent observations occur in different categories. Often this probability approaches 1.0 for both the conditional and marginal distributions as the number of response categories grows larger, in which case τ decreases toward 0.

Table 2.6 Religious Identification Now and at Age 16

Religious Identification at Age 16	Current Religious Identification				
	Protestant	Catholic	Jewish	None or other	Total
Protestant	918	27	1	70	1016
Catholic	30	351	0	37	418
Jewish	1	1	28	1	31
None or other	29	5	0	25	59
Total	978	384	29	133	1524

Source: 1978 General Social Survey; see Bohrnstedt and Knoke (1982).

2.3.6 Religious Identification Example

Table 2.6, described by Bohrnstedt and Knoke (1982), is taken from the 1978 General Social Survey. Subjects were classified by their current religious identification, and by their religious identification at age 16. When we treat current identification as the response, the sample version of Goodman and Kruskal's tau equals 0.57. The average conditional variation on religious identification is 57% smaller than the marginal variation. The sample version of the uncertainty coefficient is 0.51. There seems to be relatively strong association between religious identification now and at age 16.

In Table 2.6, both variables use the same classification. For square tables of this sort, observations often cluster on the main diagonal. Chapter 10 introduces an alternative measure, *kappa*, and special models that describe the structure of square tables.

2.4 HISTORICAL OVERVIEW

The early literature on categorical data analyses dealt primarily with summary indices of association. The subject sparked heated debate among statisticians such as Karl Pearson and G. Udny Yule about how association should be measured.

Pearson (1904, 1913) envisioned continuous bivariate distributions underlying cross-classification tables. He believed that we should describe association by approximating a measure such as the correlation for that underlying continuum. His *tetrachoric correlation* for a 2×2 table was one such measure. Suppose a bivariate normal density is collapsed to a

2×2 table having the same margins as the observed table. The tetrachoric correlation is the value of the correlation ρ in the normal density that would produce cell probabilities equal to the sample cell proportions. Pearson's *contingency coefficient* (Problem 3.24) was an attempt for $I \times J$ tables to approximate an underlying correlation.

Pearson argued that coefficients advocated by Yule (1900, 1912) for 2×2 tables, namely functions of the odds ratio such as Yule's Q, were unsuitable. He claimed they were unstable when applied to $I \times J$ tables collapsed to 2×2 tables, depending strongly on the way the table was collapsed.

Yule believed it was possible to define meaningful coefficients without assuming anything about underlying continuous distributions. He argued (1912, p. 612) that variables such as (vaccinated, unvaccinated) and (died, survived) were inherently discrete and that "at best the normal coefficient can only be said to give us in cases like these a hypothetical correlation between supposititious variables. The introduction of needless and unverifiable hypotheses does not appear to me a desirable proceeding in scientific work."

Pearson did not let this remark go unchallenged. He and D. Heron (1913) filled more than 150 pages of Pearson's journal (*Biometrika*) with a scathing reply. For instance,

> If Mr. Yule's views are accepted, irreparable damage will be done to the growth of modern statistical theory. . . . (Yule's Q) has never been and never will be used in any work done under his (Pearson's) supervision. . . . We regret having to draw attention to the manner in which Mr. Yule has gone astray at every stage in his treatment of association, but criticism of his methods has been thrust on us not only by Mr Yule's recent attack, but also by the unthinking praise which has been bestowed on a text-book (Yule's) which at many points can only lead statistical students hopelessly astray.

Pearson and Heron attacked Yule's "half-baked notions" and "specious reasoning" and concluded that he would have to withdraw his ideas "if he wishes to maintain any reputation as a statistician."

In retrospect, Pearson and Yule both had valid points. There are cases, particularly for nominal variables, where the classification is fixed and no underlying continuous distribution is apparent (e.g., religious affiliation, nationality, gender). On the other hand, in many applications it *is* possible to assume an underlying continuum, and it can be important to direct model-building and inference toward that continuum. We shall observe a kind of reconciliation between Yule and Pearson in Chapter 8, where odds ratios characterize a model that fits well when there is an underlying continuous normal distribution.

Goodman and Kruskal (1959) summarized the historical development of measures of association for contingency tables. Their book (1979) reprints four classic papers (1954, 1959, 1963, 1972) they published on this topic. The 1959 paper contains the following quote from a paper by M. H. Doolittle in 1887, which undoubtedly clarified the meaning of *association* for his contemporaries:

> Having given the number of instances respectively in which things are both thus and so, in which they are thus but not so, in which they are so but not thus, and in which they are neither thus nor so, it is required to eliminate the general quantitative relativity inhering in the mere thingness of the things, and to determine the special quantitative relativity subsisting between the thusness and the soness of the things.

CHAPTER NOTES

Section 2.2: Ways of Comparing Proportions

2.1 Yule (1912, p. 587) argued that multiplicative invariance is a desirable property for measures of association, especially when proportions sampled in various marginal categories are arbitrary. For 2×2 tables, Edwards (1963) showed that functions of the odds ratio are the only statistics that are invariant both to row/column interchange and to multiplication within rows or within columns by a constant. For $I \times J$ tables, see Altham (1970a) for related results.

Simple measures such as the difference of proportions and relative risk undoubtedly have very long history of use. Goodman and Kruskal (1959) noted that the relative risk was used by Quetelet in 1849.

Section 2.3: Summary Measures of Association

2.2 Concordance and discordance were first introduced for continuous variables. In that case, samples can be fully ranked (i.e., there are no ties), so $C + D = n(n-1)/2$ and sample gamma simplifies to $(C - D)/[n(n-1)/2]$. This is *Kendall's tau*, proposed by Kendall (1938). Kruskal (1958) and Agresti (1984, Chapters 9 and 10) surveyed ordinal measures of association.

Measures of ordinal association also apply when one variable is ordinal and the other is nominal but has only two categories. When Y is ordinal and X is nominal with $I > 2$ categories, no measure presented in Section 2.3 is very applicable. The *mean ridits* presented in Problem 8.30 can be useful for this case.

Haberman (1982a) and Magidson (1982) presented generalized versions of the concentration and uncertainty coefficients for models with nominal responses and multiple predictors, some possibly continuous.

PROBLEMS

Applications

2.1 Table 2.7, taken from the 1982 General Social Survey, was reported by Clogg and Shockey (1988).

Table 2.7

Gun Registration	Death Penalty	
	Favor	Oppose
Favor	784	236
Oppose	311	66

Source: 1982 General Social Survey; see Clogg and Shockey (1988).

a. Identify each classification as a response or explanatory variable.

b. Based on your answer in (a), choose a measure for describing the association. Interpret the direction and strength of association.

2.2 The odds ratio between treatment (A, B) and cure (yes, no) is 1.5. Explain what is wrong with the interpretation, "The probability of cure is 1.5 times higher for treatment A than for treatment B."

2.3 Let the horizontal axis of a graph be $\pi_{1|2}$, and let the vertical axis be $\pi_{1|1}$. Plot the locus of points for a 2×2 table having
a. relative risk $= 0.5$,
b. odds ratio $= 0.5$,
c. difference of proportions $= -0.5$.

2.4 Table 2.8 is based on records of accidents in 1988 compiled by the Department of Highway Safety and Motor Vehicles in the State of Florida.
a. Compute and interpret the sample odds ratio, relative risk, and difference of proportions.
b. When is the odds ratio approximately equal to the relative risk? Explain.

Table 2.8

| Safety Equipment | Injury | |
In use	Fatal	Nonfatal
None	1601	162,527
Seat belt	510	412,368

Source: Department of Highway Safety and Motor Vehicles, State of Florida.

2.5 In a 20-year cohort study of British male physicians, Doll and Peto (1976) noted that per year, the proportion who died from lung cancer was 0.00140 for cigarette smokers and 0.00010 for nonsmokers. The proportion who died from coronary heart disease was 0.00669 for smokers and 0.00413 for nonsmokers.

    a. Describe the association of smoking with each of lung cancer and heart disease, using the relative risk. Interpret.

    b. Describe the associations using the difference of proportions. Interpret.

    c. Which response is more strongly related to cigarette smoking, in terms of the reduction in number of deaths that would occur with elimination of cigarettes?

    d. Describe the associations using the odds ratio. Interpret.

2.6 A diagnostic test is used to detect whether subjects have a certain disease. A positive test outcome predicts that a subject has the disease. Given the subject has the disease, the probability the diagnostic test is positive is called the *sensitivity*. Given the subject does not have the disease, the probability the test is negative is called the *specificity*. Consider the 2×2 table in which the row variable is the true status, and the column classification is the diagnosis. If "positive" is the first level of each classification, then sensitivity is $\pi_{1|1}$ and specificity is $\pi_{2|2}$. Suppose ρ denotes the probability that a subject has the disease.

    a. Given that the diagnosis is positive, use Bayes Theorem to show that the probability a subject has the disease is

$$\pi_{1|1}\rho/[\pi_{1|1}\rho + \pi_{1|2}(1 - \rho)].$$

    b. Suppose a diagnostic test for exposure to AIDS has sensitivity and specificity both equal to 0.95. If $\rho = 0.005$, find the probability a subject has been exposed to AIDS, given that the diagnostic test is positive.

c. To better understand the answer in (b), find the four joint probabilities for the 2 × 2 table, and discuss their relative sizes.

d. Calculate the odds ratio for (b), and interpret it.

Note: A plot of y = true positive rate (sensitivity) against x = false positive rate $(1 - \text{specificity})$ for various possible ways of defining a "positive" result is called a *receiver operating characteristic* (ROC) curve. See Tosteson and Begg (1988).

2.7 Doll and Hill (1952) presented Table 2.9, from a retrospective study of lung cancer and tobacco smoking among patients in hospitals in several English cities. This table compares male lung cancer patients with control patients having other diseases, according to the average number of cigarettes smoked daily over a ten-year period preceding the onset of the disease.

Table 2.9

Daily Average Number of Cigarettes	Disease Group	
	Lung Cancer Patients	Control Patients
None	7	61
<5	55	129
5–14	489	570
15–24	475	431
25–49	293	154
50+	38	12

Source: Reprinted with permission from Doll and Hill (1952).

a. Show that the disease groups are stochastically ordered with respect to their distributions on smoking of cigarettes. Interpret.

b. Collapse the table into a 2 × 2 table according to whether daily average was less than five or at least five. Compute the odds ratio and interpret its value.

c. Refer to part (b). Suppose we wanted to estimate the difference in the proportions who got lung cancer between those who smoked fewer than five cigarettes per day and those who smoked at least five per day. Can we do so with these data? Explain.

2.8   Refer to the previous problem.

   a. Compute the sample log odds of lung cancer for each of the six levels of smoking. As level of smoking increases, is there a trend in the sample log odds?

   b. Compute the log local odds ratio for each pair of adjacent levels of smoking. Describe the nature of the association.

   c. Suppose the log odds of lung cancer were linearly related to the level of smoking, the log odds in row i satisfying $\log(\text{odds}_i) = \alpha + \beta i$. Show that this implies that the local odds ratios are identical.

2.9   Hout et al. (1987) reported Table 2.10, which summarizes responses to the questionnaire item "Sex is fun for me and my partner (a) never or occasionally, (b) fairly often, (c) very often, (d) almost always," for 91 married couples from the Tucson metropolitan area. Compute and interpret a measure of the association between wife's response and husband's response to this item.

Table 2.10

	Wife's Rating of Sexual Fun			
Husband's Rating	Never or Occasionally	Fairly Often	Very Often	Almost Always
Never or occasionally	7	7	2	3
Fairly often	2	8	3	7
Very often	1	5	4	9
Almost always	2	8	9	14

Source: Reprinted with permission from Hout et al. (1987).

2.10   Describe the association in Table 2.11, based on a sample conducted in 1965 of a probability sample of high school seniors and their parents.

Table 2.11

Parent Party Identification	Student Party Identification		
	Democrat	Independent	Republican
Democrat	604	245	67
Independent	130	235	76
Republican	63	180	252

Source: Reprinted with permission from Jennings and Niemi (1968).

2.11 Give a 3×2 table of joint probabilities for which:
   a. Goodman and Kruskal's $\tau = 1$.
   b. $\tau = 0$.
   c. $U = 1$.
   d. $U = 0$.

Theory and Methods

2.12 When two binary response variables have a joint distribution $\{\pi_{ij}\}$, show the odds ratio can be computed using the joint distribution, the within-rows conditional distributions, or the within-columns conditional distributions. Give the implications regarding sampling designs for which the odds ratio is appropriate.

2.13 **a.** Show that for a 2×2 table of cell counts $\{n_{ij}\}$, the odds ratio is invariant to (1) interchanging rows with columns, and (2) multiplication of cell counts within rows or within columns by a nonzero constant.
   b. Show that the difference of proportions and the relative risk do not have these invariance properties.

2.14 **a.** Show that $\{\alpha_{ij}\}$ in (2.8) determine all $\binom{I}{2}\binom{J}{2}$ odds ratios formed from pairs of rows and pairs of columns.
   b. Show that $\{\alpha_{ij}\}$ determine $\{\theta_{ij}\}$ in (2.7), and vice versa.

2.15 Refer to the previous problem, and suppose all rows and columns have positive probability. Show that independence between X and Y is equivalent to $\{\text{all } \alpha_{ij} = 1\}$.

2.16 Suppose the response variable in a $2 \times J$ table is ordinal. Consider the $J - 1$ *cumulative odds ratios*

$$\theta_j = \frac{F_{j|1}/(1 - F_{j|1})}{F_{j|2}/(1 - F_{j|2})}, \quad j = 1, \ldots, J - 1 .$$

   a. Suppose $\theta_j = 2.5$ for all $j = 1, \ldots, J - 1$. Interpret.
   b. Show that $\log \theta_j \geq 0$ for all $j = 1, \ldots, J - 1$ is equivalent to the conditional distribution in row 2 being stochastically higher than the one in row 1.

2.17 For the odds ratios $\{\theta_{ij}\}$ in (2.7), show that $\log \theta_{ij} \geq 0$ for $1 \leq j \leq J-1$ implies that the conditional distribution in row $i+1$ is stochastically higher than the one in row i. Show by counterexample that the converse is not true.

2.18 **a.** Show that Yule's Q falls between -1 and 1.
  b. Give conditions under which $Q = -1$ or $Q = 1$.
  c. Derive the relationship between Q and the odds ratio.

2.19 For n observations in a cross-classification of ordinal variables, the total number of pairs of observations can be partitioned into

$$n(n-1)/2 = C + D + T_X + T_Y - T_{XY}$$

where T_X is the number of pairs tied on X, T_Y is the number of pairs tied on Y, and T_{XY} is the number tied on X and Y. Find expressions for T_X, T_Y, and T_{XY} in terms of $\{n_{ij}\}$.

2.20 For each of the $n(n-1)$ ordered pairs of observations (X_a, Y_a) and (X_b, Y_b) in a sample of size n, let $X_{ab} = \text{sign}(X_a - X_b)$ and $Y_{ab} = \text{sign}(Y_a - Y_b)$.
  a. Show that the sample correlation between the $n(n-1)$ distinct (X_{ab}, Y_{ab}) pairs equals

$$(C - D)/\{[n(n-1)/2 - T_X][n(n-1)/2 - T_Y]\}^{1/2}$$

where $T_X = \Sigma \, n_{i+}(n_{i+} - 1)/2$ and $T_Y = \Sigma \, n_{+j}(n_{+j} - 1)/2$. This index of ordinal association is called *Kendall's tau-b* (Kendall 1945). Tau-*b* tends to be less sensitive than gamma to the choice of response categories (Agresti 1976).
  b. Somers' d (Somers 1962) is defined as $(C - D)/[n(n-1)/2 - T_X]$. Show that d is the difference between the proportions of concordant and discordant pairs, out of those pairs untied on X.
  c. Show that for 2×2 tables, Somers' d simplifies to the difference of proportions and tau-*b* simplifies to the Pearson correlation between X_a and Y_a.

2.21 Show that independence of X and Y implies
  a. $\tau = 0$.
  b. $U = 0$.

2.22 Goodman and Kruskal (1954) proposed an alternative measure, *lambda*, for nominal variables. It has $V(Y) = 1 - \max\{\pi_{+j}\}$ and $V(Y \mid i) = 1 - \max_j\{\pi_{j \mid i}\}$.

  a. Interpret lambda as a proportional reduction in prediction error, in which predictions are made by selecting the response category that is most likely.

  b. Show that independence implies $\lambda = 0$, but the converse is not true.

  c. Calculate lambda for Table 2.6, and compare the interpretation to that for Goodman and Kruskal's tau.

CHAPTER 3

Inference For
Two-Way Contingency Tables

This chapter presents basic inferential methods for categorical data. We introduce them here for multinomial distributions and two-way contingency tables, but the same inferential methods play a vital role in more complex analyses presented in later chapters.

One of the first applications of hypothesis testing was the problem of judging whether a set of probabilities equal certain fixed numbers. Section 3.2 presents the classical analysis of this problem, in which the test statistic is Karl Pearson's chi-squared statistic.

Another fundamental problem is testing the null hypothesis of independence between two categorical variables. Section 3.3 describes this analysis, for which test statistics also have large-sample chi-squared distributions. That section uses the partitioning property of chi-squared to extract components of test statistics that describe certain aspects of the overall association in a table.

In practice, we learn more from *estimating* descriptive parameters than from *testing hypotheses* about their values. Section 3.4 shows how to calculate large-sample standard errors and confidence intervals for parameters presented in Chapter 2. Sections 3.5 and 3.6 introduce *exact* inferential methods, which are useful when the sample size is too small to apply large-sample results.

An inferential statistical procedure is valid only to the extent that sampling assumptions upon which it is based are fulfilled. Most inferential methods for categorical data assume multinomial or Poisson sampling models. The first section of this chapter presents these sampling models.

3.1 SAMPLING DISTRIBUTIONS

Suppose we observe counts $\{n_i, i = 1, \ldots, N\}$ in the N cells of a contingency table. For instance, these might be observations for the N levels of a single categorical variable, or for $N = IJ$ cells of a two-way table. This section treats the counts as random variables. Each n_i has distribution concentrated on the nonnegative integers, with expected value denoted by $m_i = E(n_i)$. The $\{m_i\}$ are called *expected frequencies*.

3.1.1 Poisson sampling

Since n_i must be a nonnegative integer, its sampling distribution should place its mass on that range. One of the simplest such distributions is the *Poisson*. Its form depends on a single parameter, the mean m_i. The probability mass function is

$$\frac{\exp(-m_i)m_i^{n_i}}{n_i!} \quad \text{for } n_i = 0, 1, 2, \ldots . \qquad (3.1)$$

It satisfies $\text{Var}(n_i) = E(n_i) = m_i$.

The Poisson sampling model for counts $\{n_i\}$ assumes that they are independent Poisson random variables. The joint probability function for $\{n_i\}$ is then the product of the probabilities (3.1) for the N cells. The total sample size $n = \Sigma n_i$ also has a Poisson distribution, with parameter Σm_i.

The Poisson distribution is used for counts of events that occur randomly over time or space, when outcomes in disjoint periods are independent. For example, Poisson distributions might be realistic for $n_1 =$ the number of spontaneous abortions, $n_2 =$ the number of induced abortions, and $n_3 =$ the number of live births, measured in November, 1990 in London, England.

3.1.2 Multinomial Sampling

An unusual feature of Poisson sampling is that the total sample size $n = \Sigma\, n_j$ is random, rather than fixed. If we start with the Poisson model but condition on the total sample size n, $\{n_i\}$ no longer have Poisson distributions, since each n_i cannot exceed n. Conditional on n, $\{n_i\}$ are also no longer independent, since the value of one affects the possible range for the others.

Given that $\Sigma n_j = n$, the conditional probability of a set $\{n_i\}$ satisfying this condition is

$$P\left(n_i \text{ observations in cell } i, \; i = 1, \ldots, N \mid \sum n_j = n\right)$$

$$= \frac{P(n_i \text{ observations in cell } i, \; i = 1, \ldots, N)}{P\left(\sum n_j = n\right)}$$

$$= \frac{\prod_i [\exp(-m_i) m_i^{n_i}/n_i!]}{\exp\left(-\sum m_j\right)\left(\sum m_j\right)^n / n!} = \left(\frac{n!}{\prod_i n_i!}\right) \prod_i \pi_i^{n_i} \qquad (3.2)$$

where $\{\pi_i = m_i/(\sum m_j)\}$. This is the *multinomial* $(n, \{\pi_i\})$ distribution, characterized by the sample size n and the cell probabilities $\{\pi_i\}$. The *binomial* (n, π_1) distribution with index n and "success" probability π_1 is the special case of the multinomial with $N = 2$ cells. For the multinomial distribution for $\{n_1, \ldots, n_N\}$, the marginal distribution for n_i is binomial, with $E(n_i) = n\pi_i$ and $\text{Var}(n_i) = n\pi_i(1 - \pi_i)$.

The multinomial distribution for $\{n_i\}$ also applies when n independent observations are taken from a probability distribution concentrated on a set of N categories. In other words, if the same probability distribution $\{\pi_i\}$ applies to each observation, and if the n observations are independent, then the counts $\{n_i\}$ of the number of observations in each category have distribution (3.2). When cell counts have distribution (3.2), the sampling scheme is called *multinomial sampling*.

3.1.3 Independent Multinomial Sampling

Suppose we take observations on a categorical response variable Y, separately at various settings of an explanatory variable X. Let n_{ij} denote the number of observations in the jth response category, at the ith setting of X. Suppose the n_{i+} observations on Y at the ith setting of X are independent, each having probability distribution $\{\pi_{1|i}, \ldots, \pi_{J|i}\}$. Then the counts $\{n_{ij}, j = 1, \ldots, J\}$ have the multinomial distribution

$$\left(\frac{n_{i+}!}{\prod_j n_{ij}!}\right) \prod_j \pi_{j|i}^{n_{ij}}. \qquad (3.3)$$

When samples at different settings of X are independent, the joint probability function for the entire data set is the product of the multinomial probability functions (3.3) from the various settings. This sampling scheme is *independent multinomial sampling*, sometimes also called *product multinomial sampling*.

Independent multinomial sampling also results under the following conditions: Suppose $\{n_{ij}\}$ follow either independent Poisson sampling with means $\{m_{ij}\}$, or multinomial sampling with probabilities $\{\pi_{ij} = m_{ij}/n\}$. When X is an explanatory variable, it is sensible to perform statistical inference conditional on the totals $\{n_{i+}\}$, even when their values are not fixed by the sampling design. When we condition on $\{n_{i+}\}$, the cell counts $\{n_{ij}, j = 1, \ldots, J\}$ have the multinomial distribution (3.3) with response probabilities $\{\pi_{j|i} = m_{ij}/m_{i+}, j = 1, \ldots, J\}$, and cell counts from different rows are independent.

In *prospective* studies, the totals $\{n_{i+}\}$ for X are often fixed, and we regard each row of J counts as an independent multinomial sample on Y. In *retrospective* studies, the totals $\{n_{+j}\}$ for Y are often fixed, and we regard each column of I counts as an independent multinomial sample on X. In *cross-sectional* studies, the total sample size is fixed, but not the row or column totals, and we regard the IJ cell counts as a multinomial sample.

3.1.4   Seat Belt Example

Researchers in the Massachusetts Department of Transportation plan to study the relationship between seat-belt use (yes, no) and outcome of crash (fatality, non-fatality), for drivers involved in automobile accidents on the Massachusetts Turnpike. They will summarize results in Table 3.1. Suppose they catalog all accidents on the Turnpike for the next year, classifying each according to these variables. Then the total sample size is a random variable, and they might treat the numbers of observations at the four combinations of seat-belt use and outcome of crash as independent Poisson random variables with unknown means $\{m_{11}, m_{12}, m_{21}, m_{22}\}$.

Suppose, instead, the researchers randomly sample 100 police records of crashes on the Turnpike in the past year, and classify each according to

Table 3.1   Seat-Belt Use and Results of Automobile Crashes

	Result of Crash	
Seat-Belt use	Fatality	Nonfatality
Yes		
No		

seat-belt use and outcome of crash. For this cross-sectional study, the total sample size n is fixed, and they might treat the four cell counts as a multinomial random variable with $n = 100$ trials and unknown probabilities $\{\pi_{11}, \pi_{12}, \pi_{21}, \pi_{22}\}$.

Suppose police records for accidents involving fatalities were filed separately from the others. The researchers might instead randomly sample 50 records of accidents with a fatality and 50 records of accidents with no fatality. For this retrospective approach, the column totals in Table 3.1 are fixed. They might then regard each column of Table 3.1 as an independent binomial (multinomial with two levels) sample.

It is sometimes sensible to treat both margins as fixed. The appropriate sampling distribution is then the *hypergeometric*. This case, which is not common, is discussed in Section 3.5.

3.1.5 Likelihood Functions and Maximum Likelihood Estimates

The choice of sampling model is the first step toward data analysis. Given the observed data $\{n_i, i = 1, \ldots, N\}$, the *likelihood function* is the probability of $\{n_i\}$ for that sampling model, treated as a function of the unknown parameters. The maximum likelihood (abbreviated ML) estimates are parameter values that maximize this function. They are the parameter values under which the observed data would have had highest probability of occurrence.

To illustrate, we obtain ML estimates of category probabilities $\{\pi_i\}$, assuming multinomial sampling. As a function of $\{\pi_i\}$, probability function (3.2) is proportional to

$$\prod_i \pi_i^{n_i} \,. \tag{3.4}$$

The part of a probability function involving the parameters is called the *kernel*. Since we shall maximize (3.2) with respect to $\{\pi_i\}$, the factorial terms in that expression are constants that have no influence on where the maximum occurs. Thus, we ignore such terms and treat the kernel as the likelihood function. The ML estimates are values for $\{\pi_i\}$ that maximize (3.4), subject to the constraints that they are nonnegative and add to 1.0.

Parameter values that maximize (3.4) also maximize a monotone increasing function of it, such as its log. For most probability functions, including the multinomial and Poisson, it is simpler to maximize the latter. For the multinomial, the log likelihood is

$$L = \sum n_i \log \pi_i \,. \tag{3.5}$$

To eliminate redundancies in the parameters, we treat L as a function of the $N-1$ probabilities obtained by dropping π_N, since $\pi_N = 1 - (\pi_1 + \cdots + \pi_{N-1})$. Note that $\partial \pi_N / \partial \pi_i = -1$ for $i = 1, \ldots, N-1$.

Since

$$\frac{\partial \log \pi_N}{\partial \pi_i} = \frac{1}{\pi_N} \frac{\partial \pi_N}{\partial \pi_i} = -\frac{1}{\pi_N},$$

differentiating L with respect to π_i gives the likelihood equation

$$\frac{\partial L}{\partial \pi_i} = \frac{n_i}{\pi_i} - \frac{n_N}{\pi_N} = 0.$$

The ML solution satisfies $\hat{\pi}_i / \hat{\pi}_N = n_i / n_N$. Now

$$\sum \hat{\pi}_i = 1 = \frac{\hat{\pi}_N \left(\sum n_i \right)}{n_N} = \frac{\hat{\pi}_N n}{n_N}$$

so $\hat{\pi}_N = n_N / n$ and then $\hat{\pi}_i = n_i / n = p_i$. It follows from results to be presented in Section 6.1.3 that this solution does maximize the likelihood. Thus, the ML estimates of $\{\pi_i\}$ are simply the sample proportions.

For contingency tables, the ML estimates of cell probabilities are the sample cell proportions. The ML estimates of marginal probabilities are the sample marginal proportions. When two categorical variables are statistically independent, all $\pi_{ij} = \pi_{i+} \pi_{+j}$. The ML estimate of π_{ij} under this condition is

$$\hat{\pi}_{ij} = p_{i+} p_{+j} = \frac{n_{i+} n_{+j}}{n^2}. \tag{3.6}$$

This is shown by substituting $\pi_{ij} = \pi_{i+} \pi_{+j}$ in the multinomial formula and using the constraints $\sum_i \pi_{i+} = \sum_j \pi_{+j} = 1$ in the maximization process.

For multinomial sampling of size n over IJ cells of a two-way table, an individual cell count n_{ij} has the binomial distribution with index n and parameter π_{ij}. The mean of this binomial distribution is $m_{ij} = n\pi_{ij}$, which has ML estimate $\hat{m}_{ij} = n\hat{\pi}_{ij}$. Under the assumption of independence, this equals

$$\hat{m}_{ij} = np_{i+} p_{+j} = \frac{n_{i+} n_{+j}}{n}.$$

The $\{\hat{m}_{ij}\}$ are called *estimated expected frequencies*. They have the same marginal totals as do the observed data. For instance, $\hat{m}_{i+} = \sum_j \hat{m}_{ij} = (n_{i+}/n) \sum_j n_{+j} = n_{i+}$. We use $\{\hat{m}_{ij}\}$ in tests of independence in Section 3.3.

3.1.6 Overdispersion

Most categorical data analyses assume multinomial sampling. Many analyses have the pleasing result that parameter estimates are the same for Poisson, multinomial, or independent multinomial sampling schemes, because of the similarity in the likelihood functions (Section 13.2.2). In derivations of results, it is often simplest to refer to Poisson sampling, because of the lack of constraints on parameters.

Sometimes data display more variation than is predicted by Poisson or multinomial schemes. This might happen because observations on different subjects are positively correlated, rather than independent. Or, it might happen because the true sampling distribution is a mixture of different Poisson or multinomial distributions.

To illustrate, suppose we observe the number of insects in a batch that survive a certain low dose of an insecticide, where each batch has n insects. The insects are also highly sensitive to temperature change, and the temperature during the experiment varied considerably (but was not measured) from one batch to another. Then the distribution of the number of insects per batch surviving the experiment might cluster near 0 and near n, showing more dispersion than expected for binomial sampling. Extra variation could also occur when an insect in a batch has probability π of surviving, but the value of π varies for insects in the batch according to some distribution (See Problem 3.18).

The existence of greater variation than predicted by the sampling model is called *overdispersion*. One approach to dealing with overdispersion equates the variance of the counts with some constant multiple of the formula given by the sampling model. That constant, generally expected to exceed 1.0, is estimated from the data. Section 13.2.3 discusses overdispersion for models for categorical data.

3.2 TESTING GOODNESS OF FIT

This section presents a goodness-of-fit test introduced by Karl Pearson in 1900. It had a revolutionary impact on categorical data analysis, being one of the first inferential statistical methods. Pearson's test evaluates whether probabilities in a multinomial distribution equal certain hypothesized values.

3.2.1 Testing a Specified Multinomial

Consider the null hypothesis (H_0) that the N parameters $\{\pi_i\}$ of a multinomial distribution equal certain fixed values $\{\pi_{i0}\}$, where $\Sigma \pi_{i0} =$

$\Sigma \pi_i = 1$. When H_0 is true, the expected cell frequencies are $m_i = n\pi_{i0}$, $i = 1, \ldots, N$. For sample counts $\{n_i\}$, Pearson proposed the test statistic

$$X^2 = \sum \frac{(n_i - m_i)^2}{m_i}. \tag{3.7}$$

For large samples, X^2 has approximately a chi-squared null distribution, with degrees of freedom (df) equal to $N - 1$. A statistic of form (3.7) is called a *Pearson chi-squared statistic*.

Greater departures of $\{n_i\}$ from $\{m_i\}$ produce greater X^2 values, for a fixed sample size n. The P-value of the test is the null probability that X^2 takes value at least as large as the observed value. This equals the sum of the multinomial null probabilities for all cell count arrays (having a sum of n) that produce X^2 values at least as large as observed. For large samples, the P-value is approximated by the tail probability to the right of the observed X^2 value, for the chi-squared distribution with df = $N - 1$.

According to Stigler (1986), Pearson's original motivation in developing this test was to analyze whether possible outcomes on a particular Monte Carlo roulette wheel were equally likely. Pearson discussed this application in his 1900 article, as well as others involving throwing dice and testing normality.

3.2.2 Testing Mendel's Theories

Among its many applications, Pearson's test has been used in genetics to test Mendel's theories of natural inheritance. Mendel crossed pea plants of pure yellow strain with ones of pure green strain. He predicted that second-generation hybrid seeds would be 75% yellow and 25% green, yellow being the dominant strain. One experiment produced $n = 8023$ seeds, of which $n_1 = 6022$ were yellow and $n_2 = 2001$ were green. The expected frequencies for $\pi_{10} = 0.75$ are $m_1 = 6017.25$ and $m_2 = 2005.75$. The Pearson statistic is $X^2 = 0.015$, which has a P-value of $P = 0.88$, not contradictory to Mendel's hypothesis.

Mendel performed several experiments of this type. In 1936, R. A. Fisher summarized Mendel's results. He used the reproductive property of chi-squared; namely, if $X_1^2, \ldots, X_k^2$ are independent chi-squared statistics with degrees of freedom $\nu_1, \ldots, \nu_k$, then ΣX_i^2 has a chi-squared distribution with df = $\Sigma \nu_i$. Fisher obtained a summary chi-squared statistic that equaled 42, based on df = 84. A chi-squared distribution with df = 84 has mean 84 and standard deviation $(2 \times 84)^{1/2} = 13.0$, and the right-tail probability above 42 is $P = 0.99996$. In other words the chi-squared statistic was so small that the fit seemed *too* good.

Fisher commented, "the general level of agreement between Mendel's expectations and his reported results shows that it is closer than would be expected in the best of several thousand repetitions. . . . I have no doubt that Mendel was deceived by a gardening assistant, who knew only too well what his principal expected from each trial made." In a letter written at the time, he stated "Now, when data have been faked, I know very well how generally people underestimate the frequency of wide chance deviations, so that the tendency is always to make them agree too well with expectations." In summary, goodness-of-fit tests can reveal not only when a fit is inadequate, but also when it is better than random fluctuations would have us expect.

Interesting accounts of Fisher's analysis and the accompanying controversy were given by his daughter, Joan Fisher Box (1978, pp. 295–300), and by Freedman et al. (1978, pp. 420–428, 478). Despite possible difficulties with Mendel's data, subsequent work led to general acceptance of his theories.

3.2.3 Theoretical Justification

We now show why Pearson's statistic has a limiting chi-squared distribution. For a multinomial sample of size n, n_i has the binomial distribution with index n and parameter π_i. For large n, by the normal approximation to the binomial, n_i (and $p_i = n_i/n$) have approximate normal distributions. More generally, by the Central Limit Theorem, the sample proportions $\mathbf{p} = (n_1/n, \ldots, n_{N-1}/n)'$ have an approximate multivariate normal distribution (Section. 12.1.5). Let $\boldsymbol{\Sigma}_0$ denote the null covariance matrix of $\sqrt{n}\mathbf{p}$, and let $\boldsymbol{\pi}_0 = (\pi_{10}, \ldots, \pi_{N-1,0})'$. Under H_0, since $\sqrt{n}(\mathbf{p} - \boldsymbol{\pi}_0)$ converges to a $N(\mathbf{0}, \boldsymbol{\Sigma}_0)$ distribution, the quadratic form

$$n(\mathbf{p} - \boldsymbol{\pi}_0)'\boldsymbol{\Sigma}_0^{-1}(\mathbf{p} - \boldsymbol{\pi}_0) \qquad (3.8)$$

has distribution converging to a chi-squared with df $= N - 1$.

Section 12.1.5 shows that the covariance matrix of $\sqrt{n}\mathbf{p}$ has elements

$$\sigma_{ij} = -\pi_i \pi_j \qquad \text{if } i \neq j$$
$$= \pi_i(1 - \pi_i) \quad \text{if } i = j .$$

The matrix $\boldsymbol{\Sigma}_0^{-1}$ has (i, j)th element $1/\pi_{N0}$ when i $\neq j$ and $(1/\pi_{i0} + 1/\pi_{N0})$ when $i = j$ (You can verify this by showing $\boldsymbol{\Sigma}_0 \boldsymbol{\Sigma}_0^{-1}$ equals the identity matrix). With this substitution, direct calculation shows that (3.8) simplifies to X^2.

This argument is similar to the one Pearson gave in 1900. R. A. Fisher (1922) gave a simpler justification. Suppose we treat $(n_1, \ldots, n_N)$ as independent Poisson random variables, with means $(m_1, \ldots, m_N)$. For large $\{m_i\}$, the standardized values $\{z_i = (n_i - m_i)/\sqrt{m_i}\}$ have approximate standard normal distributions, and $\Sigma z_i^2 = X^2$ has an approximate chi-squared distribution with N degrees of freedom. When we add the single linear constraint $\Sigma(n_i - m_i) = 0$, thus converting the Poisson distributions to a multinomial, we lose a degree of freedom.

For $N = 2$, we can calculate the *exact* P-value using the binomial distribution for n_1. To test H_0: $\pi = 0.5$ against H_a: $\pi \neq 0.5$ based on $n = 10$ and $n_1 = 1$, for instance, $P = P(n_1 \leqslant 1$ or $n_1 \geqslant 9)$ for the binomial distribution with $\pi_{10} = 0.5$. For large samples, by the normal approximation to the binomial, we can conduct the test using the normal statistic

$$z = \frac{p_1 - \pi_{10}}{[\pi_{10}(1 - \pi_{10})/n]^{1/2}} \tag{3.9}$$

In fact, Pearson's X^2 is simply the square of this normal statistic. For Mendel's data, $p_1 = 6022/8023$, $\pi_{10} = 0.75$, $n = 8023$, and $z = 0.123$, for which $X^2 = (0.123)^2 = 0.015$.

3.2.4 Goodness-of-Fit Test with Estimated Expected Frequencies

The X^2 statistic (3.7) compares an observed distribution with a hypothetical one. In some applications, the hypothetical probabilities are functions of a smaller set of unknown parameters, and their numerical values are unknown. ML estimates of the parameters determine ML estimates of the hypothetical probabilities, and hence ML estimates of the expected frequencies for use in X^2. Replacing expected frequencies by estimates affects the distribution of X^2. When the expected frequencies are functions of t parameters, the degrees of freedom drop to $(N - 1) - t$ (Section. 12.3.3).

We now show a goodness-of-fit test with estimated expected frequencies. A sample of 156 dairy calves born in Okeechobee County, Florida, were classified according to whether they caught pneumonia within 60 days after birth. Calves that got a pneumonia infection were also classified according to whether they got a secondary infection within two weeks after the first infection cleared up. Table 3.2 shows the data. Calves that did not get a primary infection could not get a secondary infection, so no observations can fall in the cell for "no" primary infection and "yes" secondary infection. Such a cell is called a *structural zero*.

Table 3.2   Primary and Secondary Pneumonia Infections of Calves

| Primary | Secondary Infection[a] | |
Infection	Yes	No
Yes	30 (38.1)	63 (39.0)
No	0 (–)	63 (78.9)

Source: Dr. Thang Tran and Dr. G. A. Donovan, College of Veterinary Medicine, University of Florida.

[a] Values in parentheses are estimated expected frequencies.

One goal of this study was to test whether the probability of primary infection was the same as the conditional probability of secondary infection, given that the calf got the primary infection. In other words, if π_{ij} denotes the probability that a calf is classified in row i and column j of this table, the null hypothesis is

$$H_0: \pi_{11} + \pi_{12} = \pi_{11}/(\pi_{11} + \pi_{12})$$

or $\pi_{11} = (\pi_{11} + \pi_{12})^2$. Let $\pi = \pi_{11} + \pi_{12}$ denote the probability of primary infection. The null hypothesis states that the probabilities satisfy the structure shown in Table 3.3; that is, that probabilities in a trinomial for the categories (yes-yes, yes-no, no-no) for primary-secondary infection equal $[\pi^2, \pi(1 - \pi), 1 - \pi]$.

Let n_{ij} denote the number of observations in cell (i, j). The ML estimate of π is the value maximizing the kernel of the multinomial likelihood

$$(\pi^2)^{n_{11}}(\pi - \pi^2)^{n_{12}}(1 - \pi)^{n_{22}} . \tag{3.10}$$

The log likelihood is

Table 3.3   Probability Structure for Hypothesis

| Primary | Secondary Infection | | |
Infection	Yes	No	Total
Yes	π^2	$\pi(1 - \pi)$	π
No	–	$1 - \pi$	$1 - \pi$

$$L(\pi) = n_{11} \log(\pi^2) + n_{12} \log(\pi - \pi^2) + n_{22} \log(1 - \pi) .$$

Differentiation with respect to π gives the likelihood equation

$$\frac{2n_{11}}{\pi} + \frac{n_{12}}{\pi} - \frac{n_{12}}{1 - \pi} - \frac{n_{22}}{1 - \pi} = 0 .$$

The solution is

$$\hat{\pi} = (2n_{11} + n_{12})/(2n_{11} + 2n_{12} + n_{22}) .$$

For Table 3.2, the ML estimate of π is $\hat{\pi} = 0.494$. Since $n = 156$, the estimated expected frequencies are $\hat{m}_{11} = n\hat{\pi}^2 = 38.1$, $\hat{m}_{12} = n(\hat{\pi} - \hat{\pi}^2) = 39.0$, and $\hat{m}_{22} = n(1 - \hat{\pi}) = 78.9$. These values are shown in Table 3.2. The Pearson statistic for testing the null hypothesis is $X^2 = 19.7$. Since there are $N = 3$ possible responses and $t = 1$ parameter (π) determining the expected frequencies, df $= (3 - 1) - 1 = 1$. There is very strong evidence against the null hypothesis. Inspection of Table 3.2 reveals that many more calves got a primary infection but not a secondary infection than the hypothesis predicts. The researchers concluded that the primary infection had an immunizing effect that reduced the likelihood of a secondary infection.

3.3   TESTING INDEPENDENCE

In two-way contingency tables with multinomial sampling, the null hypothesis of statistical independence is H_0: $\pi_{ij} = \pi_{i+} \pi_{+j}$ for all i and j. To test H_0, we could use the Pearson X^2 statistic (3.7) with n_{ij} in place of n_i and with $m_{ij} = n\pi_{ij} = n\pi_{i+} \pi_{+j}$ in place of m_i. Here, m_{ij} is the expected value of n_{ij} under the null hypothesis. Usually, though, $\{\pi_{i+}\}$ and $\{\pi_{+j}\}$ are unknown.

3.3.1   Pearson Chi-Squared Test

We estimate the expected frequencies by $\{\hat{m}_{ij} = np_{i+}p_{+j}\}$. The X^2 statistic then equals

$$X^2 = \sum \sum \frac{(n_{ij} - \hat{m}_{ij})^2}{\hat{m}_{ij}} . \tag{3.11}$$

Pearson (1900, 1922) claimed that replacing $\{m_{ij}\}$ by the estimates $\{\hat{m}_{ij}\}$ would not affect the distribution of X^2. Since there are $N = IJ$

categories for the cross-classification, he argued that X^2 would have an asymptotic chi-squared distribution with $df = IJ - 1$. On the contrary, since $\{\hat{m}_{ij}\}$ are determined by estimating $\{\pi_{i+}\}$ and $\{\pi_{+j}\}$, the chi-squared distribution has

$$df = (IJ - 1) - (I - 1) - (J - 1) = (I - 1)(J - 1).$$

The dimensions of $\{\pi_{i+}\}$ and $\{\pi_{+j}\}$ reflect the constraints $\Sigma \pi_{i+} = \Sigma \pi_{+j} = 1$. Pearson's error was not pointed out until 1922, by R. A. Fisher, in an important article that helped to clarify geometrically the notion of *degrees of freedom*.

3.3.2 Likelihood-Ratio Chi-Squared

The likelihood-ratio test is a general-purpose way of testing a null hypothesis H_0 against an alternative hypothesis H_a. In this test, we maximize the likelihood under H_0, and also under the general condition that H_0 or H_a is true. Let Λ denote the ratio of the maximized likelihoods, which cannot exceed 1. Wilks (1935, 1938) showed that $-2 \log \Lambda$ has a limiting null chi-squared distribution, as $n \rightarrow \infty$. The degrees of freedom equal the difference in the dimensions of the parameter spaces under $H_0 \cup H_a$ and under H_0.

For multinomial sampling in a contingency table, the kernel of the likelihood is

$$\prod_i \prod_j \pi_{ij}^{n_{ij}}, \quad \text{where all } \pi_{ij} \geq 0 \quad \text{and} \quad \sum_i \sum_j \pi_{ij} = 1. \tag{3.12}$$

Under H_0: independence (all $\pi_{ij} = \pi_{i+} \pi_{+j}$), the likelihood is maximized when $\hat{\pi}_{i+} = n_{i+}/n$ and $\hat{\pi}_{+j} = n_{+j}/n$, so that $\hat{\pi}_{ij} = n_{i+}n_{+j}/n^2$. In the general case, the likelihood is maximized when $\hat{\pi}_{ij} = n_{ij}/n$. The ratio of the likelihoods equals

$$\Lambda = \frac{\displaystyle\prod_i \prod_j (n_{i+}n_{+j})^{n_{ij}}}{\displaystyle n^n \prod_i \prod_j n_{ij}^{n_{ij}}}.$$

It follows that Wilks's statistic, denoted by G^2, is

$$G^2 = -2 \log \Lambda = 2 \sum \sum n_{ij} \log(n_{ij}/\hat{m}_{ij}) \tag{3.13}$$

where $\{\hat{m}_{ij} = n_{i+}n_{+j}/n\}$ are the estimated expected frequencies under the assumption of independence. This statistic is called the *likelihood-ratio chi-squared statistic*. The larger the value of G^2, the more evidence there is against the null hypothesis.

In the general case, the parameter space consists of $\{\pi_{ij}\}$ subject to the linear restriction $\Sigma\, \Sigma\, \pi_{ij} = 1$, so the dimension is $IJ - 1$. Under H_0, the $\{\pi_{ij}\}$ are determined by $\{\pi_{i+}\}$ and $\{\pi_{+j}\}$, so the dimension is $(I - 1) + (J - 1)$. The difference in these dimensions equals $(I - 1)(J - 1)$. For large samples, G^2 has a chi-squared null distribution with df $= (I - 1)(J - 1)$.

When independence holds, the Pearson statistic X^2 and the likelihood-ratio statistic G^2 have asymptotic chi-squared distributions with df $= (I - 1)(J - 1)$. In fact, X^2 and G^2 are asymptotically equivalent in that case; $X^2 - G^2$ converges in probability to zero (Section 12.3.4). The limiting results for multinomial sampling also apply to the other sampling schemes (Roy and Mitra 1956).

It is not simple to describe the sample size needed for the chi-squared distribution to approximate well the exact distributions of X^2 and G^2. For a fixed number of cells, X^2 usually converges more quickly than G^2. The chi-squared approximation is usually poor for G^2 when $n/IJ < 5$. When I or J is large, it can be decent for X^2 for n/IJ as small as 1, if the table does not contain both very small and moderately large expected frequencies. Section 7.7.3 provides further guidelines.

3.3.3 Job Satisfaction Example

We now re-visit the data in Table 2.4 on income and job satisfaction. Table 3.4 contains the estimated expected frequencies for H_0: independence. For instance, the first cell has $\hat{m}_{11} = n_{1+}n_{+1}/n = (206 \times 62)/901 = 14.2$. The chi-squared statistics are $X^2 = 11.99$ and $G^2 = 12.03$, based on df $= (4 - 1)(4 - 1) = 9$. Each statistic yields a P-value of 0.21. These data do not show much evidence of association between income and job satisfaction. However, the discrepancy between observed and estimated expected frequencies is noticeable in the corner cells of the table, suggesting that further analysis may be fruitful.

3.3.4 Invariance of Chi-Squared to Category Orderings

The $\{\hat{m}_{ij} = n_{i+}n_{+j}/n\}$ used in X^2 and G^2 depend on the row and column marginal totals, but not on the order in which the rows and columns are listed. Thus, X^2 and G^2 do not change under permutations of rows or columns. These tests treat both classifications as nominal scales. We

Table 3.4 Observed (Estimated) Expected Frequencies for Testing Independence of Income and Job Satisfaction

	Job Satisfaction				
Income (US$)	Very Dissatisfied	Little Dissatisfied	Moderately Satisfied	Very Satisfied	Total
<6000	20 (14.2)	24 (24.7)	80 (72.9)	82 (94.2)	206
6000–15,000	22 (19.9)	38 (34.6)	104 (102.3)	125 (132.2)	289
15,000–25,000	13 (16.2)	28 (28.2)	81 (83.2)	113 (107.5)	235
>25,000	7 (11.8)	18 (20.5)	54 (60.5)	92 (78.2)	171
Total	62	108	319	412	901

ignore some information when we use them to test independence between ordinal scales.

When at least one variable is ordinal, it is usually possible to construct more powerful tests of independence. We shall see that analyses of Table 3.4 that take ordering of categories into account show strong evidence of association between income and job satisfaction.

3.3.5 Partitioning Chi-Squared

Let Z denote a random variable having a standard normal distribution. Then Z^2 has a chi-squared distribution with df = 1. A chi-squared random variable with df = ν has representation $Z_1^2 + \cdots + Z_\nu^2$, where $Z_1, \ldots, Z_\nu$ are independent standard normal random variables. If X_1^2 and X_2^2 are independent random variables having chi-squared distributions with degrees of freedom ν_1 and ν_2, the reproductive property implies that $X^2 = X_1^2 + X_2^2$ has chi-squared distribution with df = $\nu_1 + \nu_2$. Conversely, a chi-squared statistic having df = ν has partitionings into independent chi-squared components—for example, into ν components each having df = 1.

This subsection shows ways of partitioning chi-squared statistics for testing independence so the components represent certain aspects of the association. A partitioning may show that an association primarily reflects differences between certain categories or groupings of categories.

We begin with a simple partitioning of G^2 for $2 \times J$ tables, one having $J - 1$ components. The jth component is identical to G^2 for testing independence in a 2×2 table, where the first column combines columns 1 through j of the original table, and the second column is column $j + 1$. That is, G^2 for testing independence in a $2 \times J$ table equals a statistic that compares the first two columns, plus a statistic that combines the first two columns and compares them to the third column, . . . , plus a statistic that combines the first $J - 1$ columns and compares them to the last column. Each statistic has a single degree of freedom.

It might seem more natural to compute G^2 for the $(J - 1)$ 2×2 tables obtained by pairing each column with a particular one, say the last. These statistics are not asymptotically independent, and their sum does not equal G^2 for testing independence in the full table.

For an $I \times J$ table, asymptotically independent chi-squared components result when we compare columns 1 and 2, then combine them and compare them to column 3, and so forth. Each of the $J - 1$ statistics has df $= I - 1$. More refined partitions contain $(I - 1)(J - 1)$ statistics, each having df $= 1$. One such partition, suggested by Lancaster (1949), applies to the $(I - 1)(J - 1)$ 2×2 tables

$$
\begin{array}{c|c}
\displaystyle\sum_{a<i}\sum_{b<j} n_{ab} & \displaystyle\sum_{a<i} n_{aj} \\
\hline
\displaystyle\sum_{b<j} n_{ib} & n_{ij}
\end{array}
\tag{3.14}
$$

for $i = 2, \ldots, I$ and $j = 2, \ldots, J$.

3.3.6 Origin of Schizophrenia Example

Table 3.5, based on data presented by Gallagher et al. (1987), classifies a sample of psychiatrists by their school of psychiatric thought and by their opinion on the origin of schizophrenia. The independence model gives $G^2 = 23.04$, based on df $= 4$. We can better understand this association by partitioning G^2 into four independent components. The partitioning (3.14) applies to the subtables shown in Table 3.6.

The first subtable compares the eclectic and medical schools of psychiatric thought on whether the origin of schizophrenia is biogenic or environmental, given that the classification was in one of these two categories. For this subtable, $G^2 = 0.29$, based on df $= 1$. The second subtable compares these two schools in terms of the proportion of times the origin was ascribed to be a combination, rather than biogenic or environmental. This subtable gives $G^2 = 1.36$, based on df $= 1$. The sum

Table 3.5 Most Influential School of Psychiatric Thought and Ascribed Origin of Schizophrenia

School of Psychiatric Thought	Origin of Schizophrenia		
	Biogenic	Environmental	Combination
Eclectic	90	12	78
Medical	13	1	6
Psychoanalytic	19	13	50

Source: Reprinted with permission, based on Gallagher et al. (1987).

Table 3.6 Subtables Used in Partitioning Chi-Squared for Table 7.15[a]

	Bio	Env		Bio + Env	Com			Bio	Env		Bio + Env	Com
Ecl	90	12	Ecl	102	78		Ecl + Med	103	13	Ecl + Med	116	84
Med	13	1	Med	14	6		Psy	19	13	Psy	32	50

[a] Bio, biogenic; Com, combination; Ecl, eclectic; Env, environmental; Psy, psychoanalytic

of these first two components equals G^2 for the independence model applied to the first two rows of Table 3.5. There is little evidence of a difference between the eclectic and medical schools of thought on the ascribed origin of schizophrenia.

Next we combine the eclectic and medical schools and compare them to the psychoanalytic school. The third subtable in Table 3.6 compares them for the (biogenic, environmental) classification, giving $G^2 = 12.95$, with df = 1. The fourth subtable compares them for the (biogenic or environmental, combination) split, giving $G^2 = 8.43$, with df = 1.

The psychoanalytic school seems more likely than the other schools to ascribe the origins of schizophrenia as being a combination. Of those who chose either the biogenic or environmental origin, members of the psychoanalytic school were somewhat more likely than the other schools to choose the environmental origin. The sum of these four G^2 components equals the value of 23.04 for testing independence in the original table.

3.3.7 Rules for Partitioning

Goodman (1968, 1969, 1971b), Irwin (1949), Iverson (1979), and Lancaster (1949) gave rules that help in determining subtables for which components of chi-squared are independent. Among these are the following necessary conditions:

1. The degrees of freedom for the subtables must sum to the degrees of freedom for the original table.
2. Each cell count in the original table must be a cell count in one and only one subtable.
3. Each marginal total of the original table must be a marginal total for one and only one subtable.

To check empirically whether a certain partitioning gives independent components, we can note whether the G^2 values for the subtables sum to G^2 for the original table.

In this section we used the G^2 statistic, for which exact partitionings occur. The X^2 statistic does not equal the sum of the X^2 values for the separate tables in a partition. When the null hypotheses all hold, X^2 does have an asymptotic equivalence with G^2, however. In addition, when the table has small counts, it is safer to use X^2 to study the component tables.

Chi-squared tests simply indicate whether there is evidence of an association. They are rarely adequate for answering all questions we have about a data set. Many statisticians (e.g., Berkson 1938, Cochran 1954) have warned of the dangers of relying strictly on results of this test, rather than studying the nature of the association. The next section discusses estimation of parameters that describe various aspects of the association.

3.4 LARGE-SAMPLE CONFIDENCE INTERVALS

Chapter 2 presented several parameters that describe associations. The accuracy of estimators of the parameters is characterized by standard errors of their sampling distributions. For Poisson or multinomial sampling, large-sample distributions of most estimators are normal. This section presents methods for deriving large-sample standard errors and confidence intervals.

3.4.1 Estimating Odds Ratios

Let $\hat{\theta} = n_{11}n_{22}/n_{12}n_{21}$ denote the sample value of the odds ratio $\theta = \pi_{11}\pi_{22}/\pi_{12}\pi_{21}$ for a 2×2 table. The sample odds ratio equals 0 or ∞ if any $n_{ij} = 0$, and it is undefined if both entries in a row or column are zero. Since these outcomes have positive probability, the expected value and variance of $\hat{\theta}$ and $\log \hat{\theta}$ do not exist. In terms of bias and mean squared error, Gart and Zweiful (1967) and Haldane (1955) showed that the amended estimators

$$\tilde{\theta} = \frac{(n_{11} + 0.5)(n_{22} + 0.5)}{(n_{12} + 0.5)(n_{21} + 0.5)}$$

and $\log \tilde{\theta}$ behave well (see also Problem 12.4).

The estimators $\hat{\theta}$ and $\tilde{\theta}$ have the same asymptotic normal distribution around θ. The effect of adding 0.5 to cells disappears as $n \to \infty$. For small n, their distributions are highly skewed. When $\theta = 1$, for instance, $\hat{\theta}$ cannot be much smaller than θ (since $\hat{\theta} \geq 0$), but it could be much larger with nonnegligible probability. The log transform, having an additive rather than multiplicative structure, converges more rapidly to a normal distribution. For Poisson or multinomial sampling or for independent binomial sampling within the rows or within the columns, an estimated asymptotic standard error (denoted by ASE) of $\log(\hat{\theta})$ is

$$\hat{\sigma}(\log \hat{\theta}) = \left(\frac{1}{n_{11}} + \frac{1}{n_{12}} + \frac{1}{n_{21}} + \frac{1}{n_{22}} \right)^{1/2}. \tag{3.15}$$

Replacing $\{n_{ij}\}$ by $\{n_{ij} + 0.5\}$ in the ASE improves the estimator.

Let $z_{\alpha/2}$ denote the percentage point from the standard normal distribution having a right-tail probability equal to $\alpha/2$. By the large-sample normality of $\log(\hat{\theta})$,

$$\log \hat{\theta} \pm z_{\alpha/2} \hat{\sigma}(\log \hat{\theta})$$

is an approximate $100(1-\alpha)$ percent confidence interval for $\log \theta$. Exponentiating (taking antilogs of) endpoints of this confidence interval gives a confidence interval for θ.

3.4.2 Estimating Difference of Proportions and Relative Risk

Next we consider the difference of proportions and the relative risk for comparing conditional distributions of a column response variable within two rows. For these measures, we treat the rows as independent binomial samples. In row i, n_{i1} has a binomial distribution with sample size n_{i+} and probability $\pi_{1|i}$ of response in column 1. The sample proportion $p_{1|i} = n_{i1}/n_{i+}$ has expectation $\pi_{1|i}$ and variance $\pi_{1|i}(1 - \pi_{1|i})/n_{i+}$. Since the sample proportions $p_{1|1}$ and $p_{1|2}$ are independent, their difference has expectation

$$E(p_{1|1} - p_{1|2}) = \pi_{1|1} - \pi_{1|2}$$

and standard error

$$\sigma(p_{1|1} - p_{1|2}) = \left[\frac{\pi_{1|1}(1 - \pi_{1|1})}{n_{1+}} + \frac{\pi_{1|2}(1 - \pi_{1|2})}{n_{2+}} \right]^{1/2}. \tag{3.16}$$

We can estimate this by $\hat{\sigma}(p_{1|1} - p_{1|2})$, formula (3.16) with $\pi_{1|i}$ replaced by $p_{1|i}$. Then

$$(p_{1|1} - p_{1|2}) \pm z_{\alpha/2} \hat{\sigma}(p_{1|1} - p_{1|2}) \tag{3.17}$$

is a confidence interval for $\pi_{1|1} - \pi_{1|2}$.

The sample relative risk is $r = p_{1|1}/p_{1|2}$. The asymptotic standard error of $\log r$ is

$$\sigma(\log r) = \left(\frac{1 - \pi_{1|1}}{\pi_{1|1}n_{1+}} + \frac{1 - \pi_{1|2}}{\pi_{1|2}n_{2+}} \right)^{1/2}. \tag{3.18}$$

Both $\log r$ and the sample version of standard error (3.18) are undefined when $p_{1|1}$ and/or $p_{1|2}$ equal zero. A less biased estimator of the log relative risk $[\log(\pi_{1|1}) - \log(\pi_{1|2})]$ is

$$\log \tilde{r} = \log\left(\frac{n_{11} + \frac{1}{2}}{n_{1+} + \frac{1}{2}}\right) - \log\left(\frac{n_{21} + \frac{1}{2}}{n_{2+} + \frac{1}{2}}\right)$$

and a related confidence interval is

$$\log \tilde{r} \pm z_{\alpha/2}[(n_{11} + \tfrac{1}{2})^{-1} - (n_{1+} + \tfrac{1}{2})^{-1} + (n_{21} + \tfrac{1}{2})^{-1} - (n_{2+} + \tfrac{1}{2})^{-1}]^{1/2}.$$

Exponentiating endpoints gives a confidence interval for the relative risk.

3.4.3 Myocardial Infarction Examples Revisited

To illustrate inference for the odds ratio, we re-visit Table 2.2 on oral contraceptive practice and myocardial infarction. The sample value $\hat{\theta} = 2.55$ is close to $\tilde{\theta} = 2.54$, since no cell count is especially small. The ASE (3.14) of $\log \hat{\theta} = 0.937$ is 0.330. For the population this sample represents, a 95% confidence interval for $\log \theta$ is $0.937 \pm 1.96(0.330)$, or $(0.290, 1.584)$. This gives a confidence interval for θ of $[\exp(0.290), \exp(1.584)]$, or $(1.34, 4.88)$.

Since the confidence interval for θ does not contain 1.0, the odds of myocardial infarction seem higher for those using oral contraceptives. With this sample size, the estimate is rather imprecise of how much higher those odds are. Table 2.2 was based on retrospective sampling, so we cannot construct confidence intervals for the relative risk or difference of proportions for a myocardial infarction response.

Next, we re-visit Table 2.3, from a prospective study of aspirin use and myocardial infarction. The sample proportion of heart attacks was 0.0171 for physicians taking placebo, and 0.0094 for physicians taking aspirin. A 95% confidence interval for the log relative risk is $\log(0.0171/0.0094) \pm 0.237$, which translates to $(1.43, 2.30)$ for the relative risk. We are 95% confident that the risk is between 1.43 and 2.30 times higher for those taking placebo than for those taking aspirin. A 95% confidence interval for the difference of proportions is 0.008 ± 0.003, or $(0.005, 0.011)$. The diminished risk from taking aspirin is statistically significant.

3.4.4 Delta Method

We now introduce a method of deriving standard errors for large-sample inferences. Suppose an estimator is a function of statistics that are jointly asymptotically normally distributed. Then, under mild conditions, that estimator itself has a large-sample normal distribution. The method of deriving this distribution is called the *delta method*.

Suppose sample counts $\{n_i, \; i = 1, \ldots, N\}$ have a multinomial $(n, \{\pi_i\})$ distribution. The sample proportion $p_i = n_i/n$ has mean and variance

$$E(p_i) = \pi_i \quad \text{and} \quad E(p_i - \pi_i)^2 = \pi_i(1 - \pi_i)/n . \tag{3.19}$$

For $i \neq j$, the sample proportions p_i and p_j have covariance

$$\text{Cov}(p_i, p_j) = -\pi_i \pi_j/n . \tag{3.20}$$

Formulas (3.19) and (3.20) are proved in Section 12.1.5. The sample proportions $(p_1, p_2, \ldots, p_{N-1})$ have a large-sample multivariate normal distribution. For statistics that are functions of the sample proportions, the delta method implies the following result, also proved in Section 12.1.5:

Let ζ denote a differentiable function of $\{\pi_i\}$, and let $\hat{\zeta}$ denote the sample value of ζ for a multinomial sample. Let

$$\phi_i = \frac{\partial \zeta}{\partial \pi_i} , \quad i = 1, \ldots, N .$$

Then as $n \to \infty$, the distribution of $\sqrt{n}(\hat{\zeta} - \zeta)/\sigma$ converges to a standard normal, where

$$\sigma^2 = \sum \pi_i \phi_i^2 - \left(\sum \pi_i \phi_i \right)^2 . \tag{3.21}$$

The asymptotic variance depends on the cell probabilities $\{\pi_i\}$ and the partial derivatives of the measure with respect to $\{\pi_i\}$. In practice, we replace $\{\pi_i\}$ and $\{\phi_i\}$ in (3.21) by their sample values, yielding a ML estimate $\hat{\sigma}^2$ of σ^2. Then, $\hat{\sigma}/\sqrt{n}$ is an estimated asymptotic standard error for $\hat{\zeta}$. A large-sample confidence interval for ζ takes the form

$$\hat{\zeta} \pm z_{\alpha/2} \hat{\sigma}/\sqrt{n} .$$

We illustrate the delta method by applying it to the log odds ratio, taking $\zeta = \log \theta = \log \pi_{11} + \log \pi_{22} - \log \pi_{12} - \log \pi_{21}$. Since

$$\phi_{11} = \partial(\log \theta)/\partial \pi_{11} = \partial(\log \pi_{11})/\partial \pi_{11} = 1/\pi_{11}$$

$$\phi_{12} = -1/\pi_{12}, \qquad \phi_{21} = -1/\pi_{21}, \qquad \phi_{22} = 1/\pi_{22},$$

we obtain $\Sigma \Sigma \pi_{ij}\phi_{ij} = 0$ and $\sigma^2 = \Sigma \Sigma \pi_{ij}\phi_{ij}^2 = \Sigma \Sigma(1/\pi_{ij})$. The asymptotic standard error of $\log(\hat{\theta})$ for a multinomial sample $\{n_{ij}\}$ is

$$\sigma(\log \hat{\theta}) = \sigma/\sqrt{n} = \left(\sum \sum 1/n\pi_{ij}\right)^{1/2}.$$

Since $np_{ij} = n_{ij}$, we can estimate this standard error by (3.15).

3.4.5 ASE of Gamma and Other Measures

The delta method is also useful for deriving ASEs of other measures of association. For instance, for a cross-classification of ordinal variables, consider gamma (2.10). Let

$$\pi_{ij}^{(c)} = \sum_{a<i}\sum_{b<j} \pi_{ab} + \sum_{a>i}\sum_{b>j} \pi_{ab}$$

$$\pi_{ij}^{(d)} = \sum_{a<i}\sum_{b>j} \pi_{ab} + \sum_{a>i}\sum_{b<j} \pi_{ab}$$

where i and j are fixed values in the summations. The term $\pi_{ij}^{(c)}$ is the sum of probabilities for cells that are *concordant* when matched with the cell in row i and column j. The term $\pi_{ij}^{(d)}$ is the sum of the probabilities for cells that are *discordant* when matched with that cell.

Using the delta method with $\gamma = (\Pi_c - \Pi_d)/(\Pi_c + \Pi_d)$, we obtain

$$\Pi_c = \sum_i \sum_j \pi_{ij}\pi_{ij}^{(c)}, \qquad \Pi_d = \sum_i \sum_j \pi_{ij}\pi_{ij}^{(d)}$$

$$\partial\Pi_c/\partial\pi_{ij} = 2\pi_{ij}^{(c)}, \qquad \partial\Pi_d/\partial\pi_{ij} = 2\pi_{ij}^{(d)}$$

$$\phi_{ij} = [2(\Pi_c + \Pi_d)(\pi_{ij}^{(c)} - \pi_{ij}^{(d)}) - 2(\Pi_c - \Pi_d)(\pi_{ij}^{(c)} + \pi_{ij}^{(d)})]/(\Pi_c + \Pi_d)^2$$

$$= 4[\Pi_d\pi_{ij}^{(c)} - \Pi_c\pi_{ij}^{(d)}]/(\Pi_c + \Pi_d)^2$$

$$\sum_i \sum_j \pi_{ij}\phi_{ij} = 0.$$

From (3.21), sample gamma has a large-sample normal distribution, with

$$\sigma^2 = \frac{16}{(\Pi_c + \Pi_d)^4} \sum \sum \pi_{ij}[\Pi_d\pi_{ij}^{(c)} - \Pi_c\pi_{ij}^{(d)}]^2. \qquad (3.22)$$

Brown and Benedetti (1977) and Appendix A.5 of the manual for the BMDP statistical computer package (Dixon 1983) are good sources for standard error formulas for many other association measures.

When we substitute $\hat{\sigma}$ for σ in (3.21), the limiting distribution is still standard normal, but convergence may be slower. The equivalence in the large-sample distribution is justified as follows: The sample proportions converge in probability to the population proportions, by the weak law of large numbers. Since $\hat{\sigma}$ is a continuous function of the sample proportions, it converges in probability to σ, and $\sigma/\hat{\sigma}$ converges in probability to 1. Now

$$\sqrt{n}\,\frac{(\hat{\zeta} - \zeta)}{\hat{\sigma}} = \sqrt{n}\,\frac{(\hat{\zeta} - \zeta)}{\sigma} \times \left(\frac{\sigma}{\hat{\sigma}}\right).$$

The first term on the right-hand side converges in distribution to a standard normal, by (3.21), and the second term converges in probability to 1. Thus their product, the left-hand side, also has a limiting standard normal distribution.

3.4.6 Job Satisfaction Example

Sections 2.3.1 and 3.3.3 presented a cross-classification of job satisfaction with income. Using the sample version of (3.22), we obtain an ASE of 0.041 for $\hat{\gamma} = 0.127$. There is substantial evidence that $\gamma > 0$, and hence that job satisfaction and income are statistically dependent. An approximate 95% confidence interval for γ is $0.127 \pm 1.96(0.041)$, or (0.05, 0.21). The association between income and job satisfaction seems to be weak.

The conclusion that $\gamma > 0$ may be surprising, since the chi-squared analysis in Section 3.3.3 gave little indication of an association. Unlike gamma, X^2 and G^2 did not take advantage of the ordinal nature of the variables. We now have much stronger evidence of association than provided by the chi-squared tests, which were not designed to detect positive or negative trends in the association.

3.5 EXACT TESTS FOR SMALL SAMPLES

In the Preface to the first edition of *Statistical Methods for Research Workers*, R. A. Fisher stated ". . . the traditional machinery of statistical processes is wholly unsuited to the needs of practical research. Not only does it take a cannon to shoot a sparrow, but it misses the sparrow! The elaborate mechanism built on the theory of infinitely large samples is not accurate enough for simple laboratory data. Only by systematically tackling small sample problems on their merits does it seem possible to apply accurate tests to practical data."

The confidence interval and test procedures described in the previous three sections are large-sample methods. They apply as the sample size n grows, for a fixed number of cells N. As $n \to \infty$, the expected frequencies $\{m_i = n\pi_i\}$ in the cells grow. As they grow, the multinomial distribution for $\{n_i\}$ is better approximated by a multivariate normal, and X^2 and G^2 have more nearly chi-squared distributions.

There are alternatives to asymptotic procedures when the sample size is small. Using the computing power that is widespread in this modern era, we can use *exact* distributions rather than large-sample approximations. This section gives exact procedures for testing independence in two-way tables.

3.5.1 Fisher's Exact Test

We first study the 2×2 case. Under the null hypothesis of independence, an exact distribution that is free of any unknown parameters results from conditioning on the marginal frequencies in *both* margins. When we assume Poisson, multinomial, or independent multinomial sampling, and then condition on the observed marginal totals, we obtain the hypergeometric distribution

$$\frac{\binom{n_{1+}}{n_{11}} \binom{n_{2+}}{n_{+1} - n_{11}}}{\binom{n}{n_{+1}}} . \tag{3.23}$$

This formula expresses the distribution of the four cell counts in terms of only one element, n_{11}. Given the marginal totals, the value for n_{11} determines the other three cell counts. The range of possible values for n_{11} in this distribution is $m_- \leq n_{11} \leq m_+$, where $m_- = \max(0, n_{1+} + n_{+1} - n)$ and $m_+ = \min(n_{1+}, n_{+1})$.

To test independence, the P-value is the sum of hypergeometric probabilities for outcomes at least as favorable to the alternative hypothesis as the observed outcome. To illustrate, consider the alternative hypothesis H_a: $\theta > 1$ for 2×2 tables. For the given marginal totals, tables having larger n_{11} values have larger odds ratios, and hence stronger evidence in favor of this alternative. Thus, the P-value equals the hypergeometric probability that n_{11} is at least as large as the observed value. The reference set consists of tables having the same row and column totals as the observed table. This test for 2×2 tables is called *Fisher's exact test* (Fisher 1934, 1935c, Irwin 1935).

3.5.2 Fisher's Tea Drinker

R. A. Fisher (1935a) described the following experiment: When drinking tea, a British woman claimed to be able to distinguish whether milk or tea was added to the cup first. To test her claim, she was given eight cups of tea, in four of which milk was added first. She was told that there were four cups of each type, so that she should make four predictions of each order. The order of presenting the cups to her was random. Table 3.7 shows results of the experiment. We conduct Fisher's exact test of H_0: $\theta = 1$ against H_a: $\theta > 1$. The alternative hypothesis reflects the woman's claim, predicting a positive association between true order of pouring and the woman's guess.

The experimental design fixes both marginal distributions, since the woman was asked to guess which four cups had milk added first. Thus, it is completely natural to use the hypergeometric for the null distribution of n_{11}. The P-value for Fisher's exact test is the null probability of Table 3.7 and of tables that would have given even more evidence in favor of her claim. The observed table, three correct guesses of the cups having milk added first, has null probability

$$\frac{\binom{4}{3}\binom{4}{1}}{\binom{8}{4}} = 0.229 .$$

There is only one more extreme table, four correct guesses. It has $n_{11} = n_{22} = 4$ and $n_{12} = n_{21} = 0$, and a probability of $\binom{4}{4}\binom{4}{0}/\binom{8}{4} = 0.014$. The P-value equals 0.243. The experiment did not establish an association between the actual order of pouring and the woman's guess. Of course, it is difficult to do so with such a small sample.

For the one-sided alternative in Fisher's exact test, Davis (1986a)

Table 3.7 Fisher's Tea Tasting Experiment

Poured First	Guess Poured First		Total
	Milk	Tea	
Milk	3	1	4
Tea	1	3	4
Total	4	4	

Source: Based on experiment described by Fisher (1935a).

showed that the same P-value results when we order the tables using the difference of proportions, odds ratio, or n_{11}. For the two-sided alternative, the P-value is often defined as the sum of the probabilities of tables no more likely to occur than the observed table. In this case, Davis noted that other criteria for ordering the tables can yield different P-values. For instance, a different P-value might result from defining it to be the null probability that X^2 is at least as large as the observed X^2.

The exact distribution (3.23) is highly discrete for small samples, in the sense that n_{11} can assume relatively few values. When we conduct Fisher's exact test using a fixed significance level, the P-value has a small number of possible values (for the given margins), so it is usually not possible to achieve that level exactly. In the tea-tasting experiment, n_{11} can only take values 4, 3, 2, 1, 0, and the P-values are restricted to 0.014, 0.243, 0.757, 0.986, and 1.0.

It *is* possible to achieve any fixed significance level by employing randomization on the boundary of the critical region, in deciding whether to reject H_0. For the tea tasting experiment, suppose we reject the null hypothesis with probability 1.0 when $n_{11} = 4$, with probability 0.157 when $n_{11} = 3$, and with probability 0.0 otherwise. For expectation taken with respect to the null hypergeometric distribution of n_{11}, the significance level equals

$$P(\text{Reject } H_0) = E[P(\text{Reject } H_0 \mid n_{11})$$
$$= 1.0(0.014) + 0.157(0.229) + 0.0 \times P(n_{11} \leq 2) = 0.05 .$$

With this extension, Tocher (1950) showed that Fisher's test is uniformly most powerful unbiased. In practice, randomized tests are rarely (if ever) justified, and we recommend simply reporting the P-value.

3.5.3 Derivation of Exact Conditional Distribution

We now show steps that lead to the hypergeometric distribution for testing independence. Suppose there is independent multinomial sampling within rows of an $I \times J$ contingency table. For instance, we may want to compare I treatment groups, by fixing row totals $\{n_{i+}\}$ and estimating the I conditional distributions $\{\pi_{j \mid i}, j = 1, \ldots, J\}$. Under the null hypothesis of independence, $\pi_{j \mid 1} = \pi_{j \mid 2} = \cdots = \pi_{j \mid I} = \pi_{+j}$, for $j = 1, \ldots, J$. The product of the multinomial probability functions (3.3) from the I rows then simplifies to

$$\prod_i \left[\frac{n_{i+}!}{\prod_j n_{ij}!} \left(\prod_j \pi_{j|i}^{n_{ij}} \right) \right] = \frac{\left(\prod_i n_{i+}! \right)\left(\prod_j \pi_{+j}^{n_{+j}} \right)}{\prod_i \prod_j n_{ij}!} . \tag{3.24}$$

This exact distribution for $\{n_{ij}\}$ has limited use, because it depends on unknown parameters $\{\pi_{+j}\}$. These are nuisance parameters, since they do not describe the association. In statistical inference, a standard way of eliminating nuisance parameters is to condition on sufficient statistics for them. By the definition of sufficiency, the resulting conditional distribution does not depend on those parameters.

The contribution of $\{\pi_{+j}\}$ to the product multinomial distribution (3.24) depends on the data only through $\{n_{+j}\}$, which are sufficient statistics. The $\{n_{+j}\}$ have the multinomial $(n, \{\pi_{+j}\})$ distribution, namely

$$\left(\frac{n!}{\prod_j n_{+j}!} \right)\prod_j \pi_{+j}^{n_{+j}} . \tag{3.25}$$

The joint probability function of $\{n_{ij}\}$ and $\{n_{+j}\}$ is identical to the probability function of $\{n_{ij}\}$, since $\{n_{+j}\}$ are determined by $\{n_{ij}\}$. Thus, the probability function of $\{n_{ij}\}$, conditional on $\{n_{+j}\}$, equals the probability function (3.24) of $\{n_{ij}\}$ divided by the probability function (3.25) of $\{n_{+j}\}$, or

$$\frac{\left(\prod_i n_{i+}! \right)\left(\prod_j n_{+j}! \right)}{n! \prod_i \prod_j n_{ij}!} . \tag{3.26}$$

When there is a single multinomial sample for the table, the unknown parameters are $\{\pi_{ij}\}$. For testing independence ($\pi_{ij} = \pi_{i+} \pi_{+j}$ all i and j), distribution (3.26) results from conditioning on both the row and column totals. These are sufficient statistics for the unknown row and column probabilities that determine the null distribution of cell counts. For either sampling model, both sets of margins are fixed after the conditioning. The end result (3.26) does not depend on unknown parameters, and thus permits exact inference.

Probability distribution (3.26) is the *multiple hypergeometric* distribution. It is defined over the set of contingency tables $\{n_{ij}\}$ having the same row and column totals as the observed table. For 2×2 tables, it simplifies to hypergeometric distribution (3.23).

3.5.4 Other Exact Tests of Independence

Exact tests for tables of size larger than 2×2 utilize the multiple hypergeometric distribution. Freeman and Halton (1951) defined the *P*-value as the probability of the set of tables no more likely to occur than the observed table. Exact tests can just as easily order the tables using other measures than their probabilities. For instance, we could use the exact distribution of X^2, taking P to be the null probability that X^2 is at least as large as observed. For classifications having ordered categories, we should use some ordinal index. For instance, when the alternative hypothesis predicts a positive association, we could let the *P*-value be $P[C - D \geq (C - D)_o]$, where $(C - D)_o$ denotes the observed difference between the numbers of concordant and discordant pairs.

We illustrate an exact test for an ordinal response using Table 3.8, from Shapiro et al. (1979). This is a cross-classification of level of smoking and myocardial infarction, for a sample of young women. The table contains very small counts in the second row, and asymptotic approaches are inappropriate. For this table, $C = 175$ and $D = 12$, so $(C - D)_o = 163$. Given the marginal counts, the only other table having $(C - D)$ at least this large has counts (25, 26, 11) for row 1 and (0, 0, 4) in row 2. From (3.26), conditional on both sets of margins, $P[(C - D) \geq 163] = 0.0183$. Though the sample contains only four myocardial infarction patients, there is evidence of a positive association between level of smoking and incidence of myocardial infarction. The evidence is stronger than we obtain using X^2, which ignores the ordering of categories. The exact probability that X^2 is at least as large as the observed value of 6.96

Table 3.8 Example for Exact Conditional Test

	Smoking Level (Cigarettes/Day)		
	0	1–24	>25
Control	25	25	12
Myocardial infarction	0	1	3

Source: Reprinted with permission, based on Table 5 in Shapiro et al. (1979).

is 0.052. The asymptotic approximation for this using the chi-squared distribution with df = 2 is 0.031. The Freeman-Halton P-value is 0.034.

Recently developed algorithms make computations for exact tests feasible for most tables for which asymptotic approximations are invalid, even when $I > 2$ and/or $J > 2$. See Baglivo et al. (1988), Cox and Plackett (1980), Mehta and Patel (1983), and Pagano and Halvorsen (1981). Computing time can be reduced by estimating the P-value to within a satisfactory margin, such as 0.001 with probability 0.99. Mehta et al. (1988) described a fast importance sampling algorithm for doing this.

3.5.5 Conditional vs. Unconditional Tests

When we assume Poisson or multinomial sampling, we treat at most one marginal distribution as fixed. It may then seem artificial to use a method that conditions on *both* sets of marginal counts. We next describe an alternative small-sample test for 2×2 tables that conditions on only one margin.

When the rows are independent binomial samples, we regard the row totals as fixed. Suppose we want to test equality of $\pi_{1|1}$ and $\pi_{1|2}$. For fixed row totals, $p_{1|1} - p_{1|2}$ can take on a discrete set of values, one of which is the observed difference d_o. Though computationally intensive, for fixed $\pi_{1|1} = \pi_{1|2} = \pi$, we can use the independent binomial distributions to compute $P_\pi(p_{1|1} - p_{1|2} \geq d_o)$. Since π is unknown, the P-value is defined to be

$$P = \max_{0 \leq \pi \leq 1} P_\pi(p_{1|1} - p_{1|2} \geq d_o).$$

This is an *unconditional* test for comparing binomial probabilities. Originally proposed by Barnard (1945, 1947), it was later (1949) refuted by him in favor of Fisher's exact conditional test. Suissa and Shuster (1985) and Haber (1986, 1987) gave computational details and comparisons with Fisher's test. A disadvantage of the unconditional test is that computations are not currently feasible for larger tables or more complex problems.

Arguments in favor of conditioning on both sets of marginal totals are that it eliminates nuisance parameters, the margins contain little information about the association (Yates 1984, Haber 1988), and the method generalizes to give exact inference for other contingency table problems. The resulting hypergeometric distribution is highly discrete, however. There are few possible values for n_{11}, making it difficult to obtain a small P-value. In repeated use with a fixed significance level, the actual type-1 error probability may be much smaller than the nominal value.

This problem is partly unavoidable. Statistics having discrete distributions are necessarily conservative in terms of achieving nominal significance levels. Because the *unconditional* test fixes only one margin, however, it has many more tables in the reference set for its sampling distribution. That sampling distribution is less discrete, in the sense that there is a much richer array of possible *P*-values than with Fisher's exact test. When used with fixed significance levels, the test tends to be less conservative and more powerful than Fisher's exact test.

When both sets of marginal totals are naturally fixed, such as in Table 3.7, the high degree of discreteness is unavoidable and Fisher's exact test is the best procedure. Regardless of which margins are naturally fixed, a way to adjust for discreteness is to let the *P*-value be *half* the probability of the observed result plus the probability of more extreme results. Compared to the ordinary *P*-value, this *mid P*-value is more nearly uniformly distributed under the null hypothesis. It has a null expected value of 0.5, as the regular *P*-value has for continuous random variables. The mid *P*-value has been recommended (e.g., by Lancaster 1961, and by R. L. Plackett, in his discussion of Yates's 1984 article) as a good compromise between having a conservative test and using randomization on the boundary to eliminate problems from discreteness.

As *I* and/or *J* increase, the number of possible values in exact conditional sampling distributions tends to increase. Thus, the argument that Fisher-type tests are conservative for testing at fixed significance levels loses its force.

3.6 EXACT NON-NULL INFERENCE

Exact non-null distribution theory also derives from conditioning arguments. The exact distributions are used in constructing small-sample confidence intervals for measures such as the odds ratio. For simplicity, we restrict our attention here to 2×2 tables.

3.6.1 Inference for the Odds Ratio

For multinomial sampling, the distribution of $\{n_{ij}\}$ depends on the cell probabilities $\{\pi_{ij}\}$. The odds ratio is $\theta = \pi_{11}\pi_{22} / \pi_{12}\pi_{21} = \pi_{11}(1 - \pi_{1+} - \pi_{+1} + \pi_{11}) / (\pi_{1+} - \pi_{11})(\pi_{+1} - \pi_{11})$. Hence, π_{11} is a function of the odds ratio and the marginal probabilities. Since a similar argument applies to any π_{ij}, we can express the distribution of $\{n_{ij}\}$ using parameters $\{\theta, \pi_{1+}, \pi_{+1}\}$. Conditional on $\{n_{1+}, n_{+1}\}$, the distribution of $\{n_{ij}\}$ depends only on θ. Since n_{11} determines all other cell counts, given the marginal totals, the conditional distribution of $\{n_{ij}\}$ is some function $f(n_{11}; n_{1+}, n_{+1}, \theta)$. This distribution is (Fisher 1935c)

$$f(n_{11}; n_{1+}, n_{+1}, \theta) = \frac{\binom{n_{1+}}{n_{11}}\binom{n - n_{1+}}{n_{+1} - n_{11}}\theta^{n_{11}}}{\sum\limits_{u=m_-}^{m_+} \binom{n_{1+}}{u}\binom{n - n_{1+}}{n_{+1} - u}\theta^{u}} \quad \text{for } m_- \leq n_{11} \leq m_+ .$$

$$(3.27)$$

For a fixed observed n_{11}, the *conditional ML* estimate of θ is the value of θ that maximizes this probability. Differentiating the log likelihood with respect to θ shows that this estimate satisfies the equation $n_{11} = E(n_{11})$, where the expectation refers to distribution (3.27). The equation has a unique solution, and is solved using iterative methods. See Cornfield (1956), Cox (1970), and Plackett (1981, p. 45). This estimator differs from the *unconditional* ML estimator $\hat{\theta} = n_{11}n_{22}/n_{12}n_{21}$, which is a function of the ML estimates of $\{\pi_{ij}\}$ for the multinomial distribution of $\{n_{ij}\}$.

For testing H_0: $\theta = \theta_0$ against H_a: $\theta > \theta_0$, the P-value is the sum of $f(x; n_{1+}, n_{+1}, \theta_0)$ values for x at least as large as the observed n_{11}. For testing against H_0: $\theta < \theta_0$, the P-value sums the probabilities for $x \leq n_{11}$. When $\theta_0 = 1$, these tests are Fisher's exact test. We obtain a confidence interval for θ by inverting the test. The lower end point is the θ_0 value for which $P = \alpha/2$ in testing H_0: $\theta = \theta_0$ against H_a: $\theta > \theta_0$. The upper end point is the θ_0 value for which $P = \alpha/2$ in testing H_0 against H_a: $\theta < \theta_0$. The confidence coefficient is *at least* $1 - \alpha$, rather than *exactly* $1 - \alpha$, because the distribution of n_{11} is discrete.

Thomas (1971) and Baptista and Pike (1977) gave computing algorithms for calculating this interval, and Mehta et al. (1985) gave a faster algorithm for a more general problem. Section 14.1 describes a computer program, StatXact, that calculates conditional ML estimates and exact confidence intervals and tests for odds ratios.

3.6.2 Contraceptive Practice Example

To illustrate, we return to Table 2.2 on oral contraceptive practice and myocardial infarction. Section 3.4.3 obtained an unconditional ML estimate for θ of 2.55, and an approximate 95% confidence interval of (1.34, 4.88). Using a computer algorithm with the exact conditional distribution, we find that the interval (1.26, 5.10) has guaranteed confidence coefficient at least 95%. The conditional ML estimate of θ is 2.54. The sample size is relatively large, and the asymptotic approach is adequate.

CHAPTER NOTES

Section 3.1: Sampling Distributions

3.1 Breslow (1984), Cox (1983), Cox and Snell (1989, pp. 106–115), Crowder (1978), Follman and Lambert (1989), Kleinman (1975), Kupper and Haseman (1978), McCullagh and Nelder (1989), Moore (1986), Pack (1986), Paul and Plackett (1978), Paul et al. (1989), Williams (1982), and Wilson (1989) presented ways of treating overdispersed data.

  Social science applications commonly have complex sampling designs, incorporating such features as clustering and stratification. For discussion of the necessary modifications in inferential procedures for categorical data, see Altham (1976), Brier (1980), Clogg and Eliason (1987), Cohen (1976), Fay (1985), Holt et al. (1980), Koch et al. (1975), Koehler and Wilson (1986), Landis et al. (1987), Plackett and Paul (1978), Rao and Thomas (1989), Rao and Scott (1981, 1987), Rosner (1989), Shuster and Downing (1976), Tavaré (1983), and Tavaré and Altham (1983).

Section 3.2: Testing Goodness of Fit

3.2 The chi-squared distribution with df $= \nu$ has mean ν, variance 2ν, and skewness $(8/\nu)^{\frac{1}{2}}$. It is approximately normal when ν is large. See Kendall and Stuart (1979, Vol. 2) and Lancaster (1969) for other properties. Cochran (1952) gave an interesting historical survey of chi-squared tests of fit. See also Cressie and Read (1989), Koch and Bhapkar (1982), Moore (1986), and Read and Cressie (1988).

Section 3.3: Testing Independence

3.3 The $\{\hat{m}_{ij}\}$ in the test of independence are *exact* (rather than *estimated*) expected values for hypergeometric sampling, which treats both margins as fixed. Haldane (1940) gave the exact null mean and variance of X^2. Section 7.1.4 justifies the partitionings of G^2 given here. Sections 4.4 and 8.4 discuss the principle behind more powerful tests for ordered categories.

  For 2×2 tables, Yates (1934) suggested a correction to the Pearson statistic,

$$X_c^2 = \sum \sum \frac{(|n_{ij} - \hat{m}_{ij}| - 0.5)^2}{\hat{m}_{ij}} \,,$$

to adjust for using the continuous chi-squared distribution to approximate a discrete distribution. The corrected statistic gives P-values (from the chi-squared distribution) that better approximate hypergeometric probabilities obtained with Fisher's exact test. This adjustment is *not* intended to make the sampling distribution closer to the reference chi-squared distribution. In fact, the uncorrected statistic is better in that respect (Grizzle 1967, Conover 1974). Haber (1980, 1982), Plackett (1964), Upton (1982), and Yates (1984) discussed the appropriateness of the continuity correction. Since computer packages now make it possible to conduct Fisher's exact test even for fairly large samples, this correction is less important than it once was.

3.4 Pearson (1922) reacted angrily to Fisher's suggestion that he had incorrectly stated df for the test of independence. Pearson claimed that df $= 3$ for a 2×2 table that had only the total sample size fixed. He argued "I hold that such a view (Fisher's) is entirely erroneous, and that the writer has done no service to the science of statistics by giving it broad-cast circulation in the pages of the *Journal of the Royal Statistical Society*. ... I trust my critic will pardon me for comparing him with Don Quixote tilting at the windmill; he must either destroy himself, or the whole theory of probable errors, for they are invariably based on using sample values for those of the sampled population unknown to us." Fisher was unable to get his rebuttal published by the Royal Statistical Society, and ultimately resigned his membership.

A few years later, Fisher performed an empirical test that supported his df formula, and proved embarrassing for Pearson. Using 11,668 2×2 tables randomly generated under conditions of independence by Egon Pearson (Karl Pearson's son), Fisher noted the average sample value of X^2 was nearly identically 1.0, the value predicted by the formula df $= (I - 1)(J - 1)$. In the preface to his collected papers, Fisher later noted of his 1922 paper that "it had to find its way to publication past critics who, in the first place, could not believe that Pearson's work stood in need of correction, and who, if this had to be admitted, were sure that they themselves had corrected it."

Joan Fisher Box (1978) gave an entertaining presentation of this and other conflicts between Fisher and Pearson. Fienberg (1980b) summarized Fisher's contributions to categorical data analysis.

Section 3.4: Large-Sample Confidence Intervals

3.5 The delta method can be applied directly with θ to obtain $\hat{\sigma}$ ($\hat{\theta}$) and a confidence interval $\hat{\theta} \pm z_{\alpha/2}\hat{\sigma}(\hat{\theta})$. This is not recommended—$\hat{\theta}$ converges more slowly to normality than log $\hat{\theta}$, and this interval could contain negative values and does not give equivalent results to the one obtained using $1/\hat{\theta}$ and its standard error. The confidence interval based on the asymptotic distribution of log $\hat{\theta}$ also is imperfect, because the true confidence coefficient tends to be smaller than the nominal value. Gart and Thomas (1972) and Fleiss (1981, pp. 71-74) discussed various methods of constructing confidence intervals for the odds ratio. Goodman (1964b) gave asymptotic simultaneous confidence intervals for all odds ratios in an $I \times J$ table.

Hauck and Anderson (1986) noted that the interval (3.17) for the difference of proportions tends to be too narrow, and recommended an adjustment. Koopman (1984) and Miettinen and Nurminen (1985) used likelihood methods to obtain alternative confidence intervals for the relative risk that, according to Gart and Nam (1988), more accurately approximate the nominal confidence level.

Goodman and Kruskal (1963, 1972) used a formula similar to (3.21) to obtain asymptotic variances for independent multinomial sampling. Suppose each row of a two-way contingency table consists of a separate multinomial sample. A generic measure ζ is a function of $\{\pi_{i+}\}$ and $\{\pi_{j|i}\}$. Suppose we know $\{\pi_{i+}\}$, so the sample measure $\hat{\zeta}$ is a function of $\{p_{j|i}\}$. Denote the proportion sampled from the ith row by ω_i. The sampling is proportional when $\{\omega_i = \pi_{i+}\}$. The asymptotic variance of $\sqrt{n}(\hat{\zeta} - \zeta)$ equals

$$\sigma^2 = \sum_i \frac{1}{\omega_i} \left[\sum_j \pi_{j|i} \phi_{ij}^{+2} - \left(\sum_j \pi_{j|i} \phi_{ij}^{+} \right)^2 \right]$$

where $\phi_{ij}^{+} = \partial\zeta/\partial\pi_{j|i}$.

Section 3.5: Exact Tests for Small Samples

3.6 The UMPU property of Fisher's exact test follows from conditioning on a sufficient statistic that is complete and has distribution in the exponential family. Fleiss (1981), Gail and Gart (1973), and Suissa and Shuster (1985) showed how to determine minimal sample sizes for obtaining fixed power in Fisher's exact test. There has been much controversy over the appropriateness of the conditioning argument used in that test. See, for instance, Barnard (1945, 1947, 1949, 1979), Berkson (1978), Fisher (1956), Kempthorne (1979), Lloyd (1988), Pearson (1947), Rice (1988, and the following discussion), Suissa and Shuster (1984, 1985), Upton (1982), and Yates (1934, 1984 and following discussion). Yates and discussants also addressed the controversy behind the choice of P-value for the two-sided version of Fisher's test. Upton (1982) surveyed several ways of testing independence in 2 × 2 tables. Altham (1969) provided an exact Bayesian analysis for 2 × 2 tables.

The exact test for 2 × J tables using $C - D$ is equivalent to a Wilcoxon test with average ranks for the levels of the ordinal response. See Agresti and Wackerly (1977) and Mehta et al. (1984). Agresti et al. (1990) and Patefield (1982) suggested alternative exact tests for cross-classifications of ordinal variables.

Section 3.6: Exact Non-Null Inference

3.7 Suppose (θ, λ) has minimal sufficient statistic (T, U), where λ is a nuisance parameter. Cox and Hinkley (1974, pp. 35) defined U to be *ancillary* for θ if its distribution depends only on λ, and the distribution of T given U depends only on θ. For 2 × 2 tables, let θ be the odds ratio, $\lambda = (\pi_{1+}, \pi_{+1})$, $T = n_{11}$, and $U = (n_{1+}, n_{+1})$. Then U is not ancillary, because its distribution depends on θ as well as λ. Using a definition due to Godambe (1980), Bhapkar (1986) referred to the marginals U as *partial ancillary* for θ. This means that the distribution of the data, given U, depends only on θ, and that for fixed θ, the family of distributions of U for various λ is complete.

Small-sample confidence intervals for the difference of proportions and relative risk were discussed by Beal (1987), Bedrick (1987), Gart and Nam (1988), and Santner and Snell (1980). Gart and Nam reviewed "exact" and approximate methods, suggested corrections for skewness, and provided computing algorithms.

PROBLEMS

Applications

3.1 Snedecor (1937, p.156) described an experiment on chlorophyll inheritance in maize. For 1103 seedlings of self-fertilized, heterozygous green plants, 854 seedlings were green and 249 were yellow. Theory predicts the ratio of green to yellow is three to one. Test the hypothesis that 3:1 is the true ratio.

3.2 In a large city, 30% of the population is black. Prospective jurors for court trials are selected from this population. For each selection of a juror, π denotes the probability that a black person is selected.

For random sampling, $\pi = 0.3$. In a sample of 30 prospective jurors, one black is chosen.

  a. Assuming binomial sampling, find the exact P-value for testing H_0: $\pi = 0.3$ against H_a: $\pi < 0.3$.

  b. For large samples, why is the z statistic (3.9) more appropriate than X^2 for a one-sided alternative such as this?

3.3 Table 3.9 contains Ladislaus von Bortkiewicz's data on deaths of soldiers in the Prussian army from kicks by army mules. The data refer to 10 army corps, each observed for 20 years. In 109 corps-years of exposure, there were no deaths, in 65 corps-years there was one death, and so forth. Fisher (1934) and Quine and Seneta (1987) discussed these data.

Table 3.9

No. Deaths	No. Corps-Years
0	109
1	65
2	22
3	3
4	1
5 or more	0

  a. Estimate the mean number of deaths per corps year.

  b. Test whether probabilities of occurrences in these five categories follow a Poisson distribution (truncated for 4 and above).

3.4 Refer to Section 3.2.4. Let a denote the number of calves that got a primary, secondary, and tertiary infection, b the number that received a primary and secondary infection but not a tertiary one, c the number that received a primary infection but not a secondary infection, and let d the number that did not receive a primary infection. Let π be the probability of a primary infection.

  a. Consider the hypothesis that the probability of infection at time t, given infection at times $1, \ldots, t-1$, is also π, for $t = 2,3$. Show the ML estimate of π is $\hat{\pi} = (3a + 2b + c)/(3a + 3b + 2c + d)$.

  b. For the sample of calves analyzed in Section 3.2.4, $a = 5$, $b = 25$, $c = 63$, and $d = 63$. Test the hypothesis, and interpret results.

3.5 Refer to Table 2.3.

   a. Find the P-value for testing the hypothesis that the incidence of heart attacks is independent of aspirin intake, using X^2 and G^2. Interpret.

   b. Show how to partition chi-squared into two components. Interpret.

   c. Construct a 2×2 table by combining the first two columns. Calculate an approximate 95% confidence interval for the odds ratio. Compare results to those in Section 3.4.3 for the relative risk.

3.6 Helmes and Fekken (1986) reported Table 3.10, which classifies a sample of psychiatric patients by their diagnosis and by whether their treatment prescribed drugs. Partition chi-squared to describe differences and similarities among the diagnoses in terms of the relative frequency of prescribed drugs.

Table 3.10

Diagnosis	Drugs	No Drugs
Schizophrenia	105	8
Affective disorder	12	2
Neurosis	18	19
Personality disorder	47	52
Special symptoms	0	13

Source: Reprinted with permission from Helmes and Fekken (1986).

3.7 Refer to Table 2.7. Construct and interpret a 90% confidence interval for the population odds ratio between attitude toward the death penalty and attitude toward gun registration. Are the endpoints of the interval equidistant from the point estimate of the odds ratio? Why not?

3.8 Refer to Problem 2.9. Why is the Pearson chi-squared test not the optimal way to test independence between sexual fun of husbands and wives? Suggest an alternative approach.

3.9 Refer to the data in Section 6.7.1. Obtain a 95% confidence interval for gamma. Interpret the association between schooling and attitude toward abortion.

3.10 Table 3.11 contains results of a study by Mendenhall et al. (1984) to compare radiation therapy with surgery in treating cancer of the larynx. Use Fisher's exact test to test H_0: $\theta = 1$ against H_a: $\theta > 1$.

Table 3.11

	Cancer Controlled	Cancer Not Controlled
Surgery	21	2
Radiation therapy	15	3

Source: Reprinted with permission from Mendenhall et al. (1984), Pergamon Press plc.

3.11 Refer to the previous problem. Test H_0: $\theta = 1$ against H_a: $\theta \neq 1$. Explain how you formed the P-value. Using exact conditional methods, find a confidence interval for the odds ratio that has confidence coefficient at least 0.95.

3.12 For a flip of a given coin, let π denote the probability of the outcome "head." Consider the experiment of testing H_0: $\pi = 0.5$ against H_a: $\pi > 0.5$, using $n = 4$ independent flips. Show that the null probability of rejecting H_0 at the 0.05 significance level is 0.0 for the exact binomial test, and 1/16 using the approximate normal test (3.9). Since the true level for the normal test comes closer to matching the nominal level of 0.05, does this imply that the approximate test is more appropriate than the exact test? Why not?

Theory and Methods

3.13 The moment generating function of a chi-squared random variable with df $= \nu$ is $m(t) = (1 - 2t)^{-\nu/2}$, for $|t| < \frac{1}{2}$. Use this to prove the reproductive property of the chi-squared distribution.

3.14 Suppose $\{n_{ij}\}$ are independent Poisson random variables with means $\{m_{ij}\}$. Conditional on $\{n_{i+}\}$, show there is independent multinomial sampling within the rows.

3.15 For multinomial sampling, assuming statistical independence of two categorical responses, show that the ML estimator of π_{ij} is $\hat{\pi}_{ij} = n_{i+}n_{+j}/n^2$.

3.16 Consider a sequence of independent Bernoulli trials, with probability π of success on each trial. Let Y be the number of successes before the kth failure. Show that the probability mass function of Y is the *negative binomial*,

$$\frac{(y + k - 1)!}{y!(k - 1)!} \, \pi^y (1 - \pi)^k \,, \quad y = 0, 1, 2, \dots .$$

(Note: $E(Y) = k\pi/(1 - \pi)$ and $\text{Var}(Y) = k\pi/(1 - \pi)^2$; thus $\text{Var}(Y) > E(Y)$, whereas they are equal for the Poisson. The negative binomial may be useful when there is overdispersion for a Poisson model. The Poisson is the limit of the negative binomial as $k \to \infty$ and $\pi \to 0$ with $k\pi = m$ fixed.)

3.17 Conditional on m, Y has a Poisson distribution with mean m. However, values of m for the target population vary according to a gamma density

$$g(m) = \frac{1}{\Gamma(\alpha)} \, \beta^\alpha m^{\alpha - 1} \exp(-\beta m) \,, \quad m > 0$$

where $\alpha > 0$ and $\beta > 0$. Show that the unconditional distribution of Y is negative binomial, with mean $\lambda = E(m) = \alpha/\beta$ and variance $\lambda(1 + \beta)/\beta$.

3.18 Suppose Y_i is a Bernoulli random variable, $P(Y_i = 1) = 1 - P(Y_i = 0) = \pi$, $i = 1, \dots, n$. When $\{Y_i\}$ are independent, $S = \Sigma Y_i$ has a binomial distribution. There is overdispersion when $\text{Var}(S) > n\pi(1 - \pi)$, the binomial value.

   a. Show that when $\{Y_i\}$ have pairwise correlation $\rho > 0$, $\text{Var}(S) > n\pi(1 - \pi)$. (Altham (1978) discussed generalizations of the binomial that allow for correlation between Bernoulli random variables.)

   b. Suppose there is heterogeneity, in the sense that π is a random variable with mean $\bar{\pi}$ and positive variance. Show $\text{Var}(S) > n\bar{\pi}(1 - \bar{\pi})$. (McCullagh and Nelder (1989, Section 4.5), Paul and Plackett (1978) and Williams (1982) discussed related models. One approach assumes π has a beta distribution, in which case the unconditional distribution for S is the *beta-binomial*.)

3.19 Refer to quadratic form (3.8).

   a. Verify that the matrix quoted in the text for Σ_0^{-1} is the inverse of Σ_0.

b. Show that (3.8) simplifies to Pearson's statistic (3.7).

c. For z statistic (3.9), show $z^2 = X^2$.

3.20 Use the likelihood ratio approach to derive a statistic for testing that multinomial probabilities equal certain specified values.

3.21 Probabilities in a 2×2 table are hypothesized to follow the pattern $\pi_{11} = \theta^2$, $\pi_{12} = \pi_{21} = \theta(1 - \theta)$, $\pi_{22} = (1 - \theta)^2$, for some unknown θ. That is, the marginal distributions are identical (e.g., $\pi_{1+} = \pi_{+1}$), and there is independence.

   a. For a multinomial sample, show the ML estimate of θ is $\hat{\theta} = (p_{1+} + p_{+1})/2$.

   b. Explain how to test the hypothesis, and give the degrees of freedom for the test statistic.

3.22 For testing independence in an $I \times 2$ table, show that

$$X^2 = \frac{\left(\sum_i n_{i1} p_{1 \mid i}\right) - n_{+1} p_{+1}}{p_{+1} p_{+2}} = \frac{\sum_i n_{i+} (p_{1 \mid i} - p_{+1})^2}{p_{+1} p_{+2}}.$$

Fisher (1934) and Cochran (1954) attributed these formulas to A. E. Brandt and G. W. Snedecor.

3.23 For sample proportions $\{p_i\}$ and ML estimates $\{\hat{\pi}_i\}$, the likelihood-ratio statistic is

$$G^2 = -2n \sum p_i \log(\hat{\pi}_i/p_i).$$

Show that $G^2 \geq 0$, with equality if and only if $p_i = \hat{\pi}_i$ for all i. (Hint: Apply Jensen's inequality to $E[\log(X)]$, where X is a random variable that equals $\hat{\pi}_i/p_i$ with probability p_i).

3.24 For testing independence, show that $X^2 \leq n \, \min(I - 1, \, J - 1)$. Hence $V^2 = X^2/[n \, \min(I - 1, J - 1)]$ (Cramér 1946) falls between 0 and 1. (For 2×2 tables, V^2 simplifies to a measure called *phi-squared* and also equals Goodman and Kruskal's tau. Other measures based on X^2 include Pearson's *contingency coefficient* $[X^2/(X^2 + n)]^{1/2}$.)

3.25 Show the sample value of the uncertainty coefficient (2.12) satisfies $\hat{U} = -G^2/2(\Sigma \, p_{+j} \log p_{+j})$.

3.26 The *power divergence* statistic for testing goodness of fit is

$$\frac{2}{\lambda(\lambda + 1)} \sum n_i[(n_i/\hat{m}_i)^\lambda - 1], \quad \text{for } -\infty < \lambda < \infty.$$

   a. For $\lambda = 1$, show this equals the Pearson statistic X^2.
   b. Show that as $\lambda \to 0$, it converges to the likelihood ratio statistic G^2. (Hint: $\log(t) = \lim_{h \to 0}(t^h - 1)/h$.)
   c. As $\lambda \to -1$, show it converges to $2 \sum \hat{m}_i \log(\hat{m}_i/n_i)$, the *minimum discrimination information* statistic (Gokhale and Kullback 1978).
   d. For $\lambda = -2$, show it equals $\Sigma(n_i - \hat{m}_i)^2/n_i$, the *Neyman modified chi-squared* statistic (Neyman 1949).
   e. For $\lambda = -\frac{1}{2}$, show it equals $4 \Sigma(\sqrt{n_i} - \sqrt{\hat{m}_i})^2$, the *Freeman-Tukey* statistic (Freeman and Tukey 1950).

(Read and Cressie (1988) and Cressie and Read (1984) investigated properties of this family of statistics. Under regularity conditions, each member of the family has the same asymptotic distribution. They recommended $\lambda = 2/3$, because of the superiority of the chi-squared approximation to its null sampling distribution.)

3.27 Use a partitioning argument to explain why G^2 for testing independence cannot increase when we combine two rows or two columns of an $I \times J$ contingency table.

3.28 Motivate partitioning (3.14) by showing that the multiple hypergeometric distribution (3.26) for $\{n_{ij}\}$ factors as the product of hypergeometric distributions for the separate component tables (Lancaster, 1949).

3.29 For independent Bernoulli random variables $Y_1, \ldots, Y_n$, with parameter π, let $S = \Sigma Y_i$, and let $p = S/n$.
   a. For large samples, argue that $100(1 - \alpha)\%$ confidence limits for π are the two solutions to the quadratic equation in π

$$(p - \pi)^2 = (z_{\alpha/2})^2 \pi(1 - \pi)/n.$$

   b. When lower-order terms are ignored, show the interval in (a) simplifies approximately to

$$p \pm z_{\alpha/2}[p(1 - p)/n]^{1/2}.$$

c. Use the pivotal method for deriving confidence intervals to show that the roots in π of

$$\frac{\alpha}{2} = \sum_{x=0}^{S} \binom{n}{x} \pi^x (1-\pi)^{n-x} \qquad \frac{\alpha}{2} = \sum_{x=S}^{n} \binom{n}{x} \pi^x (1-\pi)^{n-x}$$

determine an interval having confidence coefficient at least $1 - \alpha$.

3.30 When data have the multinomial $(n, \{\pi_i, i = 1, \ldots, N\})$ distribution, with $N > 2$, we can obtain confidence limits for π_i as the solutions of

$$(p_i - \pi_i)^2 = (z_{\alpha/2N})^2 \pi_i (1 - \pi_i)/n, \ i = 1, \ldots, N.$$

a. Using the Bonferroni inequality, argue that these N confidence intervals hold simultaneously (for large samples) with probability at least $1 - \alpha$.

b. Show that the standard deviation of $p_i - p_j$ is $[\pi_i + \pi_j - (\pi_i - \pi_j)^2]/n$. For large n, show that the probability is at least $1 - \alpha$ that the confidence intervals

$$(p_i - p_j) \pm z_{\alpha/2a} \{ [p_i + p_j - (p_i - p_j)^2]/n \}^{1/2}$$

simultaneously contain the $a = N(N-1)/2$ differences $\{\pi_i - \pi_j\}$. See Goodman (1965), Fitzpatrick and Scott (1987), and Miller (1981).

3.31 Assuming multinomial sampling, show that $\sigma^2 = (\sum \sum \pi_{ij}^{-1})(1 - Q^2)^2/4$ in (3.21) for Yule's Q. This result was noted by Yule (1900, 1912).

3.32 Suppose a measure ζ defined for a two-way table has form $\zeta = \nu/\delta$. For multinomial sampling, show the asymptotic variance of $\sqrt{n}(\hat{\zeta} - \zeta)$ is $\sigma^2 = [\sum \sum \pi_{ij} \eta_{ij}^2 - (\sum \sum \pi_{ij} \eta_{ij})^2]/\delta^4$, where $\eta_{ij} = \delta(\partial\nu/\partial\pi_{ij}) - \nu(\partial\delta/\partial\pi_{ij})$ (Goodman and Kruskal, 1972).

3.33 Refer to Note 3.6. For independent multinomial sampling with log θ, show that ϕ_{ij}^+ equals $1/\pi_{j|i}$ when $i = j$ and $-1/\pi_{j|i}$ when $i \neq j$. Conclude that $\sigma^2 = \sum \sum 1/\omega_i \pi_{j|i}$, the same as the asymptotic variance for multinomial sampling when $\omega_i = \pi_{i+}$.

3.34 Explain why $\{n_{+j}\}$ are sufficient for $\{\pi_{+j}\}$ in (3.24).

3.35 When a test statistic has a continuous distribution, the null distribution of the P-value is uniform; that is, $P(P \leq \alpha) = \alpha$ for $0 < \alpha < 1$. For Fisher's exact test, show that under the null, $P(P \leq \alpha) < \alpha$ for $0 < \alpha < 1$. (Hint: $P(P \leq \alpha) = E[P(P \leq \alpha \mid n_{1+}, n_{+1})]$.)

3.36 Consider exact tests of independence, given the marginals, for the $I \times I$ table having $n_{ii} = 1$ for $i = 1, \ldots, I$, and $n_{ij} = 0$ otherwise.

a. Show that the test that orders tables by their probabilities has P-value equal to 1.0.

b. Show that the one-sided test that orders tables by their values of $C - D$ has P-value equal to $(1/I!)$. (Hence one can achieve significance for ordinal variables at the 0.05 level using only four observations).

3.37 For hypergeometric distribution (3.23),

$$E(n_{11}) = \frac{n_{1+}n_{+1}}{n} \quad \text{and} \quad \text{Var}(n_{11}) = \frac{n_{1+}n_{+1}n_{2+}n_{+2}}{n^2(n-1)}.$$

Letting $\rho = n_{+1}/n$, show n_{11} has the same mean as a binomial random variable for n_{1+} trials with success probability ρ, and it has the variance of that random variable multiplied by a finite population correction $(n - n_{1+})/(n - 1)$. Show the hypergeometric is similar to that binomial distribution when n_{1+} is small compared to n.

3.38 Show that the conditional ML estimate of θ satisfies $n_{11} = E(n_{11})$ for distribution (3.27).

3.39 A 2×2 contingency table, summarizing observations on two independent binomial variables, has counts $(3, 0)$ in the first row and $(0, 3)$ in the second row. In testing $H_0: \pi_{1|1} = \pi_{1|2}$ against $\pi_{1|1} > \pi_{1|2}$, let π denote the common value under H_0.

a. For given π, show that the P-value for an exact binomial test is $\pi^3(1 - \pi)^3$. Show that the maximum P-value occurs for $\pi = 0.5$, and hence the P-value for the exact unconditional test equals $1/64$.

b. Show that the P-value for Fisher's exact test is $1/20$.

c. Explain the difference in philosophy between the two tests. (For further discussion of this example, see R. J. A. Little (1989) and G. Barnard's discussion at the end of the article by Yates (1984).

CHAPTER 4

Models For Binary Response Variables

Chapters 2 and 3 presented ways of analyzing association in two-way contingency tables. But most studies have several variables. We usually want to describe effects of some of them, the explanatory variables, on one or more response variables. *Modeling* the effects helps us do this efficiently. A good-fitting model evaluates effects of explanatory variables, describes association and interaction linkages, and produces improved estimates of response probabilities. The rest of this text concerns model-building for categorical response variables. This chapter focuses on *binary* responses, response variables having only two categories.

The first section of the chapter introduces a family of *generalized linear models*. This family contains important models for categorical data, as well as standard regression and analysis of variance models for continuous response variables.

Section 4.2 introduces generalized linear models for binary response variables. The most important model of this type is the *logistic regression* model, based on the *logit* transformation of a proportion. Section 4.3 illustrates the use of logit models with categorical explanatory variables.

Section 4.4 shows that the choice of an appropriate model can improve inferential power in detecting associations. For instance, for testing independence, statistics for models that exploit category orderings can be much more powerful than standard chi-squared statistics.

Section 4.5 introduces two more models for binary data, the *probit* and *extreme-value* models. Section 4.6 introduces model diagnostics, such as residuals, which can provide detailed inspections of the quality of fit provided by a binary regression model. Section 4.7, more technical in nature, derives likelihood equations for logit models and presents the Newton-Raphson iterative method for solving the equations. The final

79

section discusses exact conditional inference for logistic regression models.

4.1 GENERALIZED LINEAR MODELS

Most models in this book are special cases of *generalized linear models*, a broad class of models introduced by Nelder and Wedderburn (1972). Generalized linear models are specified by three components: a *random component*, which identifies the probability distribution of the response variable; a *systematic component*, which specifies a linear function of explanatory variables that is used as a predictor; and a *link* describing the functional relationship between the systematic component and the expected value of the random component. We will use the acronym, GLM, as shorthand for generalized linear model.

4.1.1 Components of a Generalized Linear Model

The *random component* of a GLM consists of independent observations $Y = (Y_1, \ldots, Y_N)'$ from a distribution in the natural exponential family. That is, each observation Y_i has probability density function or mass function of the form

$$f(y_i; \theta_i) = a(\theta_i)b(y_i) \exp[y_i Q(\theta_i)] . \tag{4.1}$$

This family includes several important distributions as special cases, including the Poisson and binomial. The value of the parameter θ_i in (4.1) may vary for $i = 1, \ldots, N$, depending on values of explanatory variables. The term $Q(\theta)$ is called the *natural parameter* of the distribution. A more general representation given in Section 13.1 allows for other parameters (e.g., nuisance or scale parameters), but formula (4.1) is sufficient for our purposes.

The *systematic component* of a GLM relates a vector $\eta = (\eta_1, \ldots, \eta_N)'$ to a set of explanatory variables through a linear model

$$\eta = X\beta \tag{4.2}$$

Here X is a model matrix (sometimes called a "design" matrix) consisting of values of explanatory variables for the N observations, and β is a vector of model parameters. The vector η is called the *linear predictor*.

The third component of a GLM is a *link* between the random and systematic components. Let $\mu_i = E(Y_i)$, $i = 1, \ldots, N$. Then μ_i is linked to η_i by $\eta_i = g(\mu_i)$, where g is any monotonic differentiable function. Thus, the model links expected values of observations to explanatory variables through the formula

$$g(\mu_i) = \sum_j \beta_j x_{ij}, \quad i = 1, \ldots, N. \tag{4.3}$$

The function $g(\mu) = \mu$ gives the *identity link* $\eta_i = \mu_i$, specifying a linear model for the mean response. The link function that transforms the mean to the natural parameter is called the *canonical link*. For it, $g(\mu_i) = Q(\theta_i)$, and $Q(\theta_i) = \sum_j \beta_j x_{ij}$.

In summary, a GLM is a linear model for a transformed mean of a variable having distribution in the natural exponential family. To illustrate the three components of a GLM, we now introduce some important GLMs for categorical response variables.

4.1.2 Logit Models

Many categorical response variables have only two categories. The observation for each subject might be classified as a "success" or a "failure." Represent these possible outcomes by 1 and 0. The *Bernoulli* distribution for binary random variables specifies probabilities $P(Y = 1) = \pi$ and $P(Y = 0) = 1 - \pi$ for the two outcomes, for which $\pi = E(Y)$. When Y_i has Bernoulli distribution with parameter π_i, the probability mass function is

$$f(y_i; \pi_i) = \pi_i^{y_i}(1 - \pi_i)^{1 - y_i} = (1 - \pi_i)[\pi_i/(1 - \pi_i)]^{y_i}$$

$$= (1 - \pi_i) \exp\left[y_i \log\left(\frac{\pi_i}{1 - \pi_i}\right)\right]$$

for $y_i = 0$ and 1. This distribution is in the natural exponential family. The natural parameter $Q(\pi) = \log[\pi/(1 - \pi)]$, the log odds of response 1, is called the *logit* of π. GLMs that use the logit link are called *logit* models. We discuss logit models extensively in this chapter and throughout the text.

4.1.3 Loglinear Models

Section 3.1 noted that cell counts in a contingency table are often treated as independent Poisson random variables. Let n_i denote the count in the *i*th cell, and let $m_i = E(n_i)$ denote its expected value, $i = 1, \ldots, N$. (To be consistent with standard notation in categorical data literature, here we use n_i and m_i instead of the GLM notation y_i and μ_i.) The Poisson probability mass function for n_i is

$$f(n_i; m_i) = \frac{\exp(-m_i)(m_i)^{n_i}}{n_i!} = \exp(-m_i)\left(\frac{1}{n_i!}\right)\exp[n_i \log(m_i)]$$

for nonnegative integer values of n_i. This has natural exponential form (4.1) with $y_i = n_i$ and $\theta_i = m_i$, $a(m_i) = \exp(-m_i)$, $b(n_i) = 1/n_i!$, and $Q(m_i) = \log(m_i)$. For the Poisson distribution, a GLM links a monotone function of m_i to explanatory variables through a linear model.

Since the natural parameter is $\log(m_i)$, the canonical link function is the log link, $\eta_i = \log(m_i)$. The model using this link is

$$\log(m_i) = \sum_j \beta_j x_{ij}, \quad i = 1, \ldots, N. \tag{4.4}$$

In this GLM the cell counts, rather than individual classifications of the subjects, are the N observations. Model (4.4) is called a *loglinear model* for a contingency table. We study loglinear models in detail in Chapters 5–8.

4.1.4 Connection with Models for Continuous Variables

The class of GLMs includes not only models for discrete variables, but also continuous variables. The normal distribution is in a natural exponential family presented in Section 13.1 that includes dispersion parameters. For fixed variance, its natural parameter is the mean. Therefore, a regression model for the mean of Y is a GLM using the identity link. Table 4.1 lists this and other standard models for normal random component. The table also lists primary GLMs for categorical responses presented in the next six chapters. The distinctions between the models

Table 4.1 Types of Models for Statistical Analysis

Random Component	Link	Systematic Component	Model	Chapter
Normal	Identity	Continuous	Regression	
Normal	Identity	Categorical	Anal. of variance	
Normal	Identity	Mixed	Anal. of covariance	
Bernoulli	Logit	Mixed	Logistic regression	4
Poisson	Log	Mixed	Loglinear	5–8
Multinomial	Generalized logit	Mixed	Multinomial response	9

for categorical responses are not sharp, because many loglinear models for Poisson random component are equivalent to models using logit links for Bernoulli or multinomial random components.

A traditional way of analyzing data is to transform the response so it is approximately normal, with constant variance; then, standard regression methods are applicable. With GLMs, by contrast, the choice of link is separate from the choice of random component. If a link produces additivity of effects, it is not necessary that it also stabilize variance or produce normality. This is because the fitting process utilizes likelihood methods for the choice of random component, and we are not restricted to normality for that choice.

GLMs make possible a unified theory that encompasses important models for continuous and categorical variables. For many GLMs the log likelihood is strictly concave, and ML estimates of model parameters exist and are unique under general conditions (Wedderburn 1976). The estimates are computed with an iterative algorithm, called *Fisher scoring*, that uses a generalized version of least squares. The same algorithm applies regardless of the choice of distribution for the random component or the choice of link. For GLMs using the canonical link, this algorithm simplifies to the Newton-Raphson algorithm presented in Section 4.7.2.

Let $L(\boldsymbol{\mu}; \mathbf{y})$ denote the log likelihood function expressed in terms of the means $\boldsymbol{\mu} = (\mu_1, \ldots, \mu_N)'$, for an observed data vector $\mathbf{y} = (y_1, \ldots, y_N)'$. Let $L(\hat{\boldsymbol{\mu}}; \mathbf{y})$ denote the maximum of the log likelihood under the assumption that a given model holds. The maximum achievable log likelihood is $L(\mathbf{y}; \mathbf{y})$. This occurs for the most general model, having as many parameters as observations, so that $\hat{\boldsymbol{\mu}} = \mathbf{y}$. The (scaled) *deviance* of a GLM is defined to be

$$2[L(\mathbf{y}; \mathbf{y}) - L(\hat{\boldsymbol{\mu}}; \mathbf{y})]$$

For models in this book, the deviance has the same form as the likelihood-ratio statistic (3.13). For some GLMs, the deviance has approximately a chi-squared distribution when the model holds. The degrees of freedom equal the difference between the number of observations N and the number of parameters in the model. The deviance can then be used to test the fit of the model, and components of the deviance are diagnostic measures of lack of fit.

The statistical computer package GLIM, sponsored by the Royal Statistical Society, is a simple-to-use interactive program for fitting GLMs. It is available for the mainframe computer or PC, and it can fit nearly all models discussed in this book. Appendix A.1.2 describes GLIM more fully. Section 13.1 presents further details on the theory of GLMs.

4.2 LOGISTIC REGRESSION

Let Y denote a binary response variable. For instance, Y might indicate vote in an election (Democrat, Republican), choice of automobile (domestic, foreign import), or diagnosis of breast cancer (present, absent). Denoting the two outcomes by 0 and 1 gives the Bernoulli random variable with mean

$$E(Y) = 1 \times P(Y = 1) + 0 \times P(Y = 0) = P(Y = 1) .$$

We denote this probability by $\pi(\mathbf{x})$, reflecting its dependence on values of explanatory variables $\mathbf{X} = (X_1, \ldots, X_k)$. Since also

$$E(Y^2) = 1^2 \pi(\mathbf{x}) + 0^2 [1 - \pi(\mathbf{x})] = \pi(\mathbf{x})$$

the variance of Y is

$$V(Y) = E(Y^2) - [E(Y)]^2 = \pi(\mathbf{x})[1 - \pi(\mathbf{x})] .$$

In introducing the models in this section, for simplicity we use a single explanatory variable, X.

4.2.1 Linear Probability Model

For a binary response, the regression model

$$E(Y) = \pi(x) = \alpha + \beta x \tag{4.5}$$

is called a *linear probability model*. When observations on Y are independent, this model is a GLM with identity link function.

The linear probability model has a major structural defect. Probabilities must fall between 0 and 1, whereas linear functions take values over the entire real line. Model (4.5) predicts $\pi < 0$ and $\pi > 1$ for sufficiently large or small x values. We usually expect a *nonlinear* relationship between $\pi(x)$ and x. A fixed change in X may have less impact when π is near 0 or 1 than when π is near the middle of its range. For instance, suppose we plan to purchase an automobile, and must choose between buying a new one or used one. Let $\pi(x)$ denote the probability of selecting a new one, when annual family income $= x$. We expect an increase of \$10,000 in income to have less effect when $x = \$1,000,000$ (for which π is near 1) than when $x = \$30,000$.

Model (4.5) can be valid over a finite range of x values. There are problems with using ordinary least squares to fit the model, however, since conditions that make least squares estimators optimal are not satisfied. Since $V(Y) = \pi(x)[1 - \pi(x)]$, the variability is not constant, but rather depends on x through its influence on π. As $\pi(x)$ moves toward 0 or 1, the conditional distribution of Y is more nearly concentrated at a single point, and the variance moves toward 0. The ordinary estimators are no longer minimum variance in the class of linear unbiased estimators. Also Y, being binary, is very far from normally distributed. Thus sampling distributions for the ordinary estimators do not apply, and better estimators exist that are not linear in the observations (see Note 4.2).

4.2.2 Logistic Regression Model

Because of the structural problems with the linear probability model, it is more fruitful to study models implying a curvilinear relationship between x and $\pi(x)$. When we expect a monotonic relationship, the S-shaped curves in Figure 4.1 are natural shapes for regression curves. A function having this shape is

$$\pi(x) = \frac{\exp(\alpha + \beta x)}{1 + \exp(\alpha + \beta x)} , \qquad (4.6)$$

called the *logistic regression* function. As $x \to \infty$, $\pi(x) \downarrow 0$ when $\beta < 0$ and $\pi(x) \uparrow 1$ when $\beta > 0$. As $\beta \to 0$ the curve flattens to a horizontal straight line. When the model holds with $\beta = 0$, the binary response is independent of X.

The logistic regression curve (4.6) has $\partial \pi(x)/\partial x = \beta \pi(x)[1 - \pi(x)]$. The curve has its steepest slope at the x value where $\pi(x) = \frac{1}{2}$, which is $x = -\alpha/\beta$. The line tangent to the logistic curve at that point has slope $\beta/4$. As $|\beta|$ increases, the curve has a steeper rate of increase at the x value having any given $\pi(x)$ value. The value $1/\beta$ approximates the distance between the x values where $\pi(x) = 0.25$ or 0.75 and where $\pi(x) = 0.50$.

We now find the link function for which the logistic regression model is a GLM. For this model the odds of making response 1 are

$$\frac{\pi(x)}{1 - \pi(x)} = \exp(\alpha + \beta x) = e^{\alpha}(e^{\beta})^x .$$

This formula provides a basic interpretation for β. The odds increase

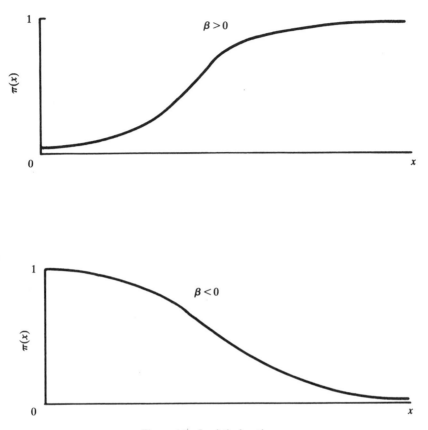

Figure 4.1 Logistic functions.

multiplicatively by e^β for every unit increase in x. The log odds has the linear relationship

$$\log\left(\frac{\pi(x)}{1 - \pi(x)}\right) = \alpha + \beta x . \qquad (4.7)$$

Thus, the appropriate link is the log odds transformation, the *logit*.

An advantage of this model over models using other links is that effects can be estimated whether the sampling design is prospective or retrospective. Effects in the logistic model refer to odds, and the estimated odds at one value of x divided by the estimated odds at another value of x is an odds ratio. Section 2.2 showed that this measure is valid with retrospective data. Problem 4.20 gives details. Cornfield (1962) provided another justification for the logistic model: Given that $Y = i$, suppose X has normal distribution $N(\mu_i, \sigma^2)$, $i = 0,1$. Then using Bayes Theorem, it follows that $P(Y = 1 | X = x)$ equals (4.6) with $\beta = (\mu_1 - \mu_0)/\sigma^2$.

To determine an appropriate form for the systematic component of a logistic regression model, it is helpful to plot sample logits against x. Suppose there are n_i observations at the ith setting of x. It is sufficient to report the total number of "1" outcomes at each setting, rather than the results of individual Bernoulli observations. Let y_i denote the outcome of this binomial random variable. The ith sample logit is then $\log[y_i/(n_i - y_i)] = \log[p_i/(1 - p_i)]$, where $p_i = y_i/n_i$. This is not defined when $y_i = 0$ or n_i, and the adjusted value

$$\log\left(\frac{y_i + \frac{1}{2}}{n_i - y_i + \frac{1}{2}}\right)$$

called an *empirical logit*, is a less biased estimator of the true logit. When X is continuous, it is necessary to group the data before calculating empirical logits. We defer to Section 4.7 the technical details regarding deriving likelihood equations and fitting logistic regression models.

4.2.3 Cancer Remission Example

Lee (1974) used logistic regression to determine characteristics associated with remission in cancer patients. Table 4.2 contains values of the most important explanatory variable, a labeling index. This index, denoted by LI, measures proliferative activity of cells after a patient receives an injection of tritiated thymidine. It represents the percentage of cells that are "labeled." The response variable was whether the patient achieved remission, a "1" indicating success. There were 27 observations at 14 distinct levels of LI. We treat these as 14 independent binomial samples.

To study the effect of LI, we first calculated empirical logits. Since the sample size is small and there are few observations at each level of LI, we grouped levels as shown in Table 4.3. The empirical logits exhibit a clear increasing trend, suggesting we model a linear effect of LI on the logit scale.

The ML parameter estimates for the logistic regression model are $\hat{\alpha} = -3.777$ and $\hat{\beta} = 0.145$ (ASE = 0.059). For a unit change in LI, the estimated odds of remission are multiplied by $\exp(0.145) = 1.16$. The predicted probability of remission equals 0.5 when $LI = -\hat{\alpha}/\hat{\beta} = 26.0$. The predicted logit $\hat{L} = \hat{\alpha} + \hat{\beta}x$ at any setting x for LI yields a predicted odds $\exp(\hat{L})$, and a predicted probability

Table 4.2 Cancer Remission and Labeling Index (LI)

LI	No. Cases	No. Remissions	$\hat{\pi}$(logistic)	$\hat{\pi}$(probit)	$\hat{\pi}$(linear)
8	2	0	0.068	0.053	−0.003
10	2	0	0.089	0.075	0.053
12	3	0	0.115	0.103	0.109
14	3	0	0.148	0.138	0.164
16	3	0	0.189	0.181	0.220
18	1	1	0.237	0.231	0.276
20	3	2	0.293	0.288	0.331
22	2	1	0.357	0.350	0.387
24	1	0	0.425	0.417	0.443
26	1	1	0.497	0.487	0.498
28	1	1	0.569	0.557	0.554
32	1	0	0.702	0.689	0.665
34	1	1	0.759	0.748	0.721
38	3	2	0.849	0.846	0.832

Source: Data reprinted with permission from Lee (1974).

$$\hat{\pi} = \exp(\hat{L})/[1 + \exp(\hat{L})] \,.$$

For instance, at the mean LI level in this sample of 20.1, the predicted logit is $-3.777 + 0.145(20.1) = -0.87$, and the predicted probability of remission is $\exp(-0.87)/[1 + \exp(-0.87)] = 0.30$. The incremental rate of change in the probability of remission at that point is $\hat{\beta}\hat{\pi}(1 - \hat{\pi}) = 0.03$.

Table 4.2 reports predicted probabilities for the logistic regression model. We obtain predicted numbers of remissions by multiplying each probability by the number of observations at that LI level. At LI = 38, for instance, the predicted number of remissions for the three cases is $3(0.849) = 2.55$. Table 4.3 reports the predicted counts, in grouped fashion. In this form, there is no evidence of lack of fit of the model.

Table 4.3 Empirical Logits for Table 4.2, and Predicted Number of Remissions for Logistic Regression Model

LI	No. Cases	No. Remissions	Empirical Logit	Predicted No. Remissions
8–12	7	0	−2.71	0.66
14–18	7	1	−1.47	1.25
20–24	6	3	0.00	2.02
26–32	3	2	0.51	1.77
34–38	4	3	0.85	3.31

4.2.4 Inference for Logistic Regression

From Wald's (1943) general asymptotic results for ML estimators, it follows that parameter estimators in logistic models have large-sample normal distributions. Thus, large-sample confidence intervals for parameters have form $\hat{\beta} \pm z_{\alpha/2}(\text{ASE})$.

Let $\gamma = (\gamma_1, \ldots, \gamma_q)'$ denote a subset of model parameters. Suppose we want to test H_0: $\gamma = 0$. Let M_1 denote the fitted model, and let M_2 denote the simpler model with $\gamma = 0$. Large-sample tests can use Wilks's (1938) likelihood-ratio approach, with test statistic based on twice the log of the ratio of maximized likelihoods for M_1 and M_2. Let L_1 denote the maximized log likelihood for M_1, and let L_2 denote the maximized log likelihood for M_2. Under H_0, the statistic

$$-2(L_2 - L_1)$$

has a large-sample chi-squared distribution with df $= q$.

Alternatively, by the large-sample normality of parameter estimators, the statistic

$$\hat{\gamma}'[\hat{\text{Cov}}(\hat{\gamma})]^{-1}\hat{\gamma}$$

has the same limiting null distribution (Wald 1943). This is called a *Wald statistic*. When γ has a single element, this chi-squared statistic with df $= 1$ is the square of the ratio of the parameter estimate to its estimated standard error. Hauck and Donner (1977) indicated that when the effect is large, the Wald test is not as powerful as the likelihood-ratio test, and can even show aberrant behavior.

To illustrate inferences for logistic regression, we continue our analysis of the cancer remission data. The statistic $z = \hat{\beta}/\text{ASE} = 2.45$ gives strong evidence of a positive association between LI and remission. The equivalent Wald chi-squared statistic, $z^2 = 5.96$, is based on df $= 1$. Even stronger evidence is provided by the likelihood-ratio statistic, $-2(L_2 - L_1) = 8.30$. In this application, M_1 is the logistic regression model, and M_2 is the independence model ($\beta = 0$).

4.2.5 Inverse CDF Links

A monotone regression curve such as the first one in Figure 4.1 has the shape of a cumulative distribution function (cdf) for a continuous random variable. The second curve also has that shape when x is replaced by $-x$. This suggests a class of models for binary responses having form

$$\pi(x) = F(\alpha + \beta x) \qquad (4.8)$$

where F is a standard continuous cdf. When F is strictly increasing over the entire real line, (4.8) has the form

$$F^{-1}[\pi(x)] = \alpha + \beta x . \qquad (4.9)$$

This model is a GLM with link function equal to the inverse of F. The link function maps the $(0,1)$ range of probabilities onto $(-\infty, \infty)$, the range of linear predictors such as $\alpha + \beta x$.

When $\beta > 0$, the logistic regression curve $\pi(x) = \exp(\alpha + \beta x)/[1 + \exp(\alpha + \beta x)]$ looks like the cdf of the *logistic* distribution. When $\beta < 0$, the formula for $1 - \pi(x)$ has that appearance. The cdf of the logistic distribution is

$$F(x) = \frac{\exp[(x - \mu)/\tau]}{1 + \exp[(x - \mu)/\tau]} , \quad -\infty < x < \infty$$

for location parameter μ and scale parameter $\tau > 0$. The distribution is symmetric and bell-shaped, with mean equal to μ and standard deviation equal to $\tau \pi/\sqrt{3}$. It looks much like the normal distribution with the same mean and standard deviation.

The logistic regression curve has form $\pi(x) = F(\alpha + \beta x)$ when F is the standardized cdf, for which $\mu = 0$ and $\tau = 1$. That curve is the cdf of a logistic distribution having mean $-\alpha/\beta$ and standard deviation $\pi/(|\beta|\sqrt{3})$. The logit transformation is the inverse function for the logistic cdf.

When F is the standard normal cdf Φ, model (4.8) is the *probit* model, which is discussed in Section 4.5. The normal tails are slightly thinner than those for the logistic, implying that $\pi(x)$ approaches 0 and 1 more quickly for the probit model than for the logistic model. For illustrative purposes, we also fitted the probit model and the linear probability model to the cancer remission data. The ML estimates for the probit model are $\hat{\alpha} = -2.317$ and $\hat{\beta} = 0.088$ (ASE = 0.033). The ordinary least squares solution for the linear probability model is $\hat{\pi}(x) = -0.225 + 0.028x$. Predicted probabilities for these models are listed in Table 4.2 and plotted in Figure 4.2. The fit for the linear probability model is quite different from the logit and probit fits. In fact, it gives negative predicted probability at the lowest recorded LI value.

Models presented in this section generalize for multiple explanatory variables. For instance, the logistic regression model for values $\mathbf{x} = (x_1, \ldots, x_k)'$ of k explanatory variables is

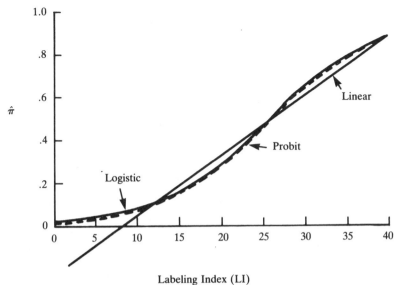

Figure 4.2 Predicted probabilities for three models.

$$\log\left(\frac{\pi(\mathbf{x})}{1 - \pi(\mathbf{x})}\right) = \alpha + \beta_1 x_1 + \beta_2 x_2 + \cdots + \beta_k x_k . \qquad (4.10)$$

4.3 LOGIT MODELS FOR CATEGORICAL DATA

Explanatory variables in the models discussed in the previous section can be continuous or categorical. When they are categorical, models with logit link are equivalent to loglinear models discussed in upcoming chapters. We consider this case separately here to help simplify the explanation of that equivalence in the next chapter.

4.3.1 Logit Model for $I \times 2$ Table

Suppose there is a single explanatory factor, having I categories. In row i of the $I \times 2$ table, the two response probabilities are $\pi_{1|i}$ and $\pi_{2|i}$, with $\pi_{1|i} + \pi_{2|i} = 1$. In the logit model

$$\log\left(\frac{\pi_{1|i}}{\pi_{2|i}}\right) = \alpha + \beta_i \ . \tag{4.11}$$

$\{\beta_i\}$ describes the effects of the factor on the response.

The right-hand side of this equation resembles the model formula for cell means in one-way ANOVA. As in the ANOVA model, identifiability requires a linear constraint on the parameters, such as $\Sigma \beta_i = 0$ or $\beta_1 = 0$. Then $I - 1$ of $\{\beta_i\}$ characterize the relationship. For the constraint $\Sigma \beta_i = 0$, α is the mean of the logits, and β_i is a deviation from the mean for row i. The higher β_i is, the higher the logit in row i, and the higher the value of $\pi_{1|i}$; that is, subjects in category i are more likely to make response 1.

Let $\{n_{ij}\}$ denote the number of times response j occurs when the factor is at level i. It is usual to treat as fixed the total counts $\{n_{i+} = n_{i1} + n_{i2}\}$ at the I factor levels. When binary responses are independent Bernoulli random variables, $\{n_{i1}\}$ are independent binomial random variables with parameters $\{\pi_{1|i}\}$.

For *any* set $\{\pi_{1|i} > 0\}$, there exist $\{\beta_i\}$ such that model (4.11) holds. That model has as many parameters as binomial observations, and it is said to be *saturated*. When a factor has *no* effect on the response variable, the simpler model

$$\log\left(\frac{\pi_{1|i}}{\pi_{2|i}}\right) = \alpha \tag{4.12}$$

holds. This is the special case of (4.11) in which $\beta_1 = \beta_2 = \cdots = \beta_I$. Since it is equivalent to $\pi_{1|1} = \cdots = \pi_{1|I}$, (4.12) is the model of statistical independence of the response and factor.

4.3.2 Logit Models for Higher Dimensions

These logit models generalize when there are several categorical factors. For instance, suppose there are two factors, A and B, for the binary response. Let I denote the number of levels of A, and J the number of levels of B. Denote by $\pi_{k|ij}$ the probability of response k, when factor A is at level i and factor B is at level j, so $\pi_{1|ij} + \pi_{2|ij} = 1$.

For the $I \times J \times 2$ table, the logit model

$$\log\left(\frac{\pi_{1|ij}}{\pi_{2|ij}}\right) = \alpha + \beta_i^A + \beta_j^B \tag{4.13}$$

represents the effects of A through I parameters $\{\beta_i^A\}$ and the effects of

B through J parameters $\{\beta_j^B\}$. It assumes the effects of each factor are the same at each level of the other factor. That is, the model assumes an absence of interaction. The right-hand side of the equation resembles the formula for cell means in the two-way ANOVA model that assumes no interaction.

This model treats $\{n_{ij+}\}$ as fixed and $\{n_{ij1}\}$ as independent binomial random variables with parameters $\{\pi_{1|ij}\}$. When a factor has *no* effect on the response variable, adjusting for the other factor, all its beta parameters are equal. A test of a hypothesis such as H_0: $\beta_1^B = \cdots = \beta_J^B$ gives a comparison of a complete and reduced model. We study ways of analyzing such models and equivalent loglinear models for multi-way tables in Chapters 6 and 7.

4.3.3 Heart Disease Example

Table 4.4 is based on data reported by Cornfield (1962). A sample of male residents of Framingham, Massachusetts, aged 40-59, were classified on several factors, including blood pressure. During a six-year follow-up period, they were classified according to whether they developed coronary heart disease. This is the response variable.

To test the null hypothesis of independence, the likelihood-ratio statistic (3.13) is $G^2 = 30.02$, based on df $= 7$. There is strong evidence that blood pressure and heart disease are statistically dependent.

Table 4.4   Cross-Classification of Framingham Men by Blood Pressure and Heart Disease

Blood Pressure	Heart Disease[a]	
	Present	Absent
<117	3 (5.2)	153 (150.8)
117–126	17 (10.6)	235 (241.4)
127–136	12 (15.1)	272 (268.9)
137–146	16 (18.1)	255 (252.9)
147–156	12 (11.6)	127 (127.4)
157–166	8 (8.9)	77 (76.1)
167–186	16 (14.2)	83 (84.8)
>186	8 (8.4)	35 (34.6)

Source: Reprinted with permission based on Cornfield (1962).

[a] Fitted values for linear logit model in parentheses.

For saturated model (4.11), regardless of the constraint chosen for $\{\beta_i\}$, the values of $\{\alpha + \beta_i\}$ are the same. Their ML estimates are the sample logits, reported in Table 4.5. For instance,

$$\hat{\alpha} + \hat{\beta}_1 = \log(3/153) = -3.93$$

Except for the reversal between the second and third categories, the logits increase as blood pressure level increases. This means that the sample proportion of subjects having heart disease increases, as also shown in Table 4.5.

4.3.4 Linear Logit Model

Models (4.11) and (4.13) treat factors as nominal, since they ignore whether there is a natural ordering of categories. When factor categories are ordered, models exist that are more parsimonious than the saturated one, yet more complex than the independence model (4.12). For instance, suppose there is an ordinal explanatory variable X, with distances between the scores $\{x_1, x_2, \ldots, x_I\}$ describing distances between its categories. When we expect X to have a monotone effect on the response, it is natural to fit the *linear logit* model

$$\log\left(\frac{\pi_{1|i}}{\pi_{2|i}}\right) = \alpha + \beta x_i . \tag{4.14}$$

For this model, X has a linear effect on Y, on the logit scale. The independence model is the special case $\beta = 0$.

We observed a monotone increase in the sample logits for Table 4.4, indicating that the linear logit model (4.14) may fit better than the

Table 4.5 Logits and Proportion of Heart Disease for Table 4.4

Blood Pressure	Sample Logit	Proportion Heart Disease	
		Observed	Fitted
<117	−3.93	0.019	0.033
117–126	−2.63	0.067	0.042
127–136	−3.12	0.042	0.053
137–146	−2.77	0.059	0.067
147–156	−2.36	0.086	0.084
157–166	−2.26	0.094	0.104
167–186	−1.65	0.162	0.144
>186	−1.48	0.186	0.194

independence model. The explanatory variable corresponds to a grouping of a naturally continuous variable, and we used the scores {111.5, 121.5, 131.5, 141.5, 151.5, 161.5, 176.5, 191.5}. The non-extreme scores are midpoints for the intervals of blood pressure. ML fitting of the linear logit model gives $\hat{\alpha} = -6.082$ and $\hat{\beta} = 0.0243$. The estimated multiplicative effect of a unit change in blood pressure on the odds of developing heart disease is $\exp(0.0243) = 1.025$. For adjacent categories of blood pressure in Table 4.4 that are 10 units apart, the estimated odds of developing heart disease are $\exp(10 \times 0.0243) = 1.28$ times higher when blood pressure falls in category $i + 1$ instead of i.

For testing $\beta = 0$, we can use

$$z = \hat{\beta} / \text{ASE} = 0.0243/0.0048 = 5.03 .$$

Equivalently, the Wald chi-squared statistic $z^2 = 25.27$ has df $= 1$. The likelihood-ratio statistic compares maximized likelihoods of the linear logit (L) and independence (I) models; that is, $M_1 = L$ and $M_2 = I$. It equals $-2(L_2 - L_1) = 24.11$. There is overwhelming evidence that the prevalence of heart disease increases with increasing blood pressure.

4.3.5 Goodness of Fit as a Likelihood-Ratio Test

For a given logit model, we can use model parameter estimates to calculate predicted logits, and hence predicted probabilities and estimated expected frequencies $\{\hat{m}_{ij} = n_{i+} \hat{\pi}_{j \mid i}\}$. When expected frequencies are relatively large, we can test goodness of fit with a Pearson or likelihood-ratio chi-squared statistic. For a model symbolized by M, we denote these statistics by $X^2(M)$ and $G^2(M)$. For instance,

$$G^2(M) = 2 \sum \sum n_{ij} \log(n_{ij}/\hat{m}_{ij}) .$$

The degrees of freedom equal the number of logits minus the number of linearly independent parameters in the model. For the independence model (4.12) for a binary response $(J = 2)$, there are I logits and 1 parameter, so df $= I - 1 = (I - 1)(J - 1)$. The linear logit model (4.14) has df $= I - 2$. The saturated model (4.11) has df $= 0$; it fits perfectly.

We used the likelihood-ratio principle to construct a statistic $-2(L_2 - L_1)$ that tests whether certain model parameters are zero, by comparing the fitted model M_1 with a simpler model M_2. When explanatory variables are categorical, we denote this statistic for testing M_2, given that M_1 holds, by $G^2(M_2 \mid M_1)$. The goodness-of-fit statistic $G^2(M)$ is a special case of that statistic, in which $M_2 = M$ and M_1 is the saturated model. In testing whether M fits, we test whether *all* parameters that are in the saturated model but not in M equal zero.

Let L_S denote the maximized log likelihood for the saturated model. The likelihood-ratio statistic for comparing models M_1 and M_2 is

$$G^2(M_2 \mid M_1) = -2(L_2 - L_1)$$

$$= -2(L_2 - L_S) - [-2(L_1 - L_S)]$$

$$= G^2(M_2) - G^2(M_1).$$

That is, the test statistic for comparing two models is identical to the difference in G^2 goodness-of-fit statistics for the two models.

We illustrate these analyses using the heart disease data in Table 4.4. For subjects at the highest blood pressure level, the linear logit model gives predicted logit $-6.082 + 0.0243(191.5) = -1.42$. The predicted probability of heart disease is $\exp(-1.42)/[1 + \exp(-1.42)] = 0.194$. Since there were 43 subjects at that blood pressure level, the estimated expected frequency of heart disease is $43(0.194) = 8.4$. Table 4.4 contains all the estimated expected frequencies. The likelihood-ratio goodness-of-fit statistic for the linear logit (L) model equals $G^2(L) = 5.91$, based on

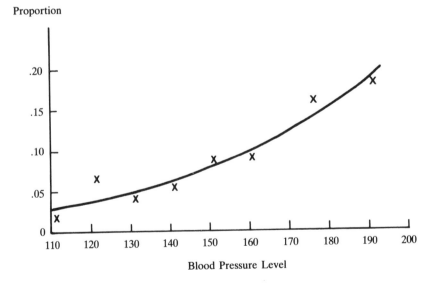

Note: × denotes *observed* proportion

Figure 4.3 Observed and predicted proportions of heart disease for linear logit model.

df = $I - 2 = 8 - 2 = 6$. The observed proportions and the predicted probabilities of heart disease are listed in Table 4.5 and plotted in Figure 4.3. The fit seems good, except in the second category. The data do not depart significantly from a linear trend on the logit scale, and more precise modeling requires a larger sample and/or finer measurement of blood pressure.

The G^2 statistic (3.13) for testing independence is simply the goodness-of-fit statistic for the independence model (4.12), $G^2(I) = 30.02$, based on df = 7. The likelihood-ratio statistic previously reported for testing $\beta = 0$ in the linear logit model is $G^2(I \mid L) = G^2(I) - G^2(L) = 30.02 - 5.91 = 24.11$, based on df = 1.

The remainder of the text uses the term *fitted values* for $\{\hat{m}_{ij}\}$ instead of the rather awkward "estimated expected frequencies."

4.4 USING MODELS TO IMPROVE INFERENTIAL POWER

The model building process should take into account the measurement scales of the variables. Proper recognition of measurement scales can have a major impact in the quality of inferences about associations.

The G^2 and X^2 tests of independence introduced in Section 3.3 treat both classifications as nominal. This section denotes those test statistics as $G^2(I)$ and $X^2(I)$. We have seen that $G^2(I)$ is a special case of the likelihood-ratio statistic $G^2(M_2 \mid M_1) = -2(L_2 - L_1)$ for comparing two models, with M_1 being the saturated model and M_2 the independence (I) model. For ordinal variables, other statistics are more powerful over important parts of the parameter space. These statistics are directed toward narrower, more relevant, alternative hypotheses than the general one of statistical dependence.

In Section 4.3, to test independence in an $I \times 2$ contingency table with ordered rows, we tested H_0: $\beta = 0$ in the linear logit model (4.14). The statistic $G^2(I \mid L) = G^2(I) - G^2(L)$ is the likelihood-ratio statistic comparing the linear logit model ($M_1 = L$) and the independence model ($M_2 = I$). This test is sensitive to departures from independence of a linear nature on the logit scale. By likelihood equations to be derived in Section 4.8, the alternative hypothesis H_a: $\beta \neq 0$ is equivalent to H_a: $\mu_1 \neq \mu_2$, where μ_1 and μ_2 are population mean scores for conditional distributions of the row variable within column 1 and column 2. As $N \rightarrow \infty$, the probability the $G^2(I \mid L)$ statistic rejects H_0 increases to 1 for tables in which $\mu_1 \neq \mu_2$. The hypothesis $\beta \neq 0$ is narrower than the general one of dependence, since unequal means implies dependence but the converse is not true when $I > 2$.

4.4.1 Noncentral Chi-Squared Distribution

For $I \times 2$ tables, both $G^2(I \mid L)$ and $G^2(I)$ have approximate large-sample chi-squared distributions—$G^2(I)$ with $df = (I - 1)(J - 1) = (I - 1)$ and $G^2(I \mid L)$ with $df = 1$. To compare power, it is necessary to compare their non-null sampling distributions. When the null hypothesis is false and the sample size is moderately large, their distributions are approximately *noncentral chi-squared*. This distribution, introduced by R. A. Fisher in 1928, arises from the following construction. If a random variable Z_i has a $N(\mu_i, 1)$ distribution, $i = 1, \ldots, \nu$, and if $Z_1, \ldots, Z_\nu$ are independent, then ΣZ_i^2 has the noncentral chi-squared distribution with $df = \nu$ and with *noncentrality parameter* $\lambda = \Sigma \mu_i^2$. Its mean is $\nu + \lambda$ and its variance is $2(\nu + 2\lambda)$. The regular (central) chi-squared distribution is the special case $\lambda = 0$. When H_0 is true, $\lambda = 0$; the further the truth is from H_0, the larger λ is.

Let $X_{\nu,\lambda}^2$ denote a noncentral chi-squared random variable with degrees of freedom ν and noncentrality λ. Let $\chi_\nu^2(\alpha)$ denote the $100(1 - \alpha)$th percentile of a central chi-squared random variable with the same degrees of freedom; that is, $\chi_\nu^2(\alpha)$ is the boundary of the critical region for a chi-squared test with degrees of freedom ν and significance level α. A fundamental result for chi-squared analyses is that, for fixed λ, $P[X_{\nu,\lambda}^2 > \chi_\nu^2(\alpha)]$ increases as ν decreases; that is, the power for rejecting H_0 at a fixed α-level increases as the df of the test decreases (Fix et al. 1959, Das Gupta and Perlman 1974). For fixed ν, the power equals α when $\lambda = 0$, and increases as λ increases.

The inverse relation between power and df suggests a basic principle for categorical data analyses: In formulating chi-squared statistics, we should try to focus the noncentrality on a statistic having a small df value. For instance, when $G^2(I \mid L)$ has the same noncentrality as $G^2(I)$, $G^2(I \mid L)$ is more powerful because it is based on fewer degrees of freedom. Section 7.5 discusses calculation of noncentrality and power in chi-squared tests.

4.4.2 Increased Power for Narrower Alternatives

Suppose we expect X to have, at least approximately, a linear effect on the logit of Y. To test independence, it is then sensible to use a statistic that has strong power over the region of the parameter space corresponding to that type of effect. This is precisely the purpose of the tests based on $G^2(I \mid L)$ and $z = \hat{\beta}/\text{ASE}$. For the region of interest, such tests can have much greater power than $G^2(I)$ and $X^2(I)$, which disperse their power over the entire parameter space.

We can be more precise concerning when $G^2(I \mid L)$ is more powerful than $G^2(I)$. The statistics are related by

$$G^2(I) = G^2(I \mid L) + G^2(L) \tag{4.15}$$

where $G^2(L)$ is the statistic for testing goodness-of-fit of the linear logit model. When the linear logit model holds, $G^2(L)$ has an asymptotic chi-squared distribution with $df = I - 2$; then if $\beta \neq 0$, $G^2(I)$ and $G^2(I \mid L)$ both have approximate noncentral chi-squared distributions with the same noncentrality. Whereas $df = I - 1$ for $G^2(I)$, $df = 1$ for $G^2(I \mid L)$. Thus, $G^2(I \mid L)$ is more powerful, since it uses fewer degrees of freedom.

When the linear logit model does not hold, $G^2(I)$ has greater noncentrality than $G^2(I \mid L)$, the discrepancy increasing as the model fits more poorly. However, when the model provides a decent approximation to reality, usually $G^2(I \mid L)$ is still more powerful. That test's single degree of freedom more than compensates for its loss in noncentrality. The closer the true relationship is to the linear logit, the more nearly $G^2(I \mid L)$ captures the same noncentrality as $G^2(I)$, and the more powerful it is compared to $G^2(I)$. For instance, Figure 4.4 shows that when the

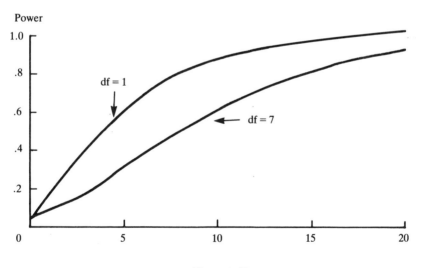

Figure 4.4 Power and noncentrality, for $df = 1$ and $df = 7$, when $\alpha = 0.05$.

noncentrality of a test having df = 1 is at least about half as large as that of a test having df = 7, the test with df = 1 is more powerful.

We obtain improved power by sacrificing power over a region of the parameter space we believe to be less relevant for the variables under study. The $G^2(I)$ and $X^2(I)$ tests of independence disperse their power over the entire parameter space, and have greater power than $G^2(I \mid L)$ over parts of that space poorly described by the linear logit model. Hence, a test that utilizes the ordinal nature of the data is not *uniformly* more powerful. The tests based on $G^2(I)$ and $X^2(I)$ are *consistent*—for *any* departure from independence, the probability of rejecting H_0 increases to 1 as $n \to \infty$. On the other hand, the $G^2(I \mid L)$ statistic is designed to detect dependencies in which the logit tends to increase or tends to decrease as X increases.

There is an analogy when Y is normally distributed, rather than binary. Suppose we want to compare I population means $(\mu_1, \ldots, \mu_I)$ for Y, obtained at I levels of X. The one-way ANOVA F test is sensitive toward any difference among the means. However, when there is approximately a linear trend in the means according to values of X, a more powerful test is the t or F test of the hypothesis that the slope β of the linear regression line is zero. The latter test describes dependence by a single parameter (β), rather than $I-1$ parameters (e.g., $\mu_1 - \mu_I, \ldots, \mu_{I-1} - \mu_I$). It is not more powerful, though, if there is actually a U-shaped pattern to the means.

4.4.3 Cochran-Armitage Trend Test

Armitage (1955) and Cochran (1954) were among the first statisticians to point out the importance of directing chi-squared tests toward narrow alternatives. For $I \times 2$ tables with ordered rows, they gave a decomposition of the Pearson statistic that is similar to the one just described for $G^2(I)$, but using the linear probability model.

In row i, let $\pi_{1 \mid i}$ denote the probability of response 1, and let $p_{1 \mid i}$ denote the sample proportion, $i = 1, \ldots, I$. Let $\{x_i\}$ be scores assigned to the rows. For the linear probability model

$$\pi_{1 \mid i} = \alpha + \beta x_i ,$$

the ordinary least squares fit gives the prediction equation

$$\hat{\pi}_{1 \mid i} = p_{+1} + b(x_i - \bar{x}) ,$$

where

$$\bar{x} = \frac{\sum n_{i+} x_i}{n} \quad \text{and} \quad b = \frac{\sum n_{i+} (p_{1|i} - p_{+1})(x_i - \bar{x})}{\sum n_{i+} (x_i - \bar{x})^2}.$$

The Pearson statistic for testing independence satisfies

$$X^2(I) = z^2 + X^2(L)$$

where

$$X^2(L) = \left(\frac{1}{p_{+1} p_{+2}}\right) \sum n_{i+} (p_{1|i} - \hat{\pi}_{1|i})^2$$

$$z^2 = \left(\frac{b^2}{p_{+1} p_{+2}}\right) \sum n_{i+} (x_i - \bar{x})^2.$$

(4.16)

When the linear probability model holds, $X^2(L)$ has an asymptotic chi-squared distribution with df $= I - 2$. It tests the goodness of fit of the model. The statistic z^2, based on df $= 1$, tests for a linear trend in the proportions.

4.4.4 Treatment of Leprosy Example

Table 4.6, from a classic article by Cochran (1954), refers to an experiment on the use of sulfones and streptomycin drugs in the treatment of leprosy. The degree of infiltration at the start of the experiment measures a certain type of skin damage. The response refers to the change in the

Table 4.6 Change in Clinical Condition by Degree of Infiltration

Clinical Change	Degree of Infiltration		Proportion High
	High	Low	
Worse	1	11	0.08
Stationary	13	53	0.20
Slight improvement	16	42	0.28
Moderate improvement	15	27	0.36
Marked improvement	7	11	0.39

Source: Reprinted with permission from the Biometric Society (Cochran 1954).

overall clinical condition of the patient after 48 weeks of treatment. We use the equal-interval scores $\{1, 2, 3, 4, 5\}$. The question of interest is whether subjects with high infiltration changed differently from those with low infiltration.

Here, the row variable is actually the response variable. It seems natural to compare the mean clinical change (for the chosen scores) for the two levels of infiltration. Cochran (1954) and Yates (1948) gave such an analysis, and noted that it is identical to the trend test based on treating the binary variable as the response. That test is sensitive to linearity between clinical change and the proportion of cases that had high infiltration.

The Pearson test of independence gives $X^2(I) = 6.88$, based on df = 4. This test does not show much evidence of association ($P = 0.14$), but it ignores the ordering of rows. The sample proportion of high infiltration increases monotonically as the clinical change improves, and the trend test gives $z^2 = 6.67$ for df = 1. Its P-value is 0.01, giving strong evidence of more positive clinical change at the higher level of infiltration. In addition,

$$X^2(I) = 6.88 = 6.67 + 0.21$$

where $X^2(L) = 0.21$, based on df = 3, reveals little departure of the proportions from linearity. Results of the trend test are similar to those we would obtain by testing that the slope is zero in the linear logit model.

4.5 PROBIT AND EXTREME VALUE MODELS

So far, our attention has focused on the logit link between a probability π and a linear predictor. This section discusses two alternative models for binary data. Like the logit model, these models have form

$$\pi(x) = F(\alpha + \beta x) \qquad (4.17)$$

for a continuous cdf F.

4.5.1 Tolerance Distributions and the Probit Model

Binary response models are used in toxicology to describe the effect of a toxic chemical dosage on whether a subject dies. In this application, the concept of a *tolerance distribution* provides justification for model (4.17).

Let x denote the dosage (or, often, the log dosage) of a toxic chemical. For a randomly selected subject, let $Y = 1$ if the subject dies. Suppose the subject has tolerance T for the dosage, with $(Y = 1)$ equivalent to $(T \leq x)$. For instance, an insect may survive exposure to a toxic spray if the dosage x is less than T, and die if the dosage is at least T. Tolerances vary among subjects, and let $G(t) = P(T \leq t)$ denote the cdf for their population distribution. For fixed dosage x, the probability a randomly selected subject dies is

$$P(Y = 1) = \pi(x) = P(T \leq x) = G(x).$$

If F is the cdf of a linear transformation of T, such as the standard cdf for the family of which G is a member, then this probability has form $F(\alpha + \beta x)$.

In many toxicological experiments, the tolerance distribution for the log dosage is approximately normal with some mean μ and standard deviation σ. If G is the cdf of that normal distribution, then

$$\pi(x) = G(x) = \Phi[(x - \mu)/\sigma]$$

where Φ is the standard normal cdf. This has form (4.17) with $F = \Phi$, $\alpha = -\mu/\sigma$ and $\beta = 1/\sigma$. The model

$$\Phi^{-1}[\pi(x)] = \alpha + \beta x \tag{4.18}$$

is the *probit* model.

For the probit model, the response curve for $\pi(x)$ (or for $1 - \pi(x)$, when $\beta < 0$) has the appearance of the normal cdf with mean $\mu = -\alpha/\beta$ and standard deviation $\sigma = 1/|\beta|$. Since 68% of the mass in a normal distribution falls within a standard deviation of the mean, $1/|\beta|$ is the distance between the x values where $\pi(x) = 0.16$ or 0.84 and where $\pi(x) = 0.50$. The rate of change in $\pi(x)$ at a particular x value is $\partial \pi(x)/\partial x = \beta \phi(\alpha + \beta x)$, where ϕ is the standard normal density function. The rate is highest when $\alpha + \beta x = 0$ (i.e., at $x = -\alpha/\beta$), where it equals $\beta/(2\pi)^{1/2} = 0.40\beta$ and at which point $\pi(x) = \frac{1}{2}$.

By comparison, the response curve for $\pi(x)$ for the logistic regression model with parameter β corresponds to the logistic cdf with standard deviation $\pi/|\beta|\sqrt{3}$. Its rate of change in $\pi(x)$ at $x = -\alpha/\beta$ is 0.25β. The rates of change where $\pi(x) = \frac{1}{2}$ are the same for the cdfs corresponding to the probit and logistic curves when the logistic β is $0.40/0.25 = 1.6$ times the probit β. The standard deviations are the same when the logistic β is $\pi/\sqrt{3} = 1.8$ times the probit β. When both models fit well, parameter estimates in logistic models are about 1.6–1.8 times those in probit models.

ML estimates for probit models can be obtained using the Fisher scoring algorithm for GLMs given in Section 13.1.3.

4.5.2 Models with Log-Log Link

The logit and probit links are symmetric about 0.5, in the sense that

$$\text{link}(\pi) = -\text{link}(1 - \pi) .$$

To illustrate,

$$\text{logit}(\pi) = \log[\pi/(1 - \pi)] = -\log[(1 - \pi)/\pi] = -\text{logit}(1 - \pi) .$$

This implies that the response curve for $\pi(x)$ has a symmetric appearance about the point where $\pi = 0.5$. In particular, $\pi(x)$ approaches 0 at the same rate that it approaches 1. Logit and probit models are not appropriate when $\pi(x)$ increases from 0 fairly slowly, but approaches 1 quite suddenly.

The response curve

$$\pi(x) = 1 - \exp[-\exp(\alpha + \beta x)] \tag{4.19}$$

has the shape shown in Figure 4.5. It is asymmetric, $\pi(x)$ departing from 1 more sharply than it departs from 0. For this model,

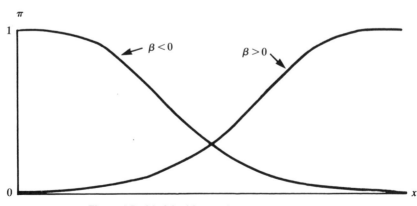

Figure 4.5 Model with complementary log–log link.

$$\log[-\log(1 - \pi(x))] = \alpha + \beta x \,.$$

The link for this GLM is called the *complementary log-log* link.

To interpret model (4.19), we note that for two values x_1 and x_2,

$$\log[-\log(1 - \pi(x_2))] - \log[-\log(1 - \pi(x_1))] = \beta(x_2 - x_1)$$

so that

$$\frac{\log(1 - \pi(x_2))}{\log(1 - \pi(x_1))} = \exp[\beta(x_2 - x_1)]$$

and

$$1 - \pi(x_2) = (1 - \pi(x_1))^{\exp[\beta(x_2 - x_1)]} \,.$$

The probability of "failure" at x_2 equals the probability of failure at x_1 raised to the power $\exp(\beta)$ higher for each unit increase in the distance $x_2 - x_1$.

A related model is

$$\pi(x) = \exp[-\exp(\alpha + \beta x)] \tag{4.20}$$

for which $\pi(x)$ departs from 1 slowly but approaches 0 sharply. As x increases, the curve is monotone decreasing when $\beta > 0$, and monotone increasing when $\beta < 0$. This model uses the *log-log* link

$$\log[-\log(\pi(x))] = \alpha + \beta x \,.$$

When the complementary log-log model holds for the probability of a success, the log-log model holds for the probability of a failure.

Model (4.20) with log-log link is the special case of (4.17) with cdf of the *extreme value* (or *Gumbel*) distribution. That cdf equals

$$G(x) = \exp\{-\exp[-(x - a)/b]\}$$

for parameters $b > 0$ and $-\infty < a < \infty$, and has mean $a + 0.577b$ and standard deviation $\pi b/\sqrt{6}$. Models with log-log links can also be fitted using the Fisher scoring algorithm for GLMs.

4.5.3 Beetle Mortality Example

Table 4.7, from Bliss (1935), reports the number of beetles killed after 5 hours exposure to gaseous carbon disulphide at various concentrations. In

Table 4.7 Beetles Killed after Exposure to Carbon Disulfide

			Fitted Values		
Log Dose	No. Beetles	No. Killed	Comp. Log-Log	Probit	Logit
1.691	59	6	5.7	3.4	3.5
1.724	60	13	11.3	10.7	9.8
1.755	62	18	20.9	23.4	22.4
1.784	56	28	30.3	33.8	33.9
1.811	63	52	47.7	49.6	50.0
1.837	59	53	54.2	53.4	53.3
1.861	62	61	61.1	59.7	59.2
1.884	60	60	59.9	59.2	58.8

Source: Data reprinted with permission from Bliss (1935).

toxicological experiments, it is usual to measure concentration in terms of log dosage. Figure 4.6 plots the proportion of beetles killed against the log dosage. The proportion killed seems to take a large jump at about $x = 1.8$, and the proportions killed are close to 1 above that concentration.

The model with complementary log-log link has ML estimates equal to $\hat{\alpha} = -39.52$ and $\hat{\beta} = 22.01$. At log dosage 1.7, the fitted probability of survival is $1 - \hat{\pi} = \exp\{-\exp[-39.52 + 22.01(1.7)]\} = 0.885$, whereas at log dosage 1.8 it is 0.332 and at log dosage 1.9 it is 5×10^{-5}. The probability of survival is the power $\exp(22.01 \times 0.1) = 9.03$ higher for each 0.1 increase in log dosage. For instance, $0.332 = (0.885)^{9.03}$.

At each log dosage value, by multiplying the estimated kill probability by the number of beetles at that level, we obtain fitted values reported in Table 4.7. These are quite close to the observed numbers killed, and the fit of the model seems adequate. The G^2 goodness-of-fit statistic equals 3.5, based on df $= 6$.

By contrast, logit and probit models fit these data poorly. The G^2 value is 11.1 for the logit model and 10.0 for the probit model. For the probit model

$$\hat{\pi}(x) = \Phi(-34.96 + 19.74x)$$

for which $\hat{\pi} = 0.5$ at $34.96/19.74 = 1.77$. This fit corresponds to a normal tolerance distribution having mean 1.77 and standard deviation $1/19.74 = 0.05$.

Proportion Killed

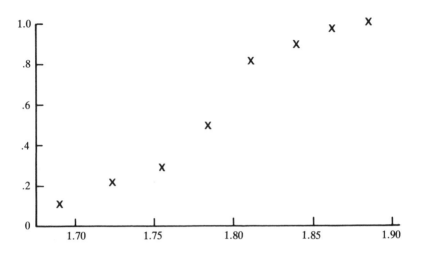

Figure 4.6   Proportion of beetles killed versus log dosage.

Table 4.7 also contains the logit and probit fitted values, which are quite similar. Their poor fit is not surprising, given the non-symmetric appearance of Figure 4.6. For further discussion of these data, see Aranda-Ordaz (1981) and Stukel (1988).

4.6   MODEL DIAGNOSTICS

Goodness-of-fit statistics such as G^2 and X^2 are summary indicators of the overall quality of fit. Additional diagnostic analyses are necessary to describe the nature of any lack of fit. Graphical displays are helpful for this purpose. We can compare the observed and fitted proportions by plotting them against each other, or by plotting both of them against explanatory variables. Residuals comparing observed and fitted counts are also useful. Such diagnostic analyses help show whether lack of fit is due to an inappropriate choice of link function or perhaps due to nonlinearity in effects of explanatory variables.

4.6.1 Residuals

Let y_i denote the number of successes for n_i trials at the ith of I settings of the explanatory variables. For a binary response model, residuals for the fits provided by the I binomial distributions are

$$e_i = \frac{y_i - n_i \hat{\pi}_{1|i}}{[n_i \hat{\pi}_{1|i}(1 - \hat{\pi}_{1|i})]^{1/2}}, \quad i = 1, \ldots, I. \tag{4.21}$$

Each residual compares the difference between an observed count and its fitted value to the estimated standard deviation of the observed count.

If $\hat{\pi}_{1|i}$ were replaced by the true value $\pi_{1|i}$ in (4.21), e_i would be the difference between a binomial random variable and its expectation, divided by its estimated standard deviation. Then if n_i were large, e_i would have an approximate standard normal distribution. The $\{\pi_{1|i}\}$ are unknown, however, so (4.21) replaces them by their estimates for the model. Because the estimates depend on $\{y_i\}$, $\{y_i - n_i \hat{\pi}_{1|i}\}$ tend to be smaller than $\{y_i - n_i \pi_{1|i}\}$. Thus, $\{e_i\}$ tend to show less variation than standard normal random variables. In fact, the Pearson statistic for testing the fit of the model is related to $\{e_i\}$ by

$$X^2 = \sum e_i^2. \tag{4.22}$$

If X^2 has df $= \nu$, it follows that the sum of squared residuals is asymptotically comparable to the sum of squares of ν (rather than I) standard normal random variables. Despite this, residuals are often treated like standard normal deviates, with absolute values larger than 2 indicating possible lack of fit.

Table 4.8 shows residuals for logit models fitted to the heart disease

Table 4.8 Residuals for Logit Models Fitted to Table 4.4

Blood Pressure	Sample Size	Observed Heart Disease	Fitted Indep. Model	Fitted Linear Logit	Residual Indep. Model	Residual Linear Logit
<117	156	3	10.8	5.2	−2.46	−0.98
117–126	252	17	17.4	10.6	−0.11	2.01
127–136	284	12	19.7	15.1	−1.79	−0.81
137–146	271	16	18.8	18.1	−0.66	−0.51
147–156	139	12	9.6	11.6	0.80	0.12
157–166	85	8	5.9	8.9	0.90	−0.30
167–186	99	16	6.9	14.2	3.62	0.51
>186	43	8	3.0	8.4	3.02	−0.14

data in Table 4.4. The residuals for the independence model are large and show an increasing trend. This trend disappears when we add a linear effect term (model 4.14), and only the second category shows evidence of lack of fit. In analyzing residual patterns, though, we must be cautious about attributing patterns to what might be chance variation from a model.

We can define *adjusted* residuals that have asymptotic standard normal distributions, using an analog of the "hat" or projection matrix in regression analysis. Sections 7.3 and 12.3 discuss these residuals.

The X^2 and G^2 goodness-of-fit statistics do *not* have approximate chi-squared distributions when applied to binary regression models with a continuous covariate, unless there are many observations at each observed level of the covariate. The asymptotic chi-squared theory applies when the number I of levels of X is fixed, and the number of observations n_i made at each level grows. This theory applies more naturally to the models of Section 4.3 for categorical covariates. Thus, to analyze lack of fit when explanatory variables are continuous, we apply goodness-of-fit statistics and related residual measures by grouping observed and fitted values for a partition of the space of explanatory variable values. Hosmer and Lemeshow (1980) and Tsiatis (1980) presented ways of doing this for logit models, though the ideas extend to any GLM (Pregibon 1982b). We can also detect lack of fit of certain types by using a likelihood-ratio test to compare the model to a more complex one, perhaps containing nonlinear or interaction terms.

4.6.2 Diagnostics for GLMs

Recently, statisticians have proposed diagnostics for GLMs that generalize diagnostics for ordinary regression models. Much work remains to be done in this area, and we briefly describe only a couple of approaches.

Separate diagnostics are useful for investigating the adequacy of choice for each of the three components of a GLM—random component, linear predictor, and link function. For instance, suppose we can specify a family $g(\mu; \alpha)$ of links, dependent on some parameter α. Pregibon (1980) showed how to estimate the α value giving the best link, and how to check the adequacy of a chosen link $g(\mu; \alpha_0)$.

We can detect influential observations by calculating residual measures and by approximating the effect on parameter estimates of deleting each single observation from the analysis. See Pregibon (1980, 1981), Pierce and Schafer (1986), and Williams (1984, 1987) for details. Section 13.1.5 discusses residuals for GLMs.

4.6.3 R-Squared Measures

In regression models, R^2 gives the proportional reduction in variation in comparing the conditional variation of the response to the marginal variation. It describes the strength of association between the model's linear predictor and the response, with $R^2 = 1$ when we can predict the response perfectly. Despite several attempts to define analogs of R^2 for models for categorical responses, no proposed measure seems as widely useful as the regression R^2. We present a couple of proposed measures in this subsection.

Let L_M denote the maximized log likelihood for the model fitted, and let L_S denote the maximized log likelihood for the saturated model, which has as many parameters as observations. Let L_0 denote the maximized log likelihood for the "null" model containing only an intercept term. Probabilities are no greater than 1.0, so log likelihoods are nonpositive. As the model complexity increases, the parameter space expands, so the value of the maximized likelihood increases. Thus, the maximized log likelihoods satisfy $L_0 \leqslant L_M \leqslant L_S \leqslant 0$, and the measure

$$\frac{L_M - L_0}{L_S - L_0} \tag{4.23}$$

falls between 0 and 1. It equals 0 when the model provides no improvement in fit over the null model, and it equals 1 when the model fits as well as the saturated model.

For the ith subject in a sample, let y_i denote the binary response, and let $\hat{\pi}_i$ denote the estimated probability of response 1, for a particular model. For N independent Bernoulli observations, the maximized log likelihood is

$$\log \prod_{i=1}^{N} [\hat{\pi}_i^{y_i}(1 - \hat{\pi}_i)^{1-y_i}] = \sum_{i=1}^{N} [y_i \log \hat{\pi}_i + (1 - y_i) \log(1 - \hat{\pi}_i)].$$

The null model gives $\hat{\pi}_i = (\Sigma \, y_i)/N = \bar{y}$, so that

$$L_0 = N[\bar{y}(\log \bar{y}) + (1 - \bar{y}) \log(1 - \bar{y})].$$

The saturated model has a dummy variable for each subject, and implies $\hat{\pi}_i = y_i$ for all i. Thus, $L_S = 0$, and (4.23) simplifies to

$$D = \frac{L_0 - L_M}{L_0}$$

proposed by McFadden (1974).

Now suppose explanatory variables are categorical and we summarize data by I binomial counts rather than N Bernoulli indicators. The saturated model then has a dummy variable for each count, and gives I fitted proportions equal to the I sample proportions for response 1. Then L_S is nonzero and (4.23) takes a different value than when we calculate it using individual subjects. For I binomial counts, the maximized likelihoods are related to the G^2 goodness-of-fit statistic by $G^2(M) = -2(L_M - L_S)$, so (4.23) becomes

$$D^* = \frac{G^2(0) - G^2(M)}{G^2(0)}.$$

Goodman (1971a) and Theil (1970) discussed this and related partial association measures.

A difficulty with D^* is that it can be large even when the strength of association is weak. For instance, a model can fit much better than the null model even though the fitted probabilities are close to 0.5 for the entire sample. In particular, $D^* = 1$ when the model fits perfectly, regardless of how well we can predict individual responses on Y with that model. Also, suppose the population satisfies the given model, but not the null model. As the sample size increases with I fixed, $G^2(M)$ behaves like a chi-squared random variable but $G^2(0)$ grows unboundedly. Thus, $D^* \to 1$ as $N \to \infty$, and its magnitude tends to depend on the sample size. This measure confounds model goodness of fit with strength of association. Similar behavior occurs for R^2 in regression analyses when we calculate it using *means* of Y values at I different x settings, rather than using individual subjects.

An alternative approach measures the association between the observed binary responses $\{y_i\}$ and their fitted values $\{\hat{\pi}_i\}$ for the model. We could use the Pearson correlation between the two sets, or a measure describing relative occurrence of concordance vs. discordance, such as gamma. Or, we could use the proportional reduction in error obtained by using $\hat{\pi}_i$ instead of $\bar{y} = \Sigma\, y_i/n$ as a predictor of y_i (Morrison, 1972), where $\bar{y}$ is the overall proportion of the sample making response 1. For squared error, this gives a measure

$$R^2 = 1 - \frac{\Sigma\,(y_i - \hat{\pi}_i)^2}{\Sigma\,(y_i - \bar{y})^2}$$

in a family developed by Efron (1978) using an axiomatic approach. When we fit the linear probability model using ordinary least squares, this

R^2 simplifies to standard R-squared for regression modeling. In that case, it also equals the square of the correlation between $\{y_i\}$ and $\{\hat{\pi}_i\}$.

A disadvantage of R^2 is its failure to incorporate in the error structure the dependence of the variance of Y_i on π. For a given model and set of data, the parameter value that gives $\{\hat{\pi}_i\}$ that maximize R^2 is not the ML estimator, and is not even an efficient estimator. When we use R^2 with ML estimates $\{\hat{\pi}_i\}$, it can decrease when we add an explanatory variable to the model. Amemiya (1981) suggested a related measure that weights squared deviations by inverse predicted variances. That measure is less simple to interpret, however.

4.7 FITTING LOGIT MODELS

We now study the mechanics of ML estimation and model fitting for logit models. We treat the N binary responses as independent Bernoulli random variables. Let $x_i = (x_{i0}, x_{i1}, \ldots, x_{ik})$ denote the ith setting of values of k explanatory variables, $i = 1, \ldots, I$, where $x_{i0} = 1$. When explanatory variables are continuous, there may be a different setting for each subject, in which case $I = N$. We express the logistic regression model (4.10) as

$$\pi(\mathbf{x}_i) = \frac{\exp\left(\sum_{j=0}^{k} \beta_j x_{ij}\right)}{\left[1 + \exp\left(\sum_{j=0}^{k} \beta_j x_{ij}\right)\right]} \tag{4.24}$$

where $\beta_0 = \alpha$.

4.7.1 Likelihood Equations

When more than one observation on Y occurs at a fixed x_i value, it is sufficient to record the number of observations n_i and the number of "1" outcomes. Thus we let Y_i refer to this "success" count rather than to individual binary responses. The $\{Y_i, i = 1, \ldots, I\}$ are independent binomial random variables with $E(Y_i) = n_i \pi(x_i)$, where $n_1 + \cdots + n_I = N$. The joint probability mass function of $(Y_1, \ldots, Y_I)$ is proportional to the product of I binomial functions,

$$\prod_{i=1}^{I} \pi(\mathbf{x}_i)^{y_i}[1 - \pi(\mathbf{x}_i)]^{n_i - y_i}$$

$$= \left\{ \prod_{i=1}^{I} [1 - \pi(\mathbf{x}_i)]^{n_i} \right\} \left\{ \prod_{i=1}^{I} \exp\left[\log\left(\frac{\pi(\mathbf{x}_i)}{1 - \pi(\mathbf{x}_i)} \right)^{y_i} \right] \right\}$$

$$= \left\{ \prod_{i=1}^{I} [1 - \pi(\mathbf{x}_i)]^{n_i} \right\} \exp\left[\sum y_i \log\left(\frac{\pi(\mathbf{x}_i)}{1 - \pi(\mathbf{x}_i)} \right) \right].$$

For model (4.24), the ith logit is $\sum_j \beta_j x_{ij}$, so the exponential term in the last expression equals $\exp[\sum_i y_i(\sum_j \beta_j x_{ij})] = \exp[\sum_j (\sum_i y_i x_{ij})\beta_j]$. Also, since $[1 - \pi(\mathbf{x}_i)] = [1 + \exp(\sum_j \beta_j x_{ij})]^{-1}$, the log likelihood equals

$$L(\boldsymbol{\beta}) = \sum_j \left(\sum_i y_i x_{ij} \right) \beta_j - \sum_i n_i \log\left[1 + \exp\left(\sum_j \beta_j x_{ij} \right) \right]. \quad (4.25)$$

This depends on the binomial counts only through the sufficient statistics $\{\sum_i y_i x_{ij}, \, j = 0, \ldots, k\}$.

We derive likelihood equations by differentiating L with respect to elements of $\boldsymbol{\beta}$ and setting the results equal to zero. Since

$$\frac{\partial L}{\partial \beta_a} = \sum_i y_i x_{ia} - \sum_i n_i x_{ia} \left[\frac{\exp\left(\sum_j \beta_j x_{ij} \right)}{1 + \exp\left(\sum_j \beta_j x_{ij} \right)} \right],$$

the likelihood equations are

$$\sum_i y_i x_{ia} - \sum_i n_i \hat{\pi}_i x_{ia} = 0, \quad a = 0, \ldots, k \quad (4.26)$$

where $\hat{\pi}_i = \exp(\sum_j \hat{\beta}_j x_{ij})/[1 + \exp(\sum_j \hat{\beta}_j x_{ij})]$ denotes the ML estimate of $\pi(\mathbf{x}_i)$. We defer solving these equations to the next subsection.

Let $\mathbf{X}$ denote the $I \times (k + 1)$ matrix of values of $\{x_{ij}\}$. The likelihood equations (4.26) have form

$$\mathbf{X}'\mathbf{y} = \mathbf{X}'\hat{\mathbf{m}}, \quad (4.27)$$

where $\hat{m}_i = n_i \hat{\pi}_i$. Similar equations apply to the least squares fit for the linear regression model; that is, $\mathbf{X}'\mathbf{y} = \mathbf{X}'\hat{\mathbf{y}}$, where $\hat{\mathbf{y}} = \mathbf{X}\hat{\boldsymbol{\beta}}$ and $\hat{\boldsymbol{\beta}} = (\mathbf{X}'\mathbf{X})^{-1}\mathbf{X}'\mathbf{y}$. Equation (4.27) illustrates a fundamental result for GLMs that use the canonical link. The likelihood equations equate the sufficient statistics to the estimates of their expected values (Nelder and Wedder-burn 1972).

The information matrix is the negative expected value of the matrix of second partial derivatives of the log likelihood. Under regularity conditions, ML estimators of parameters have a large-sample normal distribution with covariance matrix equal to the inverse of the information matrix. For the logistic regression model,

$$\frac{\partial^2 L}{\partial \beta_a \, \partial \beta_b} = -\sum_i \frac{x_{ia} x_{ib} n_i \exp\left(\sum_j \beta_j x_{ij}\right)}{\left[1 + \exp\left(\sum_j \beta_j x_{ij}\right)\right]^2} = -\sum_i x_{ia} x_{ib} n_i \pi_i (1 - \pi_i) \, . \quad (4.28)$$

Since (4.28) is not a function of $\{y_i\}$, the observed and expected second derivative matrix are identical. This happens for all GLMs that use canonical links (Nelder and Wedderburn 1972).

We estimate the covariance matrix by substituting $\hat{\beta}$ into the matrix having elements equal to the negative of (4.28), and inverting. The estimated covariance matrix has form

$$\hat{\text{Cov}}(\hat{\beta}) = \{\mathbf{X}' \, \mathbf{Diag}[n_i \hat{\pi}_i (1 - \hat{\pi}_i)]\mathbf{X}\}^{-1} \quad (4.29)$$

where $\mathbf{Diag}[n_i \hat{\pi}_i (1 - \hat{\pi}_i)]$ denotes the $I \times I$ diagonal matrix having elements $\{n_i \hat{\pi}_i (1 - \hat{\pi}_i)\}$ on the main diagonal. The square roots of the main diagonal elements of (4.29) are estimated standard errors of model parameter estimators.

At a fixed setting $\mathbf{x}$, the estimated variance of the predicted logit $\hat{L} = \mathbf{x}\hat{\beta}$ is $\hat{\sigma}^2(\hat{L}) = \mathbf{x}\hat{\text{Cov}}(\hat{\beta})\mathbf{x}'$. For large samples, $\hat{L} \pm z_{\alpha/2} \hat{\sigma}(\hat{L})$ is a confidence interval for the true logit. The endpoints invert to a corresponding interval for the probability π, using the transform $\hat{\pi} = \exp(\hat{L})/[1 + \exp(\hat{L})]$.

The log likelihood function for logistic (and probit) regression models is strictly concave, and ML estimates exist and are unique except in certain boundary cases (Wedderburn 1976). However, the likelihood equations (4.26) are nonlinear functions of the ML estimates $\hat{\beta}$, and they require iterative solution. We next study an iterative procedure for this purpose.

4.7.2 Newton–Raphson Method

The Newton–Raphson method is a method for solving nonlinear equations. It can solve equations, such as likelihood equations, that determine the location at which a function is maximized. We describe the method, and then we use it to solve likelihood equations for logistic regression models.

The method requires an initial guess for the value that maximizes the function. The function is approximated in a neighborhood of that guess by a second-degree polynomial, and the second guess is the location of that polynomial's maximum value. The function is then approximated in a neighborhood of the second guess by another second-degree polynomial, and the third guess is the location of its maximum. In this manner, the method generates a sequence of guesses. Those guesses converge to the location of the maximum, when the function is suitable and/or the initial guess is good.

In more detail, here is how the Newton-Raphson method determines the value $\hat{\boldsymbol{\beta}}$ of $\boldsymbol{\beta}$ that maximizes a function $g(\boldsymbol{\beta})$. Let $\mathbf{q}' = (\partial g/\partial \beta_1, \partial g/\partial \beta_2, \ldots)$, and let $\mathbf{H}$ denote the matrix having entries $h_{ab} = \partial^2 g/\partial \beta_a \partial \beta_b$. Let $\mathbf{q}^{(t)}$ and $\mathbf{H}^{(t)}$ be those terms evaluated at $\boldsymbol{\beta}^{(t)}$, the tth guess for $\hat{\boldsymbol{\beta}}$. At step t in the iterative process ($t = 0, 1, 2, \ldots$), $g(\boldsymbol{\beta})$ is approximated near $\boldsymbol{\beta}^{(t)}$ by the terms up to second order in its Taylor series expansion,

$$Q^{(t)}(\boldsymbol{\beta}) = g(\boldsymbol{\beta}^{(t)}) + \mathbf{q}^{(t)\prime}(\boldsymbol{\beta} - \boldsymbol{\beta}^{(t)}) + (\tfrac{1}{2})(\boldsymbol{\beta} - \boldsymbol{\beta}^{(t)})'\mathbf{H}^{(t)}(\boldsymbol{\beta} - \boldsymbol{\beta}^{(t)}) .$$

Solving $\partial Q^{(t)}/\partial \boldsymbol{\beta} = \mathbf{q}^{(t)} + \mathbf{H}^{(t)}(\boldsymbol{\beta} - \boldsymbol{\beta}^{(t)}) = \mathbf{0}$ for $\boldsymbol{\beta}$ yields the next guess,

$$\boldsymbol{\beta}^{(t+1)} = \boldsymbol{\beta}^{(t)} - (\mathbf{H}^{(t)})^{-1}\mathbf{q}^{(t)} , \tag{4.30}$$

assuming $\mathbf{H}^{(t)}$ is nonsingular.

To help you understand the process, you may want to work through these steps when β has a single element (Problem 4.31). Figure 4.7 illustrates a cycle of the method for that case.

4.7.3 Newton–Raphson for Logistic Regression

Now suppose $g(\boldsymbol{\beta})$ is the log likelihood for the logistic regression model. From (4.26) and (4.28), let

$$q_j^{(t)} = \frac{\partial L(\boldsymbol{\beta})}{\partial \beta_j}\bigg|_{\boldsymbol{\beta}^{(t)}} = \sum_i (y_i - n_i \pi_i^{(t)})x_{ij}$$

$$h_{ab}^{(t)} = \frac{\partial^2 L(\boldsymbol{\beta})}{\partial \beta_a \partial \beta_b}\bigg|_{\boldsymbol{\beta}^{(t)}} = -\sum_i x_{ia}x_{ib}n_i \pi_i^{(t)}(1 - \pi_i^{(t)}) .$$

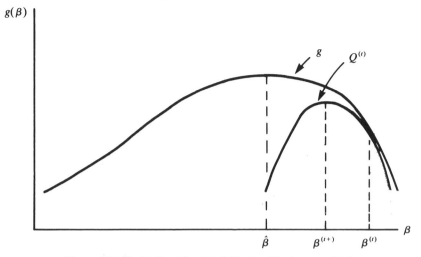

Figure 4.7   Illustration of cycle of Newton–Raphson method.

Here $\pi^{(t)}$, the tth approximation for $\hat{\pi}$, is obtained from $\beta^{(t)}$ through

$$
\pi_i^{(t)} = \frac{\exp\left(\sum\limits_{j=0}^{k} \beta_j^{(t)} x_{ij}\right)}{\left[1 + \exp\left(\sum\limits_{j=0}^{k} \beta_j^{(t)} x_{ij}\right)\right]}.
\qquad (4.31)
$$

We use $\mathbf{q}^{(t)}$ and $\mathbf{H}^{(t)}$ with formula (4.30) to obtain the next value $\beta^{(t+1)}$, which in this context is

$$
\beta^{(t+1)} = \beta^{(t)} + \{\mathbf{X}' \, \mathbf{Diag}[n_i \pi_i^{(t)}(1 - \pi_i^{(t)})]\mathbf{X}\}^{-1}\mathbf{X}'(\mathbf{y} - \mathbf{m}^{(t)})
\qquad (4.32)
$$

where $m_i^{(t)} = n_i \pi_i^{(t)}$. This is used to obtain $\pi^{(t+1)}$, and so forth.

After making an initial guess $\beta^{(0)}$, we use (4.31) to obtain $\pi^{(0)}$, and for $t > 0$ the iterations proceed as just described using (4.32) and (4.31). In the limit, $\pi^{(t)}$ and $\beta^{(t)}$ converge to the ML estimates $\hat{\pi}$ and $\hat{\beta}$ (Walker and Duncan 1967). The $\mathbf{H}^{(t)}$ matrices converge to the matrix $\hat{\mathbf{H}} = -\mathbf{X}' \, \mathbf{Diag}[n_i \hat{\pi}_i(1 - \hat{\pi}_i)]\mathbf{X}$. By (4.29) the estimated asymptotic covariance matrix of $\hat{\beta}$ is a by-product of the Newton-Raphson method, namely $-\hat{\mathbf{H}}^{-1}$.

The value $\beta^{(t+1)}$ can also be expressed as

$$\boldsymbol{\beta}^{(t+1)} = \{\mathbf{X}' \ \mathbf{Diag}[n_i \pi_i^{(t)}(1 - \pi_i^{(t)})]\mathbf{X}\}^{-1}\mathbf{X}' \ \mathbf{Diag}[n_i \pi_i^{(t)}(1 - \pi_i^{(t)})]\mathbf{z}^{(t)}$$

$$(4.33)$$

where $\mathbf{z}^{(t)}$ has elements

$$z_i^{(t)} = \log\left[\frac{\pi_i^{(t)}}{1 - \pi_i^{(t)}}\right] + \frac{y_i - n_i \pi_i^{(t)}}{n_i \pi_i^{(t)}(1 - \pi_i^{(t)})} \, . \tag{4.34}$$

In this expression, $\mathbf{z}^{(t)}$ is the linearized form of the logit link function for the sample data, evaluated at $\boldsymbol{\pi}^{(t)}$ (Problem 4.32). To interpret (4.33), it is helpful to refer to the general linear model

$$\mathbf{z} = \mathbf{X}\boldsymbol{\beta} + \boldsymbol{\epsilon} \, .$$

When the covariance matrix of $\boldsymbol{\epsilon}$ is $\mathbf{V}$, the *weighted least squares* (WLS) estimator of $\boldsymbol{\beta}$ is $(\mathbf{X}'\mathbf{V}^{-1}\mathbf{X})^{-1}\mathbf{X}'\mathbf{V}^{-1}\mathbf{z}$. Thus, $\boldsymbol{\beta}^{(t+1)}$ in (4.33) is the WLS solution for the general linear model, when $\{\epsilon_i\}$ are uncorrelated with variances $\{1/n_i \pi_i^{(t)}(1 - \pi_i^{(t)})\}$. Section 12.1.3 shows that these variances are estimated asymptotic variances of the sample logits. The ML estimate is the limit of a sequence of WLS estimates, where the weight matrix changes at each cycle. This process of calculating the ML estimate is called *iterative reweighted least squares*.

The convergence of $\boldsymbol{\beta}^{(t)}$ to $\hat{\boldsymbol{\beta}}$ for the Newton-Raphson method is usually fast. For large t, the convergence satisfies, for each j

$$|\beta_j^{(t+1)} - \hat{\beta}_j| \leqslant c \, |\beta_j^{(t)} - \hat{\beta}_j|^2 \quad \text{for some } c > 0$$

and is referred to as *second-order*. It often takes only a few iterations to get satisfactory convergence for logistic regression models.

4.8 CONDITIONAL LOGISTIC REGRESSION

ML estimators of model parameters have satisfactory properties when the sample size is large compared to the number of parameters in the model. In other cases, we can use exact conditional methods that generalize those presented for 2×2 tables in Sections 3.5 and 3.6. We will illustrate an exact analysis using the linear logit model for an $I \times 2$ table, logit $= \alpha + \beta x_i$. First we study the likelihood equations for that model.

4.8.1 Likelihood Equations for Linear Logit Model

For the linear logit model, the n_{i+} subjects in row i all have explanatory variable X equal to x_i, so the log likelihood (4.25) simplifies to

$$L(\boldsymbol{\beta}) = -\sum_{i=1}^{I} n_{i+} \log[1 + \exp(\alpha + \beta x_i)] + \sum_{i=1}^{I} n_{i+}(\alpha + \beta x_i) .$$

Differentiating, we obtain the likelihood equations

$$\frac{\partial L}{\partial \alpha} = -\sum_i n_{i+} \left[\frac{\exp(\alpha + \beta x_i)}{1 + \exp(\alpha + \beta x_i)} \right] + n_{+1} = 0 \qquad (4.35)$$

$$\frac{\partial L}{\partial \beta} = -\sum_i n_{i+} x_i \left[\frac{\exp(\alpha + \beta x_i)}{1 + \exp(\alpha + \beta x_i)} \right] + \sum_i n_{i1} x_i = 0 . \qquad (4.36)$$

These equations have simple interpretations. Let $\{\hat{m}_{ij} = n_{i+} \hat{\pi}_{j|i}\}$ denote ML estimates of the expected values $\{m_{ij} = n_{i+} \pi_{j|i}\}$ of $\{n_{ij}\}$. Then $\hat{m}_{i+} = n_{i+}$ all i, and hence $\hat{m}_{++} = N$. Since $\pi_{1|i} = \exp(\alpha + \beta x_i)/[1 + \exp(\alpha + \beta x_i)]$, (4.35) implies that

$$-\sum_i \hat{m}_{i1} + n_{+1} = 0 ,$$

so $\hat{m}_{+1} = n_{+1}$, and thus $\hat{m}_{+2} = n_{+2}$. Similarly, (4.36) implies

$$-\sum_i \hat{m}_{i1} x_i + \sum_i n_{i1} x_i = 0 \quad \text{or} \quad \sum_i x_i \left(\frac{n_{i1}}{n_{+1}} \right) = \sum_i x_i \left(\frac{\hat{m}_{i1}}{\hat{m}_{+1}} \right). \qquad (4.37)$$

The mean score across the rows in the first column is the same for the fitted data as for the observed data. It follows from (4.37) and $\{\hat{m}_{i+} = n_{i+}\}$ that $\Sigma_i(\hat{m}_{i+} - \hat{m}_{i1})x_i = \Sigma_i(n_{i+} - n_{i1})x_i$, or $\Sigma_i x_i \hat{m}_{i2}/\hat{m}_{+2} = \Sigma_i x_i n_{i2}/n_{+2}$, since $\hat{m}_{+2} = n_{+2}$. Thus, observed and fitted mean scores are also identical for the second column.

4.8.2 Exact Trend Test

When the data are I independent binomial counts, the row totals $\{n_{i+}\}$ are fixed. Sufficient statistics for the linear logit model are the total number of "1" outcomes, n_{+1}, and $T = \Sigma x_i n_{i1}$, which determines the mean score on X for those responses. To eliminate the nuisance parameter α, we must condition on n_{+1}. Conditional on n_{+1} (and hence n_{+2}) as well as the row totals, the distribution of cell counts depends on β but not α. The data enter the conditional distribution only through the remaining sufficient statistic, T. It follows from Cox (1970) that the exact conditional distribution of T has the form

$$P(T = t; \beta) = \frac{c(t)e^{\beta t}}{\sum_u c(u)e^{\beta u}} \qquad (4.38)$$

where $c(u)$ denotes the sum of $(\Pi \Pi n_{ij}!)^{-1}$ for all tables with the given marginal totals that have $T = u$.

The *conditional ML estimator* of β is the value of β that maximizes (4.38), for the given marginal totals and the observed value t of T. The estimator, which is evaluated with iterative methods, is a monotone function of T. Conditional on the marginal totals, larger values of T give evidence of larger values of β.

When $\beta = 0$, the cell counts have the multiple hypergeometric distribution (3.26). To test independence against $\beta > 0$, the P-value is the null probability that T is at least as large as the observed value. Ordering the tables with the given margins by T is equivalent to ordering them by the test statistic z or the slope b from the Cochran-Armitage trend test presented in Section 4.4.3. Thus this test for the regression parameter in the linear logit model is an exact trend test.

Computer algorithms are becoming increasingly available for conducting exact inference in logistic regression models. For Table 4.6, the P-value for the exact one-sided trend test is 0.0063. The sample size in that table is moderately large, and the result is close to that obtained with asymptotic procedures.

For the general logistic regression model (4.24), we conduct exact inference for β_j by conditioning on each sufficient statistic except $T_j = \Sigma_i y_i x_{ij}$. Inference for β_j is based on the exact conditional distribution of T_j. Exact inference for logistic regression parameters is called *conditional logistic regression*. See Breslow and Day (1980, Ch. 7), Cox (1970), and Hirji et al. (1987) for further details.

CHAPTER NOTES

Section 4.1: Generalized Linear Models

4.1 Distribution (4.1) is called a *natural* (or *linear*) exponential family to distinguish it from a more general exponential family in which y is replaced by $r(y)$ in the exponential term. GLMs have been used for other continuous random components besides the normal. For instance, survival times are usually assumed to have exponential, Weibull, or a more general extreme value distribution. (See Aitkin and Clayton 1980 and

Whitehead 1980). The gamma family, which provides a diverse collection of distributions for positive random variables, is also well suited for GLMs. For fixed shape parameter, this family has constant coefficient of variation; that is, the standard deviation is proportional to the mean. Recent books on GLMs include those by Aitkin et al. (1989), Dobson (1983), and McCullagh and Nelder (1989).

Section 4.2: Logistic Regression

4.2 A weighted least squares approach yields more efficient estimators of parameters in the linear probability model. (See Aldrich and Nelson 1984, Chap. 1, Draper and Smith 1981, pp. 108-116). We weight each observation by the inverse $1/\hat{\pi}(x)[1 - \hat{\pi}(x)]$ of the estimated variance, using $\hat{\pi}(x)$ from the ordinary least squares fit. We could use this procedure iteratively, the fit at each stage generating estimated weights for the next stage. As the number of iterations increases, the estimates in this scheme converge to the ML estimates. Complications arise when $\hat{\pi}(x)$ is outside the $[0, 1]$ range at some stage.

4.3 According to Finney (1971), Fisher and Yates (1938) first suggested the logit link for binary data. The term "logit" was introduced by Berkson (1944), who showed its similarity to the probit. Berkson (1951) argued in favor of the logistic regression model over the probit model. Dyke and Patterson (1952) used the logit in models having categorical explanatory variables. Haldane (1955) recommended the addition of $\frac{1}{2}$ to numerator and denominator given in the empirical logit. With this modification, the bias is on the order of only $1/n_i^2$, for large n_i (Problem 12.4 and Gart et al. 1985). Albert and Anderson (1984), Berkson (1951, 1953, 1955), Hodges (1958), and Walker and Duncan (1967) discussed ML estimation for logistic models. Cox's (1958a) article and (1970) book have been highly influential in advancing the use of logistic regression models.

There have been many interesting applications of logistic regression. Rosenbaum and Rubin (1983) used it to adjust for bias in comparing two groups in observational studies. They defined the *propensity* as the probability of being in one group, for a given setting of the explanatory variables x, and they used logistic regression to estimate how propensity depends on x. In comparing the groups on the response variable, they showed that one can control for differing distributions of the groups on x by adjusting for the estimated propensity. This is done by using the propensity to match samples from the groups, or to subclassify subjects into several strata consisting of intervals of propensity scores, or to adjust directly by entering the propensity in an analysis of covariance model.

Adelbasit and Plackett (1983), Chaloner and Larntz (1988), Minkin (1987), Tsutakawa (1980), and Wu (1985) discussed design problems for binary response experiments. Chaloner and Larntz presented a Bayesian approach to choosing settings for an explanatory variable to optimize various criteria in terms of estimating parameter values or estimating the setting at which the response probability equals some fixed value. For predicting classifications of subjects on binary responses, Efron (1975), Press and Wilson (1978), and Amemiya and Powell (1980) compared logistic regression to discriminant analysis, which assumes that explanatory variables have a normal distribution at each of the two levels of Y.

Stukel (1988) extended the scope of the logistic model by introducing shape parameters that modify the behavior of the curve in extreme probability regions and allow for asymmetric treatment of the two tails. Follman and Lambert (1989) presented a generalization that can account for overdispersion.

4.4 In the likelihood-ratio test of H_0: $\beta = \beta_0$, one computes twice the difference in maximized log likelihoods when $\beta = \beta_0$ and when β is unrestricted. We can obtain a confidence interval for β by inverting this test. For instance, a 95% confidence interval for β is the set of β_0 values not rejected at the 0.05 level. For small samples, the likelihood can be highly nonsymmetric about the ML estimate, and this confidence interval can be quite different from $\hat{\beta} \pm z_{\alpha/2}(\text{ASE})$. See Aitkin et al. (1989) for details.

The *score statistic*, due to R. A. Fisher and C. R. Rao, is an alternative to the likelihood-ratio and Wald statistics for testing a hypothesis $\gamma = \gamma_0$ about model parameters. It is a quadratic form based on the vector of partial derivatives of the log likelihood with respect to γ, evaluated at the H_0 estimates (i.e., assuming $\gamma = \gamma_0$). For logistic regression models with a single explanatory variable, the score test of $\beta = 0$ is equivalent to the trend test in Section 4.4.3. Cox and Hinkley (1974, Chap. 9) and Rao (1973, Section 6e) discussed, in general terms, asymptotic equivalences among the likelihood-ratio test, Wald test, and score test for model-based inference.

4.5 For continuous responses, scatter diagrams give us visual information about the dependence of Y on X. For binary responses, such diagrams are not very informative. In Section 4.2.3, we used empirical logits for grouped levels of LI to obtain preliminary information about the nature of the dependence. When X is continuous, or when it is discrete but with a small sample size at each distinct X value, nonparametric smoothing methods can describe the dependence of $\pi(x)$ on x. For instance, see Copas (1983).

Section 4.4: Using Models to Improve Inferential Power

4.6 See Armitage (1955), Chapman and Nam (1968), Cochran (1954), Mantel (1963), Wood (1978), and Yates (1948) for details and extensions of the Cochran-Armitage trend test. For testing independence, that trend test is asymptotically efficient both for linear and logistic alternatives, as the number of observations increases at each x value. Its efficiency against linear alternatives follows from the approximate normality of the sample proportions, with constant Bernoulli variance when $\beta = 0$. For the linear logit model (4.14), its efficiency follows from its equivalence with the score test. See Problem 8.29 and Cox (1958a) for related remarks.

Section 4.5: Probit and Extreme Value Models

4.7 According to Stigler (1986, p. 246), Fechner (1860) may have been the first to suggest transforming proportions by the inverse normal cdf. Articles by Gaddum (1933) and Bliss (1934, 1935) popularized the probit method for toxicological experiments. Bliss introduced the term "probit," but used the inverse normal cdf with mean 5 (rather than 0) and standard deviation 1. Chambers and Cox (1967) showed it is difficult to distinguish between the probit and logit models unless the sample size is extremely large. Ashford and Sowden (1970) generalized the probit model for multivariate binary responses. Finney (1971) gave further details about probit models.

Models with log-log link have been used for survival data (see Section 9.5) and in econometric applications (see Brennan 1949, Zellner and Lee 1965).

Section 4.6: Model Diagnostics

4.8 Diagnostics for logistic regression models were discussed by Copas (1988), Cox (1970, pp. 94–99), Fowlkes (1987), Jennings (1986), Johnson (1985), Landwehr et al. (1984), and Pregibon (1981). Copas's article noted special problems for binary data that do not occur for normal regression models.

Amemiya (1981), Efron (1978), and Maddala (1983) reviewed a variety of R^2-type measures for logistic regression. Some versions of such measures adjust for the number of parameters in the model.

Section 4.7: Fitting Logit Models

4.9 For further details on the Newton-Raphson method, see Bard (1974) and Haberman (1978). Fisher (1935b) introduced the Fisher scoring method to calculate ML estimates for probit models.

PROBLEMS

Applications

4.1 In the first nine decades of the twentieth century in baseball's National League, the percentage of times that the starting pitcher pitched a complete game were: 72.7 (1900-1909), 63.4, 50.0, 44.3, 41.6, 32.8, 27.2, 22.5, 13.3 (Source: George Will, *Newsweek*, April 10, 1989).

    a. For simplicity, suppose the number of games was the same in each decade. Fit logit and linear probability models to describe the trend in these data.

    b. Use the fitted models to predict the percentages of complete games for the next three decades.

4.2 A sample of elderly people are given a psychiatric examination to determine whether symptoms of senility are present. One explanatory variable is the score on a subtest of the Wechsler Adult Intelligence Scale. Table 4.9 shows the data.

    a. Fit a logistic regression model. For what region of WAIS scores does the estimated probability of senility exceed $\frac{1}{2}$? Estimate the probability at several WAIS values, and sketch a figure of the relationship.

Table 4.9   Data on $X = $ WAIS Score and $Y = $ Senility
(1 = Symptoms Present)

X	Y	X	Y	X	Y	X	Y	X	Y
9	1	7	1	7	0	17	0	13	0
13	1	5	1	16	0	14	0	13	0
6	1	14	1	9	0	19	0	9	0
8	1	13	0	9	0	9	0	15	0
10	1	16	0	11	0	11	0	10	0
4	1	10	0	13	0	14	0	11	0
14	1	12	0	15	0	10	0	12	0
8	1	11	0	13	0	16	0	4	0
11	1	14	0	10	0	10	0	14	0
7	1	15	0	11	0	16	0	20	0
9	1	18	0	6	0	14	0		

b. Interpret the dependence of the odds of senility on the WAIS, and test the statistical significance of the effect.

c. Show how to judge whether the model fits adequately by comparing observed to fitted values for intervals of WAIS.

d. Use least squares to fit the linear probability model. Note that, compared to the logistic regression model, predicted probabilities are quite different at the low and high ends of the WAIS scale.

4.3 Refer to Table 4.6. Analyze these data using a logit or probit model. Test independence using the Wald test and using the likelihood-ratio test, and compare results to the trend test.

4.4 Refer to Table 2.9. Fit a logit model, using scores {0, 3, 9.5, 19.5, 37, 55} for cigarette smoking. Interpret.

4.5 Table 4.10 refers to the sample discussed in Example 4.3.3, classified both by blood pressure and cholesterol level. For instance, at the lowest level of both variables, there were 53 cases, of whom 2 exhibited heart disease.

a. Fit a logit model that describes the effect of cholesterol on heart disease. Interpret.

b. Using the ordering of cholesterol levels, give two ways to test the hypothesis of independence of cholesterol and heart disease.

Table 4.10

Blood Pressure	Serum Cholesterol (mg/100 ml)						
	<200	200–209	210–219	220–244	245–259	260–284	>284
<117	2/53	0/21	0/15	0/20	0/14	1/22	0/11
117–126	0/66	2/27	1/25	8/69	0/24	5/22	1/19
127–136	2/59	0/34	2/21	2/83	0/33	2/26	4/28
137–146	1/65	0/19	0/26	6/81	3/23	2/34	4/23
147–156	2/37	0/16	0/6	3/29	2/19	4/16	1/16
157–166	1/13	0/10	0/11	1/15	0/11	2/13	4/12
167–186	3/21	0/5	0/11	2/27	2/5	6/16	3/14
>186	1/5	0/1	3/6	1/10	1/7	1/7	1/7

Source: Reprinted with permission from Cornfield (1962).

    c. Fit a logit model that simultaneously describes the effects of cholesterol and blood pressure on heart disease. Interpret effects.

4.6 Use models to analyze and interpret the data in Table 4.11 on smoking habits of students in Arizona high schools.

Table 4.11

	Student Smokes	Student Does Not Smoke
Both parents smoke	400	1380
One parent smokes	416	1823
Neither parent smokes	188	1168

By permission, S.V. Zagona, *Studies and Issues in Smoking Behavior*, Tuscon: The University of Arizona Press, Copyright 1967.

4.7 Table 4.12, reported by Clogg and Shockey (1988), is taken from the 1982 General Social Survey.

    a. Treating vote as the response, fit logit model (4.13) with nominal main effects. Does there seem to be a trend in the effects at the seven levels of political views?

    b. Fit a logit model that uses the ordinal nature of political views. Carefully interpret parameter estimates for this model.

Table 4.12

| Race | Political Views[a] | 1980 Presidential Vote | |
		Reagan	Carter or other
White	1	1	12
	2	13	57
	3	44	71
	4	155	146
	5	92	61
	6	100	41
	7	18	8
Nonwhite	1	0	6
	2	0	16
	3	2	23
	4	1	31
	5	0	8
	6	2	7
	7	0	4

Source: 1982 General Social Survey; see Clogg and Shockey (1988).
[a]Political views range from 1 = extremely liberal to 7 = extremely conservative.

4.8 Table 4.13, from Graubard and Korn (1987), refers to a prospective study of maternal drinking and congenital malformations. After the first 3 months of pregnancy, the women in the sample completed a questionnaire about alcohol consumption. Following the birth of the child, observations were recorded on presence or absence of congenital sex organ malformations.

Table 4.13

| Malformation | Alcohol Consumption (Ave. no. Drinks per Day) | | | | |
	0	<1	1–2	3–5	≥6
Absent	17,066	14,464	788	126	37
Present	48	38	5	1	1

Source: Reprinted with permission from the Biometric Society (Graubard and Korn 1987).

   a. Test independence using the trend test with midpoint scores {0.0, 0.5, 1.5, 4.0, 7.0} for alcohol consumption.
   b. Conduct the trend test using the scores {1, 2, 3, 4, 5}. Compare results to those in (a), and note that results can be sensitive to the choice of scores. (Table 4.13 has some very small counts, and exact tests have greater validity than asymptotic procedures. For the exact trend test using the scores in (a), the one-sided P-value is 0.016.)

4.9 Refer to the fit of the probit model for Table 4.2. Interpret results.

4.10 Refer to the beetle mortality data in Table 4.7. Fit the model having log-log link, and use least squares to fit the linear probability model. Why do they fit so poorly?

4.11 Calculate residuals for the linear logit model fitted to Table 4.7. Do they show evidence of lack of fit?

4.12 Calculate residuals for the independence model fitted to Table 4.6. What type of lack of fit is indicated?

4.13 Conduct the exact trend test for Table 3.8. Interpret results.

Theory and Methods

4.14 Suppose each observation is a mean $\bar{Y}_i$ of n independent Bernoulli observations; that is, $n\bar{Y}_i$ has a *binomial* distribution. Show that the same expected value and natural parameter occur as when each observation is a single Bernoulli random variable.

4.15 For given variance (say, $\sigma^2 = 1$), show that the normal distribution with mean parameter μ is in the natural exponential family, and identify the natural parameter. Formulate the usual regression model as a GLM.

4.16 Binary observations have parameter π depending on a covariate x. Consider the model $\pi(x) = \frac{1}{2} + (1/\pi)\tan^{-1}(\alpha + \beta x)$.
   a. What distribution has cdf of this form?
   b. Formulate a GLM in which $\pi(x)$ has this form. When would you expect this model to be more appropriate than the logit or probit?

4.17 Prove that the logistic regression curve (4.6) has steepest slope where $\pi(x) = \frac{1}{2}$. Generalize this result to model (4.10) having multiple explanatory variables.

4.18 In toxicological experiments in which the probability killed is modeled as a function of dose, the x value at which $\pi(x) = \frac{1}{2}$ is called the median lethal dose, and is denoted by LD 50.

  a. Show that for probit model (4.18) and logit model (4.7), LD $50 = -\alpha/\beta$.

  b. Find LD 50 for the model with complementary log-log link.

4.19 Consider the calibration problem of estimating the x value at which $\pi(x) = \pi_0$. For the linear logit model, argue that a large-sample confidence interval is the set of x values for which

$$|\hat{\alpha} + \hat{\beta}x - \text{logit}(\pi_0)| / [\text{Var}(\hat{\alpha}) + x^2 \, \text{Var}(\hat{\beta}) + 2x \, \text{Cov}(\hat{\alpha}, \hat{\beta})]^{1/2}$$

$$< z_{\alpha/2} .$$

4.20 Let D denote the event that a subject has a certain disease, let S denote the event that a subject is sampled, and let $\rho_0 = P(S \mid D)$ and $\rho_1 = P(S \mid D^c)$. Suppose logistic regression model (4.6) holds, with $\pi(x) = \pi(D \mid x)$ denoting the probability of disease. Using Bayes Theorem, show that $P(D \mid S; x)$ also follows the logistic regression model, with the same effect parameter β but with intercept $\alpha^* = \alpha + \log(\rho_0/\rho_1)$. It follows that the logistic regression model is appropriate for retrospective sampling. (See Anderson 1972, Breslow and Day 1980, p. 203, Breslow and Powers 1978, Farewell 1979, and Prentice 1976).

4.21 For the population of subjects having $Y = j$, suppose the explanatory variables $\mathbf{X}$ have a multivariate normal $N(\boldsymbol{\mu}_j, \boldsymbol{\Sigma})$ distribution, $j = 0,1$. Use Bayes Theorem to show that $P(Y = 1 \mid \mathbf{x})$ follows the logistic regression model with effect parameters $\boldsymbol{\Sigma}^{-1}(\boldsymbol{\mu}_1 - \boldsymbol{\mu}_0)$. (Cornfield 1962).

4.22 Suppose model (4.8) holds for some strictly increasing cdf F. Show there is a monotone transformation of the explanatory variable such that the logistic regression model holds. Generalize this result to alternative link functions.

4.23 For an $I \times 2$ contingency table, consider logit model (4.11). For any set $\{\pi_{1|i}\}$ with each $\pi_{1|i} > 0$, show that this model necessarily holds. Prove that the case $\beta_1 = \beta_2 = \cdots = \beta_I$ is the independence model.

4.24 For model (4.19) with complementary log-log link, show that the greatest rate of change of $\pi(x)$ occurs at $x = -\alpha/\beta$. What does $\pi(x)$ equal at that point? Give the corresponding results for the model with log-log link, and compare to the logit and probit models.

4.25 Suppose that the model (4.20) having log-log link holds. Explain how to interpret β.

4.26 Prove that the residuals for the linear logit model satisfy $X^2 = \Sigma \, e_i^2$.

4.27 When explanatory variables are categorical, show that the maximized log likelihood for an unsaturated model is the same whether we regard the data as N Bernoulli observations or I binomial observations. Show this is not true for the saturated model, which has as many parameters as observations.

4.28 Consider the log likelihood function (4.25) for the logistic regression model with a single explanatory variable taking only two values, 0 and 1. Derive the likelihood equations. Show that the ML estimate of β is the sample log odds ratio.

4.29 Derive the likelihood equations for model (4.13).

4.30 Let Y_i, $i = 1, \ldots, N$, denote N independent Bernoulli random variables.
   a. Derive the log likelihood for the probit model $\Phi^{-1}[\pi(\mathbf{x})] = \Sigma_j \beta_j x_{ij}$.
   b. Show the likelihood equations for the logistic and probit regression models are

$$\sum_i (y_i - \hat{\pi}_i) z_i x_{ij} = 0, \quad j = 0, \ldots, k,$$

where $z_i = 1$ for the logistic case and $z_i = \phi(\Sigma_j \hat{\beta}_j x_{ij})/\hat{\pi}_i(1 - \hat{\pi}_i)$ for the probit case. (When the link is not canonical, there is no reduction of the data in sufficient statistics.)

4.31 Suppose we want to find the value $\hat{\beta}$ that maximizes a function $g(\beta)$ of a single variable. Let $\beta^{(0)}$ denote an initial guess.

a. Using $g'(\hat{\beta}) = g'(\beta^{(0)}) + (\hat{\beta} - \beta^{(0)})g''(\beta^{(0)}) + \cdots$, argue that for $\beta^{(0)}$ close to $\hat{\beta}$, approximately $0 = g'(\beta^{(0)}) + (\hat{\beta} - \beta^{(0)})g''(\beta^{(0)})$. Solve this equation to obtain an approximation $\beta^{(1)}$ for $\hat{\beta}$.

b. Let $\beta^{(t)}$ denote the tth approximation for $\hat{\beta}$, $t = 0, 1, 2, \ldots$. Continuing the argument from (a), show the next approximation is

$$\beta^{(t+1)} = \beta^{(t)} - g'(\beta^{(t)})/g''(\beta^{(t)}).$$

4.32 **a.** Show that $\boldsymbol{\beta}^{(t+1)}$ can be expressed as (4.33).

b. Let p denote a proportion. For a nearby value π, show that

$$\log\left(\frac{p}{1-p}\right) \cong \log\left(\frac{\pi}{1-\pi}\right) + \frac{p-\pi}{\pi(1-\pi)}.$$

c. Show that $z_i^{(t)}$ in (4.34) is a linearized version of the ith sample logit, evaluated at the tth approximation $\pi_i^{(t)}$ for $\hat{\pi}_i$.

4.33 Use likelihood equations to show that when the linear logit model holds for an $I \times 2$ table, $\beta = 0$ is equivalent to $E(X \mid Y = 0) = E(X \mid Y = 1)$.

CHAPTER 5

Loglinear Models

Chapters 2–4 focused mainly on *bivariate* analyses—for instance, modeling the relationship between a binary response and a single explanatory variable. To help us analyze relationships among *several* variables, we now turn our attention to models for multidimensional contingency tables. We see that incorrect conclusions can result from studying variables only two at a time.

Loglinear models describe association patterns among categorical variables. With the loglinear approach, we model cell counts in a contingency table in terms of associations among the variables. When it is natural to regard one variable as a response and others as explanatory variables, certain loglinear models are equivalent to logit models for that response variable.

Section 5.1 introduces loglinear models for two-way tables, and shows their relation to logit models. Section 5.2 shows that an association between two variables can change dramatically when we control for a third variable. That section discusses types of independence in three-way tables, and introduces the concept of *three-factor interaction* among categorical variables. Section 5.3 presents loglinear and logit models for three-way tables. Section 5.4 presents models for tables of higher dimensions.

This chapter focuses on model interpretation, rather than the use of models for data analysis. Chapters 6 and 7 discuss the process of using sample data to fit the models and make inferences.

5.1 LOGLINEAR MODEL FOR TWO DIMENSIONS

Suppose there is a multinomial sample of size n over the $N = IJ$ cells of an $I \times J$ contingency table. The probabilities $\{\pi_{ij}\}$ for that multinomial

130

distribution form the joint distribution of two categorical responses. Those responses are statistically independent when $\pi_{ij} = \pi_{i+}\pi_{+j}$, $i = 1, \ldots, I$, $j = 1, \ldots, J$. The related expression for the expected frequencies $\{m_{ij} = n\pi_{ij}\}$ is $m_{ij} = n\pi_{i+}\pi_{+j}$ for all i and j. We shall construct loglinear models using $\{m_{ij}\}$ rather than $\{\pi_{ij}\}$, so they also apply for the Poisson sampling model for N cell counts with expectations $\{m_{ij}\}$.

5.1.1 Independence Model

On a logarithmic scale, independence has the additive form

$$\log m_{ij} = \log n + \log \pi_{i+} + \log \pi_{+j} . \tag{5.1}$$

The log expected frequency for cell (i, j) is an additive function of an ith row effect and a jth column effect.

Denote the row variable by X and the column variable by Y. Expression (5.1) is equivalent to

$$\log m_{ij} = \mu + \lambda_i^X + \lambda_j^Y \tag{5.2}$$

where

$$\lambda_i^X = \log \pi_{i+} - \left(\sum_h \log \pi_{h+}\right)/I$$

$$\lambda_j^Y = \log \pi_{+j} - \left(\sum_h \log \pi_{+h}\right)/J$$

$$\mu = \log n + \left(\sum_h \log \pi_{h+}\right)/I + \left(\sum_h \log \pi_{+h}\right)/J .$$

The parameters $\{\lambda_i^X\}$ and $\{\lambda_j^Y\}$ satisfy

$$\sum_i \lambda_i^X = \sum_j \lambda_j^Y = 0 .$$

Model (5.2) is called the *loglinear model of independence* in a two-way contingency table.

The zero-sum constraints are a common way to make parameters in the model identifiable, but other parameter definitions are possible. The same model results when we add a constant to one term and subtract it from another. An alternative definition sets $\lambda_i^X = \log \pi_{i+} - \log \pi_{1+}$ and $\lambda_j^Y = \log \pi_{+j} - \log \pi_{+1}$, so $\lambda_1^X = \lambda_1^Y = 0$. For either scaling, as λ_i^X increases, relatively more subjects are at the ith level of X.

5.1.2 Saturated Model

Now suppose there is dependence between the variables, with all $m_{ij} > 0$. Let $\eta_{ij} = \log m_{ij}$, let

$$\eta_{i.} = \frac{\sum\limits_{j} \eta_{ij}}{J} , \quad \eta_{.j} = \frac{\sum\limits_{i} \eta_{ij}}{I}$$

and let

$$\mu = \eta_{..} = \frac{\sum\limits_{i} \sum\limits_{j} \eta_{ij}}{IJ}$$

denote the grand mean of $\{\log m_{ij}\}$. Then letting

$$\lambda_i^X = \eta_{i.} - \eta_{..} , \qquad \lambda_j^Y = \eta_{.j} - \eta_{..}$$

$$\lambda_{ij}^{XY} = \eta_{ij} - \eta_{i.} - \eta_{.j} + \eta_{..}$$

(5.3)

it follows from substitution that

$$\log m_{ij} = \mu + \lambda_i^X + \lambda_j^Y + \lambda_{ij}^{XY} . \tag{5.4}$$

This model describes perfectly any set of positive expected frequencies. It is the *saturated* model, the most general model for two-way contingency tables. The right-hand side of (5.4) resembles the formula for cell means in two-way ANOVA, allowing interaction.

The parameters $\{\lambda_i^X\}$ and $\{\lambda_j^Y\}$ in (5.3) are deviations about a mean, and $\Sigma_i \lambda_i^X = \Sigma_j \lambda_j^Y = 0$. Thus, there are $I - 1$ linearly independent row parameters and $J - 1$ linearly independent column parameters. The $\{\lambda_i^X\}$ and $\{\lambda_j^Y\}$ pertain to the relative number of cases in cells at various levels of X and Y, on a log scale. If $\lambda_i^X > 0$, for instance, the average log expected frequency for cells in row i exceeds the average log expected frequency over the entire table.

The $\{\lambda_{ij}^{XY}\}$ satisfy

$$\sum_i \lambda_{ij}^{XY} = \sum_j \lambda_{ij}^{XY} = 0 . \tag{5.5}$$

Given $\{\lambda_{ij}^{XY}\}$ in the $(I-1)(J-1)$ cells in the first $I-1$ rows and $J-1$ columns, these constraints determine the parameters for cells in the last column or the last row. Thus, $(I-1)(J-1)$ of these terms are linearly independent.

The independence model (5.2) is the special case of the saturated model (5.4) in which all $\lambda_{ij}^{XY} = 0$. The additional $\{\lambda_{ij}^{XY}\}$ parameters in (5.4) are association parameters that reflect departures from independence of X and Y. The number of linearly independent parameters equals $1 + (I - 1) + (J - 1) = I + J - 1$ for the independence model and $1 + (I - 1) + (J - 1) + (I - 1)(J - 1) = IJ$ for the saturated model. For tables of any number of dimensions, the number of parameters in the saturated loglinear model equals the number of cells in the table.

5.1.3 Interpretation of Parameters

Loglinear model formulas make no distinction between response and explanatory variables. Most studies have such a distinction, however, and this can influence our choice of model and interpretation of parameters.

The interpretation is simplest for binary responses. For instance, consider the independence model for an $I \times 2$ table. The logit of the binary variable equals

$$\log(\pi_{1\,|\,i}/\pi_{2\,|\,i}) = \log(m_{i1}/m_{i2}) = \log m_{i1} - \log m_{i2}$$

$$= (\mu + \lambda_i^X + \lambda_1^Y) - (\mu + \lambda_i^X + \lambda_2^Y)$$

$$= \lambda_1^Y - \lambda_2^Y .$$

The logit is the same in every row. For zero-sum constraints on parameters, the I logits equal $2\lambda_1^Y$, since $\lambda_2^Y = -\lambda_1^Y$. In each row, $\exp(2\lambda_1^Y)$ is the odds that the column classification is category 1 rather than category 2.

Direct relationships exist between the odds ratio and association parameters in loglinear models. The relationship is simplest for 2×2 tables. For the saturated model,

$$\log \theta = \log\left(\frac{m_{11}m_{22}}{m_{12}m_{21}}\right) = \log m_{11} + \log m_{22} - \log m_{12} - \log m_{21}$$

$$= (\mu + \lambda_1^X + \lambda_1^Y + \lambda_{11}^{XY}) + (\mu + \lambda_2^X + \lambda_2^Y + \lambda_{22}^{XY})$$

$$- (\mu + \lambda_1^X + \lambda_2^Y + \lambda_{12}^{XY}) - (\mu + \lambda_2^X + \lambda_1^Y + \lambda_{21}^{XY})$$

$$= \lambda_{11}^{XY} + \lambda_{22}^{XY} - \lambda_{12}^{XY} - \lambda_{21}^{XY} .$$

For the constraints $\Sigma_i \lambda_{ij}^{XY} = \Sigma_j \lambda_{ij}^{XY} = 0$,

$$\lambda_{11}^{XY} = \lambda_{22}^{XY} = -\lambda_{12}^{XY} = -\lambda_{21}^{XY},$$

so that

$$\log \theta = 4\lambda_{11}^{XY}. \tag{5.6}$$

The odds ratio for a 2×2 table equals the antilog of four times the association parameter in the saturated loglinear model.

Alternative constraints on parameters lead to alternative interpretations. For instance, instead of (5.3), we could set

$$\mu = \eta_{11}, \qquad \lambda_i^X = \eta_{i1} - \eta_{11}, \qquad \lambda_j^Y = \eta_{1j} - \eta_{11}$$

$$\lambda_{ij}^{XY} = \eta_{ij} - \eta_{i1} - \eta_{1j} + \eta_{11}. \tag{5.7}$$

Then (5.4) holds but $\lambda_1^X = \lambda_1^Y = \lambda_{1j}^{XY} = \lambda_{i1}^{XY} = 0$ for all i and j, and $\lambda_{ij}^{XY} = \log[(m_{11}m_{ij})/(m_{1j}m_{i1})]$.

5.1.4 Models for Cell Probabilities

For the Poisson sampling model, cell counts are independent Poisson random variables with means $\{m_{ij}\}$. The natural parameter of the Poisson distribution is the log mean, so it is natural to model $\{\log m_{ij}\}$. For Poisson sampling, a loglinear model is a generalized linear model using the canonical link.

Conditional on the sample size n, Poisson loglinear models for $\{m_{ij}\}$ have equivalent expressions as multinomial models for the cell probabilities $\{\pi_{ij} = m_{ij}/(\Sigma \Sigma m_{ab})\}$. To illustrate, for the saturated loglinear model (5.4),

$$m_{ij} = \exp(\mu + \lambda_i^X + \lambda_j^Y + \lambda_{ij}^{XY})$$

$$\pi_{ij} = \frac{\exp(\mu + \lambda_i^X + \lambda_j^Y + \lambda_{ij}^{XY})}{\displaystyle\sum_a \sum_b \exp(\mu + \lambda_a^X + \lambda_b^Y + \lambda_{ab}^{XY})}. \tag{5.8}$$

This representation implies the usual constraints for probabilities, $\{\pi_{ij} \geq 0\}$ and $\Sigma \Sigma \pi_{ij} = 1$. The μ parameter cancels in the multinomial model (5.8). This parameter relates purely to the total sample size, which is random in the Poisson model but not the multinomial model.

5.2 TABLE STRUCTURE FOR THREE DIMENSIONS

An important part of any research study is the choice of predictor and control variables. Unless we include relevant variables in the analysis, results will have limited usefulness.

In studying the relationship between a response variable and an explanatory variable, we should control covariates that can influence that relationship. For instance, suppose we are studying effects of passive smoking—the effects on a nonsmoker of living with a smoker. We might compare lung cancer rates between nonsmokers whose spouses smoke and nonsmokers whose spouses do not smoke. In doing so, we should control for age, work environment, socioeconomic status, or other factors that might relate both to whether one's spouse smokes and to whether one has lung cancer. This section discusses statistical control and other concepts important in the study of multivariate relationships.

5.2.1 Partial Association

When all variables are categorical, a multidimensional contingency table displays the data. We illustrate ideas using the three-variable case. Denote the variables by X, Y, and Z. We display the distribution of X–Y cell counts at different levels of Z using cross sections of the three-way contingency table. These cross-sections are called *partial tables*. In the partial tables, Z is controlled; that is, its value is held constant. The two-way contingency table obtained by combining the partial tables is called the X–Y *marginal table*. That table, rather than controlling Z, ignores it.

Partial tables can exhibit quite different associations than marginal tables. In fact, it can be misleading to analyze only the marginal tables of a multi-way table, as the following example illustrates.

5.2.2 Death Penalty Example

Table 5.1, presented by Radelet (1981), is a $2 \times 2 \times 2$ contingency table. Radelet's article studied effects of racial characteristics on whether individuals convicted of homicide receive the death penalty. The variables in Table 5.1 are "death penalty verdict," having categories (yes, no), and "race of defendant" and "race of victim," each having categories (white, black). The 326 subjects were defendants in homicide indictments in 20 Florida counties during 1976–1977.

Table 5.2 is the marginal table for defendant's race and the death penalty verdict. We obtain it by summing the cell counts in Table 5.1 over

Table 5.1 Death Penalty Verdict by Defendant's Race and Victim's Race

Defendant's Race	Victim's Race	Death Penalty		Percentage Yes
		Yes	No	
White	White	19	132	12.6
	Black	0	9	0.0
Black	White	11	52	17.5
	Black	6	97	5.8

Source: Reprinted with permission from Radelet (1981).

the levels of victim's race. About 12% of white defendants and about 10% of black defendants received the death penalty. *Ignoring* victim's race, the percentage of "yes" death penalty verdicts was lower for blacks than for whites.

For each combination of defendant's race and victim's race, Table 5.1 lists the percentage of subjects who received the death penalty. These are displayed in Figure 5.1. Consider the association between defendant's race and the death penalty verdict, controlling for victim's race. When the victim was white, the death penalty was imposed about 5 percentage points more often for black defendants than for white defendants. When the victim was black, the death penalty was imposed over 5 percentage points more often for black defendants than for white defendants. *Controlling* for victim's race, the percentage of "yes" death penalty verdicts was higher for blacks than for whites. The direction of the association is the reverse of that in the marginal table.

Why does the association between death penalty verdict and defendant's race change direction when we control victim's race? Let us study Table 5.3. For each pair of variables, it lists the marginal odds ratio and also the partial odds ratio at each level of the third variable. The

Table 5.2 Frequencies for Death Penalty Verdict and Defendant's Race

Defendant's Race	Death Penalty		Total
	Yes	No	
White	19	141	160
Black	17	149	166
Total	36	290	326

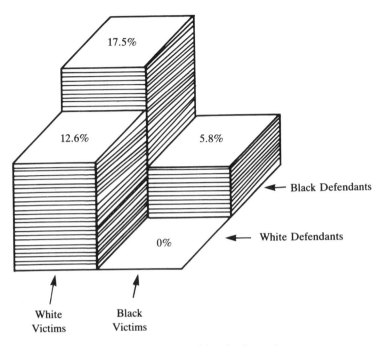

Figure 5.1 Percent receiving death penalty.

marginal odds ratios describe the association when the third variable is ignored (i.e., when we sum the counts over the levels of the third variable to obtain a marginal two-way table). The partial odds ratios describe the association when the third variable is controlled. Since one cell count in the three-dimensional table equals zero and since several of them are small, we added 0.5 to each cell count before computing these odds ratios.

Table 5.3 Odds Ratios for Death Penalty (P), Victim's Race (V), and Defendant's Race (D)[a]

Association		Variables		
		$P-D$	$P-V$	$D-V$
Marginal		1.18	2.71	25.99
Partial	Level 1	0.67	2.80	22.04
	Level 2	0.79	3.29	25.90

[a] The value 0.5 was added to each cell frequency before calculation of odds ratios.

The marginal odds ratio for death penalty verdict and defendant's race is 1.18; the estimated odds of the death penalty were 1.18 times as high for white defendants as for black defendants. But, when the victim was white, the estimated odds of the death penalty were 0.67 times as high for white defendants as for black defendants; when the victim was black, the estimated odds were 0.79 times as high for white defendants as for black defendants. To understand this reversal in association, it helps to study the other odds ratios in Table 5.3. The association between victim's race and defendant's race is strong. The odds of having killed a white are estimated to be 25.99 times higher for white defendants than for black defendants. The odds ratios relating death penalty verdict and victim's race indicate the death penalty was more likely when the victim was white than when the victim was black. So, whites are tending to kill whites, and killing a white is more likely to result in the death penalty. This suggests that the marginal association should show more of a tendency for white defendants to receive the death penalty than do the partial associations. In fact, this is the result observed in Table 5.2.

For each defendant's race, Figure 5.2 plots the % receiving the death penalty at each level of victim's race. Each observation is represented by a letter giving the level of victim's race. Surrounding each observation is a circle having area proportional to the number of observations at that combination of defendant's race and victim's race. The largest circles occur when whites kill whites or blacks kill blacks, and these cause the marginal result whereby whites are more likely to receive the death penalty.

The result that a pair of variables can have marginal association of different direction from their partial associations is called *Simpson's paradox* (Problem 5.2).

5.2.3 Types of Independence

Next we introduce types of independence for cell probabilities in a three-way cross-classification of response variables X, Y, and Z. Denote the cell probabilities by $\{\pi_{ijk}, i = 1, \ldots, I, j = 1, \ldots, J, k = 1, \ldots, K\}$, where $\Sigma_i \Sigma_j \Sigma_k \pi_{ijk} = 1.0$.

The three variables are *mutually independent* when

$$\pi_{ijk} = \pi_{i++} \pi_{+j+} \pi_{++k} \text{ for all } i, j, \text{ and } k . \qquad (5.9)$$

On a log scale, mutual independence is the loglinear model

$$\log m_{ijk} = \mu + \lambda_i^X + \lambda_j^Y + \lambda_k^Z . \qquad (5.10)$$

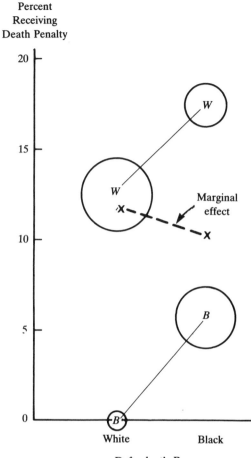

Note: Victim's race at center of circles.

x = marginal effect of defendant's race, ignoring victim's race.

Figure 5.2 Percent receiving death penalty by defendant's race, controlling and ignoring victim's race.

Variable Y is *jointly independent* of X and Z when

$$\pi_{ijk} = \pi_{i+k}\pi_{+j+} \quad \text{for all } i, j, \text{ and } k . \tag{5.11}$$

This is ordinary two-way independence for Y and a new variable composed of the IK combinations of levels of X and Z. The loglinear model is

$$\log m_{ijk} = \mu + \lambda_i^X + \lambda_j^Y + \lambda_k^Z + \lambda_{ik}^{XZ} \ . \tag{5.12}$$

Similarly, X could be jointly independent of Y and Z, or Z could be jointly independent of X and Y. When condition (5.9) holds, then $\pi_{i+k} = \pi_{i++} \pi_{++k}$, so that (5.11) also holds. Thus, mutual independence implies joint independence of any one variable from the others.

Next consider the relationship between X and Y, controlling for Z. If X and Y are independent in the partial table for the kth category of Z, then X and Y are said to be *conditionally independent at level k of Z*. Let $\{\pi_{ij \mid k} = \pi_{ijk}/\pi_{++k}, \ i = 1, \ldots, I, \ j = 1, \ldots, J\}$ denote the joint distribution of X and Y at level k of Z. Then conditional independence at level k of Z is

$$\pi_{ij \mid k} = \pi_{i+ \mid k} \pi_{+j \mid k} \quad \text{for all } i \text{ and } j \ . \tag{5.13}$$

More generally, X and Y are *conditionally independent given Z* when they are conditionally independent at every level of Z; that is, when (5.13) holds for all k, or equivalently, when

$$\pi_{ijk} = \pi_{i+k} \pi_{+jk}/\pi_{++k} \quad \text{for all } i, j, \text{ and } k \ . \tag{5.14}$$

Conditional independence of X and Y is the loglinear model

$$\log m_{ijk} = \mu + \lambda_i^X + \lambda_j^Y + \lambda_k^Z + \lambda_{ik}^{XZ} + \lambda_{jk}^{YZ} \ . \tag{5.15}$$

Suppose Y is jointly independent of X and Z, so $\pi_{ijk} = \pi_{i+k} \pi_{+j+}$. Then $\pi_{ij \mid k} = \pi_{ijk}/\pi_{++k} = \pi_{i+k} \pi_{+j+}/\pi_{++k}$, and summing both sides over i, we obtain $\pi_{+j \mid k} = \pi_{+j+}$. Therefore $\pi_{ij \mid k} = (\pi_{i+k}/\pi_{++k}) \pi_{+j+} = \pi_{i+ \mid k} \pi_{+j \mid k}$, so X and Y are also conditionally independent. In summary, mutual independence of the variables implies that Y is jointly independent of X and Z, which itself implies that X and Y are conditionally independent. Table 5.4 summarizes these three types of independence.

Table 5.4 Summary of Independence Models

Model	Probabilistic Form for π_{ijk}	Association Terms in Loglinear Model	Interpretation
(5.10)	$\pi_{i++} \pi_{+j+} \pi_{++k}$	None	Variables mutually independent
(5.12)	$\pi_{i+k} \pi_{+j+}$	λ_{ik}^{XZ}	Y independent of X and Z
(5.15)	$\pi_{i+k} \pi_{+jk}/\pi_{++k}$	$\lambda_{ik}^{XZ} + \lambda_{jk}^{YZ}$	X and Y independent, given Z

5.2.4 Marginal vs. Conditional Independence

Section 5.2.2 showed that partial associations can be quite different from marginal associations. For further illustration, we now see that conditional independence of X and Y, given Z, does not imply marginal independence of X and Y.

The joint probabilities in Table 5.5 show a hypothetical relationship among three variables for new graduates of a university. The association between Y = income at first job (high, low) and X = gender (female, male) at the two levels of Z = major discipline (liberal arts, science or engineering) is described by the odds ratios

$$\theta_{\text{Lib}} = \frac{0.18 \times 0.08}{0.12 \times 0.12} = 1.0, \qquad \theta_{\text{Sci}} = \frac{0.02 \times 0.32}{0.08 \times 0.08} = 1.0.$$

Income and gender are conditionally independent, given major. The odds ratio for the (income, gender) marginal table equals $(0.20 \times 0.40)/(0.20 \times 0.20) = 2.0$, so the variables are not independent when we ignore major.

Why are the odds of a high income twice as high for males as females, when we ignore major? The conditional $X - Z$ and $Y - Z$ odds ratios, all of which equal 6.0, give a clue. The conditional odds (given income) of majoring in a science or engineering discipline are six times higher for males than for females, and the conditional odds (given gender) of having a high income are six times higher for those majoring in a science or engineering discipline than for those in a liberal arts discipline. Science and engineering majors have relatively more males, and those majors also have relatively more graduates at high incomes.

Conditional independence and marginal independence both hold when one of the stronger types of independence studied in the previous

Table 5.5 Conditional Independence Does Not Imply Marginal Independence

Major	Gender	Income	
		Low	High
Liberal Arts	Female	0.18	0.12
	Male	0.12	0.08
Science or Engineering	Female	0.02	0.08
	Male	0.08	0.32
Total	Female	0.20	0.20
	Male	0.20	0.40

subsection applies. Suppose Y is jointly independent of X and Z, that is $\pi_{ijk} = \pi_{i+k}\pi_{+j+}$. We have seen that this implies conditional independence of X and Y. Summing over k on both sides, we obtain $\pi_{ij+} = \pi_{i++}\pi_{+j+}$. Thus, X and Y also exhibit marginal independence. So, joint independence of Y from X and Z (or of X from Y and Z) implies X and Y are both marginally and conditionally independent. Since mutual independence of X, Y, and Z implies that Y is jointly independent of X and Z, mutual independence also implies that X and Y are both marginally and conditionally independent. However, when we know only that X and Y are conditionally independent, $\pi_{ijk} = \pi_{i+k}\pi_{+jk}/\pi_{++k}$. Summing over k on both sides, we obtain $\pi_{ij+} = \Sigma_k(\pi_{i+k}\pi_{+jk}/\pi_{++k})$. All three terms in the summation involve k, and this does not simplify to $\pi_{i++}\pi_{+j+}$, marginal independence.

Figure 5.3 summarizes relationships among the four types of independence.

5.2.5 Three-Factor Interaction

In loglinear models (5.10), (5.12), and (5.15), there are three, two, and one pair of conditionally independent variables, respectively. In the latter two models, the doubly-subscripted terms (such as λ_{ij}^{XY}) pertain to conditionally dependent variables. To permit all three pairs of variables to be conditionally dependent, we introduce the generalization of these models

$$\log m_{ijk} = \mu + \lambda_i^X + \lambda_j^Y + \lambda_k^Z + \lambda_{ij}^{XY} + \lambda_{ik}^{XZ} + \lambda_{jk}^{YZ}. \qquad (5.16)$$

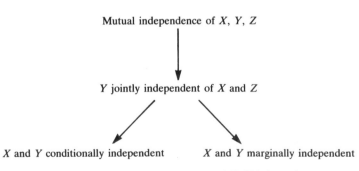

Mutual independence of X, Y, Z

Y jointly independent of X and Z

X and Y conditionally independent X and Y marginally independent

Figure 5.3 Relationships among types of X–Y independence.

From exponentiating both sides of (5.16), we see the cell probabilities have form

$$\pi_{ijk} = \psi_{ij}\phi_{jk}\omega_{ik} \; .$$

There is no closed-form expression for the three components in terms of $\{\pi_{ijk}\}$ except in certain special cases (see Note 5.3).

The next section shows that this model implies the conditional odds ratios between any two variables are identical at each level of the third variable. The term "no interaction" is used in Statistics to refer to situations in which the effect of an explanatory variable on the response is the same at all levels of other explanatory variables. Model (5.16) is called the loglinear model of *no three-factor interaction*.

5.3 LOGLINEAR MODELS FOR THREE DIMENSIONS

This section takes a closer look at loglinear models for three-way contingency tables. We begin by formally defining model parameters.

5.3.1 Hierarchical Loglinear Models

Let $\{m_{ijk}\}$ denote expected frequencies. Suppose all $m_{ijk} > 0$, and let $\eta_{ijk} = \log m_{ijk}$. A dot in a subscript denotes the average with respect to that index; for instance, $\eta_{.jk} = (\Sigma_i \eta_{ijk})/I$. We set

$$\mu = \eta_{...}$$
$$\lambda_i^X = \eta_{i..} - \eta_{...}, \qquad \lambda_j^Y = \eta_{.j.} - \eta_{...}, \qquad \lambda_k^Z = \eta_{..k} - \eta_{...}$$
$$\lambda_{ij}^{XY} = \eta_{ij.} - \eta_{i..} - \eta_{.j.} + \eta_{...}$$
$$\lambda_{ik}^{XZ} = \eta_{i.k} - \eta_{i..} - \eta_{..k} + \eta_{...}$$
$$\lambda_{jk}^{YZ} = \eta_{.jk} - \eta_{.j.} - \eta_{..k} + \eta_{...}$$
$$\lambda_{ijk}^{XYZ} = \eta_{ijk} - \eta_{ij.} - \eta_{i.k} - \eta_{.jk} + \eta_{i..} + \eta_{.j.} + \eta_{..k} - \eta_{...} \; .$$

The sum of the parameters for any index equals zero. That is,

$$\sum_i \lambda_i^X = \sum_j \lambda_j^Y = \sum_k \lambda_k^Z = \sum_i \lambda_{ij}^{XY} = \sum_j \lambda_{ij}^{XY} = \cdots = \sum_k \lambda_{ijk}^{XYZ} = 0 \; .$$

The general loglinear model for a three-way table is

$$\log m_{ijk} = \mu + \lambda_i^X + \lambda_j^Y + \lambda_k^Z + \lambda_{ij}^{XY} + \lambda_{ik}^{XZ} + \lambda_{jk}^{YZ} + \lambda_{ijk}^{XYZ} . \tag{5.17}$$

Doubly-subscripted terms pertain to partial associations, and the triply-subscripted term pertains to three-factor interaction.

Setting certain parameters equal to zero in (5.17) yields models introduced in the previous section. Table 5.6 lists some of these models. The linear predictors for the models have structure analogous to models for a three-way factorial ANOVA. The singly-subscripted terms are analogous to main effects, and the doubly-subscripted terms are analogous to two-factor interactions. For simplicity, Table 5.6 also assigns each model a symbol that lists the highest-order term(s) for each variable.

The loglinear models in Table 5.6 are called *hierarchical* models. This means that whenever the model contains higher-order effects, it also incorporates lower-order effects composed from the variables. For instance, when the model contains λ_{ij}^{XY}, it also must contain λ_i^X and λ_j^Y. An example of a nonhierarchical model is

$$\log m_{ij} = \mu + \lambda_i^X + \lambda_{ij}^{XY} .$$

This model permits association between X and Y but forces the average log expected frequency across rows to be the same in every column. Nonhierarchical models are sensible in very few applications. Using such models is analogous to using ANOVA or regression models with interaction terms but without corresponding main effects.

5.3.2 Interpretation of Model Parameters

To interpret loglinear models, we describe their marginal and partial associations using odds ratios. The X–Y marginal table $\{\pi_{ij+}\}$ uses a set of $(I-1)(J-1)$ odds ratios, such as

Table 5.6 Some Loglinear Models for Three-Dimensional Tables

Loglinear Model	Symbol
$\log m_{ijk} = \mu + \lambda_i^X + \lambda_j^Y + \lambda_k^Z$	(X, Y, Z)
$\log m_{ijk} = \mu + \lambda_i^X + \lambda_j^Y + \lambda_k^Z + \lambda_{ij}^{XY}$	(XY, Z)
$\log m_{ijk} = \mu + \lambda_i^X + \lambda_j^Y + \lambda_k^Z + \lambda_{ij}^{XY} + \lambda_{jk}^{YZ}$	(XY, YZ)
$\log m_{ijk} = \mu + \lambda_i^X + \lambda_j^Y + \lambda_k^Z + \lambda_{ij}^{XY} + \lambda_{jk}^{YZ} + \lambda_{ik}^{XZ}$	(XY, YZ, XZ)
$\log m_{ijk} = \mu + \lambda_i^X + \lambda_j^Y + \lambda_k^Z + \lambda_{ij}^{XY} + \lambda_{jk}^{YZ} + \lambda_{ik}^{XZ} + \lambda_{ijk}^{XYZ}$	(XYZ)

$$\theta_{ij}^{XY} = \frac{\pi_{ij+}\,\pi_{i+1,j+1,+}}{\pi_{i+1,j,+}\,\pi_{i,j+1,+}}, \quad 1 \leq i \leq I-1,\ 1 \leq j \leq J-1.$$

Within a fixed level k of Z, the corresponding odds ratios

$$\theta_{ij(k)} = \frac{\pi_{ijk}\,\pi_{i+1,j+1,k}}{\pi_{i,j+1,k}\,\pi_{i+1,j,k}}, \quad 1 \leq i \leq I-1,\quad 1 \leq j \leq J-1 \qquad (5.18)$$

describe the *conditional* X–Y association. Similarly, conditional association between X and Z is described by $(I-1)(K-1)$ odds ratios $\{\theta_{i(j)k}\}$ at each of the J levels of Y, and conditional association between Y and Z is described by $(J-1)(K-1)$ odds ratios $\{\theta_{(i)jk}\}$ at each of the I levels of X.

Loglinear model parameters are functions of conditional odds ratios. The relationships are simplest for binary variables. For $2 \times 2 \times 2$ tables, substituting (5.17) for $\log m_{ijk}$ in the log conditional odds ratios, we obtain

$$\lambda_{111}^{XYZ} = \frac{1}{8}\log\!\left(\frac{\theta_{11(1)}}{\theta_{11(2)}}\right) = \frac{1}{8}\log\!\left(\frac{\theta_{1(1)1}}{\theta_{1(2)1}}\right) = \frac{1}{8}\log\!\left(\frac{\theta_{(1)11}}{\theta_{(2)11}}\right) \qquad (5.19)$$

for zero-sum constraints on $\{\lambda_{ijk}^{XYZ}\}$. Each λ_{ijk}^{XYZ} term is zero when the odds ratio between two variables is the same at each level of the third variable. More generally, $\{\lambda_{ijk}^{XYZ} = 0\}$ in a $2 \times 2 \times K$ table when $\theta_{11(1)} = \cdots = \theta_{11(K)}$, and then

$$\lambda_{11}^{XY} = \frac{1}{4}\log\theta_{11(k)} \quad \text{for } k = 1, \ldots, K. \qquad (5.20)$$

As in the two-dimensional case (see (5.6)), the association parameter is proportional to the log odds ratio.

Loglinear models can be characterized using conditional odds ratios. For instance, conditional independence of X and Y is equivalent to $\{\theta_{ij(k)} = 1,\ i = 1, \ldots, I-1,\ j = 1, \ldots, J-1,\ k = 1, \ldots, K\}$.

5.3.3 Conditions for Identical Marginal and Partial Associations

We next give sufficient conditions for X–Y odds ratios to be the same in partial tables as in the marginal table. When they are the same, we can study the X–Y association in a simplified manner by collapsing over the Z dimension. In the following conditions, Z may be a single variable or multidimensional.

Collapsibility conditions

$$\theta_{ij}^{XY} = \theta_{ij(1)} = \theta_{ij(2)} = \cdots = \theta_{ij(K)} \, , \quad 1 \le i \le I-1 \, , \, 1 \le j \le J-1$$

$$(5.21)$$

if either or both of the following hold:

$$\theta_{i(j)k} = 1, \quad 1 \le i \le I-1, \, 1 \le j \le J, \, 1 \le k \le K-1$$

$$\theta_{(i)jk} = 1, \quad 1 \le i \le I, \, 1 \le j \le J-1 \, , \, 1 \le k \le K-1 \, .$$

In other words, $X-Y$ marginal and partial associations are identical if either Z and X are conditionally independent (i.e., the model symbolized by (XY, YZ) holds), or if Z and Y are conditionally independent (i.e., model (XY, XZ) holds).

The proof of the collapsibility conditions is straightforward (Bishop, 1971). One shows that when model (XY, YZ) or (XY, XZ) holds, the loglinear model for $\{m_{ij+}\}$ has the same association parameters as $\{\lambda_{ij}^{XY}\}$ in those models, hence the same odds ratios (Problem 5.27).

5.3.4 Interpretation of Models

We now use the collapsibility conditions to show properties of the loglinear models listed in Table 5.6. We also interpret the models using corresponding logit models.

(X, Y, Z)

For the loglinear model

$$\log m_{ijk} = \mu + \lambda_i^X + \lambda_j^Y + \lambda_k^Z \tag{5.22}$$

the three variables are mutually independent. Section 5.2 showed that each pair of variables is also conditionally independent and marginally independent. The result that marginal independence implies conditional independence (or vice versa) also follows from the collapsibility conditions. For instance, the $X-Y$ marginal association is identical to the $X-Y$ partial association (given Z), because Z is conditionally independent of X (given Y), or also because Z is conditionally independent of Y (given X).

Suppose Y is a binary response. For (5.22) the logit on Y equals

$$\log(m_{i1k}/m_{i2k}) = \lambda_1^Y - \lambda_2^Y \, .$$

For zero-sum constraints, this equals $2\lambda_1^Y$, and $\exp(2\lambda_1^Y)$ is the odds

m_{i1k}/m_{i2k} for each combination of settings of X and Z. When Y has $J > 2$ levels, $\log(m_{iak}/m_{ibk}) = \lambda_a^Y - \lambda_b^Y$. The higher the value of λ_j^Y, relatively more subjects are classified in the jth level of Y.

(XZ, Y) or (YZ, X) or (XY, Z)

There are three models in which only one pair of variables is conditionally dependent. The symbol (XZ, Y) denotes model

$$\log m_{ijk} = \mu + \lambda_i^X + \lambda_j^Y + \lambda_k^Z + \lambda_{ik}^{XZ} \tag{5.23}$$

whereby Y is jointly independent of X and Z. By Section 5.2, Y and X are conditionally independent (given Z), and Y and Z are conditionally independent (given X). The model symbol reflects the conditional dependence of X and Z, and the parameters $\{\lambda_{ik}^{XZ}\}$ pertain to that dependence.

For this model, by Section 5.2.4, Y is also independent of X and Z in the X–Y and Y–Z marginal tables. In addition, the collapsibility conditions imply that the X–Z marginal odds ratios are identical to corresponding X–Z partial odds ratios, since Y is independent of X (given Z), or because Y is independent of Z (given X).

The logit on Y is the same for model (XZ, Y) as for model (X, Y, Z). The joint independence of Y from X and Z means that the logit on Y is identical at all combinations of levels of X and Z.

(XY, YZ) or (XY, XZ) or (XZ, YZ)

There are three models in which only one pair of variables is conditionally independent. The symbol (XZ, YZ) denotes the model of conditional independence of X and Y, given Z,

$$\log m_{ijk} = \mu + \lambda_i^X + \lambda_j^Y + \lambda_k^Z + \lambda_{ik}^{XZ} + \lambda_{jk}^{YZ} . \tag{5.24}$$

The parameters $\{\lambda_{ik}^{XZ}\}$ and $\{\lambda_{jk}^{YZ}\}$ pertain to $X - Z$ and $Y - Z$ partial associations.

Section 5.2.4 noted that X and Y may be marginally dependent, even though they are conditionally independent. This follows also from the collapsibility conditions, because Z is conditionally dependent with both X and Y. By those same conditions, however, the $X - Z$ and $Y - Z$ marginal tables display the same odds ratios as do their corresponding partial tables, because X and Y are conditionally independent.

For model (XZ, YZ) with binary Y, the logit is

$$\log(m_{i1k}/m_{i2k}) = (\lambda_1^Y - \lambda_2^Y) + (\lambda_{1k}^{YZ} - \lambda_{2k}^{YZ}) .$$

For zero-sum constraints, this equals $2\lambda_1^Y + 2\lambda_{1k}^{YZ}$. The logit has form $\alpha + \beta_k^Z$, when we identify $2\lambda_{1k}^{YZ}$ as the kth effect of Z on the logit of Y (i.e., β_k^Z), and $2\lambda_1^Y = \alpha$. The logit depends on the level k of Z through the association term for Y and Z, but it does not depend on the level i of X.

Model (XZ, YZ) is a very important one. An association observed between X and Y may be spurious if we find a third variable Z such that the association disappears when we control Z.

(XY, XZ, YZ)

For the model

$$\log m_{ijk} = \mu + \lambda_i^X + \lambda_j^Y + \lambda_k^Z + \lambda_{ij}^{XY} + \lambda_{ik}^{XZ} + \lambda_{jk}^{YZ} \qquad (5.25)$$

partial association terms appear for each pair of variables, so no pair is conditionally independent. This is the model of *no three-factor interaction*. No collapsibility conditions are fulfilled, so for each pair of variables, marginal odds ratios may differ from partial odds ratios.

Substituting (5.25) into the expression for $\log \theta_{ij(k)}$, we obtain

$$\log \theta_{ij(k)} = \lambda_{ij}^{XY} + \lambda_{i+1,j+1}^{XY} - \lambda_{i,j+1}^{XY} - \lambda_{i+1,j}^{XY}.$$

Since the right-hand side is the same for all k, an absence of three-factor interaction is equivalent to

$$\theta_{ij(1)} = \theta_{ij(2)} = \cdots = \theta_{ij(K)} \quad \text{for all } i \text{ and } j.$$

The same argument for the other partial odds ratios shows that this model is also equivalent to

$$\theta_{i(1)k} = \theta_{i(2)k} = \cdots = \theta_{i(J)k} \quad \text{for all } i \text{ and } k,$$

and to

$$\theta_{(1)jk} = \theta_{(2)jk} = \cdots = \theta_{(I)jk} \quad \text{for all } j \text{ and } k.$$

When there is an absence of three-factor interaction, the association between two variables is identical at each level of the third variable.

For model (XY, XZ, YZ) with binary Y and zero-sum parameter constraints, Section 5.4.3 shows that the logit simplifies to

$$\log(m_{i1k}/m_{i2k}) = 2\lambda_1^Y + 2\lambda_{i1}^{XY} + 2\lambda_{1k}^{YZ}$$

The logit has form $\alpha + \beta_i^X + \beta_k^Z$. It depends both on the level of X and the level of Z, but in an additive manner. The effect of X on the logit is the same at each level of Z, and the effect of Z is the same at each level of X. Additivity on the logit scale has become the generally accepted definition of "no interaction" for categorical variables. However, this is only one possible type of "no interaction." We could just as legitimately define "no interaction" as additivity on some other scale, such as the probit or linear probability scales. Interaction can occur on one scale when there is no interaction on another scale. In some applications, another definition of no interaction may be more sensible. For instance, the theory underlying some biological model might predict that the probability of success (rather than its logit) is an additive function of effects of predictors.

(XYZ)

The general model (5.17) for three variables allows for three-factor interaction. Each pair of variables may be conditionally dependent, and an odds ratio for any pair may vary across levels of the third variable. This model describes the entire set of $\{m_{ijk}\}$ having positive values. Since $\sum_i \lambda_i^X = 0$, there are $I - 1$ linearly independent $\{\lambda_i^X\}$ parameters. Since $\sum_i \lambda_{ij}^{XY} = \sum_j \lambda_{ij}^{XY} = 0$, there are $(I - 1)(J - 1)$ linearly independent $\{\lambda_{ij}^{XY}\}$ parameters. Similar formulas apply to the other parameters. The total number of linearly independent parameters (including μ) is

$$
\begin{aligned}
&1 + (I - 1) + (J - 1) + (K - 1) \\
&+ (I - 1)(J - 1) + (I - 1)(K - 1) + (J - 1)(K - 1) \\
&+ (I - 1)(J - 1)(K - 1) = IJK
\end{aligned}
$$

which equals the total number of cells in the table. The model is saturated.

5.3.5 Alternative Constraints

This section defined parameters using zero-sum constraints. Alternative schemes give equivalent models, but different parameter interpretations. For instance, if we let

$$\mu = \log(m_{111})$$

$$\lambda_i^X = \log(m_{i11}/m_{111}), \qquad \lambda_j^Y = \log(m_{1j1}/m_{111}), \qquad \lambda_k^Z = \log(m_{11k}/m_{111})$$

$$\lambda_{ij}^{XY} = \log(m_{ij1}m_{111}/m_{i11}m_{1j1})$$

$$\lambda_{ik}^{XZ} = \log(m_{i1k}m_{111}/m_{i11}m_{11k})$$

$$\lambda_{jk}^{YZ} = \log(m_{1jk}m_{111}/m_{1j1}m_{11k})$$

$$\lambda_{ijk}^{XYZ} = \log[(m_{ijk}m_{11k}/m_{i1k}m_{1jk})/(m_{ij1}m_{111}/m_{i11}m_{1j1})]$$

then (5.17) still holds. The two-factor terms are log odds ratios for the cross product with the cell at the first level of each variable. A three-factor term is a log of ratios of odds ratios. The parameters satisfy the constraints

$$\lambda_1^X = \lambda_1^Y = \lambda_1^Z = \lambda_{1j}^{XY} = \lambda_{i1}^{XY} = \cdots = \lambda_{ij1}^{XYZ} = 0 .$$

The computer package GLIM uses these types of constraints.

5.4 LOGLINEAR MODELS FOR HIGHER DIMENSIONS

Models are more complex for three-way tables than for two-way tables, because of the variety of potential partial association and three-factor interaction patterns. Once we understand loglinear models for three-way tables, however, we can readily extend the concepts to multi-way tables.

As the number of dimensions increases, there are some complicating factors. One complication is the tremendous increase in the number of possible interaction patterns. Another difficulty is caused by the dramatic increase in number of cells. Unless the sample size is large, there may be many zero cell counts. Later chapters show this can cause difficulties with existence of estimates, with the appropriateness of standard asymptotic theory, and with feasibility of computing routines for fitting the models.

5.4.1 Four-Dimensional Tables

To illustrate models for higher dimensions, we study four-way cross-classifications of variables W, X, Y, and Z. For simplicity of interpretation, it helps to have few higher-order terms.

The model of mutual independence, denoted by (W, X, Y, Z), is so simple that it rarely has much practical application. Models having no three-factor interaction terms are nested within the model

$$\log m_{hijk} = \mu + \lambda_h^W + \lambda_i^X + \lambda_j^Y + \lambda_k^Z + \lambda_{hi}^{WX} + \lambda_{hj}^{WY} + \lambda_{hk}^{WZ} + \lambda_{ij}^{XY} + \lambda_{ik}^{XZ}$$
$$+ \lambda_{jk}^{YZ}$$

denoted by (WX, WY, WZ, XY, XZ, YZ). For this model, each pair of variables is conditionally dependent, given the other two variables. Conditional independences correspond to the absence of certain two-factor terms. If $\lambda_{jk}^{YZ} = 0$ for all j and k, for instance, then Y and Z are conditionally independent at each combination of levels of W and X.

A variety of models exhibit some form of three-factor interaction. For model (WXY, WZ, XZ, YZ), for instance, each pair of variables is conditionally dependent, but at each level of Z the association between W and X or between W and Y or between X and Y varies across the levels of the remaining variable. The partial association between Z and another variable is the same at each combination of levels of the other two variables.

5.4.2 Collapsibility

Analysis is simplified for multi-way tables when we can collapse over one or more dimensions. For three-way tables we studied sufficient conditions under which partial and marginal associations are identical. The following result, based on a theorem in Bishop et al. (1975, p. 47), gives collapsibility conditions for multi-way tables.

Suppose variables in a T-dimensional table are divided into three mutually exclusive groups. If all model terms linking variables from the first group with variables from the second group equal zero, then model terms among variables for the first group are unchanged when the table is collapsed over the second group.

We illustrate this result using a four-dimensional table, with W and X forming the first group of variables, Y the second group, and Z the third group. If the λ_{hj}^{WY} and λ_{ij}^{XY} terms are zero, the W–X partial association remains the same when the table is collapsed over Y. This happens for model (WXZ, YZ) or simpler models. Next suppose W and X are in the first group, Y and Z are in the second group, and the third group is empty. For model (WX, YZ), the W–X partial association is the same as the W–X marginal association.

This result also implies that when any variable is independent of all other variables, the table can be collapsed over it without affecting any other model terms. For instance, associations among W, X, and Y in model (WX, WY, XY, Z) are the same as in model (WX, WY, XY).

When all two-factor effects are present in a model, collapsing over any variable may cause those effects to change. For model (WX, WY, WZ, XY, XZ, YZ), each pair of variables may have differing partial and marginal associations.

5.4.3 Correspondence Between Loglinear and Logit Models

In many applications, the potentially useful models are a small subset of all loglinear models. For instance, most studies distinguish between explanatory and response variables. Modeling effects of explanatory variables on the response is then more important than modeling relationships among explanatory variables. A subset of loglinear models, equivalent to logit models for the response variable, serve this purpose. The loglinear models contain the same structure as the logit models for associations between the response variable and the explanatory variables, and they contain the most general interaction term for relationships among the explanatory variables.

To illustrate, for an $I \times J \times 2$ table, we construct the loglinear model corresponding to the logit model

$$\log\left(\frac{m_{ij1}}{m_{ij2}}\right) = \alpha + \beta_i^A + \beta_j^B . \tag{5.26}$$

In this model, the response Y is associated with factors A and B, but the effect of each factor is the same at each level of the other factor. The loglinear model includes the association terms λ_{ik}^{AY} and λ_{jk}^{BY} and the general term λ_{ij}^{AB} for the relationship between the factors. The resulting model is (AB, AY, BY).

To show that loglinear model (AB, AY, BY) implies logit model (5.26), we note that for the loglinear model,

$$\log\left(\frac{m_{ij1}}{m_{ij2}}\right) = \log(m_{ij1}) - \log(m_{ij2})$$

$$= [\mu + \lambda_i^A + \lambda_j^B + \lambda_1^Y + \lambda_{ij}^{AB} + \lambda_{i1}^{AY} + \lambda_{j1}^{BY}]$$

$$- [\mu + \lambda_i^A + \lambda_j^B + \lambda_2^Y + \lambda_{ij}^{AB} + \lambda_{i2}^{AY} + \lambda_{j2}^{BY}]$$

$$= (\lambda_1^Y - \lambda_2^Y) + (\lambda_{i1}^{AY} - \lambda_{i2}^{AY}) + (\lambda_{j1}^{BY} - \lambda_{j2}^{BY}) .$$

For zero-sum constraints $\Sigma_k \lambda_k^Y = \Sigma_k \lambda_{ik}^{AY} = \Sigma_k \lambda_{jk}^{BY} = 0$, we have $\lambda_1^Y = -$

λ_2^Y, $\lambda_{i1}^{AY} = -\lambda_{i2}^{AY}$, and $\lambda_{j1}^{BY} = -\lambda_{j2}^{BY}$, since Y has two levels. Thus the logit simplifies to

$$\log\left(\frac{m_{ij1}}{m_{ij2}}\right) = 2\lambda_1^Y + 2\lambda_{i1}^{AY} + 2\lambda_{j1}^{BY}.$$

This is precisely the form of logit model (5.26), when we identify $2\lambda_{i1}^{AY}$ as the ith effect of A on the logit of Y (i.e., β_i^A), $2\lambda_{j1}^{BY}$ as the jth effect of B on the logit of Y (i.e., β_j^B), and $2\lambda_1^Y = \alpha$.

The λ_{ij}^{AB} terms for association among explanatory variables cancel in the difference in logarithms defined by the logit. The logit model does not contain information about this association.

Next, consider a four-way classification in which Y is a binary response variable. The logit model

$$\log\left(\frac{m_{hij1}}{m_{hij2}}\right) = \alpha + \beta_h^A + \beta_i^B + \beta_j^C \tag{5.27}$$

contains main effect terms for the factors, but no interaction terms. This model corresponds to the loglinear model that contains the fullest interaction term among the factors, and associations between each factor and the response Y; that is, model (ABC, AY, BY, CY). The special case of (5.27) in which all $\beta_j^C = 0$ corresponds to loglinear model (ABC, AY, BY), for which there is conditional independence between Y and C, given A and B. The generalization of (5.27) that also contains the term β_{hi}^{AB} corresponds to the loglinear model (ABC, ABY, CY) in which there is interaction between A and B in their effects on Y.

In summary, when there is a response variable, relevant loglinear models correspond to logit models for that response. When the response has more than two categories, relevant loglinear models correspond to generalized logit models presented in Section 9.1.

CHAPTER NOTES

Section 5.1: Loglinear Model for Two Dimensions

5.1 Loglinear models have a geometric representation. For $I \times J$ tables, for instance, the possible $\{\log m_{ij}\}$ for positive cell means is IJ-dimensional Euclidean space. This is the space mapped out by the saturated model for all its possible parameter values. Unsaturated models constrain $\{\log m_{ij}\}$ to a T-dimensional linear manifold in that space, with $T < IJ$. For the independence model, for instance, $T = (I + J - 1)$. More complex models provide greater flexibility for fitting data, since the linear manifold of potential $\{\log m_{ij}\}$ values is larger. Haberman (1974a) presented a theoretical development of loglinear models based largely on geometric ideas.

Section 5.2: Table Structure for Three Dimensions

5.2 Roy and Mitra (1956) wrote an important article on this topic. They described types of independence for three-way tables, and gave Pearson-type statistics for large-sample tests of those types.

Paik (1985) suggested circle diagrams like Figure 5.2 for presenting results from three-way tables.

Section 5.3: Loglinear Models for Three Dimensions

5.3 For $I \times J \times 2$ tables, the collapsibility conditions in Section 5.3.3 are necessary as well as sufficient (Simpson 1951, Whittemore 1978). For $I \times J \times K$ tables with $K > 2$, they are sufficient but not necessary. In that case, Ducharme and Lepage (1986) showed the conditions are necessary and sufficient for the odds ratios to remain the same no matter how the levels of Z are pooled (i.e., no matter how Z is partially collapsed).

Darroch (1962) defined a *perfect* table as one for which

$$\sum_i \pi_{ij+} \pi_{i+k} / \pi_{i++} = \pi_{+j+} \pi_{++k} \quad \text{for all } j \text{ and } k \,,$$

$$\sum_j \pi_{+jk} \pi_{ij+} / \pi_{+j+} = \pi_{i++} \pi_{++k} \quad \text{for all } i \text{ and } k \,,$$

and

$$\sum_k \pi_{i+k} \pi_{+jk} / \pi_{++k} = \pi_{i++} \pi_{+j+} \quad \text{for all } i \text{ and } j \,.$$

He showed that, for perfect tables, no three-factor interaction implies that $\{\pi_{ijk} = \pi_{ij+} \pi_{i+k} \pi_{+jk} / \pi_{i++} \pi_{+j+} \pi_{++k}\}$ and that partial odds ratios are identical to marginal odds ratios. Perfect tables do not exhibit paradoxes; for instance, conditional independence is equivalent to marginal independence. Whittemore (1978) used perfect tables to illustrate that for $I \times J \times K$ tables with $K > 2$, partial and marginal odds ratios can be identical even when no pair of variables is conditionally independent. See Darroch (1962, 1976), Davis (1986b, 1990), and Whittemore (1978) for further discussion of perfect tables.

Shapiro (1982) gave the following result about collapsibility of tables for the difference of proportions or relative risk: Suppose one of these measures, computed for a response Y and explanatory factor A, has the same value at every level of factor B. If B is independent of A in the marginal A–B table or if B is conditionally independent of Y given A, then the measure has the same value in the marginal A–Y table. Thus, whenever factors are marginally independent, such as in factorial designs with the same number of observations at each combination of levels, the difference of proportions and relative risk are collapsible. See also Wermuth (1987).

5.4 The definition for three-factor interaction originated in an article by Bartlett (1935). For $2 \times 2 \times 2$ tables, he defined no three-factor interaction as $\theta_{11(1)} = \theta_{11(2)}$, attributing the idea to R. A. Fisher. Roy and Kastenbaum (1956) extended this to multi-way tables. See also Darroch (1962), Good (1963), Goodman (1964c), and Plackett (1962).

These articles and a fundamental article by M. W. Birch (1963) on *ML* estimation for loglinear models in three-way tables stimulated a great deal of research on loglinear models in the 1960s. Mantel (1966) discussed early results, and made the loglinear model formula explicit. Several articles by Leo Goodman at the University of Chicago popularized loglinear models in the social sciences. His 1968 and 1970 articles summarize much of his early work. Goodman's developments led to outstanding theoretical work on loglinear models by Shelby Haberman, in his PhD dissertation which is the basis of a 1974a monograph. Simultaneously, related research at Harvard University by students of Frederick Mosteller (such as Yvonne Bishop and Stephen Fienberg) and William Cochran and also at the University of North Carolina by Gary Koch and several students and co-workers was highly influential in biomedical sciences. Research at North Carolina focused on weighted least squares methods for regression-type models for categorical data, whereas that at Chicago and Harvard focused on ML methods for loglinear and logit models.

The research at Harvard (e.g. Bishop 1967) was largely inspired by problems arising in analyses of large, multivariate data sets in the National Halothane Study. That study investigated whether halothane was more likely than other anesthetics to cause death due to liver damage. Mosteller's (1968) presidential address to the American Statistical Association described that and other early applications of loglinear models for smoothing multidimensional discrete data sets. The landmark book by Bishop et al. (1975) was largely responsible for the introduction of loglinear models to the general statistical world.

Since loglinear models use odds ratios as the building blocks of association and interaction, and since G. Udny Yule was the first strong proponent of that measure, he may be regarded as the grandfather of the loglinear model approach.

PROBLEMS

Applications

5.1 Construct a 2×3 table of $\{m_{ij}\}$ that satisfies the loglinear model of independence. Calculate μ, $\{\lambda_i^X\}$, and $\{\lambda_j^Y\}$, indicating the constraints that the parameters satisfy.

5.2 *Simpson's paradox* states that for three events E_1, E_2, E_3, and their complements, it is possible that $P(E_1 | E_2) > P(E_1 | E_2^c)$ even if both $P(E_1 | E_2 E_3) < P(E_1 | E_2^c E_3)$ and $P(E_1 | E_2 E_3^c) < P(E_1 | E_2^c E_3^c)$.

   a. For $2 \times 2 \times 2$ contingency tables, show this implies that the X–Y marginal association can have a different direction than the X–Y partial association.

   b. Letting $E_1 = \{$death penalty verdict "yes"$\}$, $E_2 = \{$"white" defendant$\}$, and $E_3 = \{$"white" victim$\}$, show the sample proportions in Table 5.1 satisfy Simpson's paradox. (Though its name refers to an article by E. H. Simpson in 1951, this paradox was discussed well before that date, for instance by Yule 1903.)

5.3 Smith and Jones are baseball players. Smith has a higher batting average than Jones in each of K years. Is it possible that when we combine the data from the K years, Jones has the higher batting average? Explain.

5.4 Two balanced coins are flipped, independently of each other. Let X (yes, no) indicate whether the first flip resulted in a head, let Y (yes, no) indicate whether the second flip resulted in a head, and let Z (yes, no) indicate whether both flips had the same result. Show that each pair of variables is marginally independent, but the three variables are not mutually independent.

5.5 Give three "real-world" variables X, Y, and Z, for which you expect a marginal association between X and Y, but conditional independence controlling for Z.

5.6 Give a $2 \times 2 \times 2$ table showing it is possible for X and Y to be conditionally dependent at each level of Z, yet marginally independent. (Variable Z is a *suppressor* variable.) Show this can happen

   a. when there is no three-factor interaction,

   b. when the association has opposite direction in the partial tables.

5.7 Based on 1987 murder rates in the United States, an Associated Press story reported that the probability a newborn child has of eventually being a murder victim is 0.0263 for nonwhite males, 0.0049 for white males, 0.0072 for nonwhite females, and 0.0023 for white females.

   a. Find the conditional odds ratios between race and whether a murder victim, given gender. Interpret.

   b. Assume half the newborns are of each gender, for each race. Find the marginal odds ratio between race and whether a murder victim.

   c. Do these variables exhibit three-factor interaction? Explain.

5.8 A $2 \times 2 \times 2$ table satisfies $\pi_{i++} = \pi_{+j+} = \pi_{++k} = \frac{1}{2}$ for all i, j, and k. Give an example of cell probabilities that satisfy loglinear model

   a. (X, Y, Z)

   b. (XY, Z)

   c. (XY, YZ)

   d. (XY, XZ, YZ)

   e. (XYZ)

5.9 Table 5.7 gives expected frequencies for whether a boy scout (B), delinquency (D), and socioeconomic status (S).

   a. Which loglinear model describes these expected frequencies?

Table 5.7

Socioeconomic Status	Boy Scout	Delinquent	
		Yes	No
Low	Yes	10	40
	No	40	160
Medium	Yes	18	132
	No	18	132
High	Yes	8	192
	No	2	48

b. Construct the marginal $B-D$ table. Why is it misleading to claim that scouting leads to lower delinquency rates? Using this example, explain why it is difficult to make causal inferences from statistical associations.

5.10 Table 5.8 appeared in a national study of 15–16 year old adolescents described by Morgan and Teachman (1988). The event of interest is ever having sexual intercourse.

Table 5.8

Race	Gender	Intercourse	
		Yes	No
White	Male	43	134
	Female	26	149
Black	Male	29	23
	Female	22	36

Source: Reprinted with permission from Morgan and Teachman (1988); copyrighted 1988 by the National Council on Family Relations.

a. Calculate the conditional odds ratios between gender and intercourse and between race and intercourse. Interpret.

b. We plan to construct a loglinear model that corresponds to a logit model in which intercourse is the response. Based on the odds ratios and allowing for sampling error, which loglinear model seems appropriate?

5.11 In a $2 \times 2 \times K$ table, suppose the $X-Y$ conditional odds ratio is the same at each level of Z, but it is different from the $X-Y$ marginal odds ratio. Is there three-factor interaction? Explain.

5.12 Suppose X is independent of Y, and Y is independent of Z. Does it follow that X is independent of Z? Explain.

5.13 Opposition to the legal availability of abortion is stronger among the religious than the nonreligious, and stronger among those with conservative sexual attitudes than those with more permissive attitudes. Does this imply that the religious are more likely than the nonreligious to have conservative sexual attitudes? Use sample tables in your answer.

5.14 Table 5.9 describes association between smoking status and a breathing test result, by age, for Caucasians in certain industrial plants in Houston in 1974–1975. Describe the associations, and indicate which three-variable relationship these data seem to satisfy.

Table 5.9

Age	Smoking	Breathing Test Results	
		Normal (%)	Not Normal (%)
<40	Never smoked	577 (94.4)	34 (5.6)
	Current smoker	682 (92.3)	57 (7.7)
40–59	Never smoked	164 (97.6)	4 (2.4)
	Current smoker	245 (76.8)	74 (23.2)

Source: From *Public Program Analysis* by R. N. Forthofer and R. G. Lehnen. Copyright © 1981 by Lifetime Learning Publications, Belmont, CA 94002, a division of Wadsworth, Inc. Reprinted by permission of Van Nostrand Reinhold. All rights reserved.

5.15 Explain what is meant by "no statistical interaction" in modeling response Y and explanatory X and Z in the following cases. Use graphs or tables to illustrate.

a. All variables are continuous (multiple regression).

b. Y and X are continuous, Z is categorical (analysis of covariance).

c. Y is continuous, X and Z are categorical (two-way ANOVA).

d. All variables are categorical.

5.16 Let m_{hijk} denote the expected frequency for gender h ($h = 1$, male; $h = 2$, female), religious affiliation i ($i = 1$, Protestant; $i = 2$, Catholic; $i = 3$, Jewish), political party affiliation j ($j = 1$, Democrat; $j = 2$, Republican; $j = 3$, Independent), and opinion about current laws legalizing abortion k ($k = 1$, oppose; $k = 2$, support). The logit model

$$\log(m_{hij1}/m_{hij2}) = \alpha + \beta_h^G + \beta_i^R + \beta_j^P$$

has values $\alpha = -0.62$, $\beta_1^G = 0.08$, $\beta_2^G = -0.08$, $\beta_1^R = 0.16$, $\beta_2^R = 0.25$, $\beta_3^R = -0.41$, $\beta_1^P = -0.87$, $\beta_2^P = 1.27$, $\beta_3^P = -0.40$.

   a. Use the logit model to interpret how the odds of supporting legalized abortion depends on gender.

   b. Find the probability of supporting legalized abortion for (i) Male Catholic Republicans, (ii) Female Jewish Democrats.

   c. Give the symbol for the loglinear model that is equivalent to this logit model.

   d. Give the logit model corresponding to the loglinear model (AR, AP, GRP), where A denotes the response variable (opinion on abortion).

5.17 Refer to the previous problem. Suppose we define parameters in this model using constraints $\beta_1^G = \beta_1^R = \beta_1^P = 0$. Give the values of the other parameters, and explain how to interpret them.

5.18 Fowlkes et al. (1988) reported Table 5.10, from a 1981 survey of employees of a large national corporation.

   a. Which loglinear model is equivalent to the logit model relating S = job satisfaction to main effects for the demographic variables R = race, G = gender, A = age, and L = regional location?

   b. Which logit model is equivalent to loglinear model $(AGLR, AS, LS, GRS)$? How would you interpret the model? (We further analyze these data in Problems 6.12 and 7.15.)

Theory and Methods

5.19 For $I \times J$ tables, the computer package GLIM defines parameters for the general loglinear model as in (5.7).

   a. Show this is a valid way of expressing the model, and show that $\lambda_1^X = \lambda_1^Y = \lambda_{1j}^{XY} = \lambda_{i1}^{XY} = 0$ for all i and j.

   b. For 2×2 tables, show that λ_{22}^{XY} is the log odds ratio.

Table 5.10

Region	White Less than 35 Male	White Less than 35 Female	White 35–44 Male	White 35–44 Female	White Greater than 44 Male	White Greater than 44 Female	Other Less than 35 Male	Other Less than 35 Female	Other 35–44 Male	Other 35–44 Female	Other Greater than 44 Male	Other Greater than 44 Female
Northeast												
Satisfied	288	60	224	35	337	70	38	19	32	22	21	15
Not satisfied	177	57	166	19	172	30	33	35	11	20	8	10
Mid-Atlantic												
Satisfied	90	19	96	12	124	17	18	13	7	0	9	1
Not Satisfied	45	12	42	5	39	2	6	7	2	3	2	1
Southern												
Satisfied	226	88	189	44	156	70	45	47	18	13	11	9
Not satisfied	128	57	117	34	73	25	31	35	3	7	2	2
Midwest												
Satisfied	285	110	225	53	324	60	40	66	19	25	22	11
Not satisfied	179	93	141	24	140	47	25	56	11	19	2	12
Northwest												
Satisfied	270	176	215	80	269	110	36	25	9	11	16	4
Not satisfied	180	151	108	40	136	40	20	16	7	5	3	5
Southwest												
Satisfied	252	97	162	47	199	62	69	45	14	8	14	2
Not satisfied	126	61	72	27	93	24	27	36	7	4	5	0
Pacific												
Satisfied	119	62	66	20	67	25	45	22	15	10	8	6
Not satisfied	58	33	20	10	21	10	16	15	10	8	6	2

Source: Reprinted with permission from Fowlkes et al. (1988).

5.20 Suppose $\{\pi_{ij}\}$ satisfy the loglinear model of independence.
   a. Show that $\lambda_a^Y - \lambda_b^Y = \log(\pi_{+a}/\pi_{+b})$.
   b. Show that $\lambda_a^Y > \lambda_b^Y$ is equivalent to $\pi_{+a} > \pi_{+b}$.
   c. Show that $\{$all $\lambda_j^Y = 0\}$ is equivalent to $\pi_{+j} = 1/J$ for $j = 1, \ldots, J$.

5.21 Refer to the previous problem. Show each part is true when we use parameterization (5.7).

5.22 For the saturated loglinear model for a $2 \times J$ table with parameters defined as in (5.3), show that $\lambda_{11}^{XY} = (\Sigma_j \log \alpha_j)/2J$, where $\alpha_j = (m_{11}m_{2j})/(m_{21}m_{1j})$, $j = 2, \ldots, J$.

5.23 X and Y are conditionally independent, given Z, and X and Z are marginally independent.
   a. Show X is jointly independent of Y and Z.
   b. Show X and Y are marginally independent.
   c. Conclude that if X and Y are conditionally independent, and if Z is marginally or conditionally independent of X or Y, then X and Y are also marginally independent.

5.24 Show that $\theta_{ij(1)} = \cdots = \theta_{ij(K)}$ for all i and j implies that $\theta_{i(1)k} = \cdots = \theta_{i(J)k}$ for all i and k, and $\theta_{(1)jk} = \cdots = \theta_{(I)jk}$ for all j and k. (An argument for defining "no three-factor interaction" as equality of conditional odds ratios is that the odds ratio exhibits this symmetry, unlike measures of association that are not functions of the odds ratio. See Altham 1970b for other such properties.)

5.25 Consider a $2 \times 2 \times 2$ cross-classification table, with parameters defined as in Section 5.3.1.
   a. For model (XY, XZ, YZ), show $\lambda_{11}^{XY} = (1/4) \log \theta_{11(k)}$, and use this result to interpret the loglinear partial association parameters.
   b. For model (XYZ), show $\lambda_{111}^{XYZ} = (1/8) \log[\theta_{11(1)}/\theta_{11(2)}]$. Thus, $\lambda_{ijk}^{XYZ} = 0$ is equivalent to $\theta_{11(1)} = \theta_{11(2)}$.

5.26 Consider the general loglinear model for a three-way table with parameters constrained to equal zero at the first level of each variable (Section 5.3.5).
   a. For the $2 \times 2 \times 2$ table, show $\lambda_{22}^{XY} = \log[\theta_{11(1)}]$ and $\lambda_{222}^{XYZ} = \log[\theta_{11(1)}/\theta_{11(2)}]$.

    b. Suppose all $\lambda_{ijk}^{XYZ} = 0$. Explain how to interpret the partial association parameters.

5.27 Suppose loglinear model (XY, XZ) holds.

    a. Using the expression for m_{ijk}, calculate m_{ij+} and $\log m_{ij+}$.

    b. Show that when (XY, XZ) holds, the loglinear model for the X–Y marginal table has the same association parameters as $\{\lambda_{ij}^{XY}\}$ in model (XY, XZ). Deduce that odds ratios are the same in the X–Y marginal table as in the partial tables. Using an analogous result for model (XY, YZ), deduce the collapsibility conditions (5.21).

    c. Calculate $\log m_{ij+}$ for model (XY, XZ, YZ), and explain why marginal associations need not be the same as partial associations.

5.28 Consider a three-way contingency table.

    a. When any pair of variables is conditionally independent, explain why there is no three-factor interaction.

    b. When three-factor interaction exists, explain why no pair of variables can be conditionally independent.

    c. When there is no three-factor interaction, but X–Y partial associations differ from the X–Y marginal association, explain why Z cannot be conditionally independent of X or Y.

5.29 For an $I \times J \times 2$ table, suppose logit model

$$\log(m_{ij1}/m_{ij2}) = \alpha + \beta_i^A + \beta_j^B$$

holds. Do $\{\beta_i^A\}$ take the same values as in the model

$$\log(m_{i1}/m_{i2}) = \alpha + \beta_i^A$$

for the $I \times 2$ collapsed table? Explain.

5.30 Consider the general two-way loglinear model between X and Y at a fixed level k of Z,

$$\log m_{ijk} = \mu(k) + \lambda_i^X(k) + \lambda_j^Y(k) + \lambda_{ij}^{XY}(k), \quad 1 \leq i \leq I, \ 1 \leq j \leq J.$$

Using zero-sum constraints as in Sections 5.1.2 and 5.3.1, show that parameters in the general three-way loglinear model satisfy

    a. $\mu = [\Sigma \, \mu(k)]/K$.

b. $\lambda_i^X = [\Sigma_k \lambda_i^X(k)]/K$

c. $\lambda_{ij}^{XY} = [\Sigma_k \lambda_{ij}^{XY}(k)]/K$

d. $\lambda_k^Z = \mu(k) - \mu$

e. $\lambda_{ik}^{XZ} = \lambda_i^X(k) - \lambda_i^X$

f. $\lambda_{ijk}^{XYZ} = \lambda_{ij}^{XY}(k) - \lambda_{ij}^{XY}$.

5.31 Show that the general loglinear model in T dimensions has 2^T terms. (Hint: It has a grand mean, $\binom{T}{1}$ single-factor terms, $\binom{T}{2}$ two-factor terms, and so forth.)

5.32 Consider a four-way cross-classification of W, X, Y, Z.
  a. Define parameters for the saturated loglinear model, indicating constraints that the parameters satisfy.
  b. Show that model (WXZ, WYZ) assumes only that X and Y are conditionally independent, given W and Z.
  c. Give the model in which X and Y are conditionally independent *and* there is no three-factor interaction.

5.33 For a four-way table, is the $W–X$ partial association the same as the $W–X$ marginal association, for the loglinear model symbolized by
  a. (WX, XYZ)?
  b. (WX, WZ, XY, YZ)?

5.34 For a three-way table with binary response Y, give the equivalent loglinear and logit models for which:
  a. Y is jointly independent of A and B.
  b. Y is conditionally independent of B.
  c. There is no interaction between A and B in their effects on Y.
  d. There is interaction between A and B in their effects on Y.

5.35 Show that logit model (5.27) is implied by loglinear model (ABC, AY, BY, CY).

5.36 For a four-way table with binary response Y, give the equivalent loglinear and logit models for which
  a. Y is jointly independent of A, B, and C.
  b. Y is jointly independent of B and C, given A.
  c. There are main effects of A and B on Y, but Y is conditionally independent of C, given A and B.

d. There are main effects of A, B, and C on Y.

e. There is interaction between A and B in their effects on Y, and C has main effects.

f. There is interaction between A and B and between A and C in their effects on Y.

g. There is interaction between A and B, between A and C, and between B and C, in their effects on Y.

h. There is interaction among A, B, and C in their effects on Y.

5.37 Prove the collapsibility conditions in Section 5.4.2.

5.38 Suppose model (XY, XZ, YZ) holds in a $2 \times 2 \times 2$ table. Suppose the common X–Y conditional log odds ratio at the two levels of Z is positive. Show that if the $X - Z$ and $Y - Z$ conditional log odds ratios are both positive or both negative, then the X–Y marginal log odds ratio is larger than the X–Y conditional log odds ratio. (Hence, Simpson's paradox cannot occur for the X–Y association.)

5.39 Let Π denote an $I \times J$ matrix of cell probabilities for the joint distribution of X and Y. Suppose there exist $I \times 1$ column vectors π_{1k} and $J \times 1$ column vectors π_{2k} of probabilities, $k = 1, \ldots, K$, and a set of probabilities $\{\rho_k\}$ such that

$$\Pi = \sum \rho_k \pi_{1k} \pi'_{2k} .$$

Show there is a hypothetical variable Z such that X and Y are conditionally independent, given Z. (The variable Z is called a *latent* variable, and the model for Π is a *latent-class* model. Clogg and Goodman (1984, 1985), Gilula (1983, 1984), Goodman (1974), Haberman (1979, Chap. 10), Henry (1983), Lazarsfeld and Henry (1968), McCutcheon (1987), and Palmgren and Ekholm (1987) discussed fitting and interpretation of such models.)

CHAPTER 6

Fitting Loglinear and Logit Models

Chapter 5 introduced loglinear models for contingency tables. This chapter shows how to fit the models and use them for categorical data analysis.

First, we derive sufficient statistics and likelihood equations for log-linear models fitted to three-way tables. Section 6.2 gives ML estimates of cell probabilities and expected frequencies. Section 6.3 uses chi-squared statistics to check whether a model fits adequately. Section 6.4 presents large-sample normal distributions for estimates of model parameters and cell probabilities. That section shows that a benefit of model-fitting is improved precision for estimating functions of the cell probabilities.

Many loglinear models do not have closed-form expressions for ML estimates. Section 6.5 presents two iterative algorithms for obtaining estimates, iterative proportional fitting and the Newton–Raphson method. Section 6.6 uses the algorithms to analyze loglinear models for two closely related types of data—rates and survival times. The final section applies the algorithms to table standardization, a process of smoothing cell counts so that a contingency table matches a specified set of marginal distributions.

When explanatory variables are categorical, logit models have equivalent loglinear representations. Thus, results in this chapter also apply to logit models.

6.1 SUFFICIENCY AND LIKELIHOOD FOR LOGLINEAR MODELS

After selecting a loglinear model, we use sample data to estimate model parameters and thus cell probabilities and expected frequencies. The ML estimates depend on the data only through sufficient statistics. In this section we obtain the minimal form of such statistics, illustrating with loglinear models for three-way tables. We then derive likelihood equations.

6.1.1 Minimal Sufficient Statistics

For ease of exposition, we assume the simplest sampling model—namely, that sample counts $\{n_{ijk}\}$ for the cross-classification of X, Y, and Z are independent Poisson random variables with expected values $\{m_{ijk}\}$. The same results occur when the cell counts are a multinomial sample with cell probabilities $\{\pi_{ijk} = m_{ijk}/(\Sigma \Sigma \Sigma m_{abc})\}$.

The joint Poisson probability mass function of $\{n_{ijk}\}$ is

$$\prod_i \prod_j \prod_k \frac{e^{-m_{ijk}} m_{ijk}^{n_{ijk}}}{n_{ijk}!}$$

where $\Pi_i \Pi_j \Pi_k$ gives a product over all cells of the table. The kernel of the log likelihood is

$$L(\mathbf{m}) = \sum_i \sum_j \sum_k n_{ijk} \log(m_{ijk}) - \sum_i \sum_j \sum_k m_{ijk} . \tag{6.1}$$

Now consider the general loglinear model for $\{m_{ijk}\}$,

$$\log(m_{ijk}) = \mu + \lambda_i^X + \lambda_j^Y + \lambda_k^Z + \lambda_{ij}^{XY} + \lambda_{ik}^{XZ} + \lambda_{jk}^{YZ} + \lambda_{ijk}^{XYZ} .$$

The log likelihood (6.1) equals

$$L(\mathbf{m}) = n\mu + \sum_i n_{i++}\lambda_i^X + \sum_j n_{+j+}\lambda_j^Y + \sum_k n_{++k}\lambda_k^Z$$

$$+ \sum_i \sum_j n_{ij+}\lambda_{ij}^{XY} + \sum_i \sum_k n_{i+k}\lambda_{ik}^{XZ} + \sum_j \sum_k n_{+jk}\lambda_{jk}^{YZ}$$

$$+ \sum_i \sum_j \sum_k n_{ijk}\lambda_{ijk}^{XYZ}$$

$$- \sum_i \sum_j \sum_k \exp[\mu + \lambda_i^X + \lambda_j^Y + \lambda_k^Z + \lambda_{ij}^{XY} + \lambda_{ik}^{XZ} + \lambda_{jk}^{YZ} + \lambda_{ijk}^{XYZ}] .$$
$$\tag{6.2}$$

Since the Poisson distribution is in the exponential family, the coefficients of the parameters in this log likelihood are sufficient statistics. For this saturated model, the $\{n_{ijk}\}$ are coefficients of $\{\lambda_{ijk}^{XYZ}\}$, so there is no reduction of the data. For simpler models, certain lambda parameters are zero, and (6.2) simplifies. For instance, for the model of mutual independence,

$$\log m_{ijk} = \mu + \lambda_i^X + \lambda_j^Y + \lambda_k^Z$$

sufficient statistics are the coefficients in (6.2) of $\{\lambda_i^X\}$, $\{\lambda_j^Y\}$, and $\{\lambda_k^Z\}$, namely $\{n_{i++}\}$, $\{n_{+j+}\}$, and $\{n_{++k}\}$.

Table 6.1 lists minimal sufficient statistics for several loglinear models. We determine them by inspecting, for each variable, the coefficient of the highest-order term(s) in which that variable appears. Simpler models use sample information in more condensed form. For instance, the simple model (X, Y, Z) requires only the single-factor marginal distributions, whereas the model (XY, Z) requires also the X–Y two-factor marginal distribution.

The minimal sufficient statistics are simply the marginal distributions corresponding to the terms in the model symbol. For the model (XY, XZ, YZ) having all two-factor dependencies, for instance, the minimal sufficient statistics are the two-way marginal tables $\{n_{ij+}\}$, $\{n_{i+k}\}$, and $\{n_{+jk}\}$.

6.1.2 Likelihood Equations for Loglinear Models

The fitted values $\{\hat{m}_{ijk}\}$ for a model are solutions to a set of likelihood equations. We illustrate the derivation of likelihood equations using loglinear model (XZ, YZ), for which X and Y are conditionally independent, given Z. For this model, log likelihood (6.2) simplifies to

$$L(\mathbf{m}) = n\mu + \sum_i n_{i++}\lambda_i^X + \sum_j n_{+j+}\lambda_j^Y + \sum_k n_{++k}\lambda_k^Z + \sum_i \sum_k n_{i+k}\lambda_{ik}^{XZ}$$

$$+ \sum_j \sum_k n_{+jk}\lambda_{jk}^{YZ} - \sum_i \sum_j \sum_k \exp[\mu + \lambda_i^X + \lambda_j^Y + \lambda_k^Z + \lambda_{ik}^{XZ} + \lambda_{jk}^{YZ}].$$

$$(6.3)$$

Table 6.1 Minimal Sufficient Statistics for Fitting Loglinear Models

Model	Minimal Sufficient Statistics
(X, Y, Z)	$\{n_{i++}\}$, $\{n_{+j+}\}$, $\{n_{++k}\}$
(XY, Z)	$\{n_{ij+}\}$, $\{n_{++k}\}$
(XY, YZ)	$\{n_{ij+}\}$, $\{n_{+jk}\}$
(XY, XZ, YZ)	$\{n_{ij+}\}$, $\{n_{i+k}\}$, $\{n_{+jk}\}$

We obtain each likelihood equation by differentiating $L(\mathbf{m})$ with respect to a parameter and setting the result equal to zero. For instance,

$$\frac{\partial L}{\partial \mu} = n - \sum_i \sum_j \sum_k \exp[\mu + \lambda_i^X + \lambda_j^Y + \lambda_k^Z + \lambda_{ik}^{XZ} + \lambda_{jk}^{YZ}]$$

$$= n - \sum \sum \sum m_{ijk} \, .$$

Setting this equal to zero gives the likelihood equation $\hat{m}_{+++} = n$. So, the fitted values have the same total as do the observed frequencies. Next,

$$\frac{\partial L}{\partial \lambda_i^X} = n_{i++} - \sum_j \sum_k m_{ijk} = n_{i++} - m_{i++} \, ,$$

from which we obtain equations $\{\hat{m}_{i++} = n_{i++}, \ i = 1, \ldots, I\}$. Differentiating with respect to λ_j^Y yields equations $\{\hat{m}_{+j+} = n_{+j+}, \ j = 1, \ldots, J\}$, and differentiating with respect to λ_k^Z yields equations $\{\hat{m}_{++k} = n_{++k}, \ k = 1, \ldots, K\}$.

In a similar manner, the derivatives

$$\frac{\partial L}{\partial \lambda_{ik}^{XZ}} = n_{i+k} - m_{i+k} \quad \text{and} \quad \frac{\partial L}{\partial \lambda_{jk}^{YZ}} = n_{+jk} - m_{+jk}$$

yield the likelihood equations

$$\hat{m}_{i+k} = n_{i+k} \quad \text{for all } i \text{ and } k \tag{6.4}$$

$$\hat{m}_{+jk} = n_{+jk} \quad \text{for all } j \text{ and } k \, . \tag{6.5}$$

Since (6.4) and (6.5) imply the equations previously obtained, these last two sets of equations determine the ML estimates. In summary, for model (XZ, YZ), the fitted values have the same X–Z and Y–Z marginal totals as do the observed data.

6.1.3 Birch's Results

For model (XZ, YZ), we see from (6.4), (6.5), and Table 6.1 that the minimal sufficient statistics are the ML estimates of the corresponding marginal distributions of expected frequencies. The same result occurs when we calculate likelihood equations for any other loglinear model studied so far. In fact, Birch (1963) showed that likelihood equations for loglinear models match minimal sufficient statistics to their expected values.

Birch showed there is a unique solution $\{\hat{m}_{ijk}\}$ that both satisfies the model and matches the sample data in their minimal sufficient statistics. Hence, if we produce such a solution, it must be the ML solution. To illustrate, the independence model for a two-way table

$$\log m_{ij} = \mu + \lambda_i^X + \lambda_j^Y$$

has minimal sufficient statistics $\{n_{i+}\}$ and $\{n_{+j}\}$. The likelihood equations are

$$\hat{m}_{i+} = n_{i+}, \qquad \hat{m}_{+j} = n_{+j}, \qquad \text{for all } i \text{ and } j .$$

The fitted values $\{\hat{m}_{ij} = n_{i+}n_{+j}/n\}$ satisfy these equations and also satisfy the model, so Birch's result implies they are the ML estimates. In summary, fitted values for loglinear models are smoothed versions of the sample counts that match them in certain marginal distributions but have associations and interactions satisfying certain model-implied patterns.

Birch showed that ML estimates are the same for multinomial sampling as for independent Poisson sampling. He showed estimates are also the same for independent multinomial sampling, as long as the model contains a term for the marginal distribution fixed by the sampling design. To illustrate, suppose at each of the IJ combinations of levels of X and Y, there is an independent multinomial sample for the cell counts at the K levels of Z. Then, $\{n_{ij+}\}$ are fixed. The model must contain the term λ_{ij}^{XY}, so the fitted values satisfy $\hat{m}_{ij+} = n_{ij+}$ for all i and j.

We derive Birch's results in Section 13.2. His derivations assumed that all cell counts are strictly positive. Section 7.7.2 presents results on the existence of ML estimates that show this strong an assumption is unnecessary.

6.2 ESTIMATING EXPECTED FREQUENCIES

We now derive fitted values $\{\hat{m}_{ijk}\}$ for loglinear models in three-way tables. By Birch's results, $\{\hat{m}_{ijk}\}$ both satisfy the model and are solutions to the likelihood equations.

6.2.1 Solving Likelihood Equations

To illustrate how to solve likelihood equations, we continue the analysis from the previous section of model (XZ, YZ). From (5.14), the model satisfies

$$\pi_{ijk} = \frac{\pi_{i+k}\pi_{+jk}}{\pi_{++k}} \quad \text{for all } i, j, \text{ and } k \, .$$

For Poisson sampling, we use the related formula for expected frequencies. Setting $\pi_{ijk} = m_{ijk}/n$, this is $\{m_{ijk} = m_{i+k}m_{+jk}/m_{++k}\}$. The likelihood equations (6.4) and (6.5) specify that ML estimates satisfy $\hat{m}_{i+k} = n_{i+k}$ and $\hat{m}_{+jk} = n_{+jk}$ for all $i, j,$ and k, and thus also $\hat{m}_{++k} = n_{++k}$ for all k. Since ML estimates of functions of parameters are simply the same functions of the ML estimates of those parameters, we have

$$\hat{m}_{ijk} = \frac{\hat{m}_{i+k}\hat{m}_{+jk}}{\hat{m}_{++k}} = \frac{n_{i+k}n_{+jk}}{n_{++k}} \, . \tag{6.6}$$

This solution satisfies the model and matches the data in the sufficient statistics. By Birch's results, it is the unique ML solution.

Similar reasoning produces ML estimates for all except one model listed in Table 6.1. Table 6.2 contains formulas for $\{\hat{m}_{ijk}\}$. That table also expresses $\{\pi_{ijk}\}$ in terms of marginal probabilities. We can use these expressions and the likelihood equations to determine the ML formulas, using the approach just described.

6.2.2 Direct versus Iterative Estimation

For the first three models in Table 6.2, there are explicit formulas for $\hat{m}_{ijk}$, and the estimates are said to be *direct*. Many loglinear models do

Table 6.2 Fitted Values for Loglinear Models in Three-Way Tables

Model[a]	Probabilistic Form	Fitted Value
(X, Y, Z)	$\pi_{ijk} = \pi_{i++}\pi_{+j+}\pi_{++k}$	$\hat{m}_{ijk} = \dfrac{n_{i++}n_{+j+}n_{++k}}{n^2}$
(XY, Z)	$\pi_{ijk} = \pi_{ij+}\pi_{++k}$	$\hat{m}_{ijk} = \dfrac{n_{ij+}n_{++k}}{n}$
(XY, XZ)	$\pi_{ijk} = \dfrac{\pi_{ij+}\pi_{i+k}}{\pi_{i++}}$	$\hat{m}_{ijk} = \dfrac{n_{ij+}n_{i+k}}{n_{i++}}$
(XY, XZ, YZ)	$\pi_{ijk} = \psi_{ij}\phi_{jk}\omega_{ik}$	Iterative methods (Section 6.5)
(XYZ)	No restriction	$\hat{m}_{ijk} = n_{ijk}$

[a] Formulas for models not listed are obtained by symmetry. For example, for (XZ, Y), $\hat{m}_{ijk} = n_{i+k}n_{+j+}/n$.

not have direct estimates. For such models, ML estimation requires iterative methods. Of models we have studied for two and three variables, the only one not having direct estimates is (XY, XZ, YZ), the model of no three-factor interaction. Though the two-way marginal tables are minimal sufficient statistics for this model, it is not possible to express $\{\pi_{ijk}\}$ directly in terms of $\{\pi_{ij+}\}$, $\{\pi_{i+k}\}$, and $\{\pi_{+jk}\}$.

Many models for four-way tables do not have direct estimates. Table 6.22 for Problem 6.26 lists types having direct estimates. In practice, it is not essential to know which models have direct estimates. Iterative methods used for models not having direct estimates can also be used with models that have direct estimates. Most statistical computer packages for loglinear models use such iterative routines for *all* cases.

6.2.3 Death Penalty Example

Section 5.2.2 discussed a $2 \times 2 \times 2$ table relating effects of defendant's race and victim's race on imposition of the death penalty for individuals convicted of homicide. The last column of Table 6.3 shows the cell counts. We now use the data to illustrate loglinear model-fitting. Widely available statistical computer packages can calculate $\{\hat{m}_{ijk}\}$ for loglinear models (Appendix A.2). Because of this, our discussion emphasizes interpretive aspects more than calculations.

Table 6.3 lists fitted values for several models. In the model symbols, D is defendant's race, V is victim's race, and P is death penalty verdict. Let $\hat{m}_{ijk}$ denote the fitted value at level i of defendant's race, level j of victim's race, and level k of death penalty verdict. For instance, for model (D, V, P), $\hat{m}_{111} = (n_{1++}n_{+1+}n_{++1})/n^2 = (160 \times 214 \times 36)/(326)^2 = 11.60$. The fitted values for models (VP, DV) and (VP, DP, DV) are uniformly quite close to the observed data.

Table 6.4 illustrates various patterns of association by presenting estimated odds ratios for marginal and partial associations for the models listed in Table 6.3. For example, the entry 1.0 for the partial association for model (VP, DV) is the common value of the D–P conditional odds ratios of $\{\hat{m}_{ijk}\}$ at the two levels of V: that is,

$$1.0 = \frac{21.17 \times 54.17}{129.83 \times 8.83} = \frac{0.48 \times 97.48}{8.52 \times 5.52}.$$

The entry 1.65 for the marginal D–P association for that same model is the odds ratio of $\{\hat{m}_{ijk}\}$ for the marginal D–P table; that is,

$$1.65 = \frac{(21.17 + 0.48)(54.17 + 97.48)}{(129.83 + 8.52)(8.83 + 5.52)}.$$

Table 6.3 Fitted Values for Death Penalty Data

Defendant's Race	Victim's Race	Death Penalty	(D, V, P)	(P, DV)	(VP, DV)	(DP, VP, DV)	(DVP)
						Model	
White	White	Yes	11.60	16.68	21.17	18.67	19
		No	93.43	134.32	129.83	132.33	132
	Black	Yes	6.07	0.99	0.48	0.33	0
		No	48.90	8.01	8.52	8.67	9
Black	White	Yes	12.03	6.96	8.83	11.33	11
		No	96.94	56.04	54.17	51.67	52
	Black	Yes	6.30	11.37	5.52	5.67	6
		No	50.73	91.63	97.48	97.33	97
		G^2	137.9	8.1	1.9	0.7	0
		df	4	3	2	1	0
		P-value	0.00	0.04	0.39	0.40	–

Table 6.4 Summary of Estimated Odds Ratios

Model	Partial Association			Marginal Association		
	$D{-}P$	$V{-}P$	$D{-}V$	$D{-}P$	$V{-}P$	$D{-}V$
(D, V, P)	1.0	1.0	1.0	1.0	1.0	1.0
(P, DV)	1.0	1.0	27.4	1.0	1.0	27.4
(VP, DV)	1.0	2.9	27.4	1.65	2.9	27.4
(DP, VP, DV)	0.6	3.7	28.7	1.2	2.9	27.4
$(DVP)^a$ Level 1	0.68	∞	∞	1.2	2.9	27.4
	(0.67)	(2.80)	(22.04)	(1.18)	(2.71)	(25.99)
Level 2	0	3.42	27.36			
	(0.79)	(3.29)	(25.90)			

[a] Values in parentheses for model (DVP) are obtained after adding 0.5 to each cell.

The odds ratios for the observed data are those reported for model (DVP), for which $\hat{m}_{ijk} = n_{ijk}$. For that model, the entry $\hat{m}_{121} = 0$ causes some partial odds ratios to equal 0 or ∞. Hence, for model (DVP) we also report the odds ratios obtained after adding 0.5 to each cell count.

For each model in Table 6.4, the estimated odds ratios reflect the connection between marginal and partial associations described by the collapsibility conditions in Section 5.3.3. For instance:

(D, V, P)
All partial odds ratios equal 1.0 and are necessarily the same as the marginal odds ratios, since the collapsibility conditions are fulfilled.

(P, DV)
Only the $D{-}V$ odds ratios do not equal 1.0. All partial odds ratios are necessarily the same as the marginal odds ratios. Since $\{n_{ij+}\}$ is in the set of minimal sufficient statistics for this model, all $\hat{m}_{ij+} = n_{ij+}$. Thus, the marginal $D{-}V$ odds ratio is the same for $\{\hat{m}_{ijk}\}$ as for the observed data.

(VP, DV)
Only the $D{-}P$ partial odds ratios equal 1.0. The $V{-}P$ and $D{-}V$ partial odds ratios are necessarily the same as the marginal odds ratios. This is not true of the $D{-}P$ partial and marginal odds ratios, since V is not conditionally independent of P or D. The marginal $V{-}P$ and $D{-}V$ odds ratios are the same for $\{\hat{m}_{ijk}\}$ as for the observed data, since all $\hat{m}_{ij+} = n_{ij+}$ and $\hat{m}_{+jk} = n_{+jk}$.

(DP, VP, DV)

All pairs of variables are conditionally dependent. No partial odds ratios equal 1.0, and none of them need be the same as the related marginal odds ratios. The estimated marginal odds ratios equal those for the observed data, since all $\hat{m}_{ij+} = n_{ij+}$, $\hat{m}_{i+k} = n_{i+k}$, and $\hat{m}_{+jk} = n_{+jk}$.

(DVP)

Since this model allows three-factor interaction, the two partial odds ratios for a given pair of variables are no longer equal. They *are* close (after 0.5 is added to each cell), however, which indicates why model (DP, VP, DV) fits well.

The values of the estimated odds ratios in Table 6.4 are useful for interpreting the associations. For instance, consider model (DP, VP, DV), which fits well but provides a smoothing of the data. The D–P odds ratios mean that the estimated odds of the death penalty verdict "yes" are (a) 1.2 times as high for a white defendant as for a black defendant, (b) 0.6 times as high for a white defendant as for a black defendant, within each level of victim's race. Section 5.2.2 explained the reversal in the D–P association when V is controlled (Simpson's paradox).

6.3 TESTING GOODNESS OF FIT

Having obtained fitted cell counts, we can assess model goodness of fit by comparing them to the observed cell counts. We use chi-squared statistics to test the hypothesis that population expected frequencies satisfy a given model.

6.3.1 Chi-Squared Statistics and Degrees of Freedom

For three-way tables, the likelihood-ratio statistic is

$$G^2 = 2 \sum_i \sum_j \sum_k n_{ijk} \log\left(\frac{n_{ijk}}{\hat{m}_{ijk}}\right) \tag{6.7}$$

and the Pearson statistic is

$$X^2 = \sum_i \sum_j \sum_k \frac{(n_{ijk} - \hat{m}_{ijk})^2}{\hat{m}_{ijk}} . \tag{6.8}$$

When the model holds, these statistics have large-sample chi-squared distributions, as a consequence of results proved in Section 12.3.

The degrees of freedom (df) for goodness-of-fit tests equal the difference in dimension between the alternative and null hypotheses. This equals the difference between the number of parameters in the general case and when the model holds. For instance, consider the mutual independence model (X, Y, Z), for multinomial sampling with probabilities $\{\pi_{ijk}\}$. In the general case, the only constraint is $\Sigma_i \Sigma_j \Sigma_k \pi_{ijk} = 1$, so there are $IJK - 1$ linearly independent parameters. For model (X, Y, Z), $\{\pi_{ijk} = \pi_{i++} \pi_{+j+} \pi_{++k}\}$ are determined by $I - 1$ of $\{\pi_{i++}\}$ (since $\Sigma_i \pi_{i++} = 1$), $J - 1$ of $\{\pi_{+j+}\}$, and $K - 1$ of $\{\pi_{++k}\}$, a total of $I + J + K - 3$ parameters. Thus,

$$df = (IJK - 1) - (I + J + K - 3) = IJK - I - J - K + 2 .$$

The same formula applies for Poisson sampling, for which there are IJK $\{m_{ijk}\}$ parameters in the general case. For model (X, Y, Z), $\{m_{ijk} = m_{i++}m_{+j+}m_{++k}/(m_{+++})^2\}$ are determined by m_{+++}, $I - 1$ of $\{m_{i++}\}$ (since $\Sigma_i m_{i++} = m_{+++}$), $J - 1$ of $\{m_{+j+}\}$, and $K - 1$ of $\{m_{++k}\}$, so $df = IJK - [1 + (I - 1) + (J - 1) + (K - 1)]$. Table 6.5 contains df formulas for testing three-way loglinear models.

Degrees of freedom for testing a loglinear model are identical to the residual df for the model. That is, they equal the number of cells in the table minus the number of linearly independent parameters in the model. For instance, the model

$$\log m_{ijk} = \mu + \lambda_i^X + \lambda_j^Y + \lambda_k^Z$$

Table 6.5   Residual Degrees of Freedom for Loglinear Models for Three-Way Tables

Model	Degrees of Freedom
(X, Y, Z)	$IJK - I - J - K + 2$
(XY, Z)	$(K - 1)(IJ - 1)$
(XZ, Y)	$(J - 1)(IK - 1)$
(YZ, X)	$(I - 1)(JK - 1)$
(XY, YZ)	$J(I - 1)(K - 1)$
(XZ, YZ)	$K(I - 1)(J - 1)$
(XY, XZ)	$I(J - 1)(K - 1)$
(XY, XZ, YZ)	$(I - 1)(J - 1)(K - 1)$
(XYZ)	0

has residual df $= IJK - [1 + (I - 1) + (J - 1) + (K - 1)]$. This equals the number of linearly independent parameters equated to zero in the saturated model to obtain the given model. In summary,

df = number cells − number linearly independent parameters in model

= number parameters equated to zero in saturated model

A saturated model has as many parameters as there are cells in the table, so its residual df $= 0$. The saturated model is the most general model, and it gives a perfect fit.

6.3.2 Death Penalty Example

We now return to the analysis, begun in Section 6.2.3, of the death penalty data. Table 6.6 contains the likelihood-ratio statistic and P-value for testing goodness of fit of several models. For a given df, larger G^2 values give smaller right-tail probabilities (P-values), and represent poorer fits. Table 6.6 indicates that models (D, V, P), (VP, D), (DP, V), and (VP, DP) fit the data very poorly. The common feature of these models is omission of the D–V association. This suggests an important association between defendant's race and victim's race.

Of the remaining four unsaturated models, (VP, DV) and (VP, DP, DV) are the only ones fitting adequately according to a formal

Table 6.6 Goodness-of-Fit Tests for Loglinear Models Relating Death Penalty Verdict (P), Defendant's Race (D), and Victim's Race (V)

Model	G^2	df	P-value
(D, V, P)	137.93	4	0.000
(VP, D)	131.68	3	0.000
(DP, V)	137.71	3	0.000
(DV, P)	8.13	3	0.043
(DP, VP)	131.46	2	0.000
(DP, DV)	7.91	2	0.019
(VP, DV)	1.88	2	0.390
(DP, VP, DV)	0.70	1	0.402
(DVP)	0	0	−

Source: Data from Radelet (1981). See Table 5.1

0.05-level significance test. In model (VP, DP, DV), all pairs of variables are conditionally dependent, but there is no three-factor interaction. According to the simpler model (VP, DV), the death penalty verdict is independent of defendant's race, given victim's race.

6.3.3 Corresponding Logit Models

For the death penalty data, it seems sensible to treat death penalty verdict as a response variable and defendant's race and victim's race as explanatory variables. In fitting logit models, we normally condition on the cell counts for the marginal table consisting of combinations of levels of the explanatory variables. For these data, we would treat the marginal D–V table $\{n_{ij+}\}$ as fixed and regard $\{n_{ij1}\}$ as four independent binomial samples on the P response. Logit models of interest are loglinear models containing the DV term, since all such models force the fitted values to satisfy $\{\hat{m}_{ij+} = n_{ij+}\}$.

Let $L_{ij} = \log(m_{ij1}/m_{ij2})$ denote the logit for P, at level i of D and level j of V. Loglinear model (P, DV) is equivalent to logit model

$$L_{ij} = \alpha$$

in which neither D nor V has an effect on P. Loglinear model (VP, DV) is equivalent to logit model

$$L_{ij} = \alpha + \beta_j^V$$

in which V has an effect on P, but D is conditionally independent of P given V. The term β_j^V in the logit model equals $\lambda_{j1}^{VP} - \lambda_{j2}^{VP}$ in the loglinear model.

Loglinear model (DP, VP, DV) is equivalent to logit model

$$L_{ij} = \alpha + \beta_i^D + \beta_j^V$$

in which both D and V have effects on P, but there is no three-factor interaction. The parameters in the two models are related by $\beta_i^D = \lambda_{i1}^{DP} - \lambda_{i2}^{DP}$ and $\beta_j^V = \lambda_{j1}^{VP} - \lambda_{j2}^{VP}$. Finally, the saturated model (DVP) corresponds to the general logit model

$$L_{ij} = \alpha + \beta_i^D + \beta_j^V + \beta_{ij}^{DV}$$

having interaction in the effects of D and V on P.

We can fit logit models using the methods in this chapter for equivalent loglinear models. For instance, fitted values, goodness-of-fit statistics, and residual df for the logit models just presented are identical to those given in Tables 6.3 and 6.6 for the equivalent loglinear models. For zero-sum constraints, the estimates of effect parameters are twice those of corresponding loglinear parameters.

Often, more than one loglinear or logit model fits a data set well. For the death penalty data, models (DP, VP, DV) and (VP, DV) both fit well. Chapter 7 discusses issues involved in comparing models and selecting "best" models.

6.4 ESTIMATING MODEL PARAMETERS[*]

Let $(n_1, \ldots, n_N)$ and $(m_1, \ldots, m_N)$ denote observed counts and expected frequencies for cells in a contingency table, with $n = \Sigma\, n_i$ the total sample size. For simplicity we use a single index, though the table may be multidimensional. We estimate parameters in loglinear or logit models by substituting fitted values into formulas defining the parameters. For instance, μ is the grand mean of $\{\log m_i\}$, so the ML estimate $\hat{\mu}$ of μ is the grand mean of $\{\log \hat{m}_i\}$. For models having direct estimates of $\{m_i\}$, the parameter estimates are linear combinations of logarithms of the minimal sufficient statistics.

6.4.1 Distribution of Parameter Estimators

Let **m** and **n** denote column vectors of $\{m_i\}$ and $\{n_i\}$. Loglinear models for positive Poisson cell means have form

$$\log \mathbf{m} = \mathbf{X}\boldsymbol{\beta} \tag{6.9}$$

where **X** is a model matrix containing known constants and $\boldsymbol{\beta}$ is a column vector of t parameters. We illustrate with the independence model, $\log m_{ij} = \mu + \lambda_i^X + \lambda_j^Y$. For the 2×2 table with constraints $\Sigma\, \lambda_i^X = \Sigma\, \lambda_j^Y = 0$, it is

$$\begin{bmatrix} \log m_{11} \\ \log m_{12} \\ \log m_{21} \\ \log m_{22} \end{bmatrix} = \begin{bmatrix} 1 & 1 & 1 \\ 1 & 1 & -1 \\ 1 & -1 & 1 \\ 1 & -1 & -1 \end{bmatrix} \begin{bmatrix} \mu \\ \lambda_1^X \\ \lambda_1^Y \end{bmatrix}.$$

For instance, $\log m_{12} = \mu + \lambda_1^X + \lambda_2^Y = \mu + \lambda_1^X - \lambda_1^Y$, so the second row of **X** is $(1, 1, -1)$.

The model matrix $\mathbf{X}$ occurs in expressions for likelihood equations and the asymptotic covariance matrix of parameter estimators for loglinear models. For independent Poisson sampling, the log likelihood involves the parameters of the loglinear model $\log(m_i) = \Sigma_j x_{ij}\beta_j$ through

$$L(\mathbf{m}) = \sum_i n_i \log(m_i) - \sum_i m_i = \sum_i n_i\left(\sum_j x_{ij}\beta_j\right) - \sum_i \exp\left(\sum_j x_{ij}\beta_j\right).$$
(6.10)

Since $m_i = \exp(\Sigma_j x_{ij}\beta_j)$,

$$\frac{\partial L(\mathbf{m})}{\partial \beta_j} = \sum_i n_i x_{ij} - \sum_i m_i x_{ij}, \quad j = 1, 2, \ldots, t.$$

The likelihood equations equate these derivatives to zero, and have form

$$\mathbf{X'n} = \mathbf{X'\hat{m}}.$$
(6.11)

These equations equate sufficient statistics to their expected values, which is one of Birch's (1963) results. For models considered so far, these sufficient statistics are certain marginal tables.

The matrix of second partial derivatives of the log likelihood has elements

$$\frac{\partial^2 L(\mathbf{m})}{\partial \beta_j \, \partial \beta_k} = -\sum_i x_{ij} \frac{\partial m_i}{\partial \beta_k} = -\sum_i x_{ij}\left\{\frac{\partial}{\partial \beta_k}\left[\exp\left(\sum_h x_{ih}\beta_h\right)\right]\right\} = -\sum_i x_{ij} x_{ik} m_i.$$

Loglinear models, like logistic regression models, are generalized linear models using the canonical link; thus this matrix does not depend on the observed data $\mathbf{n}$. We estimate the information matrix, the negative of this matrix, by

$$\mathbf{X'} \, \mathbf{Diag(\hat{m})X}$$

where $\mathbf{Diag(\hat{m})}$ has the elements of $\hat{\mathbf{m}}$ on the main diagonal.

For a fixed number of cells, as $n \to \infty$, the ML estimator $\hat{\boldsymbol{\beta}}$ has an asymptotic normal distribution with mean $\boldsymbol{\beta}$ and covariance matrix equal to the inverse of the information matrix. Thus, for Poisson sampling, the estimated covariance matrix of $\hat{\boldsymbol{\beta}}$ is

$$\hat{\mathrm{Cov}}(\hat{\boldsymbol{\beta}}) = [\mathbf{X'Diag(\hat{m})X}]^{-1}.$$
(6.12)

For multinomial sampling, the parameter μ is not relevant, and $\mathbf{X}$ deletes the column pertaining to it. The estimated covariance matrix is then

$$\hat{\text{Cov}}(\hat{\boldsymbol{\beta}}) = \{\mathbf{X}'[\text{Diag}(\hat{\mathbf{m}}) - \hat{\mathbf{m}}\hat{\mathbf{m}}'/n]\mathbf{X}\}^{-1}. \tag{6.13}$$

(Section 12.4.) The third standard case is independent multinomial sampling. Palmgren (1981) showed that conditional on observed marginal totals for explanatory variables, the asymptotic covariances for estimators of parameters involving the response are the same as for Poisson sampling. Section 13.2.2 outlines the argument. Palmgren's result implies (6.13) is identical to (6.12) with the row and column referring to the parameter μ deleted. Birch (1963) and Goodman (1970) gave related results.

6.4.2 Death Penalty Example

For the death penalty data, Table 6.7 contains parameter estimates and estimated standard errors for the saturated model and for model (VP, DV), the simplest one to fit the data adequately. For the saturated model, $\hat{m}_{121} = n_{121} = 0$. Since parameter estimates are contrasts of $\{\log \hat{m}_{ijk}\}$, we added $\frac{1}{2}$ to each cell count before calculating the estimates and standard errors. The values of $\hat{\lambda}_{11}^{DP}$ and $\hat{\lambda}_{111}^{DVP}$ for the saturated model are small compared to their standard errors, so it is not surprising that model (VP, DV) fits well.

The positive value (0.264) of $\hat{\lambda}_{11}^{VP}$ for model (VP, DV) indicates that when the victim was white, the death penalty was imposed more often than would be expected if those variables were independent, given defendant's race. From (5.20), $\exp(4\hat{\lambda}_{11}^{VP}) = 2.9$ is the estimated V–P odds ratio at each level of D.

Table 6.7   Estimates of Parameters and Standard Errors for Models Fitted to Death Penalty Data[a]

	Model (DVP)		Model (VP, DV)	
Parameter	Estimate	ASE	Estimate	ASE
λ_{11}^{DV}	0.793	0.195	0.828	0.095
λ_{11}^{VP}	0.277	0.195	0.264	0.116
λ_{11}^{DP}	−0.079	0.195	0.0	−
λ_{111}^{DVP}	−0.020	0.195	0.0	−

[a] Other parameter estimates are determined by zero-sum parameter constraints. For the model (DVP), 0.5 was added to each cell count before the estimates were calculated.

The estimated standard error of $\hat{\lambda}_{11}^{VP}$ for model (VP, DV) is 0.116, and an approximate 95% confidence interval for λ_{11}^{VP} is $0.264 \pm 1.96 \times 0.116$, or $(0.037, 0.492)$. A 95% confidence interval for the partial V–P odds ratio is $[\exp(4 \times 0.037), \exp(4 \times 0.492)]$, or $(1.16, 7.15)$. Though there seems to be an association between V and P, the sample is too small to judge whether it is a strong or weak one.

6.4.3 Distribution of Probability Estimates

Consider a given loglinear model with fitted values $\hat{m}$. For multinomial sampling, the ML estimates of cell probabilities are $\hat{\pi} = \hat{m}/n$. We next give the asymptotic covariance matrix for $\hat{\pi}$.

The saturated model has $\hat{\pi} = p$, the sample proportions. Let $\text{Cov}(p)$ denote their covariance matrix. Under multinomial sampling, from (3.19) and (3.20),

$$\text{Cov}(p) = [\text{Diag}(\pi) - \pi\pi']/n \qquad (6.14)$$

where $\text{Diag}(\pi)$ has the elements of π on the main diagonal. When there are I independent multinomial samples on a response variable with J categories, π and p consist of I sets of proportions, each having $J - 1$ nonredundant elements. Then, $\text{Cov}(p)$ is a block diagonal matrix. Each of the independent samples has a $(J - 1) \times (J - 1)$ block of form (6.14), and the matrix contains zeroes off the main diagonal of blocks.

Now consider an unsaturated model. Using the delta method, Sections 12.2.2 and 12.4 show that when the model holds, $\hat{\pi}$ has an asymptotic normal distribution about the true values π. The estimated covariance matrix equals

$$\hat{\text{Cov}}(\hat{\pi}) = \{\hat{\text{Cov}}(p)X[X' \, \hat{\text{Cov}}(p)X]^{-1}X' \, \hat{\text{Cov}}(p)\}/n .$$

For multinomial sampling, this expression equals

$$\hat{\text{Cov}}(\hat{\pi}) =$$
$$\{[\text{Diag}(\hat{\pi}) - \hat{\pi}\hat{\pi}']X[X'(\text{Diag}(\hat{\pi}) - \hat{\pi}\hat{\pi}')X]^{-1}X'[\text{Diag}(\hat{\pi}) - \hat{\pi}\hat{\pi}']\}/n .$$
$$(6.15)$$

For the death penalty example, we treat the data as four independent binomial samples on the death penalty response. Table 6.8 shows two sets of estimates of the probability the death penalty is imposed. One set is the sample proportions and the other set is based on model (DV, VP).

Table 6.8 Estimated Probability of Death Penalty and Asymptotic Standard Error (ASE) for Two Loglinear Models[a]

Defendant	Victim	Model (DV, VP)		Model (DVP)	
		Estimate	ASE	Estimate	ASE
White	White	0.140	0.024	0.126	0.027
	Black	0.054	0.021	0.000	0.000
Black	White	0.140	0.024	0.175	0.048
	Black	0.054	0.021	0.058	0.023

[a] Estimates for (DVP) are sample proportions. Estimated standard errors obtained using PROC CATMOD in SAS.

The table also reports estimated standard errors. The estimated covariance matrix of the sample proportions is a 4×4 diagonal matrix with elements $p_{1|ij}(1 - p_{1|ij})/n_{ij+}$ for $\{(i, j) = \{(1, 1), (1, 2), (2, 1), (2, 2)\}$. The unappealing estimated standard error of 0 for the sample proportion for the nine cases with white defendants and black victims is caused by $p_{1|12} = 0$, for which $p_{1|12}(1 - p_{1|12})/9 = 0$.

6.4.4 Model Smoothing Improves Precision of Estimation

When the sampled population satisfies a certain model, both $\hat{\pi}$ for that model and the sample proportion vector **p** are consistent estimators of the true probabilities π. The model-based estimator is better, in the sense that its true asymptotic standard error cannot exceed that of the sample proportion. This result follows from there being fewer parameters in the unsaturated model (on which $\hat{\pi}$ is based) than in the saturated model (on which **p** is based). The following result suggests that model-based estimators are also more efficient in estimating *functions* $g(\pi)$ of cell probabilities:

Suppose a certain model holds. Let $\hat{\pi}$ denote the ML estimator of π for the model, and let **p** denote the sample proportions. For any differentiable function g,

$$\text{Asymp. Var}[\sqrt{n}g(\hat{\pi})] \leq \text{Asymp. Var}[\sqrt{n}g(\mathbf{p})].$$

Section 12.2.2 proves this result.

Models that are more complex than necessary produce poorer inferences than adequate simpler models. For instance, when a given model holds, parameter estimators for that model have smaller asymptotic variances than estimators of the same parameters in more complex models (see, e.g., Lee 1977). When model (XY, XZ) holds, we obtain better estimators of λ^{XY} terms by fitting that model than by fitting model (XY, XZ, YZ) or the saturated model (compare, for instance, ASE values in Table 6.7 for models (VP, DV) and (DVP)). This is one reason that, in model selection, statisticians prefer parsimonious models. Altham (1984) showed that such results hold more generally than for models for categorical data.

When the model does not hold, $\hat{\pi}$ is inconsistent. As n gets larger, $\hat{\pi}$ does not converge to the true value π. The total mean squared error, MSE $= \Sigma\, E(\hat{\pi}_i - \pi_i)^2$, does not converge to zero. Since any unsaturated model is unlikely to describe reality *exactly*, we might expect MSE to be smaller for **p** than $\hat{\pi}$, since **p** is always consistent. Suppose, though, that a model gives a good approximation to the true probabilities. Then for small to moderate-sized samples, $\hat{\pi}$ is still better than **p**. The model smooths the sample counts, somewhat damping the random sampling fluctuations. The resulting estimators are better until the sample size is quite large.

When the model does not hold but n is small to moderate, the formula

$$\text{MSE} = \text{Variance} + (\text{Bias})^2$$

explains why model-based estimators can have smaller MSE than sample proportions. When the model does not hold, model-based estimators are biased. However, since the model producing $\hat{\pi}$ has fewer parameters than the model producing **p**, $\hat{\pi}_i$ has smaller variance than p_i. The MSE of $\hat{\pi}_i$ is smaller than that for p_i when its variance plus squared bias is smaller than the variance of p_i. This cannot happen for large samples. As the sample size increases, the variance goes to zero for both estimators, but the bias does not go to zero when the model does not hold.

To illustrate, consider Table 6.9. The cell probabilities are $\{\pi_{ij} = \pi_{i+}\pi_{+j}(1 + \delta u_i u_j)\}$ for $\{\pi_{i+} = \pi_{+j} = 1/3\}$ and $\{u_i = i - 2\}$. The value δ is a constant between -1 and 1, with 0 corresponding to independence. The independence model provides a good approximation to the relationship when δ is close to zero. The total MSE values of the two estimators are

$$\text{MSE}(\mathbf{p}) = \sum\sum E(p_{ij} - \pi_{ij})^2 = \sum\sum \text{Var}(p_{ij})$$

$$= \sum\sum \pi_{ij}(1 - \pi_{ij})/n = \left(1 - \sum\sum \pi_{ij}^2\right)/n$$

$$\text{MSE}(\hat{\pi}) = \sum\sum E(\hat{\pi}_{ij} - \pi_{ij})^2$$

Table 6.9 Cell Probabilities for Comparison of Estimators

$(1+\delta)/9$	$1/9$	$(1-\delta)/9$
$1/9$	$1/9$	$1/9$
$(1-\delta)/9$	$1/9$	$(1+\delta)/9$

for $\{\hat{\pi}_{ij} = n_{i+}n_{+j}/n^2\}$ in the independence model. For Table 6.9,

$$MSE(\mathbf{p}) = \frac{1}{n}\left\{\frac{8}{9} - \frac{4\delta^2}{81}\right\}$$

and it follows from Problem 12.26 that

$$MSE(\hat{\pi}) = \frac{1}{n}\left\{\frac{4}{9} + \frac{4}{9n}\right\} + \frac{4\delta^2}{81}\left\{1 - \frac{2}{n} + \frac{2}{n^2} - \frac{2}{n^3}\right\}.$$

Table 6.10 lists the total MSE values for various values of δ and n. When there is independence $(\delta = 0)$, $MSE(\mathbf{p}) = 8/9n$ whereas $MSE(\hat{\pi}) \doteq 4/9n$ for large n. The independence-model estimator is then considerably better than the sample proportion. When the table is close to independence and the sample size is not large, MSE is only about half as large for the model-based estimator. When $\delta \neq 0$, the inconsistency of $\hat{\pi}$ is reflected by $MSE(\hat{\pi}) \rightarrow 4\delta^2/81$ (whereas $MSE(\mathbf{p}) \rightarrow 0$) as $n \rightarrow \infty$. When the table is close to independence, however, the model-based estimator has smaller total MSE even for moderately large sample sizes.

6.5 ITERATIVE MAXIMUM LIKELIHOOD ESTIMATION

Many loglinear models do not have direct ML estimates. For instance, direct estimates do not exist for unsaturated models containing all

Table 6.10 Comparison of Total MSE($\times 10,000$) for Sample Proportion and Independence Estimators

	$\delta = 0$		$\delta = 0.1$		$\delta = 0.2$		$\delta = 0.6$		$\delta = 1.0$	
n	p	$\hat{\pi}$	p	$\hat{\pi}$	p	$\hat{\pi}$	p	$\hat{\pi}$	p	$\hat{\pi}$
10	889	489	888	493	887	505	871	634	840	893
50	178	91	178	95	177	110	174	261	168	565
100	89	45	89	50	89	65	87	220	84	529
500	18	9	18	14	18	28	17	186	17	500
∞	0	0	0	5	0	20	0	178	0	494

two-factor associations. For models not having direct estimates, iterative algorithms for solving likelihood equations yield fitted values and parameter estimates. This section discusses two important iterative procedures—*iterative proportional fitting*, and the *Newton-Raphson* method.

6.5.1 Iterative Proportional Fitting

The *iterative proportional fitting* (IPF) algorithm, originally presented by Deming and Stephan (1940), is a simple method for calculating $\{\hat{m}_i\}$ for hierarchical loglinear models. The procedure has the following steps:

1. Start with any initial estimates $\{\hat{m}_i^{(0)}\}$ having association and interaction structure no more complex than the model being fitted. For instance, $\{\hat{m}_i^{(0)} \equiv 1.0\}$ satisfies all models discussed so far.
2. By multiplying by appropriate scaling factors, successively adjust $\{\hat{m}_i^{(0)}\}$ so they match each marginal table in the set of minimal sufficient statistics.
3. Continue the process until the maximum difference between the sufficient statistics and their fitted values is sufficiently close to zero.

We illustrate the IPF algorithm using loglinear model (XY, XZ, YZ). For it, the minimal sufficient statistics are $\{n_{ij+}\}$, $\{n_{i+k}\}$ and $\{n_{+jk}\}$. Initial estimates must satisfy the model. The first cycle of the adjustment process has three steps:

$$\hat{m}_{ijk}^{(1)} = \hat{m}_{ijk}^{(0)}\left(\frac{n_{ij+}}{\hat{m}_{ij+}^{(0)}}\right)$$

$$\hat{m}_{ijk}^{(2)} = \hat{m}_{ijk}^{(1)}\left(\frac{n_{i+k}}{\hat{m}_{i+k}^{(1)}}\right)$$

$$\hat{m}_{ijk}^{(3)} = \hat{m}_{ijk}^{(2)}\left(\frac{n_{+jk}}{\hat{m}_{+jk}^{(2)}}\right)$$

Summing both sides of the first expression over k, we have $\hat{m}_{ij+}^{(1)} = n_{ij+}$ for all i and j. So, after the first step, all observed and fitted frequencies match in the X–Y marginal table. After the second step, all $\hat{m}_{i+k}^{(2)} = n_{i+k}$, but the X–Y marginal tables no longer match. After the third step, all $\hat{m}_{+jk}^{(3)} = n_{+jk}$, but the X–Y and X–Z marginal tables no longer match. A new cycle begins by again matching the X–Y marginal tables, using $\hat{m}_{ijk}^{(4)} = \hat{m}_{ijk}^{(3)}(n_{ij+}/\hat{m}_{ij+}^{(3)})$, and so forth.

At each step of the process, the updated estimates continue to satisfy the model. For instance, in step 1, we use the same adjustment factor $[n_{ij+}/\hat{m}_{ij+}^{(0)}]$ at different levels k of Z. Thus, X–Y odds ratios from different levels of Z have ratio equal to 1; that is, the no three-factor interaction pattern continues at each step.

As the cycles progress, the G^2 statistic using the updated estimates is monotone decreasing, and the process must converge (Andersen 1974, Fienberg 1970a, Haberman 1974a). That is, the fitted values come simultaneously closer to matching the observed data in all sufficient statistics, and they exhibit those associations allowed in the model. Table 6.11 contains the first four steps of IPF for fitting model (DP, VP, DV) to the death penalty data, using initial estimates $\{\hat{m}_{ijk}^{(0)} \equiv 1\}$. After four complete cycles (12 steps), $\{\hat{m}_{ijk}^{(12)}\}$ have all two-way marginal frequencies within 0.02 of the observed sufficient statistics. The limiting ML estimates are also shown in Table 6.11.

The reason the IPF algorithm produces ML estimates is that it generates a sequence of fitted values converging to a solution that both satisfies the model and matches the sufficient statistics. By Birch's results quoted in Section 6.1.4, there is only one such solution, and it is ML.

The IPF method works even for models having direct estimates. In that case, when the number of dimensions is no greater than six, IPF yields ML estimates within one cycle (Haberman 1974a, p. 197). To illustrate, we use IPF to fit the independence model (X, Y) for two variables. The minimal sufficient statistics are $\{n_{i+}\}$ and $\{n_{+j}\}$. For the initial estimates $\{\hat{m}_{ij}^{(0)} \equiv 1.0\}$, the first cycle gives

Table 6.11　Iterative Proportional Fitting of Model (DP, VP, DV) to Death Penalty Data[a]

| Cell | | | | | Fitted Marginal | | | |
| | | | | DV | DP | VP | DV | |
D	V	P	$\hat{m}^{(0)}$	$\hat{m}^{(1)}$	$\hat{m}^{(2)}$	$\hat{m}^{(3)}$	$\hat{m}^{(4)}$	$\hat{m}$
W	W	Y	1.0	75.5	17.93	22.06	22.03	18.67
		N	1.0	75.5	133.07	129.13	128.97	132.33
	B	Y	1.0	4.5	1.07	0.55	0.56	0.33
		N	1.0	4.5	7.93	8.38	8.44	8.67
B	W	Y	1.0	31.5	6.45	7.94	7.96	11.33
		N	1.0	31.5	56.55	54.87	55.04	51.67
	B	Y	1.0	51.5	10.55	5.45	5.44	5.67
		N	1.0	51.5	92.45	97.62	97.56	97.33

[a] W = White, B = Black, Y = Yes, N = no.

$$\hat{m}_{ij}^{(1)} = \hat{m}_{ij}^{(0)}\left(\frac{n_{i+}}{\hat{m}_{i+}^{(0)}}\right) = \frac{n_{i+}}{J}$$

$$\hat{m}_{ij}^{(2)} = \hat{m}_{ij}^{(1)}\left(\frac{n_{+j}}{\hat{m}_{+j}^{(1)}}\right) = \frac{n_{i+}n_{+j}}{n}.$$

The IPF algorithm then gives $\hat{m}_{ij}^{(t)} = n_{i+}n_{+j}/n$ for all $t > 2$. Hence, the ML estimates appear at the end of the first cycle.

6.5.2 Newton–Raphson Method

Section 4.7 introduced the Newton–Raphson method and applied it to ML estimation for logistic regression models. Referring to notation introduced there, we now identify $g(\boldsymbol{\beta})$ as the log likelihood for Poisson loglinear models.

From (6.10), let

$$L(\boldsymbol{\beta}) = \sum_i n_i\left(\sum_h x_{ih}\beta_h\right) - \sum_i \exp\left(\sum_h x_{ih}\beta_h\right).$$

Then

$$q_j = \frac{\partial L(\boldsymbol{\beta})}{\partial \beta_j} = \sum_i n_i x_{ij} - \sum_i m_i x_{ij}$$

$$h_{jk} = \frac{\partial^2 L(\boldsymbol{\beta})}{\partial \beta_j \, \partial \beta_k} = -\sum_i m_i x_{ij} x_{ik}$$

so that

$$q_j^{(t)} = \sum_i (n_i - m_i^{(t)}) x_{ij} \quad \text{and} \quad h_{jk}^{(t)} = -\sum_i m_i^{(t)} x_{ij} x_{ik}.$$

Now $\mathbf{m}^{(t)}$, the tth approximation for $\hat{\mathbf{m}}$, is obtained from $\boldsymbol{\beta}^{(t)}$ through $\mathbf{m}^{(t)} = \exp(\mathbf{X}\boldsymbol{\beta}^{(t)})$. It generates the next value $\boldsymbol{\beta}^{(t+1)}$ using formula (4.30), which in this context is

$$\boldsymbol{\beta}^{(t+1)} = \boldsymbol{\beta}^{(t)} + [\mathbf{X}'\mathbf{Diag}(\mathbf{m}^{(t)})\mathbf{X}]^{-1}\mathbf{X}'(\mathbf{n} - \mathbf{m}^{(t)})$$

where $\mathbf{Diag}(\mathbf{m}^{(t)})$ has elements $\{m_i^{(t)}\}$ on the main diagonal. This in turn produces $\mathbf{m}^{(t+1)}$, and so forth.

Alternatively, $\boldsymbol{\beta}^{(t+1)}$ can be expressed as

$$\boldsymbol{\beta}^{(t+1)} = -(\mathbf{H}^{(t)})^{-1}\mathbf{r}^{(t)} \tag{6.16}$$

where
$$r_j^{(t)} = \sum m_i^{(t)} x_{ij} \left\{ \log m_i^{(t)} + \frac{(n_i - m_i^{(t)})}{m_i^{(t)}} \right\}.$$

The expression in braces is the first term in the Taylor series expansion of $\log n_i$ at $\log m_i^{(t)}$.

The iterative process begins with all $m_i^{(0)} = n_i$, or all $m_i^{(0)} = n_i + 0.5$ if any $n_i = 0$. Then (6.16) produces $\boldsymbol{\beta}^{(1)}$, and for $t > 0$ the iterations proceed as just described. In the limit as t increases, $\mathbf{m}^{(t)}$ and $\boldsymbol{\beta}^{(t)}$ usually converge rapidly to the ML estimates $\hat{\mathbf{m}}$ and $\hat{\boldsymbol{\beta}}$. The $\mathbf{H}^{(t)}$ matrix converges to the matrix $\hat{\mathbf{H}} = -\mathbf{X}'\mathbf{Diag}(\hat{\mathbf{m}})\mathbf{X}$. By (6.12), the estimated large-sample covariance matrix of $\hat{\boldsymbol{\beta}}$ is $-\hat{\mathbf{H}}^{-1}$, a by-product of the Newton–Raphson method.

Formula (6.16) has the form

$$\boldsymbol{\beta}^{(t+1)} = (\mathbf{X}'\hat{\mathbf{V}}^{-1}\mathbf{X})^{-1}\mathbf{X}'\hat{\mathbf{V}}^{-1}\mathbf{y}$$

where $\mathbf{y}$ has elements $y_i = \log m_i^{(t)} + (n_i - m_i^{(t)})/m_i^{(t)}$ and where $\hat{\mathbf{V}} = [\mathbf{Diag}(\hat{\mathbf{m}}^{(t)})]^{-1}$. In other words, $\boldsymbol{\beta}^{(t+1)}$ is the weighted least squares solution for the model

$$\mathbf{y} = X\boldsymbol{\beta} + \boldsymbol{\epsilon}$$

where $\{\epsilon_i\}$ are uncorrelated with variances $\{1/m_i^{(t)}\}$. The ML estimate is the limit of a sequence of iteratively reweighted least squares estimates. When we take $\{m_i^{(0)} = n_i\}$, the first approximation for the ML estimate is simply the weighted least squares estimate for model

$$\log \mathbf{n} = X\boldsymbol{\beta} + \boldsymbol{\epsilon}.$$

6.5.3 Comparison of Iterative Methods

The IPF algorithm has the advantage of simplicity. For loglinear models discussed so far, it is conceptually simple and easy to implement. It converges to correct values even when the likelihood is poorly behaved, for instance when there are zero fitted counts and estimates on the boundary of the parameter space. The Newton–Raphson method is more complex, requiring solving a system of equations at each step. Though computing routines do not require formal matrix inversion, Newton–Raphson routines are not feasible with many computing systems when the model is of high dimensionality—for instance, when the contingency table has several dimensions and the parameter vector is large.

However, IPF also has disadvantages. It is primarily applicable to models for which likelihood equations equate observed and fitted counts in certain marginal tables. By contrast, Newton–Raphson is a general-purpose method that can solve more complex systems of likelihood equations. IPF sometimes converges very slowly compared to the Newton–Raphson method. The rate of convergence is second order for Newton–Raphson and only first order for IPF, though each cycle takes less time with IPF. Unlike Newton–Raphson, IPF does not produce the estimated covariance matrix of the parameter estimators as a by-product. Of course, fitted values produced by IPF can be substituted into (6.12) to obtain that matrix.

Because the Newton–Raphson method is applicable to a wide variety of models, it is the iterative routine used by most statistical computer packages for categorical data.

6.6 ANALYZING RATES AND SURVIVAL TIMES USING LOGLINEAR MODELS

To further illustrate the use of loglinear models, this section discusses their application to the analysis of *rates* and the analysis of *survival-time* data. For survival-time data, the response is the length of time until some event—for instance, the number of months until death for subjects diagnosed as having AIDS, or the number of hours before failure for hard disks in IBM personal computers. For rate data, the response is the number of events of some type divided by a relevant baseline measure—for instance, the number of hard disks that fail, divided by the number of hours of operation, for a test sample of IBM personal computers.

Data sets of these types usually contain *censored* observations. We cannot observe the response for some subjects, but we do know that response exceeds a certain value. If Jones is diagnosed as having AIDS, and is still alive nine months later when the study ends, the researchers know only that survival time ≥ 9 for Jones. This observation would also result if Jones withdrew from the study after 9 months. Such observations are said to be *right censored.*

When explanatory variables are categorical, loglinear models are useful both for modeling survival times until an event and for modeling rates at which the event occurs. Under the assumption that survival times have an exponential distribution, we obtain the same results whether we analyze the survival times or the rate of occurrence. The presentation in this section is based on articles by Aitkin and Clayton (1980), Holford (1976, 1980), and Laird and Olivier (1981).

6.6.1 Modeling Rates for Heart Valve Operations

Laird and Olivier (1981) described a study of patient survival after heart valve replacement operations. A sample of 109 patients were classified by type of heart valve replaced (aortic, mitral) and by age (<55, ≥ 55). Follow-up observations were made for each patient until the patient died or until the study ended. Heart valve operations were performed throughout the study period, and follow-up observations covered varying lengths of time, ranging from 3 to 97 months.

Table 6.12 lists the numbers of deaths during the follow-up period, by valve type and age. These counts are the first layer of a three-way contingency table that classifies valve type, age, and whether died (yes, no). Even if we had that complete table, it would be inappropriate to analyze it using standard methods, since subjects had differing lengths of time at risk. A subject who died after 3 months should not be treated the same as one who died after 8 years. To use age and valve type as predictors in a logit model for frequency of death, the proper baseline is not the frequency of nondeath but rather the total time subjects were at risk. That is, we should model the *rate* of death.

We define the *time at risk*, or *exposure time*, for a subject as follows: For subjects observed to die, it is the follow-up time until death occurred; for censored subjects, it is the follow-up time until the study ended or the subject withdrew from it. For a given age and valve type, the total exposure time is the sum of the exposure times for all subjects in that cell. Table 6.12 gives the total exposure times in months. We define the *risk* to be the number of deaths divided by total exposure time. Table 6.12 also contains the sample risks. For instance, the sample risk for replacement

Table 6.12 Deaths, Months of Exposure, and Risk, by Age of Patient and Type of Heart Valve

Age		Types of Heart Valve	
		Aortic	Mitral
<55	Deaths	4	1
	Exposure	1259	2082
	Risk	0.0032	0.0005
$55+$	Deaths	7	9
	Exposure	1417	1647
	Risk	0.0049	0.0055

Source: Reprinted with permission, based on data in Laird and Olivier (1981).

of the aortic valve for younger subjects is $4/1259 = 0.0032$. The risk is the monthly rate at which death occurs.

We now model effects on risk of $A =$ age and $V =$ valve type. Let n_{ij} denote the number of deaths for age i and valve type j, with expected value m_{ij}. Let E_{ij} denote the total exposure time for that cell. Given E_{ij}, the expected risk is m_{ij}/E_{ij}. The model

$$\log(m_{ij}/E_{ij}) = \mu + \lambda_i^A + \lambda_j^V \qquad (6.17)$$

assumes a lack of interaction in the effects of age and valve type on risk.

Model (6.17) has the appearance of a loglinear model for a two-way table, except the left-hand side is $\log(m_{ij}) - \log(E_{ij})$ instead of $\log(m_{ij})$. The adjustment term $\log(E_{ij})$ is called an *offset*. We fit the model conditional on the exposures $\{E_{ij}\}$. Assuming $\{n_{ij}\}$ are independent Poisson random variables with means $\{m_{ij}\}$, we can fit model (6.17) using Newton–Raphson methods or iterative proportional fitting.

Since $m_{ij} = E_{ij} \exp(\mu + \lambda_i^A + \lambda_j^V)$, the cell means are proportional to the exposures and have the same odds ratios. The IPF routine uses $\{E_{ij}\}$ as initial values, and then scales them to successively match the sufficient statistics. For model (6.17), these are the marginal death totals $\{n_{i+}\}$ and $\{n_{+j}\}$. The steps are

$$\hat{m}_{ij}^{(0)} = E_{ij}$$

$$m_{ij}^{(t)} = \hat{m}_{ij}^{(t-1)}(n_{i+}/\hat{m}_{i+}^{(t-1)}), \qquad \hat{m}_{ij}^{(t+1)} = \hat{m}_{ij}^{(t)}(n_{+j}/\hat{m}_{+j}^{(t)})$$

for $t = 1, 3, 5, \ldots$. The limiting fitted values match the observed counts $\{n_{ij}\}$ in their row and column totals, and have the same odds ratio patterns as the total exposures. Thus, the fitted risks $\{\hat{m}_{ij}/E_{ij}\}$ have odds ratios of 1, and satisfy the independence model. The ordinary loglinear model of independence is (6.17) with $\{E_{ij} = 1\}$. Its IPF routine is this one with initial values identically equal to 1.0.

Table 6.13 presents the fitted numbers of deaths and fitted risks for model (6.17). The effects of age on risk satisfy

$$\hat{\lambda}_2^A - \hat{\lambda}_1^A = 1.221 \quad (\text{ASE} = 0.514) .$$

Given valve type, the risk is estimated to be $\exp(1.221) = 3.39$ times higher for the older age group than for the younger age group. The effects of valve type satisfy

$$\hat{\lambda}_2^V - \hat{\lambda}_1^V = -0.330 \quad (\text{ASE} = 0.438)$$

Table 6.13 Fitted Values for Model (6.17)

Age		Aortic	Mitral
		Type of Heart Valve	
<55	Deaths	2.28	2.72
	Risk	0.0018	0.0013
55+	Deaths	8.72	7.28
	Risk	0.0062	0.0044

The study contains much censored data. Of the 109 patients, we observe deaths for only 21. We cannot estimate the effects of either factor very precisely. Note, though, that the analysis uses all 109 patients through their contributions to the exposure times.

We check model goodness of fit by comparing the death counts $\{n_{ij}\}$ to the fitted values $\{\hat{m}_{ij}\}$. For Tables 6.12 and 6.13, $G^2 = 3.22$ and $X^2 = 3.12$. The residual df $= 1$, since there are four logits and three model parameters. The null hypothesis that the model holds corresponds to a lack of interaction between valve type and age in their effects on risk. There is some evidence of interaction, though the model without valve-type effects fits nearly as well, with $G^2 = 3.79$ and $X^2 = 3.85$, based on df $= 2$. Models omitting age effects fit poorly.

Let t_i denote the time at risk for the ith subject. This is the time until an observed death or until censoring occurred. Let $w_i = 1$ if we observed a death for subject i, and $w_i = 0$ if subject i was censored. For subjects at given levels of age and valve type, Σw_i is the observed number of deaths and Σt_i is the total exposure time. These sums, computed by age and valve type, are the totals in the four cells of Table 6.12.

6.6.2 Modeling Survival Times

An alternative way of modeling survival data focuses on *times* until death, rather than on *numbers* of deaths. Let $f(t)$ denote the probability density function for the time to death T, and let $F(t)$ denote the cdf. More generally, these could refer to the time until occurrence of any event, such as the failure time of some product. We now observe that results of ML estimation are identical using a Poisson likelihood for numbers of deaths or a negative exponential likelihood for survival times.

A subject whose survival time until death equals t contributes a factor $f(t)$ to the likelihood. A subject whose censoring time equals t contributes a factor $1 - F(t)$, the probability of survival for at least time t. Using the indicator w_i of 1 for death and 0 for censoring, the survival-time likelihood for n independent observations is

$$\prod_{i=1}^{n} f(t_i)^{w_i}[1 - F(t_i)]^{1-w_i} .$$

The log likelihood equals

$$\sum w_i \log[f(t_i)] + \sum(1 - w_i)\log[1 - F(t_i)] . \qquad (6.18)$$

Further analysis requires assuming a parametric form for f and a model for the dependence of the parameters on explanatory variables.

It is common practice to model the *rate* at which death occurs, rather than the expected value of the survival time distribution. The function

$$h(t) = f(t)/[1 - F(t)]$$

is called the *hazard function* (Cox and Oakes 1984). It represents the instantaneous rate of death for subjects who have survived to time t; that is, it is the limit as $\epsilon \downarrow 0$ of $P[t < T < t + \epsilon \mid T > t]/\epsilon$.

A simple density for survival modeling is the negative exponential. The pdf is

$$f(t) = \lambda e^{-\lambda t} , \quad t > 0 ,$$

the cdf is $F(t) = 1 - e^{-\lambda t}$ for $t > 0$, and $E(T) = \lambda^{-1}$. The hazard function is constant,

$$h(t) = \lambda , \quad t > 0 .$$

The rate of death is identical for all t, and the expected survival time is the inverse of that rate.

Let $\mathbf{x}$ denote a set of explanatory variables, with $\mathbf{x}_i$ the value for subject i. Suppose the hazard for a negative exponential survival distribution is related to $\mathbf{x}$ by

$$h(t; \mathbf{x}) = \lambda \exp(\boldsymbol{\beta}'\mathbf{x}) . \qquad (6.19)$$

That is, the distribution for T has parameter $\lambda(\mathbf{x})$ depending on $\mathbf{x}$ through (6.19). The choice of functional form (6.19) for effects of explanatory variables is a simple way of ensuring that the hazard is nonnegative at all $\mathbf{x}$. For instance, loglinear model (6.17) for the log risk corresponds to a multiplicative model of type (6.19) for the risk.

Now we consider the log likelihood (6.18) with $f(t)$ equal to the negative exponential density with parameter $\lambda \exp(\boldsymbol{\beta}'\mathbf{x})$. For subject i, the product

$$m_i = t_i \lambda \exp(\boldsymbol{\beta}'\mathbf{x}_i)$$

of time at risk and hazard rate gives an "expected number of deaths." With this substitution, the log likelihood simplifies to

$$\sum w_i \log m_i - \sum m_i - \sum w_i \log(t_i) .$$

The first two terms involve the parameters $\boldsymbol{\beta}$. This part is identical to the log likelihood for independent Poisson variates $\{w_i\}$ with expected values $\{m_i\}$. In this application the $\{w_i\}$ are binary rather than Poisson, but that is irrelevant to the process of maximizing the likelihood with respect to $\boldsymbol{\beta}$. This process is equivalent to maximizing the likelihood for the Poisson loglinear model

$$\log m_i - \log(t_i) = \log(\lambda) + \boldsymbol{\beta}'\mathbf{x}_i$$

with offset $\log(t_i)$, using observed data $\{w_i\}$. When we sum terms in the log likelihood for subjects having a common value of $\mathbf{x}$, the observed data are the numbers of deaths $(\sum w_i)$ at each setting of $\mathbf{x}$, and the offset is the log of the total exposure time $(\sum t_i)$ at each setting.

The assumption of constant hazard over time implied by the negative exponential density is not sensible in many applications. As products wear out, their failure rate increases. To generalize this approach, we could divide the time scale into disjoint time intervals and assume the hazard is constant in each, namely

$$h(t; \mathbf{x}) = \lambda_k \exp(\boldsymbol{\beta}'\mathbf{x})$$

for t in interval k. There is a separate exponential hazard rate for each piece of the time scale. Consider the contingency table for numbers of deaths, in which one dimension is a discrete time scale and other dimensions represent categorical explanatory variables. Holford (1980) and Laird and Olivier (1981) showed that Poisson loglinear models and likelihoods for this table are equivalent to loglinear hazard models and likelihoods that assume piecewise exponential hazards for the survival times.

For very short time intervals, the piecewise exponential approach is essentially nonparametric, making no assumption about the dependence of the hazard on time. This suggests the generalization of model (6.19),

$$h(t; \mathbf{x}) = \lambda(t) \exp(\boldsymbol{\beta}'\mathbf{x}) \tag{6.20}$$

called a *proportional hazards* model (Cox 1972). This model has the property that the ratio of hazards $h(t; \mathbf{x}_1)/h(t; \mathbf{x}_2) = \exp[\boldsymbol{\beta}'(\mathbf{x}_1 - \mathbf{x}_2)]$ is the same for all t.

6.6.3 Lung Cancer Survival Example

Table 6.14, part of a data set presented by Holford (1980), describes survival for 539 males diagnosed as having lung cancer in 1973. The prognostic factors are histology and stage of disease, each having three levels. The period of follow-up was divided into 2-month intervals.

Let m_{ijk} denote the expected number of deaths for histology i and stage of disease j, in follow-up time interval k. Let E_{ijk} denote the total exposure for that cell. The model

$$\log(m_{ijk}/E_{ijk}) = \mu + \lambda_i^H + \lambda_j^S + \lambda_k^T \tag{6.21}$$

Table 6.14 Number of Deaths and Total Follow-Up (in Parentheses) by Follow-Up Time Interval, Histology, and Stage of Disease

Time Interval	Stage:	I			II			III		
		1	2	3	1	2	3	1	2	3
0–		9	12	42	5	4	28	1	1	19
		(157	134	212	77	71	130	21	22	101)
2–		2	7	26	2	3	19	1	1	11
		(139	110	136	68	63	72	17	18	63)
4–		9	5	12	3	5	10	1	3	7
		(126	96	90	63	58	42	14	14	43)
6–		10	10	10	2	4	5	1	1	6
		(102	86	64	55	42	21	12	10	32)
8–		1	4	5	2	2	0	0	0	3
		(88	66	47	50	35	14	10	8	21)
10–		3	3	4	2	1	3	1	0	3
		(82	59	39	45	32	13	8	8	14)
12–		1	4	1	2	4	2	0	2	3
		(76	51	29	42	28	7	6	6	10)

Source: Reprinted with permission from the Biometric Society, based on Holford (1980).

Table 6.15   Results of Fitting Models to Table 6.14

Effects	G^2	df
T	170.7	56
T, H	143.1	54
T, S	45.8	54
T, S, H	43.9	52
$T, S, H, S \times H$	41.5	48

has residual $G^2 = 43.9$, with df $= 52$. All models assuming no interaction between follow-up time interval and either prognostic factor are proportional hazards models, since the effects of histology and stage of disease are then the same for each time interval. Table 6.15 contains results of fitting several such models. Though stage of disease is an important prognostic factor, histology did not contribute significant additional information.

For model (6.21), the effects of stage of disease satisfy

$$\hat{\lambda}_2^S - \hat{\lambda}_1^S = 0.470 \quad (\text{ASE} = 0.174) \,,$$

$$\hat{\lambda}_3^S - \hat{\lambda}_1^S = 1.324 \quad (\text{ASE} = 0.152) \,.$$

For instance, at a fixed follow-up time for a given histology, we estimate the risk to be $\exp(1.324) = 3.76$ times higher at the third stage of disease than at the first stage.

6.7   TABLE STANDARDIZATION

In fitting models for rates in the previous section, we scaled exposure times to match the marginal distributions of the data that are sufficient statistics for the model. Using the same methodology, it is possible to adjust contingency tables so they match *any* set of fixed values in certain marginal distributions. We can use IPF or Newton–Raphson to generate a set of fitted values that have the desired marginal totals, but maintain an observed association or interaction structure. The following example illustrates such an analysis, referred to as *table standardization*.

6.7.1   Abortion and Education Example

Table 6.16, relating education and attitudes toward abortion, is from the 1972 General Social Survey. To make patterns of association more clearly

Table 6.16 Marginal Standardization of Attitudes Toward Abortion by Years of Schooling

	Attitude toward Abortion			
Schooling	Generally Disapprove	Middle Position	Generally Approve	Total
Less than high school	209 (49.4)	101 (32.0)	237 (18.6)	(100)
High school	151 (32.8)	126 (36.6)	426 (30.6)	(100)
More than high school	16 (17.8)	21 (31.3)	138 (50.9)	(100)
Total	(100)	(100)	(100)	

Source: Smith (1976).

visible, Smith (1976) adjusted the table so that all row and column marginal totals would equal 100. He used an IPF routine with the sample cell counts $\{n_{ij}\}$ as initial values $\{\hat{m}_{ij}^{(0)}\}$, so that the adjusted table would preserve the observed odds ratios.

The IPF routine is

$$\hat{m}_{ij}^{(0)} = n_{ij}$$

and then for $t = 1, 3, 5, \ldots,$

$$\hat{m}_{ij}^{(t)} = \hat{m}_{ij}^{(t-1)} \left(\frac{100}{\hat{m}_{i+}^{(t-1)}} \right),$$

$$\hat{m}_{ij}^{(t+1)} = \hat{m}_{ij}^{(t)} \left(\frac{100}{\hat{m}_{+j}^{(t)}} \right).$$

At the end of each odd-numbered step, all row totals equal 100. At the end of each even-numbered step, all column totals equal 100. Odds ratios are unchanged at each odd (even) step, since all counts in a given row (column) are multiplied by the same constant. The process converges to the entries shown in parentheses in Table 6.16. The association is clearer in the standardized table. There is a ridge down the main diagonal, with higher levels of education having more favorable attitudes about abortion. The other counts fall away smoothly on both sides.

Table standardization is useful for comparing tables that have quite different marginal structures. For instance, Mosteller (1968) compared intergenerational occupational mobility tables from Britain and Denmark. Though the original tables appeared quite different, because of differing occupational distributions in the two societies, the standardized tables had very similar appearance. In an early application, Yule (1912) used it to compare three hospitals on vaccination and recovery for smallpox patients. A common modern application is adjusting sample data to match marginal distributions specified by census results.

The use of IPF to standardize a table resembles its use for fitting rate model (6.17) in the previous section. Here, the data (rather than the total exposures) are the initial values, and the uniform marginal counts (rather than the observed marginal counts) are the targets of each iterative step. Let $\{m_{ij}\}$ denote expected frequencies for the observed data, $\{E_{ij}\}$ expected frequencies for the standardized table, and $\{\hat{E}_{ij}\}$ fitted values in the standardized table. The standardization process corresponds to fitting the model

$$\log(E_{ij}/m_{ij}) = \mu + \lambda_i^E + \lambda_j^A .$$

To fit the model, an alternative to IPF is to use the Newton–Raphson method with a table of pseudo data that satisfy independence and have the desired margins, taking $\log(n_{ij})$ as an offset. For Table 6.16 the pseudo data would be counts uniformly equal to 100/3. When the ratios $\{\hat{E}_{ij}/n_{ij}\}$ satisfy the independence model, the standardized table necessarily has the same odds ratios as the observed table.

The process of table standardization is sometimes called *raking* the table. Imrey et al. (1981) gave the asymptotic covariance matrix for raked sample proportions.

6.7.2   Analyzing Weighted Data

The process of standardizing a table or of modeling rates using exposure times corresponds to fitting a loglinear model with an offset. This process is also useful in other applications.

Let $\{m_i\}$ be expected frequencies in a contingency table, let $\{E_i\}$ be a set of fixed constants for the cells, and consider a model of form

$$\log(m_i/E_i) = \alpha + \beta_1 x_{i1} + \beta_2 x_{i2} + \cdots . \tag{6.22}$$

The standard hierarchical loglinear models introduced in Chapter 5 have this form with $\{E_i = 1\}$. The survival models of Section 6.6 have this form with $\{E_i\}$ equal to cell exposures.

Another application is the analysis of categorical data for complex sampling designs. Most data sets collected in the social sciences have sampling designs more complex than simple random sampling. Such designs commonly employ stratification and/or clustering. Case weights can be assigned to inflate or deflate the influence of each observation according to features of the sampling design. By adding the case weights for subjects classified in a particular cell i in a contingency table, we obtain a total weighted frequency for that cell. The average cell weight z_i is defined to be the total weighted frequency divided by the original frequency n_i. Conditional on the average cell weights $\{z_i\}$, we can fit loglinear models for the weighted expected frequencies $\{z_i m_i = m_i / z_i^{-1}\}$ by expressing the model as a standard loglinear model for $\{\log m_i\}$, with offset $\{\log z_i^{-1}\}$ and data $\{n_i\}$. In this way, we obtain appropriate parameter estimates and estimated standard errors for the given sampling design. See Clogg and Eliason (1987) for details.

CHAPTER NOTES

Section 6.2: Estimating Expected Frequencies

6.1 Goodman (1970, 1971b), Haberman (1974a, Chapter 5), and Sundberg (1975) discussed families of loglinear models that have direct ML estimates and can be interpreted in terms of conditional independence, independence, or equiprobability. Such models are called *decomposable*, or *Markov-type*. Haberman proved conditions under which loglinear models have direct estimates.

Darroch et al. (1980) used mathematical graph theory to represent loglinear models that contain only single-factor and two-factor terms. Each graph has as many vertices as dimensions of the table. Each vertex represents a variable, and each edge an association. For instance, the graph

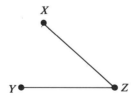

represents model (XZ, YZ). Darroch et al. defined a class of *graphical models* that contains the family of decomposable models.

Section 6.4: Estimating Model Parameters

6.2 Goodman (1970) gave asymptotic variances of parameter estimators for saturated loglinear models. Lee (1977) used the delta method to derive asymptotic variances for estimators of parameters in unsaturated models having direct estimates. Lee provided variance formulas for models for three-way and four-way tables.

Parameter estimates and standard errors must be adjusted when there is misclassification error. For ways of dealing with errors in measurement, see Assakul and Proctor (1967), Chen (1979), Ekholm and Palmgren (1987), Espeland and Odoroff (1985), Fleiss (1981, Chap. 11, 12), Goldberg (1975), Hochberg (1977), Mote and Anderson (1965), Palmgren (1987), and Tenenbein (1970).

Section 6.5: Iterative ML Estimation

6.3 Darroch (1962) and Bishop (1967) seem to have been the first statisticians to use IPF to obtain ML estimates in contingency tables. Darroch and Ratcliff (1972) gave a generalized iterative scaling method useful for models in which sufficient statistics are more complex than marginal distributions. Fienberg (1970a) and Fienberg and Meyer (1983) presented other applications of the IPF method.

Section 6.6: Analyzing Rates and Survival Times Using Loglinear Models

6.4 Log likelihood (6.18) actually applies only for "noninformative" censoring mechanisms. It does not make sense, for example, if subjects tend to withdraw from the study because of factors related to it—for instance, because of health effects related to one of the treatments. See Lagakos (1979).

Other articles dealing with analysis of rate data include Breslow and Day (1975), Clogg (1978), Clogg and Eliason (1988), Freeman and Holford (1980), Gail (1978), and Hoem (1987). Breslow and Day (1987, Section 4.5) discussed the choice between additive and multiplicative models. Articles dealing with grouped survival data, particularly loglinear models and logit models for survival probabilities, include Brown (1975), Johnson and Koch (1978), Larson (1984), Mantel and Hankey (1978), Mitchell and Turnbull (1979), Prentice and Gloeckler (1978), Schluchter and Jackson (1989), and Thompson (1977). Allison (1984) presented related models for social science applications. Aitkin and Clayton (1980) discussed exponential survival models, and also presented similar models that have hazard functions corresponding to Weibull or extreme value survival distributions. See also Whitehead (1980).

Section 6.7: Table Standardization

6.5 For further discussion of table standardization methods, see Bishop et al. (1975, pp. 76–102), Clogg and Eliason (1988), Fleiss (1981, Chap. 14), Gail (1978), Haberman (1979, Chap. 9), and Hoem (1987).

PROBLEMS

Applications

6.1 Using a statistical computer package, conduct the analyses described in this chapter for the death penalty data.

6.2 Refer to Problem 5.10. Find a good fitting loglinear model corresponding to a logit model for ever having sexual intercourse as the response. Use the parameter estimates to interpret the effects of race and gender.

6.3 Refer to Table 6.17, based on automobile accident records in 1988, supplied by the State of Florida Department of Highway Safety and Motor Vehicles. Subjects are classified by whether they were wearing a seat belt, whether ejected, and whether killed.

Table 6.17

Safety Equipment In Use	Whether Ejected	Injury	
		Nonfatal	Fatal
Seat belt	Yes	1105	14
	No	411,111	483
None	Yes	4624	497
	No	157,342	1008

Source: Florida Department of Highway Safety and Motor Vehicles.

a. Though this table refers to the population of accidents in Florida in 1988, one can use loglinear models to analyze associations among these variables. Do so, and find a loglinear model that describes the data well. Interpret the model.

b. Suppose we regard whether wearing a seat belt and whether ejected as explanatory variables. Which loglinear models are then appropriate? For a model of this type, explain the nature of the effects on whether killed.

6.4 Table 6.18, from DiFrancisco and Critelman (1984), refers to effects of nationality and education level on whether one follows politics regularly. Analyze these data.

Table 6.18

Follow Politics Regularly	USSR		USA		UK		Italy		Mexico	
	Yes	No	Yes	No	Yes	No	Yes	No	Yes	No
Primary	94	84	227	112	356	144	166	526	447	430
Secondary	318	120	371	71	256	76	142	103	78	25
College	473	72	180	8	22	2	47	7	22	2

Source: Reprinted with permission from DiFrancesco and Critelman (1984).

6.5   Table 6.19, from Demo and Parker (1987), refers to the effect of academic achievement on self-esteem among black and white college students.

Table 6.19

		Black		White	
Gender	Cumulative GPA	High Self-Esteem	Low Self-Esteem	High Self-Esteem	Low Self-Esteem
Males	High	15	9	17	10
	Low	26	17	22	26
Females	High	13	22	22	32
	Low	24	23	3	17

Source: Reprinted with permission of the Helen Dwight Reid Educational Foundation from Demo and Parker (1987). Published by Heldref Publications, copyright © 1987.

a. Which loglinear model would you pick to represent these data? Interpret the data using this model.

b. Treating self-esteem as a response variable, find a logit model that adequately describes the data. Interpret.

6.6   Refer to Table 11.1. Use loglinear models to analyze these data.

6.7   Table 6.20 refers to applicants to graduate school at the University of California at Berkeley, for the fall 1973 session. Admissions decisions are presented by gender of applicant, for the six largest

Table 6.20

	Whether Admitted			
	Male		Female	
Department	Yes	No	Yes	No
A	512(529.3)	313(295.7)	89(71.7)	19(36.3)
B	353(353.6)	207(206.4)	17(16.4)	8(8.6)
C	120(109.3)	205(215.7)	202(212.8)	391(380.2)
D	138(137.2)	279(279.8)	131(131.8)	244(243.2)
E	53(45.7)	138(145.3)	94(101.3)	299(291.7)
F	22(23.0)	351(350.0)	24(23.0)	317(318.0)
Total	1198	1493	557	1278

Source: Data from Freedman et al. (1978, p. 14). See also Bickel et al. (1975).

graduate departments. Denote the three variables by $A =$ whether admitted, $G =$ gender, and $D =$ department.

a. Test independence in the marginal A–G table. Calculate and interpret the sample A–G odds ratio.

b. The parenthesized values in Table 6.20 are fitted values for model (AD, AG, DG). Using these, estimate the A–G partial odds ratio, and interpret.

c. Compare the estimated A–G partial odds ratio from (b) to the marginal odds ratio in (a). Explain why they give such different indications of the A–G association.

6.8 Refer to the preceding problem.

a. Fit loglinear models (A, DG), (AG, DG), (AD, DG), and (AD, AG, DG), or the related logit models having A as the response. Report G^2 and df values, and comment on the quality of fit.

b. Refer to Problem 6.24, and consider the following six single-degree-of-freedom components of G^2 for model (AD, DG):

Department	G^2
A	19.05
B	0.26
C	0.75
D	0.30
E	0.99
F	0.38
Total	21.73

Comment on the fit of (AD, DG). Study the fitted values, and describe the lack of fit for Department A. How do you interpret the data for the other departments?

6.9 Refer to the previous two problems. Use only the data for Departments B–F.

a. Fit models (AD, DG) and (AD, AG, DG). Calculate the model parameter estimates and their estimated standard errors, for each model. Interpret $\hat{\lambda}_{11}^{AG}$ for model (AD, AG, DG), and construct a 95% confidence interval for the conditional A–G odds ratio.

b. Which departments seem to have more male applicants than would be expected if D and G were conditionally independent? Which departments seem to admit more students than would be expected if D and A were conditionally independent?

6.10 Refer to the previous problem. Treating A as the response variable, use models (AD, DG) and (ADG) to estimate the probability of being admitted, for each D–G combination. Find the estimated standard errors of these estimates. Comment on the sizes of the estimated standard errors for the model estimators compared to the sample proportions.

6.11 Refer to Problem 6.2. For the model you chose, obtain estimated probabilities of ever having sexual intercourse, by race and gender. Report the standard errors of the estimates.

6.12 Refer to Problem 5.18. Fit model $(AGLR, AS, LS, GRS)$, and carefully interpret the parameter estimates for the effects on S. Fowlkes et al. (1988) reported "The least-satisfied employees are less than 35 years of age, female, other (race), and work in the Northeast; . . . The most satisfied group is greater than 44 years of age, male, other, and working in the Pacific or Mid-Atlantic regions; the odds of such employees being satisfied are about 3.5 to 1." Show how to obtain these interpretations from the fit of this model.

6.13 One hundred subjects are randomly assigned to each of two treatments for leukemia. During the study, 10 subjects receiving treatment A died, and 18 subjects receiving treatment B died. The total exposure time was 170.4 years for treatment A and 147.3 years for treatment B. Test whether the rates of death are the same for the two treatments, by fitting a model that assumes equal rates.

6.14 For Table 6.12, fit the model in which risk depends on age but not on valve type. Report the estimated risk and fitted number of deaths in each cell. Interpret the age effect.

6.15 Consider Table 6.14.
   a. Fit the survival model that deletes the histology effects. Obtain and interpret the ML estimates of the stage of disease effects for this model.
   b. Show how to check whether the hazard depends on time.

6.16 Table 6.21, from Buckley (1988), gives the frequency of all reported game-related concussions for players on 49 college football teams, between 1975 and 1982. The total exposure for these data was 216,690 athlete-games. Suppose the total exposure was identical for offense and defense, six times higher for blocking than for tackling, and 2.2 times higher for rushing plays than for passing plays.

Table 6.21

Team	Situation	Activity	
		Tackle	Block
Offense	Rushing	125	129
	Passing	85	31
Defense	Rushing	216	61
	Passing	62	16

Source: Reprinted with permission from Buckley (1988).

 a. Calculate the total exposure per cell, and the sample rates of concussion. Which activity has greatest sample risk?
 b. Use loglinear models to analyze these risks.

6.17 Standardize Table 2.11, and describe the association between student party identification and parent party identification.

6.18 Standardize Table 10.2, in Chapter 10. Describe the pattern of migration.

Theory and Methods

6.19 Refer to (6.1). When $\{n_{ijk}\}$ has a multinomial distribution with probabilities $\{\pi_{ijk} = m_{ijk}/(\Sigma \Sigma \Sigma m_{abc})\}$, show that the part of the log likelihood dealing with both the data and parameters is $\Sigma \Sigma \Sigma n_{ijk} \log(m_{ijk})$, the same as for Poisson sampling.

6.20 For a two-way table, consider loglinear model

$$\log m_{ij} = \mu + \lambda_i^X$$

Show that $\hat{m}_{ij} = n_{i+}/J$ and that residual df $= I(J-1)$.

6.21 For the independence model for a two-way table, derive minimal sufficient statistics, likelihood equations, fitted values, and residual df.

6.22 For model (i) (X, Y, Z) and (ii) (XY, Z), derive
   a. minimal sufficient statistics,
   b. likelihood equations,
   c. fitted values, and
   d. residual df for testing goodness of fit.

6.23 Repeat the previous problem for model (WXY, WZ, XZ).

6.24 Consider model (XZ, YZ).

   a. For fixed k, show that $\{\hat{m}_{ijk}, i = 1, \ldots, I, j = 1, \ldots, J\}$ equal the fitted values for testing independence between X and Y within level k of Z.
   b. Show the Pearson and likelihood-ratio statistics for testing fit of this model have form $X^2 = \Sigma\, X_k^2$, where X_k^2 is the statistic for testing independence between X and Y within level k of Z.

6.25 Verify the df values given in Table 6.5 for models (XY, Z), (XY, YZ), and (XY, XZ, YZ).

6.26 Table 6.22 contains formulas for fitted values for the types of models for four-way tables that have direct estimates.

Table 6.22 Model Types Having Direct Estimates in Four Dimensions[a]

Model	Expected Frequency Estimate	Residual DF
(W, X, Y, Z)	$n_{h+++}n_{+i++}n_{++j+}n_{+++k}/n^3$	$HIJK - H - I - J - K + 3$
(WX, Y, Z)	$n_{hi++}n_{++j+}n_{+++k}/n^2$	$HIJK - HI - J - K + 2$
(WX, WY, Z)	$n_{hi++}n_{h+j+}n_{+++k}/n_{h+++}n$	$HIJK - HI - HJ - K + H + 1$
(WX, YZ)	$n_{hi++}n_{++jk}/n$	$(HI - 1)(JK - 1)$
(WX, WY, XZ)	$n_{hi++}n_{h+j+}n_{+i+k}/n_{h+++}n_{+i++}$	$HIJK - HI - HJ - IK + H + I$
(WX, WY, WZ)	$n_{hi++}n_{h+j+}n_{h++k}/(n_{h+++})^2$	$HIJK - HI - HJ - HK + 2H$
(WXY, Z)	$n_{hij+}n_{+++k}/n$	$(HIJ - 1)(K - 1)$
(WXY, WZ)	$n_{hij+}n_{h++k}/n_{h+++}$	$H(IJ - 1)(K - 1)$
(WXY, WXZ)	$n_{hij+}n_{hi+k}/n_{hi++}$	$HI(J - 1)(K - 1)$

[a] Number of levels of W, X, Y, Z, denoted by H, I, J, K. Estimates for other models of each type are obtained by symmetry.

a. Use Birch's results to verify that the entry is correct for model (W, X, Y, Z).

b. Repeat part (a) for model (WX, Y, Z).

c. Verify the residual df values for models $(W, X, Y, Z))$ and (WX, Y, Z).

d. Motivate the estimate and df formulas given for (WX, YZ) and (WXY, Z), using composite variables.

e. Motivate the formulas given for (WXY, WZ) by noting that, given W, Z is independent of the composite XY variable.

f. Motivate the formulas given for (WXY, WXZ) by noting that Y and Z are independent, given W and X.

6.27 Consider a T-dimensional table, having I_i categories in the ith dimension.

a. Find minimal sufficient statistics, ML estimates of cell probabilities, and residual df for the mutual independence model.

b. Find the minimal sufficient statistics and the residual df for the hierarchical model having all two-factor associations but no three-factor interaction terms. Does this model have direct estimates?

6.28 Consider loglinear model (X, Y, Z) for a $2 \times 2 \times 2$ table.

a. Express the model in the form $\log \mathbf{m} = \mathbf{X}\boldsymbol{\beta}$.

b. Show that the likelihood equations $\mathbf{X}'\mathbf{n} = \mathbf{X}'\hat{\mathbf{m}}$ equate $\{n_{ijk}\}$ and $\{\hat{m}_{ijk}\}$ in the one-dimensional margins.

6.29 Refer to the example in Section 6.4.4.

a. Suppose $\delta = 0.1$. For what range of n is $\text{MSE}(\hat{\boldsymbol{\pi}}) < \text{MSE}(\mathbf{p})$?

b. How would you expect the range of such values to change as δ increases? To illustrate, answer part (a) assuming $\delta = 0.2$.

6.30 Suppose the independence model holds, and let $p_{ij} = n_{ij}/n$ and $\hat{\pi}_{ij} = p_{i+}p_{+j}$.

a. Show that both p_{ij} and $\hat{\pi}_{ij}$ are unbiased for $\pi_{ij} = \pi_{i+}\pi_{+j}$.

b. Show $\text{Var}(p_{ij}) = \pi_{i+}\pi_{+j}(1 - \pi_{i+}\pi_{+j})/n$.

c. Show

$$\text{Var}(\hat{\pi}_{ij}) = \{\pi_{i+}\pi_{+j}[\pi_{i+}(1 - \pi_{+j}) + \pi_{+j}(1 - \pi_{i+})]\}/n$$
$$+ \pi_{i+}(1 - \pi_{i+})\pi_{+j}(1 - \pi_{+j})/n^2.$$

(Hint: $E(p_{i+}p_{+j})^2 = E(p_{i+}^2)E(p_{+j}^2)$, and $E(p_{i+}^2) = \text{Var}(p_{i+}) + [E(p_{i+})]^2$.)

d. As $n \to \infty$, show $\lim \text{Var}(\sqrt{n}\hat{\pi}_{ij}) \leq \lim \text{Var}(\sqrt{n}p_{ij})$, with equality only if $\pi_{ij} = 1$ or 0. Hence, if the model holds or if it "nearly" holds, the model estimator is better than the sample proportion. proportion.

6.31 For a multinomial distribution, let $\gamma = \Sigma\, b_i \pi_i$, and suppose $\pi_i = f_i(\theta) > 0$, $i = 1, \ldots, I$. Let $S = \Sigma\, b_i p_i$ and $T = \Sigma\, b_i \hat{\pi}_i$, where $\hat{\pi}_i = f_i(\hat{\theta})$, for the ML estimator $\hat{\theta}$ of θ.

a. Show that $\text{Var}(S) = [\Sigma\, b_i^2 \pi_i - (\Sigma\, b_i \pi_i)^2]/n$.

b. Using a Taylor series expansion, show that $\text{Var}(T)$ is approximately $[\text{Var}(\hat{\theta})][\Sigma\, b_i f_i'(\theta)]^2$.

c. By computing the information for $L(\theta) = \Sigma\, n_i \log[f_i(\theta)]$, show that $\text{Var}(\hat{\theta})$ is approximately $[n\Sigma\, (f_i'(\theta))^2/f_i(\theta)]^{-1}$.

d. Show that the asymptotic variance of $\sqrt{n}[T - \gamma]$ is less than or equal to the asymptotic variance of $\sqrt{n}[S - \gamma]$. (Hint: Consider $\text{Var}(T)/\text{Var}(S)$, and note that it is a squared correlation between two random variables, where with probability π_i the first equals b_i and the second equals $f_i'(\theta)/f_i(\theta)$.)

6.32 Apply IPF to loglinear model (X, YZ), and show the ML estimates are obtained within one cycle.

6.33 Show how to use IPF to fit model (WX, WY, WZ, XY, XZ, YZ) to a four-way table.

6.34 Show (6.16).

6.35 Consider loglinear model

$$\log m_{ij} = \mu + \lambda_i^X + \lambda_j^Y + \delta I(a)$$

where $I(a) = 1$ in cell a and equals 0 otherwise.

a. Find the likelihood equations, and note that there is a perfect fit in cell a.

b. Show that residual df $= IJ - I - J$.

c. Give an IPF algorithm for finding fitted values that satisfy the model. (Hint: Replace the entry in cell a by 0. Apply IPF for the independence model, with a starting value of 0 in cell a, to obtain other fitted values. See also Section 10.2.3)

6.36 Consider model (6.17). What, if any, effect is there on the model parameter estimates, their standard errors, and the goodness-of-fit statistics in the following situations:

    a. The exposure times are doubled, but the numbers of deaths stay the same.

    b. The exposure times stay the same, but the numbers of deaths double.

    c. The exposure times and the numbers of deaths both double.

6.37 Refer to model (6.17). Given the exposures $\{E_{ij}\}$, show that sufficient statistics are $\{n_{i+}\}$ and $\{n_{+j}\}$. What are sufficient statistics for the simpler model having effects only for age?

6.38 Let t_i be the survival or censoring time for the ith subject, and let $w_i = 1$ for a death and 0 for a censoring. Let $T = \Sigma\, t_i$ and $W = \Sigma\, w_i$. Suppose survival times have an exponential distribution with parameter λ.

    a. Using log likelihood (6.18), show the ML estimator of λ is W/T, the number of deaths divided by the total exposure time.

    b. Conditional on T, show that W has a Poisson distribution with mean $T\lambda$. Using the Poisson likelihood, show the ML estimator of λ is W/T.

6.39 Suppose we fix marginal row proportions $\{r_i\}$ and column proportions $\{c_j\}$ for an $I \times J$ table, all of them being positive.

    a. Show how to adjust cell proportions $\{p_{ij}\}$ so they have those totals, but maintain the same odds ratios as $\{p_{ij}\}$.

    b. Show how to find cell proportions that have these marginal totals and for which the local odds ratios all take fixed value $\theta > 0$. (Hint: Take initial values of 1.0 in all cells in the first row and in the first column. This determines all other initial cell entries such that all local odds ratios equal θ.)

    c. Show that cell proportions are determined by the marginal proportions and by the values of the local odds ratios.

6.40 Suppose we fit a multiplicative model M to a table, except for certain cells where we constrain $m_a = 0$. We express the model as $m_i = E_i M$, where $E_a = 0$ for those cells and all other $E_i = 1$. Explain how to fit this using the model-with-offset representation (6.22). (In practice, E_a must be a very small constant, such as 10^{-8}, so that its logarithm exists. Some computer packages allow the user to fit this model by assigning zero *weights* to certain cells.)

CHAPTER 7

Building and Applying
Loglinear Models

The previous chapter gave the mechanics of fitting loglinear models. Next we discuss model selection. The first section shows how to analyze whether certain effects are present by checking whether a model containing the effects fits better than a simpler model that omits them. Such comparisons are the basis of exploratory model selection strategies, using stepwise algorithms discussed in Section 7.2.

Section 7.3 presents cell residuals for loglinear models. Residuals highlight cells in which the fit is poor, and often direct us to more appropriate models.

Section 7.4 presents tests of conditional independence: A test based on loglinear models, the Cochran–Mantel–Haenszel test, and an exact test. Section 7.5 discusses related estimation and modeling of conditional association.

Section 7.6 discusses effects of sample size on model analyses. That section shows how to calculate power for tests using loglinear models. Such calculations help us determine the sample size needed to detect effects of a certain size.

Many studies encounter *sparse* contingency tables—tables in which cells have small frequencies, often with lots of zero cell counts. Section 7.7 discusses effects of sparse tables on the existence of ML estimates for loglinear models and on the use of large-sample procedures.

7.1 PARTITIONING CHI-SQUARED TO COMPARE MODELS

Section 3.3.5 showed that a chi-squared statistic with df = ν has partitionings into independent chi-squared components—for example, into ν

components each having df = 1. This section uses partitionings to compare loglinear models.

7.1.1 Partitioning for Nested Models

Partitioning of chi-squared is algebraically simple and natural using the likelihood-ratio statistic, G^2. Let $G^2(M)$ denote the value of G^2 for testing the fit of a model M, based on comparing fitted values $\hat{\mathbf{m}}$ to data $\mathbf{n}$. This statistic equals

$$G^2(M) = -2[L(\hat{\mathbf{m}}; \mathbf{n}) - L(\mathbf{n}; \mathbf{n})]$$

where $L(\hat{\mathbf{m}}; \mathbf{n})$ is the maximized log likelihood under the assumption that the model holds, and $L(\mathbf{n}; \mathbf{n})$ is the maximized log likelihood in the unrestricted (saturated) case. The difference in maximized log likelihoods is simply the log of the ratio of maximized likelihoods.

Consider two parametric models, M_1 and M_2, such that model M_2 is a special case of model M_1; that is, M_2 is simpler than M_1, so when M_2 holds, necessarily M_1 also holds. Model M_2 is said to be *nested* within M_1. Let ν_1 and ν_2 denote their residual degrees of freedom. Since M_2 is simpler than M_1, $\nu_1 < \nu_2$, and a smaller set of parameter values satisfies M_2 than satisfies M_1. Thus, $L(\hat{\mathbf{m}}_2; \mathbf{n})$ cannot exceed $L(\hat{\mathbf{m}}_1; \mathbf{n})$, and it follows that

$$G^2(M_1) \leqslant G^2(M_2).$$

Assuming model M_1 holds, the likelihood-ratio approach for testing that M_2 holds uses test statistic

$$-2[L(\hat{\mathbf{m}}_2; \mathbf{n}) - L(\hat{\mathbf{m}}_1 : \mathbf{n})]$$
$$= -2[L(\hat{\mathbf{m}}_2; \mathbf{n}) - L(\mathbf{n}; \mathbf{n})] - \{-2[L(\hat{\mathbf{m}}_1; \mathbf{n}) - L(\mathbf{n}; \mathbf{n})]\}$$
$$= G^2(M_2) - G^2(M_1).$$

This statistic, which we denote by $G^2(M_2 \mid M_1)$, is large when M_2 fits poorly compared to M_1.

We can partition $G^2(M_2)$ into

$$G^2(M_2) = G^2(M_1) + G^2(M_2 \mid M_1).$$

When model M_2 holds, $G^2(M_1)$ has an asymptotic chi-squared distribution with df $= \nu_1$, $G^2(M_2)$ has an asymptotic chi-squared distribution with df = ν_2, and $G^2(M_2 \mid M_1)$ has an asymptotic chi-squared distribution with df = $\nu_2 - \nu_1$. When M_1 holds but M_2 does not, $G^2(M_1)$ still has its asymptotic chi-squared distribution, but the other two statistics tend to grow unboundedly as n increases.

In many studies, a certain set of models has special importance. We may want to compare models that differ only by the inclusion of certain association terms, for instance. When the models form a nested set, we can partition G^2 for the simplest model in order to test a sequence of hypotheses.

To illustrate, consider the nested set of loglinear models $\{(XYZ), (XY, XZ, YZ), (XZ, YZ), (X, YZ), (X, Y, Z)\}$ and the decomposition

$$G^2(X, Y, Z) = G^2(XYZ) + \{G^2(XY, XZ, YZ) - G^2(XYZ)\}$$

$$+ \{G^2(XZ, YZ) - G^2(XY, XZ, YZ)\}$$

$$+ \{G^2(X, YZ) - G^2(XZ, YZ)\}$$

$$+ \{G^2(X, Y, Z) - G^2(X, YZ)\}$$

$$= G^2(XY, XZ, YZ)$$

$$+ G^2[(XZ, YZ) \mid (XY, XZ, YZ)]$$

$$+ G^2[(X, YZ) \mid (XZ, YZ)]$$

$$+ G^2[(X, Y, Z) \mid (X, YZ)] .$$

The component $G^2(XY, XZ, YZ)$ tests whether (XY, XZ, YZ) fits adequately. Given that model holds, the second component tests the goodness of fit of (XZ, YZ). Given that (XZ, YZ) holds, the third component tests the fit of (X, YZ). Finally, given that (X, YZ) holds, the final component tests the fit of (X, Y, Z).

Before accepting a model using a conditional breakdown, it is a good idea to check its fit according to the unconditional test. Each of several conditional components may be fairly large but not quite statistically significant, yet their sum may be significant. P-values or α-levels for these tests are not strictly accurate when the nested set is not an *a priori* choice, but is based on viewing results of fitting several models.

We can use simultaneous test procedures to lessen the probability of attributing importance to sample effects that simply reflect chance variation. These procedures use adjusted significance levels. For instance, suppose we compare s nested models. The $s - 1$ separate tests for comparing pairs of models are asymptotically independent. Goodman (1969) noted that when each test has level $1 - (1 - \alpha)^{1/(s-1)}$, the overall asymptotic Type-I error probability cannot exceed α.

7.1.2 Comparing Models for Death Penalty Data

We fitted several loglinear models to the death penalty data in Section 6.3.2. Consider the nested set $\{(DVP), (DP, VP, DV), (VP, DV), (P, DV), (D, V, P)\}$. To ensure the overall Type-I error probability for the four comparisons does not exceed $\alpha = 0.10$, we perform each test at the level $1 - (0.90)^{0.25} = 0.026$. When df $= 1$, the critical value for a chi-squared statistic is then 4.96 (i.e., a standard normal random variable exceeds $\sqrt{4.96} = 2.23$ in absolute value with probability 0.026).

Table 7.1 displays results for comparing models. The four single-degree-of-freedom components for $G^2(D, V, P)$ are 0.70, $1.88 - 0.70 = 1.18$, $8.13 - 1.88 = 6.25$, and $137.93 - 8.13 = 129.80$. We first accept (DP, VP, DV), and then accept (VP, DV), given (DP, VP, DV). At the third step the component 6.25 is large for df $= 1$, and we reject (P, DV). Model (VP, DV) also fits well according to the unconditional test, and we select it to represent these data.

The final model selected depends on the choice of nested set. For example, the set $\{(DP, VP, DV), (DP, DV), (P, DV), (D, V, P)\}$ results in the selection (DP, VP, DV).

Table 7.1 Components of G^2 for Death Penalty Data

Model	G^2	Difference	df	
(D, V, P)	137.93		4	
		129.80		1
(P, DV)	8.13		3	
		6.25		1
(VP, DV)	1.88		2	
		1.18		1
(DP, VP, DV)	0.70		1	
		0.70		1
(DVP)	0.00		0	

7.1.3   Expressions for $G^2(M_2 | M_1)$ and $X^2(M_2 | M_1)$

Let $\{\hat{m}_{1i}\}$ and $\{\hat{m}_{2i}\}$ denote fitted values for models M_1 and M_2. Then,

$$
\begin{aligned}
G^2(M_2) &= 2 \sum n_i \log(n_i/\hat{m}_{2i}) \\
&= 2 \sum n_i \log(n_i/\hat{m}_{1i}) + 2 \sum n_i \log(\hat{m}_{1i}/\hat{m}_{2i}) \\
&= G^2(M_1) + G^2(M_2 | M_1) \, .
\end{aligned}
\tag{7.1}
$$

A general result due to Simon (1973) implies that

$$
G^2(M_2 | M_1) = 2 \sum \hat{m}_{1i} \log(\hat{m}_{1i}/\hat{m}_{2i}) \, .
\tag{7.2}
$$

Simon showed that when observations have distribution in the natural exponential family, the difference in likelihood-ratio statistics takes this form whenever models are linear in the natural parameters. Statistic (7.2) has the same form as the likelihood-ratio statistic $G^2(M_2)$, but with $\{\hat{m}_{1i}\}$ playing the role of the observed data. In fact, $G^2(M_2)$ is the special case of $G^2(M_2 | M_1)$ with M_1 being the saturated model.

Just as $G^2(M)$ is a measure of the distance of $\hat{\mathbf{m}}$ from the observed data, (7.2) measures the distance of the best fit of M_2 from the best fit of M_1. In this sense, the decomposition $G^2(M_2) = G^2(M_1) + G^2(M_2 | M_1)$ expresses a certain orthogonality: The distance of the data from the best fit of M_2 equals the distance of the data from the best fit of M_1 plus the distance of the best fit of M_1 from the best fit of M_2.

For the Pearson statistic, the difference $X^2(M_2) - X^2(M_1)$ for nested models does not have Pearson form. This difference is not even necessarily nonnegative. A more appropriate statistic of Pearson form for comparing two models is

$$
X^2(M_2 | M_1) = \sum (\hat{m}_{1i} - \hat{m}_{2i})^2 / \hat{m}_{2i}
\tag{7.3}
$$

which has the usual form with $\{\hat{m}_{1i}\}$ in place of $\{n_i\}$. Statistics (7.2) and (7.3) depend on the data only through the fitted values, and thus only through sufficient statistics for model M_1. Under the hypothesis that M_1 holds, Haberman (1977a) showed that $G^2(M_2 | M_1)$ and $X^2(M_2 | M_1)$ have identical large-sample behavior, even for fairly sparse tables.

7.1.4   Partitioning in a $2 \times J$ Table

We next use results of this section to justify the partitioning of G^2 given in Section 3.3.5 for $2 \times J$ tables. For $j = 2, \ldots, J$, let M_j denote the model that satisfies

$$\theta_i = (m_{1i}m_{2,i+1})/(m_{1,i+1}m_{2i}) = 1, \quad i = 1, \ldots, j-1 .$$

For M_j, expected frequencies in the $2 \times j$ table consisting of columns 1 through j satisfy independence. Model M_h is a special case of M_j whenever $h > j$. Model M_J represents independence in the complete $2 \times J$ table. By decomposition (7.1),

$$G^2(M_J) = G^2(M_J \mid M_{J-1}) + G^2(M_{J-1})$$

$$= G^2(M_J \mid M_{J-1}) + G^2(M_{J-1} \mid M_{J-2}) + G^2(M_{J-2})$$

$$= \cdots = G^2(M_J \mid M_{J-1}) + \cdots + G^2(M_3 \mid M_2) + G^2(M_2) .$$

From (7.2), $G^2(M_j \mid M_{j-1})$ has the G^2 form with the fitted values for model M_{j-1} playing the role of the observed data. By substituting fitted values for the two models into (7.2), one can show that $G^2(M_j \mid M_{j-1})$ is identical to G^2 for testing independence in a 2×2 table; the first column combines columns 1 through $j-1$ of the original table, and the second column is column j of the original table.

For tables with ordered classifications, a useful partitioning of chi-squared has one component that tests for a linear trend in the data, and another component that tests for independence, given that a linear trend model is appropriate. The trend test in Section 4.4.3 is one such decomposition. We present other partitions of this type in the next chapter.

7.2 STRATEGIES IN MODEL SELECTION

This section describes strategies for selecting a loglinear model. The selection process becomes harder as the number of variables increases, because of the rapid increase in possible associations and interactions. Fitting all possible models is impractical when the number of dimensions exceeds three, and it helps to have guidelines. We balance two competing goals: We want a model complex enough to provide a good fit to the data. On the other hand, we want a model that is simple to interpret, one that smooths rather than overfits the data.

Most studies are designed to answer certain questions. Those questions guide the choice of potential terms for a model. Confirmatory analyses then respond to project goals using a restricted set of models. For instance, a study's hypotheses about associations and interactions may be tested by comparing nested models. For studies that are exploratory rather than confirmatory, a search among a large set of models may provide clues about important facets of the dependence structure.

7.2.1 Sampling Considerations

Regardless of the strategy used to select a model, the sampling design often dictates the inclusion of certain terms. For independent multinomial sampling, certain marginal totals are fixed. Any potential model should include those totals as sufficient statistics, so that the fitted values are identical in those margins. To illustrate, suppose at each combination of levels of W = clinic and X = treatment, there is independent multinomial sampling on two responses, Y and Z. Then $\{n_{hi++}\}$ are fixed, and we should fit models that are more complex than model (WX, Y, Z), to ensure that $\{\hat{m}_{hi++} = n_{hi++}\}$. If 29 people received treatment 1 in clinic A, then the fitted counts will have a total of 29 people at that combination.

Related to this point, models should recognize distinctions between response and explanatory variables. It is normally sensible to include the fullest interaction term among explanatory variables, and concentrate the modeling process on terms linking responses and terms linking explanatory variables to responses. If W and X are both explanatory variables, models assuming conditional independence between W and X are not of interest.

7.2.2 Exploratory Analyses

In exploratory analyses, it helps to have a rough guide to the complexity of linkages among the variables. One approach to investigating this fits the model having only single-factor terms, then the model having only single-factor and two-factor terms, then the model having only three-factor and lower order terms, and so forth. Fitting such models often reveals a restricted range of good-fitting models. For example, when the mutual independence model fits poorly but the model having all two-factor associations fits well, we can then study intermediate models containing some subset of those associations.

This strategy must recognize the sampling design and the distinction between response and explanatory variables. In a four-way table, suppose there is independent multinomial sampling on responses Y and Z at each combination of levels of explanatory variables W and X. Then we could fit models (WX, Y, Z), (WX, WY, WZ, XY, XZ, YZ), (WXY, WXZ, WYZ, XYZ) to determine the likely complexity of a good-fitting model.

Goodman (1971a) gave an alternative strategy for determining a starting point for model-fitting. He suggested fitting the saturated model and noting which estimates $\hat{\lambda}$ of association and interaction parameters are large compared to their estimated standard errors $\hat{\sigma}(\hat{\lambda})$. For large samples $(\hat{\lambda} - \lambda)/\hat{\sigma}(\hat{\lambda})$ has an approximate standard normal distribution. The ratio $\hat{\lambda}/\hat{\sigma}(\hat{\lambda})$ is called a *standardized parameter estimate*. A large standardized estimate provides evidence that the true value λ is nonzero. A natural starting point is a hierarchical model containing terms that have at least one standardized estimate exceeding a certain number, say 2.0, in absolute value. When that model fits well, it may be possible to simplify it and maintain an adequate fit. When it does not fit well, we can add additional parameters having moderate standardized estimates until the fit is adequate. A weakness of this strategy is that an estimator's precision depends on whether parameters of higher order are in the model. For instance, the standardized estimate $\hat{\lambda}_{ij}^{XY}/\hat{\sigma}(\hat{\lambda}_{ij}^{XY})$ may be quite different when λ_{ijk}^{XYZ} is in the model than when it is not. See Goodman (1970, 1971a) for other cautionary remarks on this approach.

Brown (1976) and Benedetti and Brown (1978) suggested a third strategy for preliminary fitting. They used two tests to screen the importance of each possible term. In one test the term is the most complex parameter in a simple model, whereas in the other test all parameters of its level of complexity are included. For instance, suppose we want to gauge whether λ^{XY} is needed in a model for variables X, Y, and Z. The first test is $G^2[(X, Y, Z)\,|\,(XY, Z)]$, which (Section 7.4.2 shows) tests the X–Y *marginal* association. The second test is $G^2[(XZ, YZ)\,|\,(XY, XZ, YZ)]$, which tests for the X–Y *partial* association. This strategy includes the term in a preliminary model when either test indicates it may be needed.

To illustrate these three strategies, we re-consider the death penalty data. First we study models having terms of uniform order. From Table 7.1, $G^2(D, V, P) = 137.73$ and $G^2(DP, VP, DV) = 0.70$. These suggest that a good-fitting model must be more complex than the mutual independence model, but may be simpler than the no three-factor interaction model. Second, we study standardized parameter estimates. Table 6.7 presented lambda estimates and their standard errors for the saturated model. Table 7.2 contains the standardized parameter estimates. These point toward model (P, DV). Since Table 7.2 suggests that the three-factor interaction term is not important, we could also inspect standardized estimates for model (DP, VP, DV). Table 7.2 contains these estimates, which point toward model (DV, VP). Finally, Brown's strategy also highlights the λ^{DV} and λ^{VP} terms. For instance, consider the λ^{VP} term. From Table 7.1, $G^2[(D, V, P)\,|\,(VP, D)] = 137.93 - 131.68 = 6.25$

Table 7.2 Standardized Parameter Estimates for
Models Fitted to Death Penalty Data[a]

| Term | Standardized Estimate | |
	(DVP)	(DV, DP, VP)
DV	4.07	8.79
VP	1.42	2.55
DP	-0.41	-1.10
DVP	-0.10	—

[a]Since $n_{121} = 0$, 0.5 was added to each cell count before
calculating estimates for model (DVP).

and $G^2[(DP, DV)\,|\,(DP, VP, DV)] = 7.91 - 0.70 = 7.21$, both based on
df $= 1$, for which P-values are near 0.01.

7.2.3 Stepwise Procedures

In exploratory studies, it can be informative to use an algorithmic search
method to help select a model. Goodman (1971a) proposed methods
analogous to forward selection and backward elimination procedures
employed in multiple regression analysis (Draper and Smith 1981, Chap.
6).

The forward selection process adds terms sequentially to the model
until further additions do not improve the fit. At each stage, we select the
term giving the greatest improvement in fit. The maximum P-value for
the resulting model is a sensible criterion, since reductions in G^2 for
different terms may have different degrees of freedom. A stepwise
variation of this procedure re-tests, at each stage, terms added at previous
stages to see if they are still needed.

The backward elimination process begins with a complex model and
sequentially removes terms. At each stage, we remove the term for which
there is the least damaging effect on the model. The process stops when
any further deletion leads to a significantly poorer-fitting model.

It is reasonable to start either selection procedure with a model having
terms of uniform order. For instance, for a four-way table, suppose $(W,
X, Y, Z)$ fits poorly but (WX, WY, WZ, XY, XZ, YZ) fits well. Then we
could begin forward selection with the former model or backward elimination with the latter. It is usually safer to delete terms from an
overspecified model than to add terms to an underspecified one. For
either approach, the sampling design may dictate including certain terms.

If we want to treat W and X as explanatory variables, we would begin forward selection with model (WX, Y, Z).

There is no guarantee that either strategy will lead to a meaningful model. Algorithmic selection procedures are no substitute for careful theory construction in guiding the formulation of models.

7.2.4 Pre/Extramarital Sex and Marital Status Example

Table 7.3 refers to a British study reported by Thornes and Collard (1979) and described by Gilbert (1981). A sample of men and women who had petitioned for divorce and a similar number of married people were asked (a) "Before you married your (former) husband/wife, had you ever made love with anyone else?", (b) "During your (former) marriage, (did you have) have you had any affairs or brief sexual encounters with another man/woman?" The table has size $2 \times 2 \times 2 \times 2$, with variables $G = $ gender, $E = $ whether reported extramarital sex, $P = $ whether reported premarital sex, and $M = $ marital status.

Table 7.4 displays the forward selection process for these data, starting with the mutual independence model (E, G, M, P). Since each variable has two levels, each possible term has a single degree of freedom. Thus, at each stage we add the term that gives the greatest decrease in G^2. Of models containing a single pairwise association, (GP, E, M) gives the best fit. Next we consider models having two pairwise associations, including the GP term. The best fitting model is (GP, EM). Similarly, the third stage of the process produces the model (GP, EM, EP), and the fourth stage produces (GP, EM, EP, MP). The next candidates for inclusion are the EG and GM associations, and the EMP three-factor interaction. The latter choice gives the best fit, and the resulting model (GP, EMP) is the

Table 7.3 Marital Status by Gender by Report of Pre- and Extramarital Sex (PMS and EMS)

		Gender							
		Women				Men			
	PMS:	Yes		No		Yes		No	
Marital Status	EMS:	Yes	No	Yes	No	Yes	No	Yes	No
Divorced		17	54	36	214	28	60	17	68
Still Married		4	25	4	322	11	42	4	130

Source: UK Marriage Research Centre; see Gilbert (1981). Reprinted with permission from Unwin Hyman Ltd.

Table 7.4 Forward Selection Process for Table 7.3

Stage	Model	G^2	df	Best Model
Initial	(E, G, M, P)	232.14	11	
1	(GM, E, P)	232.11	10	
	(EG, M, P)	219.38	10	
	(MP, E, G)	189.59	10	
	(EP, G, M)	186.12	10	
	(EM, G, P)	167.73	10	
	(GP, E, M)	156.88	10	*
2	(GP, GM, E)	156.85	9	
	(GP, EG, M)	144.12	9	
	(GP, MP, E)	114.33	9	
	(GP, EP, M)	110.86	9	
	(GP, EM)	92.47	9	*
3	(GP, EM, GM)	92.44	8	
	(GP, EM, EG)	79.71	8	
	(GP, EM, MP)	49.92	8	
	(GP, EM, EP)	46.46	8	*
4	(GP, EM, EP, GM)	46.36	7	
	(GP, EM, EP, EG)	43.55	7	
	(GP, EM, EP, MP)	21.07	7	*
5	(GP, EM, EP, MP, EG)	18.16	6	
	(GP, EM, EP, MP, GM)	18.13	6	
	(GP, EMP)	8.15	6	*
6	(GP, EG, EMP)	5.25	5	
	(GP, GM, EMP)	5.21	5	*
7	(GMP, EMP)	4.84	4	
	(GP, GM, EG, EMP)	0.76	4	*

first to give a decent fit. Two more steps produce model (GP, GM, EG, EMP), which gives an outstanding fit, and further refinement is unnecessary.

Considering the large number of models compared in this process, the evidence that (GP, GM, EG, EMP) is better than (GP, EMP) is not strong. We would not expect much association between G and M in the marginal table (i.e., similar proportions of men and women should be divorced), and there is only weak evidence that we need this partial association term for these variables.

Table 7.5 displays the backward elimination process, beginning with the model having all three-factor interaction terms. At each stage we delete the term for which the resulting increase in G^2 is smallest. The first stage gives (EGM, EMP, GMP), though two other models are nearly as good. Subsequent stages produce (EGM, EMP, GP), (GP, GM, EG, EMP), (GP, GM, EMP), and (GP, EMP). Further simplifications are not fruitful, and (GP, EMP) is the simplest model to fit decently. Like the forward selection process, the backward elimination process suggests that this model or (GP, GM, EG, EMP) is reasonable.

The EMP interaction seems vital to explaining relationships in Table 7.3. To describe it, we obtained estimated odds ratios for model (EMP, EG, GM, GP). Given gender, for those who reported pre-marital sex, the odds of a divorce are estimated to be 1.82 times higher for those who reported extra-marital sex than for those who did not; for those who did

Table 7.5 Backward Elimination Process for Table 7.3

Stage	Model	G^2	df	Best Model
Initial	(EMP, EGM, EGP, GMP)	0.15	1	
1	(EMP, EGM, GMP)	0.19	2	*
	(EMP, EGM, EGP)	0.29	2	
	(EMP, EGP, GMP)	0.44	2	
	(EGM, EGP, GMP)	10.33	2	
2	(GP, EMP, EGM)	0.37	3	*
	(EG, EMP, GMP)	0.46	3	
	(EP, EGM, GMP)	10.47	3	
3	(EG, GM, GP, EMP)	0.76	4	*
	(EP, GP, MP, EGM)	10.80	4	
	(EMP, EGM)	67.72	4	
4	(GM, GP, EMP)	5.21	5	*
	(EG, GP, EMP)	5.25	5	
	(EG, GM, GP, EM, EP, MP)	13.63	5	
	(EG, GM, EMP)	70.10	5	
5	(GP, EMP)	8.15	6	*
	(GM, GP, EM, EP, MP)	18.13	6	
	(GM, EMP)	83.38	6	
6	(GP, EM, EP, MP)	21.07	7	*
	(G, EMP)	83.41	7	

not report premarital sex, the odds of a divorce are estimated to be 10.94 times higher for those who reported extramarital sex than for those who did not. The effect of extramarital sex on divorce is greater for subjects who had no premarital sex. Given gender, the *MP* estimated odds ratio is 0.50 for those who reported extramarital sex, and 3.00 for those who did not report extramarital sex. Those who had premarital sex were less likely to be divorced in the first case, and more likely to be divorced in the second case. Given gender, the *EP* estimated odds ratio is 1.82 for divorced subjects and 10.95 for married subjects.

7.2.5 Using Causal Hypotheses to Guide Model Building

Though selection procedures are helpful exploratory tools, statisticians should try to relate the model-building process to theoretical constructs. Often there is a time ordering among the variables that suggests possible causal relationships. We can then analyze a certain sequence of models to investigate those relationships.

We use the previous example to illustrate. The points in time at which the four variables are determined suggests the ordering of the variables,

G	*P*	*E*	*M*
Gender	Premarital sex	Extra-marital sex	Marital status

We could regard any of these as an explanatory variable, when a variable listed to its right is the response variable. Figure 7.1 shows one possible causal structure. In this figure, a variable at the tip of an arrow is a response for a model at some stage. The explanatory variables have arrows pointing to the response, directly or indirectly.

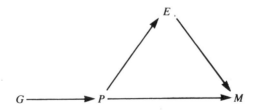

Figure 7.1 A Causal Diagram for Table 7.3.

We begin by treating P as a response. Figure 7.1 predicts that G has a direct effect on P, so that the model (G, P) of independence of these variables is inadequate. At the second stage, E is the response. Figure 7.1 predicts that P and G have direct effects on E. It also suggests that G has an indirect effect on E, through its effect on P. These effects on E can be analyzed using the logit model for E corresponding to loglinear model (GP, EP, EG). If G has only an indirect effect on E, then (GP, EP) is adequate; that is, controlling for P, E and G are conditionally independent. At the third stage, Figure 7.1 predicts that E has a direct effect on M, P has direct effects and indirect effects (through its effects on E), and G has indirect effects (through its effects on P and E). This suggests the logit model for M corresponding to loglinear model (EGP, EM, PM). For this model, G and M are independent, given P and E.

For the first stage of analysis, $G^2[(G, P)] = 75.26$, based on df $= 1$. The sample odds ratio for the G-P marginal table is 0.27; the estimated odds of premarital sex is 0.27 times as high for females as males.

For the second stage, with G and P explanatory and E the response, relevant models are logit models for E corresponding to loglinear models (GP, E), (GP, EP), and (GP, EP, EG). The G^2 values are 48.92 (df $= 3$), 2.91 (df $= 2$), and 0.00 (df $= 1$). For the second model, the estimated E–P partial odds ratio is 3.99, the same as the E–P marginal odds ratio. For each gender, the estimated odds of extra-marital sex are 3.99 times higher for those who had premarital sex than for those who did not. The third model gives weak evidence that G had a direct as well as an indirect effect on E.

The third stage regards E, G, and P as explanatory variables and M as the response variable. Figure 7.1 specifies model (EGP, EM, PM), which fits poorly, with $G^2 = 18.16$ based on df $= 5$. The lack of fit is not surprising, since the previous sub-section revealed an important interaction between E and P in their effects on M. The model (EGP, EMP) that allows this interaction but assumes conditional independence of G and M fits much better, with $G^2 = 5.25$ based on df $= 4$. The model (EGP, EMP, GM) that adds a main effect for G also fits well, with $G^2 = 0.70$ based on df $= 3$. Either model is more complicated than Figure 7.1 predicted, since the effects of E on M vary according to the level of P. However, some preliminary thought about causal relationships suggested a model similar to one that gives a good fit. We leave it to the reader to estimate and interpret effects in these models for the third stage.

This exposition just scratches the surface of types of analyses for investigating causal hypotheses using a sequence of models. For further discussion, see Goodman (1973a,b) and Fienberg (1980a, Chap. 7).

7.3 ANALYSIS OF RESIDUALS

Goodness-of-fit statistics provide only broad summaries of how models fit data. After selecting a preliminary model, we obtain further insight by switching to a microscopic mode of analysis. The pattern of lack of fit revealed in cell-by-cell comparisons of observed and fitted frequencies may suggest an alternative model that provides a better fit.

7.3.1 Standardized Residuals

For cell i in a contingency table, the simple residual $n_i - \hat{m}_i$ has limited usefulness, since a difference of fixed size is more important for smaller samples. For Poisson sampling, for instance, the standard deviation of n_i is $\sqrt{m_i}$, so we expect larger differences when m_i is larger. The *standardized residual*, defined for cell i by

$$e_i = \frac{n_i - \hat{m}_i}{\hat{m}_i^{1/2}} \tag{7.4}$$

adjusts for this. Standardized residuals are related to the Pearson statistic by $\sum e_i^2 = X^2$.

When the model holds, $\{e_i\}$ are asymptotically normal with mean 0. Like the logistic regression residuals (4.21), their asymptotic variances are less than 1.0, averaging (residual df)/(number of cells). For a complex model, the variance of e_i may be much less than 1.0. In comparing standardized residuals to standard normal percentage points (such as ± 2.0), we obtain conservative indications of cells having lack of fit.

7.3.2 Adjusted Residuals

The standardized residual divided by its estimated standard error is an alternative residual, one having an asymptotic standard normal distribution. This ratio, called an *adjusted residual*, was defined by Haberman (1973a). The general formula for the standard error of e_i is derived in Section 12.4.1.

The formula for the adjusted residual simplifies for many loglinear models having direct estimates. The independence model in a two-way table has adjusted residuals

$$r_{ij} = \frac{n_{ij} - \hat{m}_{ij}}{[\hat{m}_{ij}(1 - p_{i+})(1 - p_{+j})]^{1/2}} \tag{7.5}$$

where $\hat{m}_{ij} = np_{i+}p_{+j}$. Table 7.6 presents adjusted residuals for some models for three-way tables.

Suppose a loglinear model has fitted value of form

$$\hat{m} = n_{T(1)}\left(\frac{n_{T(2)} \cdots n_{T(b)}}{n_{V(2)} \cdots n_{V(b)}}\right)$$

where $\{n_{T(a)}\}$ and $\{n_{V(a)}\}$ are marginal totals of the table. Haberman (1978, p. 275) noted that the adjusted residual is $(n_i - \hat{m}_i)$ divided by the square root of

$$\hat{m}\left[1 - \hat{m}\left(\sum_{a=1}^{b} \frac{1}{n_{T(a)}} - \sum_{a=2}^{b} \frac{1}{n_{V(a)}}\right)\right]. \tag{7.6}$$

For instance, model (XY, Z) has $\hat{m}_{ijk} = n_{ij+}n_{++k}/n$, for which $b = 2$, $n_{T(1)} = n_{ij+}$, $n_{T(2)} = n_{++k}$, and $n_{V(2)} = n$. Thus, (7.6) equals

$$\hat{m}_{ijk}\left[1 - \frac{n_{ij+}n_{++k}}{n}\left(\frac{1}{n_{ij+}} + \frac{1}{n_{++k}} - \frac{1}{n}\right)\right]$$

which simplifies to the square of the denominator in the adjusted residual reported in Table 7.6.

7.3.3 Graduate Admissions Example

Table 7.7, taken from Bickel et al. (1975) and Freedman et al. (1978, p. 14), displays the effect of applicant's gender on whether admitted into

Table 7.6   Adjusted Residuals for Loglinear Models for Three-Way Tables

Model	Adjusted Residual[a]
(X, Y, Z)	$\dfrac{n_{ijk} - \hat{m}_{ijk}}{[\hat{m}_{ijk}(1 - p_{i++}p_{+j+} - p_{i++}p_{++k} - p_{+j+}p_{++k} + 2p_{i++}p_{+j+}p_{++k})]^{1/2}}$
(XY, Z)	$\dfrac{n_{ijk} - \hat{m}_{ijk}}{[\hat{m}_{ijk}(1 - p_{ij+})(1 - p_{++k})]^{1/2}}$
(XY, XZ)	$\dfrac{n_{ijk} - \hat{m}_{ijk}}{\left[\hat{m}_{ijk}\left(1 - \dfrac{n_{ij+}}{n_{i++}}\right)\left(1 - \dfrac{n_{i+k}}{n_{i++}}\right)\right]^{1/2}}$

[a] $\{\hat{m}_{ijk}\}$ have the formulas given in Table 6.2.

Table 7.7 Whether Admitted into Graduate School by Gender and Department[a]

Department	Whether admitted, male		Whether admitted, female	
	Yes	No	Yes	No
A	512 (531.4)	313 (293.6)	89 (69.6)	19 (38.4)
B	353 (354.2)	207 (205.8)	17 (15.8)	8 (9.2)
C	120 (114.0)	205 (211.0)	202 (208.0)	391 (385.0)
D	138 (141.6)	279 (275.4)	131 (127.7)	244 (247.6)
E	53 (48.1)	138 (142.9)	94 (98.9)	299 (294.1)
F	22 (24.0)	351 (349.0)	24 (22.0)	317 (319.0)
Total	1198	1493	557	1278

[a]Fitted Values for Model (*AD*, *DG*) in parentheses.

graduate school at the University of California at Berkeley, for the fall 1973 session. Applicants were classified by A = whether admitted, G = gender, and D = department, for the six largest graduate departments at Berkeley. Table 7.7 also reports fitted values for the loglinear model (AD, DG) of conditional independence of whether admitted and gender, given department. This model fits poorly, with $G^2 = 21.7$ based on df = 6. However, no standard unsaturated model fits these data well, the more complex model (AD, DG, AG) having $G^2 = 20.2$. An analysis of residuals explains this lack of fit.

Table 7.8 contains adjusted residuals for model (AD, DG). Let r_{ijk} denote the adjusted residual for level i of G, j of A, k of D. Each level of D has only a single nonredundant residual, because of marginal constraints for this model. For instance, since $n_{111} + n_{121} = n_{1+1} = \hat{m}_{1+1}$, it follows that $n_{111} - \hat{m}_{111} = -(n_{121} - \hat{m}_{121})$. Thus, standard errors of these differences are identical, and the adjusted residuals are identical in absolute value.

The fitted values in Table 7.7 and the adjusted residuals in Table 7.8 reveal that model (AD, DG) actually fits well for all departments except the first. This suggests the slightly more general model,

$$\log m_{ijk} = \mu + \lambda_i^G + \lambda_j^A + \lambda_k^D + \lambda_{ik}^{GD} + \lambda_{jk}^{AD} + \lambda_{ij}^{GA} I(k = 1) \quad (7.7)$$

where $I(\cdot)$ is the indicator function; that is, $I(k = 1)$ equals 1 when $k = 1$ (Department A) and equals 0 otherwise. This model assumes conditional independence between whether admitted and gender for all departments except the first. It gives a perfect fit for Department A, and the same fit as the conditional independence model for the other departments. The model has residual $G^2 = 2.7$, based on df = 5. This is identical to the sum of five components, where the kth is G^2 for testing independence between whether admitted and gender for Department $k + 1$.

Table 7.8   Adjusted Residuals for Table 7.7

Department	Whether admitted, male		Whether admitted, female	
	Yes	No	Yes	No
A	−4.15	4.15	4.15	−4.15
B	−0.50	0.50	0.50	−0.50
C	0.87	−0.87	−0.87	0.87
D	−0.55	0.55	0.55	−0.55
E	1.00	−1.00	−1.00	1.00
F	−0.62	0.62	0.62	−0.62

Though there is strong evidence of three-factor interaction, much simplification in interpretation is possible. Whether admitted seems independent of gender for all departments except the first. For that department, relatively more women were admitted than men. Incidentally, the marginal $A-G$ table gives the misleading indication that women were admitted *less* frequently than men (refer back to Problems 6.7 and 6.8). This happens because women applied relatively more frequently than men to departments having low admissions rates.

7.3.4 Cell Deletion Diagnostics

Fitting model (7.7) corresponds to deleting the data for the first department and fitting the conditional independence model to the remaining data. Whenever a residual indicates that a model fits a cell poorly, it can be informative to delete the cell and re-fit the model to remaining cells. This is equivalent to adding a parameter for the cell to the loglinear model, forcing a perfect fit in that cell. For instance, suppose we add the term $\delta I(a)$ to a model, where $I(a) = 1$ in cell a and equals 0 otherwise. We obtain the additional likelihood equation $\hat{m}_a = n_a$, implying a perfect fit in cell a, and the residual df reduces by 1. We can fit the resulting model to the other cells using IPF by replacing the entry in cell a by 0, and using a starting value of 0 in cell a in the IPF process.

In summary, studying residuals helps us understand either why a model fits poorly, or where there is lack of fit in a generally good-fitting model. When there are many cells in the table, we should realize that some residuals may be large purely by chance. Other diagnostic tools useful in regression modeling can also be helpful in assessing the fit. These include plots of ordered residuals against normal percentiles (Haberman 1973a), and other influence analyses, such as deleting single observations and noting effects on parameter estimates and goodness of fit statistics.

7.4 TESTING CONDITIONAL INDEPENDENCE

The hypothesis of conditional independence is an important one. For instance, many epidemiological studies investigate whether an association exists between a risk factor and a disease. They analyze whether an observed association persists when other factors that might influence that association are controlled. This involves testing conditional independence of the disease and the risk factor, controlling for the other factors. This section presents three approaches to testing conditional independence for categorical data: Tests based directly on loglinear models, a non-model-based test introduced by Mantel and Haenszel (1959), and an exact test.

7.4.1 Using Models to Test Independence

There are two ways of using loglinear models to test conditional independence. The loglinear model (XZ, YZ) represents conditional independence of X and Y. The goodness-of-fit test of that model gives a direct test of the hypothesis. An alternative test is given by $G^2[(XZ, YZ)|(XY, XZ, YZ)]$, which is the test of model (XZ, YZ) under the assumption that model (XY, XZ, YZ) holds. We now contrast these two tests.

The test using $G^2[(XZ, YZ)|(XY, XZ, YZ)]$ assumes that the model

$$\log m_{ijk} = \mu + \lambda_i^X + \lambda_j^Y + \lambda_k^Z + \lambda_{ij}^{XY} + \lambda_{ik}^{XZ} + \lambda_{jk}^{YZ}$$

of no three-factor interaction holds. It is a test of H_0: $\lambda_{ij}^{XY} = 0$ for all i and j, for this model. On the other hand, the statistic $G^2(XZ, YZ)$ makes no such assumption. It trivially has the conditional form $G^2[(XZ, YZ)|(XYZ)] = G^2(XZ, YZ) - G^2(XYZ)$. In other words, the null hypothesis for the direct test is H_0: $\lambda_{ij}^{XY} = 0$ and $\lambda_{ijk}^{XYZ} = 0$ for all i, j, and k, in the saturated model. So, this test statistic tests both for an absence of three-factor interaction and for conditional independence. The statistic could be large if there is three-factor interaction, or if there is no three-factor interaction but conditional dependence.

Suppose Y is a binary response. The statistic $G^2[(XZ, YZ)|(XY, XZ, YZ)]$ assumes logit model

$$\log(m_{i1k}/m_{i2k}) = \alpha + \beta_i^X + \beta_k^Z$$

holds, and tests H_0: all $\beta_i^X = 0$. By contrast, $G^2(XZ, YZ)$ tests H_0: all β_i^X and $\beta_{ik}^{XZ} = 0$ in the saturated logit model.

When there is no three-factor interaction, both tests have the same noncentrality. From results in Sections 4.4.1 and 4.4.2, $G^2[(XZ, YZ)|(XY, XZ, YZ)]$ is then more powerful than $G^2(XZ, YZ)$, since the former statistic has fewer degrees of freedom, $(I-1)(J-1)$ vs. $K(I-1)(J-1)$. When there is interaction of minor substantive importance, the conditional test is still likely to be more powerful, especially when K is large. However, when the direction of the $X-Y$ association varies across the levels of Z, $G^2(XZ, YZ)$ tends to be more powerful. When there is *any* association in *any* of the population partial tables, this test will detect it, given enough data.

7.4.2 Identical Marginal and Partial Tests of Independence

We refer to a test of form $G^2(M_2 \mid M_1)$ as *model-based*, since it tests whether certain parameters in a model equal zero. The test simplifies dramatically when both models have direct estimates. For instance, consider $G^2[(X, Y, Z) \mid (XY, Z)]$. This statistic, which tests $\lambda^{XY} = 0$ in model (XY, Z), is a test of X–Y conditional independence under the assumption that X and Y are jointly independent of Z. From (7.2), it equals

$$2 \sum \sum \sum \frac{n_{ij+}n_{++k}}{n} \log\left(\frac{n_{ij+}n_{++k}/n}{n_{i++}n_{+j+}n_{++k}/n^2} \right)$$

$$= 2 \sum \sum n_{ij+} \log\left(\frac{n_{ij+}}{n_{i++}n_{+j+}/n} \right).$$

This is identical to $G^2[(X, Y)]$ for testing independence in the marginal X–Y table. This is not surprising. The collapsibility conditions in Section 5.3.3 imply that, for model (XY, Z), the marginal X–Y association is the same as the partial X–Y association.

We now give a sufficient condition for a test of conditional independence to give the same result as the test of independence applied to the marginal table. The key condition is that *direct* estimates exist for both models, which implies the models have independence linkages necessary to ensure collapsibility.

> When two direct models are identical except for an association term for a pair of variables, the difference in G^2 statistics for the models is identical to the G^2 statistic applied to test independence in the marginal table for that pair of variables.

Sundberg (1975) proved this condition. For related discussion, see Bishop (1971) and Goodman (1970, 1971b).

Models (X, Y, Z) and (XY, Z) are direct and are identical except for the X–Y association term, so the result just observed follows from this condition. However, the test $G^2[(XZ, YZ) \mid (XY, XZ, YZ)]$ of X–Y conditional independence does not simplify to $G^2[(X, Y)]$ for the marginal table. Direct estimates do not exist for model (XY, XZ, YZ), and the X–Y partial association is not the same as the X–Y marginal association.

7.4.3 Cochran–Mantel–Haenszel Test

Mantel and Haenszel (1959) gave a test comparing two groups on a binary response, adjusting for control variables. This test of conditional independence applies to K strata of 2×2 tables. Focusing on retrospective studies of disease, they treated response totals as fixed. They analyzed the data by conditioning, within each stratum, on both the group totals and the response totals. In stratum k, given the marginal totals $\{n_{+1k}, n_{+2k}, n_{1+k}, n_{2+k}\}$, the sampling model for cell counts is the hypergeometric (Section 3.5). We can express that distribution in terms of count n_{11k} alone, since it determines $\{n_{12k}, n_{21k}, n_{22k}\}$, given the marginal totals.

Assuming conditional independence, the hypergeometric mean and variance of n_{11k} are

$$m_{11k} = E(n_{11k}) = n_{1+k}n_{+1k}/n_{++k}$$
$$V(n_{11k}) = n_{1+k}n_{2+k}n_{+1k}n_{+2k}/n_{++k}^2(n_{++k} - 1).$$

Given the strata totals, cell counts from different strata are independent. Thus, $\Sigma\, n_{11k}$ has mean $\Sigma\, m_{11k}$ and variance $\Sigma\, V(n_{11k})$.

Mantel and Haenszel proposed the summary statistic

$$M^2 = \frac{\left(\left|\Sigma\, n_{11k} - \Sigma\, m_{11k}\right| - \frac{1}{2}\right)^2}{\Sigma\, V(n_{11k})}. \tag{7.8}$$

Under the null hypothesis of conditional independence, this statistic has approximately a chi-squared distribution with df $= 1$. Cochran (1954) proposed a similar statistic. He did not use the continuity correction ($\frac{1}{2}$) and conditioned only on the group totals in each stratum, treating each 2×2 table as two independent binomials. Cochran's statistic is (7.8) with the $\frac{1}{2}$ deleted and with $V(n_{11k})$ replaced by

$$\hat{V}(n_{11k}) = n_{1+k}n_{2+k}n_{+1k}n_{+2k}/n_{++k}^3.$$

Because of the basic similarity in their approaches, we call (7.8) the *Cochran–Mantel–Haenszel* statistic.

Formula (7.8) indicates that M^2 is larger when $n_{11k} - m_{11k}$ is consistently positive or consistently negative for all strata, rather than positive for some and negative for others. This test is inappropriate when the association changes dramatically across the strata. The test is analogous to model-based test $G^2[(XZ, YZ)|(XY, XZ, YZ)]$. Both are most appropriate when there is no three-factor interaction. For $2 \times 2 \times K$ tables, the model-based statistic also is a chi-squared statistic with df $= 1$. If the sample sizes in the strata are moderately large, the two statistics usually take similar values.

An advantage of M^2 is its applicability for sparse data sets for which asymptotic theory does not hold for $G^2[(XZ, YZ)\,|\,(XY, XZ, YZ)]$. For instance, suppose each stratum has a single matched pair of subjects, one of whom is in each group. Then $n_{1+k} = n_{2+k} = 1$, for each k. When both subjects in stratum k make the same response, $n_{+1k} = 0$ or $n_{+2k} = 0$. Given the marginal counts in stratum k, the internal counts are then completely determined, and $m_{11k} = n_{11k}$ and $V(n_{11k}) = 0$. When the subjects make differing responses, $n_{+1k} = n_{+2k} = 1$, so that $m_{11k} = 0.5$ and $V(n_{11k}) = 0.25$. Thus, a matched pair contributes to M^2 only when the subjects' responses differ. The distribution of M^2 is approximately chi-squared when the total number of pairs of subjects with differing responses is relatively large. This can happen when there are few observations in each strata, as long as the number of strata is large. Section 7.7 shows that an approximate chi-squared distribution holds for $G^2[(XZ, YZ)\,|\,(XY, XZ, YZ)]$ only when every set of two-way marginal totals is fairly large. The X–Z and Y–Z marginal totals are the strata marginal totals, which should mostly exceed about 5–10 for the asymptotic theory to hold. Thus, the model-based test requires many observations in each stratum.

7.4.4 Exact Test of Conditional Independence

For $2 \times 2 \times K$ tables, the Cochran-Mantel-Haenszel test is based on $\Sigma_k\, n_{11k}$. When there is no three-factor interaction (i.e., the conditional X–Y odds ratios satisfy $\theta_{11(1)} = \cdots = \theta_{11(K)}$), Birch (1964b) showed that uniformly most powerful unbiased tests of conditional independence are those based on this sum.

Conditional on the strata margins, exact tests can be based on $\Sigma_k\, n_{11k}$, just as they use n_{11} in Fisher's exact test for 2×2 tables. The exact test uses hypergeometric probabilities in each stratum to determine the independent null distributions of $\{n_{11k}, k = 1, \ldots, K\}$. The product of the K hypergeometric mass functions gives the null joint distribution of $\{n_{11k}, k = 1, \ldots, K\}$. This determines the null distribution of their sum. For the one-sided alternative $\theta_{11(k)} > 1$, the P-value is the null probability that $\Sigma_k\, n_{11k}$ is at least as large as observed, for the fixed strata marginal totals. When the sample size is small, this test is preferred over the Cochran–Mantel–Haenszel test or the model-based test. Mehta et al. (1985) presented a fast algorithm for executing this exact test.

7.4.5 Penicillin and Rabbits Example

Table 7.9, taken from Mantel (1963), refers to the effectiveness of immediately injected or $1\frac{1}{2}$-hour-delayed penicillin in protecting rabbits

Table 7.9 Example of Cochran–Mantel–Haenszel Test

Penicillin Level	Delay	Response Cured	Died	m_{11k}	$V(n_{11k})$
1/8	None	0	6	0	0
	$1\frac{1}{2}$ h	0	5		
1/4	None	3	3	1.5	27/44
	$1\frac{1}{2}$ h	0	6		
1/2	None	6	0	4.0	32/44
	$1\frac{1}{2}$ h	2	4		
1	None	5	1	5.5	11/44
	$1\frac{1}{2}$ h	6	0		
4	None	2	0	2	0
	$1\frac{1}{2}$ h	5	0		

Source: Reprinted with permission from Mantel (1963).

against lethal injection with β-hemolytic streptococci. The table contains the expected value and variance of the number cured when there is immediate injection, under the hypothesis of conditional independence at each penicillin level. These are summarized by $\Sigma\, n_{11k} = 16$, $\Sigma\, m_{11k} = 13$, and $\Sigma\, V(n_{11k}) = 70/44$. The Cochran–Mantel–Haenszel statistic is $M^2 = 3.93$, with df $= 1$, which gives a P-value of 0.047. There is some evidence of a higher cure rate for immediate injection.

Let P = penicillin level, D = delay, and C = whether cured. The ML estimates for loglinear models (PD, PC) and (CD, PD, PC) do not exist, because of zeroes in the P–C marginal table. The estimates exist when all cell counts are positive. Table 7.10 shows the effect of adding constants of

Table 7.10 Results of Fitting Models to Table 7.9

Constant Added to Cells	$G^2(PD, PC)$	$G^2(CD, PD, PC)$	Common log $\hat{\theta}$	ASE
10^{-8}	14.29	7.49	2.55	1.18
10^{-6}	14.29	7.49	2.55	1.18
10^{-4}	14.29	7.49	2.55	1.18
0.01	13.99	7.31	2.50	1.15
0.1	12.56	6.85	2.11	1.00
0.2	11.57	6.66	1.80	0.90
0.3	10.83	6.53	1.58	0.82
0.4	10.22	6.41	1.41	0.77
0.5	9.71	6.29	1.27	0.72

various sizes to the cells. Results are stable as long as a very small constant is added. This results in $G^2[(PD, PC)|(CD, PD, PC)] = 14.29 - 7.49 = 6.80$, based on df $= 1$. We also calculated $X^2[(PD, PC)|(CD, PD, PC)]$ (not the *difference* in Pearson statistics, which is negative here, but statistic 7.3), which equals 2.86. Since the data are so sparse, these model-based tests are not reliable.

The exact test of conditional independence has a *P*-value of 0.020 for the one-sided alternative of a higher cure rate for immediate injection, and 0.040 for a two-sided alternative. The exact result is close to that given by the Cochran–Mantel–Haenszel test.

7.4.6 Generalized Cochran–Mantel–Haenszel Tests for $I \times J \times K$ Tables

Birch (1965), Landis et al. (1978), and Mantel and Byar (1978) generalized the Cochran–Mantel–Haenszel statistic to handle more than two groups or more than two responses. Suppose there are I groups and a J-category response variable, with observations at each of K strata. The test given here treats rows and columns as unordered. Section 8.4 presents analogous tests for ordered classifications.

Conditional on the row and column totals in each stratum, there are $(I - 1)(J - 1)$ nonredundant cell counts in each stratum. Let

$$\mathbf{n}_k = (n_{11k}, n_{12k}, \ldots, n_{1,J-1,k}, \ldots, n_{I-1,J-1,k})'$$

denote the counts for cells in the first $I - 1$ rows and $J - 1$ columns for stratum k. Let $\mathbf{m}_k$ denote the expected values under the hypothesis of conditional independence, namely

$$\mathbf{m}_k = (n_{1+k}n_{+1k}, n_{1+k}n_{+2k}, \ldots, n_{I-1,+,k}n_{+,J-1,k})'/n_{++k}.$$

Let $\mathbf{V}_k$ denote the null covariance matrix of $\mathbf{n}_k$, where

$$\mathrm{Cov}(n_{ijk}, n_{i'j'k}) = \frac{n_{i+k}(\delta_{ii'}n_{++k} - n_{i'+k})n_{+jk}(\delta_{jj'}n_{++k} - n_{+j'k})}{n_{++k}^2(n_{++k} - 1)}$$

with $\delta_{ab} = 1$ when $a = b$ and $\delta_{ab} = 0$ otherwise.

Summing over the K strata, we let

$$\mathbf{n} = \sum \mathbf{n}_k, \quad \mathbf{m} = \sum \mathbf{m}_k, \quad \mathbf{V} = \sum \mathbf{V}_k.$$

The generalized Cochran–Mantel–Haenszel statistic is

$$M^2 = (\mathbf{n} - \mathbf{m})'\mathbf{V}^{-1}(\mathbf{n} - \mathbf{m}) . \tag{7.9}$$

Under conditional independence, this statistic has a large-sample chi-squared distribution with df $= (I - 1)(J - 1)$. The df value equals that for the statistic $G^2[(XZ, YZ) | (XY, XZ, YZ)]$. Both statistics are sensitive to detecting conditional associations when the association is similar in each stratum. The M^2 statistic is preferred when the sample size is small for each stratum. For $K = 1$ stratum with n observations, M^2 simplifies to the multiple $(n - 1)/n$ of the Pearson chi-squared statistic for testing independence.

7.5 ESTIMATING AND COMPARING CONDITIONAL ASSOCIATIONS

When we reject the hypothesis of conditional independence, we should estimate the strength of association. If the association seems to be stable across the strata, we can estimate an assumed common value of the K odds ratios.

7.5.1 Estimation of Common Odds Ratios

In a $2 \times 2 \times K$ table, suppose $\theta_{11(1)} = \cdots = \theta_{11(K)}$. One estimate of the common odds ratio uses the ML estimator $\hat{\lambda}^{XY}$ of λ^{XY} for loglinear model (XY, XZ, YZ). For zero-sum parameter constraints, this estimator is $\exp(4\hat{\lambda}^{XY})$. The same estimator results from fitting the logit model for Y,

$$\log\left(\frac{\pi_{1|ik}}{\pi_{2|ik}}\right) = \alpha + \beta_i^X + \beta_k^Z$$

and calculating $\exp(\hat{\beta}_1^X - \hat{\beta}_2^X)$. When the number of strata is large and the data are sparse, this estimator tends to overestimate the degree of association. Standard asymptotic properties of unconditional ML estimators do not hold when the number of parameters grows at the same rate as the sample size. In the extreme case in which each stratum consists of only a single matched pair, the unconditional estimator converges to the square of the true common odds ratio (see Breslow and Day 1980, p. 250).

Woolf (1955) estimated the common odds ratio using the antilog of a weighted average of $\{\log(n_{11k}n_{22k}/n_{12k}n_{21k}), k = 1, \ldots, K\}$. Mantel and Haenszel (1959) proposed the estimator

$$\hat{\theta}_{MH} = \frac{\sum\limits_{k} (n_{11k}n_{22k}/n_{++k})}{\sum\limits_{k} (n_{12k}n_{21k}/n_{++k})}.$$

Compared to the model-based estimator and Woolf's estimator, it is less affected by strata with small sample sizes. Robins et al. (1986) gave an estimated variance for $\log(\hat{\theta}_{MH})$ of

$$\hat{\sigma}^2[\log(\hat{\theta}_{MH})] = \frac{\sum (n_{11k} + n_{22k})(n_{11k}n_{22k})/n_{++k}^2}{2\left(\sum n_{11k}n_{22k}/n_{++k}\right)^2}$$

$$+ \frac{\sum [(n_{11k} + n_{22k})(n_{12k}n_{21k}) + (n_{12k} + n_{21k})(n_{11k}n_{22k})]/n_{++k}^2}{2\left(\sum n_{11k}n_{22k}/n_{++k}\right)\left(\sum n_{12k}n_{21k}/n_{++k}\right)}$$

$$+ \frac{\sum (n_{12k} + n_{21k})(n_{12k}n_{21k})/n_{++k}^2}{2\left(\sum n_{12k}n_{21k}/n_{++k}\right)^2}.$$

Assuming no three-factor interaction and conditioning on the strata totals, the distribution of the data depends only on the common odds ratio. Birch (1964b) proposed a conditional ML estimator of the common odds ratio, based on maximizing the hypergeometric likelihood. This approach is computationally more complex. Mehta et al. (1985) provided a fast network algorithm that utilizes the exact conditional distribution to obtain an exact confidence interval for the common odds ratio.

In a recent survey article, Hauck (1989) recommended using the conditional ML estimator or the Mantel-Haenszel estimator of the common odds ratio. These estimators have good asymptotic properties for both asymptotic cases: (1) The number of strata is fixed, and the sample size within each stratum becomes large; (2) The stratum sizes are fixed, but the number of strata becomes large. The unconditional ML and Woolf estimators do not behave well in the second case.

For Table 7.9 on penicillin and rabbits, the Mantel-Haenszel estimate of a common log odds ratio is $\log \hat{\theta}_{MH} = \log(7.0) = 1.95$, with estimated standard error 0.98. An approximate 95% confidence interval for the common odds ratio is $\exp(1.95 \pm 1.96 \times 0.98)$, or $(1.03, 47.85)$. Though there is evidence of an effect, the estimate of the strength of association is very imprecise. Birch's conditional ML estimate of the common odds ratio is 10.36. The unconditional ML estimator does not exist. Table 7.10 shows that when we add a very small constant to the cells, that estimate exists and is stable, equaling $\exp(2.55) = 12.8$.

7.5.2 Multiple Response and Explanatory Variables

Some studies distinguish between response and explanatory variables, yet have more than one response variable. Section 7.2.5 showed one such example, in which a response variable for one model serves as an explanatory variable in other models.

Some studies treat two variables simultaneously as responses. For such data, we could use a separate logit model to describe effects of explanatory variables on each response. When X and Y are binary, we could also directly model the conditional X-Y log odds ratio in terms of the explanatory variables. For instance, consider Table 7.11, taken from Ashford and Sowden (1970). The data refer to a sample of coalminers, measured on B = breathlessness, W = wheeze, and A = age. B and W are binary response variables. To study whether the B-W association varies across levels of age, we fit the model (BW, AB, AW). It has residual $G^2 = 26.69$, based on df = 8. Table 7.11 reports the adjusted residuals.

The fit is not so poor, considering the large size of the sample, but the adjusted residuals show a decreasing tendency as age increases. This suggests the model

Table 7.11 Coalminers Classified by Breathlessness, Wheeze, and Age

| | Breathlessness | | | | |
| | Yes | | No | | |
Age	Wheeze Yes	Wheeze No	Wheeze Yes	Wheeze No	Adjusted Residual[a]
20–24	9	7	95	1841	0.75
25–29	23	9	105	1654	2.20
30–34	54	19	177	1863	2.10
35–39	121	48	257	2357	1.77
40–44	169	54	273	1778	1.13
45–49	269	88	324	1712	−0.42
50–54	404	117	245	1324	0.81
55–59	406	152	225	967	−3.65
60–64	372	106	132	526	−1.44

Source: Reprinted with permission from Ashford and Sowden (1970).
[a]Residual refers to yes-yes and no-no cells; reverse sign for yes-no and no-yes cells.

$$\log m_{ijk} = (BW, AB, AW) + kI(i = j = 1)\delta \qquad (7.10)$$

where I is the indicator function; that is, we amend the no-three-factor interaction model by adding δ in the cell for m_{111}, 2δ in the cell for $m_{112}, \ldots, K\delta$ in the cell for m_{11K}. In this model, the B–W log odds ratio changes linearly across the levels of age. The ML estimate of δ is -0.131 (ASE $= 0.029$), and the estimated B–W log odds ratio at level k of age is $3.676 - 0.131k$. The model has residual $G^2 = 6.80$, based on df $= 7$. It recognizes the ordinal nature of age, and is a special case of a *uniform interaction* model discussed in Section 8.3.3.

7.5.3 Testing Homogeneity of Odds Ratios

The estimators in Section 7.5.1 of a common odds ratio θ in K strata assume no three-factor interaction. To test this assumption, the unconditional ML approach is based on the fit of loglinear model (XY, XZ, YZ). Breslow and Day (1980, p. 142) proposed an alternative large-sample test. For $2 \times 2 \times K$ tables, both these tests have asymptotic chi-squared distributions with df $= K - 1$.

Zelen (1971) presented an exact test of homogeneity of the odds ratios. The exact distribution, which is free of θ under the null hypothesis, results from conditioning on n_{11+} as well as the strata marginal totals. In other words, we condition on all two-factor marginal tables, under the assumption of a common odds ratio. All $2 \times 2 \times K$ tables having the observed two-factor margins are ordered according to their exact conditional probabilities. The P-value is the sum of probabilities of all $2 \times 2 \times K$ tables that are no more probable than the observed table. Zelen's exact test requires a computer program for practical implementation, but Mehta et al. (1988) gave a fast network algorithm for executing it. For Table 7.9 on penicillin and rabbits, the exact test has a P-value of 0.074.

These tests treat the strata as levels of a nominal variable. When the strata are ordered, the directed alternative of a monotone trend in the odds ratios is often of interest. We could compare (XY, XZ, YZ) to model (7.10) (i.e., test $\delta = 0$ in (7.10)) to test homogeneity of odds ratios against the alternative that the log odds ratios change linearly across the strata. For such an alternative, Zelen (1971) gave an exact test and Breslow and Day (1980, p. 142) gave an asymptotic chi-squared statistic that applies when there are many strata but few observations per strata.

Table 7.11 has large counts, and asymptotic approaches are valid. From the fit of (BW, AB, AW), it is clear that we can reject the hypothesis of common odds ratio. Comparing to model (7.10), the ratio $z = \hat{\delta}/\text{ASE} = -0.131/0.029 = -4.5$ and the difference $26.69 - 6.80 =$

19.89 in G^2 statistics give strong evidence against the null hypothesis, relative to the alternative that the association decreases with age.

7.6   SAMPLE SIZE AND POWER CONSIDERATIONS

7.6.1   Effects of Sample Size

In any statistical procedure, the sample size can strongly influence the results. When there is a strong effect, we are likely to detect it even when the sample size is small. By contrast, detection of weak effects requires larger samples. As a consequence, to pass goodness-of-fit tests, larger samples usually require more complex models.

These facts suggest some cautionary remarks. For small data sets, the true picture may be more complex than indicated by the most parsimonious model accepted in a goodness-of-fit test. By contrast, for very large sample sizes, some effects that are statistically significant may be weak and substantively unimportant. It may then be adequate to use a model that is simpler than the simplest model that passes a goodness-of-fit test. An analysis that focuses solely on goodness-of-fit tests is incomplete. It is also necessary to estimate model parameters and describe strengths of effects.

These remarks simply reflect limitations of statistical hypothesis testing. Sharp null hypotheses are rarely true in practice, and with large enough samples they will be rejected. A more relevant concern is whether the difference between true parameter values and null hypothesis values is sufficient to be important.

Most statisticians believe that users of inferential methods overemphasize hypothesis testing, and underutilize estimation procedures such as confidence intervals. When we reject a null hypothesis, a confidence interval specifies the extent to which H_0 may be false, thus helping us determine whether its rejection is of practical importance. When we do not reject H_0, the confidence interval indicates whether there are plausible parameter values very far from H_0. A wide confidence interval containing the null hypothesis value indicates that the test had weak power at important alternatives.

7.6.2   Power for Comparing Proportions

It can be misleading to use statistical tests without studying their power. When we fail to reject a null hypothesis, we should be wary of concluding that no effect exists, unless there is high power for detecting effects of substantive size.

Many statistics, such as model parameter estimators, have large-sample normal distributions. For them, we can calculate power using ordinary methods for normal test statistics. To illustrate, suppose we want to test whether two medical treatments differ in their effectiveness in treating some disease. We plan an experiment with n subjects, independent samples of size $n/2$ receiving each treatment. A success rate of about 60% is expected for each treatment, with a difference of at least 6% between the treatments considered important.

For treatments A and B, let π_A and π_B denote the probabilities of successful response. We can test for equality of π_A and π_B with a test statistic given by the difference $p_A - p_B$ in sample proportions divided by an estimated standard error. The variance of $p_A - p_B$ is $\pi_A(1 - \pi_A)/(n/2) + \pi_B(1 - \pi_B)/(n/2) \doteq 0.6 \times 0.4 \times (4/n) = 0.96/n$. In particular,

$$z = \frac{(p_A - p_B) - (\pi_A - \pi_B)}{(0.96/n)^{1/2}}$$

has approximately a standard normal distribution for π_A and π_B near 0.6.

When the true difference is $\pi_A - \pi_B = 0.06$, the power of an α-level test for H_0: $\pi_A = \pi_B$ is approximately

$$P\left[\frac{|p_A - p_B|}{(0.96/n)^{1/2}} \geq z_{\alpha/2}\right].$$

For significance level $\alpha = 0.05$, $z_{0.025} = 1.96$, and this equals

$$P\left[\frac{(p_A - p_B) - 0.06}{(0.96/n)^{1/2}} > 1.96 - 0.06(n/0.96)^{1/2}\right]$$

$$+ P\left[\frac{(p_A - p_B) - 0.06}{(0.96/n)^{1/2}} < -1.96 - 0.06(n/0.96)^{1/2}\right]$$

$$= P[z > 1.96 - 0.06(n/0.96)^{1/2}]$$

$$+ P[z < -1.96 - 0.06(n/0.96)^{1/2}]$$

$$= 1 - \Phi[1.96 - 0.06(n/0.96)^{1/2}]$$

$$+ \Phi[-1.96 - 0.06(n/0.96)^{1/2}]$$

where Φ is the standard normal cdf. The power equals approximately 0.07 when the total sample size is 50, and 0.14 when it is 200. It is difficult to attain significance when effects are small and the sample is not very large.

7.6.3   Power for Chi-Squared Tests

When hypotheses are false, the test statistics X^2 and G^2 have approximate large-sample noncentral chi-squared distributions (Section 4.4.1). Suppose we test a hypothesis that is equivalent to model M. Let π_i denote the true probability in cell i, and let $\pi_i(M)$ denote the value to which the ML estimate based on model M converges, where $\Sigma \pi_i = \Sigma \pi_i(M) = 1$. For a multinomial sample of size n, the noncentrality parameter for X^2 equals

$$\lambda = n \sum \frac{[\pi_i - \pi_i(M)]^2}{\pi_i(M)} . \tag{7.11}$$

This has the same form as the Pearson statistic, with π_i in place of the sample proportion p_i and $\pi_i(M)$ in place of the ML estimate $\hat{\pi}_i$. The noncentrality parameter for G^2 equals

$$\lambda = 2n \sum \pi_i \log\left(\frac{\pi_i}{\pi_i(M)} \right) . \tag{7.12}$$

When the hypothesis is true, all $\pi_i = \pi_i(M)$. Then, for either statistic, $\lambda = 0$ and the ordinary (central) chi-squared distribution applies.

The approximate power for a chi-squared test with df $= \nu$ and significance level α is determined by (1) choosing a hypothetical set of true values $\{\pi_i\}$, (2) calculating $\{\pi_i(M)\}$ by fitting to $\{\pi_i\}$ the model M for the null hypothesis, (3) calculating the noncentrality parameter λ, and (4) using a table to determine $P[X^2_{\nu,\lambda} > \chi^2_\nu(\alpha)]$. Haynam et al. (1970) reported power values for λ between 0 and 100 and df between 1 and 100. Table 7.12 shows an excerpt from their tables, for $\alpha = 0.05$.

To illustrate, we could use a chi-squared statistic to compare the medical treatments in the example just discussed, by testing whether response is independent of treatment. Suppose the success probability is 0.63 for treatment A and 0.57 for treatment B. When sample sizes are equal for the two treatments, the joint probabilities in the 2×2 table are 0.315 and 0.185 in the row for treatment A, and 0.285 and 0.215 in the row for treatment B. For the model of independence, the fitted values are 0.30 for success and 0.20 for failure, in each row. The Pearson statistic has noncentrality parameter

$$n[(0.315 - 0.3)^2/0.3 + (0.185 - 0.2)^2/0.2 + (0.285 - 0.3)^2/0.3$$
$$+ (0.215 - 0.2)^2/0.2] = 0.00375n$$

Table 7.12 Power of Chi-Squared Test, for $\alpha = 0.05$

df	0.0	0.2	0.4	0.6	0.8	1.0	1.4	2.0	3.0	4.0	5.0	6.0	7.0	8.0	9.0	10.0	12.0	15.0	18.0	21.0	25.0	30.0
												Noncentrality										
1	0.050	0.073	0.097	0.121	0.146	0.170	0.220	0.293	0.410	0.516	0.609	0.688	0.754	0.807	0.851	0.885	0.934	0.972	0.989	0.996	0.998	1.000
2	0.050	0.065	0.081	0.098	0.115	0.133	0.169	0.226	0.322	0.415	0.504	0.584	0.655	0.718	0.771	0.815	0.883	0.944	0.974	0.989	0.996	0.999
3	0.050	0.062	0.075	0.088	0.102	0.116	0.145	0.192	0.275	0.358	0.440	0.518	0.590	0.654	0.711	0.761	0.840	0.917	0.959	0.981	0.993	0.998
4	0.050	0.060	0.071	0.082	0.093	0.106	0.131	0.172	0.244	0.320	0.396	0.470	0.540	0.605	0.664	0.716	0.802	0.891	0.943	0.972	0.989	0.997
6	0.050	0.058	0.066	0.075	0.084	0.094	0.114	0.146	0.206	0.270	0.336	0.403	0.468	0.531	0.590	0.644	0.738	0.843	0.911	0.952	0.980	0.994
8	0.050	0.057	0.064	0.071	0.079	0.087	0.104	0.131	0.182	0.238	0.296	0.357	0.417	0.477	0.534	0.588	0.685	0.799	0.879	0.930	0.968	0.989
10	0.050	0.056	0.062	0.068	0.075	0.082	0.097	0.121	0.166	0.215	0.268	0.323	0.379	0.435	0.490	0.542	0.640	0.760	0.848	0.908	0.956	0.984
20	0.050	0.053	0.056	0.060	0.063	0.066	0.074	0.096	0.125	0.158	0.193	0.232	0.273	0.315	0.358	0.402	0.489	0.611	0.716	0.801	0.883	0.944
50	0.050	0.052	0.054	0.056	0.059	0.061	0.066	0.076	0.092	0.110	0.129	0.150	0.173	0.198	0.223	0.250	0.307	0.398	0.490	0.580	0.687	0.796

Source: Reprinted with permission from Haynam, Govindarajulu, and Leone (1970).

and df $= 1$. For a total sample size of $n = 200$, the noncentrality is $\lambda = 0.75$. From Table 7.12, for an $\alpha = 0.05$ level test, the power is approximately 0.14.

7.6.4 Power for Testing Conditional Independence

The next example is taken from an article by O'Brien (1986) on power analyses for loglinear models. A standard fetal heart rate monitoring test predicts whether a fetus will require nonroutine care following delivery. The standard test has categories "worrisome" and "reassuring." The outcome of the delivery is classified "perfect" or "newborn required some nonroutine medical care during first week after birth." A new fetal heart rate monitoring test is developed, having categories "very worrisome," "somewhat worrisome," and "reassuring." A physician plans to study whether this new test can help make predictions about the outcome; for instance, given the result of the standard test, is there an association between delivery outcome and the result of the new test? A relevant statistic is $G^2[(NS, SO) \mid (NS, SO, NO)]$, where $N =$ new test, $S =$ standard test, $O =$ outcome. To help select the sample size for the study, a statistician approximated the power for various sample sizes and scenarios for the $N-O$ partial association.

The statistician asked the physician to conjecture about the joint distribution of the explanatory variables N and S; for instance, the statistician asked "What proportion of the cases do you think will be scored 'reassuring' by both tests?" Then for each $N-S$ combination, the physician guessed the probability the newborn would require nonroutine medical care. Table 7.13 contains one scenario for the marginal and conditional probabilities. These probabilities produce a joint distribution $\{\pi_{ijk}\}$, which has fitted probabilities $\pi(M_1)$ and $\pi(M_2)$ for the models $M_1 = (NS, SO, NO)$ and $M_2 = (NS, SO)$. The likelihood-ratio test com-

Table 7.13 Scenario for Power Computation

Standard	New	Joint Prob.	P(Nonroutine Care)
Worrisome	Very worrisome	0.04	0.40
	Somewhat worrisome	0.08	0.32
	Reassuring	0.04	0.27
Reassuring	Very worrisome	0.02	0.30
	Somewhat worrisome	0.18	0.22
	Reassuring	0.64	0.15

Source: Reprinted with permission from O'Brien (1986).

paring these models has form (7.2). Its noncentrality is (7.12) with $\pi(M_1)$ playing the role of π and $\pi(M_2)$ playing the role of $\pi(M)$.

For the scenario in Table 7.13, the noncentrality equals $0.00816n$, based on df $= 2$. For sample sizes 400, 600, 800, and 1000, the approximate powers for a 0.05-level test are 0.35, 0.49, 0.62, and 0.73. This scenario predicts 64% of the observations to occur at only one combination (reassuring, reassuring) of the factors. The lack of dispersion for the factors weakens the power.

7.7 EMPTY CELLS AND SPARSENESS IN CONTINGENCY TABLES

We refer to contingency tables as *sparse* when many cells have small frequencies. If N is the number of cells and n is the total sample size, then n/N is an index of sparseness, smaller values referring to more sparse tables. Sparse tables occur when the sample size is small. They also occur when the sample size is large but there is a large number of cells. They are common when there are many variables, or when some classifications have several categories. This section discusses effects of sparseness on inference procedures for loglinear models.

7.7.1 Sampling vs. Structural Zeroes

Sparse tables often contain cells having zero counts. Such cells are called *empty* cells. Suppose an empty cell has $m_i = n\pi_i > 0$; that is, the true cell probability is positive. Then $n_i = 0$ is due to the small size of the sample, and it is called a *sampling zero*.

A zero for a cell in which it is theoretically impossible to have observations is called a *structural zero*. For such cells $m_i = 0$ and necessarily $\hat{m}_i = 0$. For instance, suppose professors working at a certain university for at least 5 years were cross-classified on their current rank (assistant professor, associate professor, professor) and their rank five years ago. If it is impossible for professors to be demoted in rank, then three of the nine cells in the table contain structural zeroes. Contingency tables containing at least one structural zero are called *incomplete tables*.

A sampling zero is an observation having value 0, and we regard it as one of the observed counts. A structural zero is not an observation, however. A structural zero is not part of the data set, and models are fitted without such cells making any contribution to the likelihood. In some computer packages for analyzing categorical data, it is necessary to explicitly identify structural zeroes. For instance, packages that use iterative proportional fitting (such as BMDP-4F) enter initial values of 0 in cells having structural zeroes, and the iterative process preserves the 0 values. In other packages (such as GLIM), structural zeroes are simply not included as part of the data set.

7.7.2 Existence of Estimates

Haberman (1973b, 1974a) studied the influence of empty cells on the existence and uniqueness of ML estimates of parameters for loglinear models. He generalized work by Birch (1963) and Fienberg (1970b). We now list several of Haberman's results. Let $\mathbf{n}$ denote the vector of cell counts, and $\mathbf{m}$ their expected values. Haberman showed results (1)–(5) for Poisson sampling, but by result (6) they apply also to multinomial sampling.

1. The log likelihood function is a strictly concave function of log $\mathbf{m}$.
2. If a ML estimate of $\mathbf{m}$ exists, it is unique and satisfies the likelihood equations $\mathbf{X}'\mathbf{n} = \mathbf{X}'\hat{\mathbf{m}}$. Conversely, if $\hat{\mathbf{m}}$ satisfies the model and also the likelihood equations, then it is the ML estimate of $\mathbf{m}$.
3. If all $n_i > 0$, then ML estimates of loglinear model parameters exist.
4. Suppose ML parameter estimates exist for a loglinear model that equates observed and fitted counts in certain marginal tables. Then those marginal tables have uniformly positive counts.
5. Suppose model M_2 is a special case of model M_1. If ML estimates exist for M_1, then they also exist for M_2.
6. For any loglinear model, the ML estimates $\hat{\mathbf{m}}$ are identical for multinomial and independent Poisson sampling, and those estimates exist in the same situations.

To illustrate, consider the saturated model. By (2) and (3), when all $n_i > 0$, the ML estimate of $\mathbf{m}$ is $\mathbf{n}$. By (4), parameter estimates do not exist when any $n_i = 0$. Model parameter estimates are contrasts of $\{\log \hat{m}_i\}$, and since $\hat{\mathbf{m}} = \mathbf{n}$ for the saturated model, the estimates exist only when all $n_i > 0$.

Next, consider unsaturated models. From (3) and (4), ML estimates exist when all $n_i > 0$ and do not exist when any count is zero in the set of sufficient marginal tables. This does not tell us what happens when at least one $n_i = 0$ but the sufficient marginal counts are all positive. ML estimates for independence models exist if and only if all sufficient marginal totals are positive. For model (XY, XZ, YZ), however, ML estimates exist when only one $n_i = 0$, but may not exist when at least two cells are empty. For instance, ML estimates do not exist for Table 7.14, even though all sufficient statistics (the two-way marginal totals) are positive (Problem 7.40).

Haberman showed the supremum of the likelihood function is finite, even when ML estimates of loglinear model parameters do not exist. This motivated him to define "extended" ML estimators of $\mathbf{m}$, which always exist but need not have the same properties as regular ML estimators. A sequence of estimates satisfying the model that converges to the extended estimate has log likelihood approaching its supremum. In this extended sense, $\hat{m}_i = 0$ is the ML estimate of m_i for the saturated model when $n_i = 0$.

Baker et al. (1985) noted that the problem with existence of ML estimates occurs only when we require $\mathbf{m} > \mathbf{0}$ (as did Haberman), so that model parameters exist on a logarithmic scale. ML estimates of $\mathbf{m}$ always exist and are unique when models are formulated using constraint specifications rather than loglinear equations. (Section 13.5.1 introduces the constraint specification approach.)

7.7.3 Effect of Small Samples on X^2 and G^2

The true sampling distributions of "chi-squared" statistics presented in this text converge to chi-squared as the sample size $n \rightarrow \infty$, for a fixed number of cells N. The adequacy of the chi-squared approximation depends both on n and N.

Cochran studied the chi-squared approximation for X^2 in a series of articles. In 1954, he suggested that to test independence with $\mathrm{df} > 1$, a minimum expected value of 1 is permissible as long as no more than about 20% of the cells have expected values below 5. Larntz (1978), Koehler and Larntz (1980), and Koehler (1986) showed that X^2 is valid with smaller sample sizes and more sparse tables than G^2. They showed that the distribution of G^2 is usually poorly approximated by chi-squared when n/N is less than 5. When most expected frequencies are smaller

Table 7.14 Data for which ML Estimates Do Not Exist for Model (XY, XZ, YZ)[a]

Z		1		2	
Y		1	2	1	2
X 1		0	*	*	*
2		*	*	*	0

[a]Cells containing * may contain any positive numbers.

than 0.5, treating G^2 as chi-squared gives a highly conservative test; that is, when H_0 is true, the probability that G^2 exceeds the percentage point $\chi^2(\alpha)$ is much less than α, and reported P-values are larger than true ones. When most expected frequencies are between 0.5 and 4, G^2 tends to give too liberal a test.

The first two panels of Table 7.15 compare nominal and actual α-levels for X^2 and G^2 for testing independence, for sample sizes 50 and 100 and various table sizes. We calculated these levels for tables having marginal probabilities (0.5, 0.5) when there are two categories, (0.27, 0.46, 0.27) for three categories, (0.21, 0.29, 0.29, 0.21) for four categories, (0.12, 0.16, 0.22, 0.22, 0.16, 0.12) for six categories, and (0.05, 0.06, 0.10, 0.13, 0.16, 0.16, 0.13, 0.10, 0.06 0.05) for ten categories. The G^2 statistic behaves adequately at both sample sizes for 2×3 tables, but deteriorates dramatically as I and J increase for fixed n (i.e., as the degree of sparseness increases). The X^2 statistic is adequate here when n/N exceeds 1.

The size of n/N that produces adequate approximations for X^2 tends to decrease as N increases. For instance, Koehler and Larntz (1980) suggested the guideline $n > (10N)^{1/2}$ (i.e., $n/N > (10/N)^{1/2}$ for using X^2 in goodness-of-fit testing of the uniform multinomial probabilities $(1/N, \ldots, 1/N)$. They noted, however, that the chi-squared approximation tends to be poor for sparse tables containing both small and moderately large expected frequencies. It seems hopeless to expect a single rule to cover all cases.

7.7.4 Alternative Asymptotics and Alternative Statistics

When cell counts are too small to permit chi-squared approximations, there are various remedies. Researchers often combine categories of variables to obtain larger cell frequencies. Pooling categories is usually not a good idea unless there is a natural way to combine them and little information is lost in defining the variable more crudely. Given recent advances in computer power and in sophistication of algorithms, it is becoming increasingly feasible to use *exact* methods. When exact methods are infeasible, it is often possible to approximate exact distributions precisely using Monte Carlo methods (Mehta et al. 1988). Another remedy is to use an alternative approximation for the distribution of the statistic.

Current research is producing asymptotic approximations that apply when the number of cells N increases as the sample size n increases. For this approach, expected frequencies in the cells need not increase, as they must do in the usual (fixed N, $n \to \infty$) large-sample theory. For goodness-

Table 7.15 Proportion of Times Statistic Exceeds Chi-Squared Percentage Point, When There is Independence

Statistic	I	J	α:	$n = 50$				$n = 100$			
				0.01	0.05	0.10	0.25	0.01	0.05	0.10	0.25
$G^2(I)$	2	3		0.013	0.051	0.107	0.261	0.011	0.058	0.108	0.258
	4	4		0.027	0.102	0.173	0.371	0.016	0.075	0.131	0.304
	6	6		0.024	0.126	0.236	0.501	0.031	0.129	0.224	0.445
	10	10		0.000	0.000	0.002	0.028	0.013	0.086	0.188	0.464
$X^2(I)$	2	3		0.008	0.046	0.100	0.255	0.009	0.055	0.106	0.255
	4	4		0.010	0.052	0.104	0.265	0.008	0.048	0.103	0.267
	6	6		0.008	0.043	0.093	0.255	0.010	0.048	0.095	0.253
	10	10		0.010	0.036	0.070	0.188	0.010	0.047	0.096	0.245
$X^2(I)^a$	2	3		0.006	0.042	0.088	0.238	0.007	0.048	0.100	0.251
	4	4		0.004	0.027	0.066	0.196	0.006	0.037	0.084	0.235
	6	6		0.001	0.007	0.024	0.089	0.004	0.023	0.054	0.166
	10	10		0.000	0.000	0.000	0.003	0.001	0.005	0.011	0.040

Source: Reprinted with permission from Agresti and Yang (1987).
[a]Based on adding $1/J$ to each cell count.

of-fit testing of a specified multinomial, Koehler and Larntz (1980) showed that a standardized version of G^2 has an approximate normal distribution for very sparse tables. Koehler (1986) presented limiting normal distributions for G^2 for use in testing models having direct ML estimates. His results apply when the probability of each cell becomes small as N increases, when none of the probabilities converge to zero too rapidly relative to the rate of increase of n, and when the number of estimated parameters in the null hypothesis increases at a slower rate than N.

McCullagh (1986) reviewed ways of handling sparse tables and presented an alternative normal approximation for G^2. It assumes, however, that the dimension of the parameter vector in H_0 is fixed as the size of the table increases. Further research is needed to determine the usefulness of these alternative asymptotics for sparse tables.

Recent research has also produced alternative statistics for use in testing goodness of fit. Cressie and Read (1984) introduced a family of statistics

$$\frac{2}{\lambda(\lambda + 1)} \sum n_i \left[\left(\frac{n_i}{\hat{m}_i} \right)^{\lambda} - 1 \right], \quad \text{for } -\infty < \lambda < \infty$$

called *power divergence* statistics. This equals X^2 for $\lambda = 1$ and G^2 as $\lambda \to 0$ (Problem 3.26). Cressie and Read recommended the statistic with $\lambda = 2/3$, which they found less susceptible to effects of sparseness than X^2 and G^2. Other adaptations of chi-squared tests have been proposed by Berry and Mielke (1988) and Zelterman (1987).

7.7.5 Adding Constants to Cells

Empty cells and sparse tables can cause problems with existence of estimates for loglinear model parameters, problems with severe bias in estimation of descriptive measures such as odds ratios, problems with the performance of computational algorithms, as well as problems with asymptotic approximations of chi-squared statistics (see, e.g., Brown and Fuchs 1983). Because of this, many researchers add a small constant to cell counts before conducting an analysis. Some routinely add $\frac{1}{2}$, as Goodman (1964c, 1970, 1971a) recommended for saturated models. For unsaturated models, this usually smooths the data too much. Adding a constant to each cell or to each empty cell can cause havoc with sampling distributions. This operation smooths the data towards the simple model in which cells are equally probable, resulting in a conservative influence on test statistics.

To illustrate, the third panel of Table 7.15 show the effect on X^2 of adding $1/J$ to every cell in an $I \times J$ table. This makes the statistic far too conservative when n/N is less than about 5. The effect is very severe when there is a large number of cells.

It is difficult to give general recommendations on the matter of adding constants. We recommend performing a sensitivity analysis. Repeat the analysis by adding constants of various sizes, in order to gauge the effect on parameter estimates and goodness-of-fit statistics. The total count added should be no more than a very small percentage of the total sample size. Indeed, for many ML analyses it is unnecessary to add any constant. When there is a problem with existence or computations, it is often adequate to add an extremely small constant, such as 10^{-8}. This alleviates the problem but avoids over-smoothing the data before the fitting process.

7.7.6 Model-Based Tests and Sparseness

For a model M, the statistics $G^2(M)$ and $X^2(M)$ for testing fit have null distributions converging to chi-squared as cell expected frequencies grow unboundedly. The true sampling distributions may be far from chi-squared when $\{m_i\}$ are small. From (7.2) and (7.3), the model-based statistics $G^2(M_2 \mid M_1)$ and $X^2(M_2 \mid M_1)$ depend on the data only through the fitted values, and hence only through minimal sufficient statistics for the more complex model. These statistics have null distributions converging to chi-squared as the expected values of the minimal sufficient statistics grow. For most loglinear models, these expected values refer to marginal tables. Since marginal expected values are larger than cell expected values, $G^2(M_2 \mid M_1)$ converges to its limiting chi-squared distribution more quickly than does $G^2(M_2)$; similarly, $X^2(M_2 \mid M_1)$ converges more quickly than $X^2(M_2)$. Haberman (1977a) provided theoretical justification for these comments.

The sample size in sparse tables is often sufficiently large to use the model-based statistics but not the others. When fitted cell counts are small but the sufficient marginal totals for M_1 are mostly in at least the 5–10 range, the chi-squared approximation is likely to be adequate for the model-based statistics. When we test conditional independence of X and Y in a sparse table, this is another reason for preferring $G^2[(XZ, YZ) \mid (XY, XZ, YZ)]$ over $G^2(XZ, YZ)$ as the test statistic.

CHAPTER NOTES

Section 7.1: Partitioning Chi-Squared to Compare Models

7.1 Lancaster (1951) and Kullback et al. (1962) were among the first to partition chi-squared statistics in multi-way tables. Plackett (1962), Goodman (1970), and Bishop et al. (1975) pointed out difficulties with their approaches.

Section 7.2: Strategies in Model Selection

7.2 When a certain model holds, the index G^2/df has an asymptotic mean of 1 as $n \to \infty$. Goodman (1971a) recommended this index for comparing fits provided by different models to a data set. Smaller values represent better fits. The Akaike information criterion orders models according to

$$\text{AIC} = -2(\text{maximized likelihood—no. parameters in model}).$$

For loglinear models, this ordering is equivalent to one using $[G^2 - 2(\text{df})]$. Schwarz (1978) and Raftery (1986) used a Bayesian argument to motivate the alternative criterion $\text{BIC} = [G^2 - (\log n)(\text{df})]$. This takes sample size directly into account. These criteria support model parsimony, imposing a penalty for increasing the number of parameters in the model. Compared to AIC, the BIC criterion gravitates less quickly toward more complex models as the sample size increases.

Aitkin (1979, 1980) and Whittaker and Aitkin (1978) discussed model-selection methods, and suggested a simultaneous testing procedure to control the overall error rate in Brown's approach for comparing models of various orders. For further discussion of model selection, see Benedetti and Brown (1978), Goodman (1971a), Jolayemi and Brown (1984), and Wermuth (1976a).

Section 7.3: Analysis of Residuals

7.3 Pregibon (1982b) showed that the squared adjusted residual is the score test statistic (Note 4.4) for the hypothesis regarding whether the count in the ith cell is an outlier. Brown (1974) and Simonoff (1988) gave other useful residual analyses for categorical data.

Section 7.4: Testing Conditional Independence

7.4 Day and Byar (1979) noted that the Cochran–Mantel–Haenszel test is equivalent to the score test for the logit model for the $2 \times 2 \times K$ table. Kuritz et al. (1988) reviewed Mantel–Haenszel-type methods.

Section 7.5: Estimating and Comparing Conditional Associations

7.5 For further discussion of comparison of odds ratios and estimation of a common value, see Breslow and Day (1980, Section 4.4), Breslow (1976, 1981), Breslow and Liang (1982), Donner and Hauck (1986), Gart (1970), Greenland (1989), Hauck (1989),

Hauck and Donner (1988), Jones et al. (1989), Liang and Self (1985), Mehta et al. (1985), Tarone et al. (1983), Thomas (1975), and Zelen (1971). Jones et al. (1989) reviewed and compared several large-sample tests of homogeneity of odds ratios.

Breslow (1976), Breslow and Cologne (1986), Breslow and Day (1980, Section 7.5), Prentice (1976), and Zelen (1971) discussed direct modeling of the dependence of the odds ratio on explanatory variables. McCullagh and Nelder (1989, Section 6.6) described the study that produced Table 7.11, and fitted model (7.10) while simultaneously fitting linear or quadratic logit models for the marginal effects of age on breathlessness and wheeze (see also Problem 13.28).

Section 7.6: Sample Size and Power Considerations

7.6 For comparing proportions, Fleiss (1981, Section 3.2) provided a formula and tables giving the sample size necessary to achieve a certain power at a fixed significance level. Kraemer and Thiemann (1987) and Rochon (1989) gave alternative approaches to this and more general problems.

O'Brien (1986) studied the quality of the noncentral chi-squared approximation for the nonnull distribution of G^2. His simulations, conducted for tables of dimensions ranging from $2 \times 2 \times 2$ to $3 \times 3 \times 3$ with sample sizes ranging from 125 to 1000, showed that the approximation held well for a wide range of powers. Read and Cressie (1988, pp. 147–148) listed several other articles that have studied the non-null behavior of X^2 and G^2.

Chapman and Meng (1966), Drost et al. (1989), Haberman (1974a, pp. 109–112), Harkness and Katz (1964), Mitra (1958), and Patnaik (1949) derived theory for the asymptotic nonnull behavior of the chi-squared statistics. See Section 12.3.5.

Section 7.7: Empty Cells and Sparseness in Contingency Tables

7.7 For hierarchical loglinear models, Glonek et al. (1988) showed that positivity of the marginal sufficient statistics implies the existence of ML estimates if and only if the model is decomposable (see Note 6.1).

Hutchinson (1979) gave a bibliography of articles that have studied the adequacy of the chi-squared approximation for statistics such as X^2 and G^2. Cressie and Read (1989), Haberman (1988), Hosmane (1987), Lawal (1984), and Lawal and Upton (1984) also studied this issue.

PROBLEMS

Applications

7.1 Decompose chi-squared to compare a nested set of models for Table 5.8.

7.2 Using a nested set of loglinear models for a four-way table, give a decomposition of $G^2(W, X, Y, Z)$.

7.3 Use Brown's strategy to screen all two-factor associations for the death penalty data. Which model does this screening suggest?

7.4 Wermuth (1976b) reported Table 7.16. The variables are age of mother (A), length of gestation (G) in days, infant survival (I), and number of cigarettes smoked per day during the prenatal period (S).

Table 7.16

Age	Smoking	Gestation	Infant Survival	
			No	Yes
<30	<5	≤260	50	315
		>260	24	4012
	5+	≤260	9	40
		>260	6	459
30+	<5	≤260	41	147
		>260	14	1594
	5+	≤260	4	11
		>260	1	124

Source: Reprinted with permission from the Biometric Society (Wermuth 1976b).

a. Fit the models $(AGIS)$, (AGI, AIS, AGS, GIS), (AG, AI, AS, GI, GS, IS), and (A, G, I, S), and identify a subset of models that should be further studied.

b. Use Brown's (1976) method to screen each two-factor association. Which model does this analysis suggest?

c. Use standardized parameter estimates for the saturated model to suggest a starting point for model-building.

d. Compare results of the strategies.

7.5 Refer to the previous problem. Suppose you wish to treat A and S as explanatory variables, and G and I as responses. For part (a), give four models that help identify useful models for further study. Fit those models, and using results, construct a model that fits these data well. Interpret.

7.6 Refer to the previous problem. Repeat the analysis, treating A, S, and G as explanatory variables and I as the response.

7.7   Refer to Problem 7.4.

     a. Use forward selection to build a model.

     b. Use backward elimination to build a model.

     c. Compare results of the two strategies.

7.8   Refer to the previous two problems. Use forward selection to build a model, treating A, S, and G as explanatory variables.

7.9   Refer to Table 7.3.

     a. Starting with model (EG, EM, EP, GM, GP, MP), use the forward selection process to build a model.

     b. Find the standardized parameter estimates for the saturated model. What model do they suggest?

7.10   Refer to the previous problem. Compute adjusted or standardized residuals for your final model, and comment on any lack of fit.

7.11   Refer to Problem 6.3. Analyze these data, using methods of this chapter.

7.12   Table 7.17 was discussed by Sewell and Orenstein (1965) and Elliott (1988). Analyze these data.

7.13   Refer to the previous problem. Find a loglinear model that fits well and is equivalent to a logit model that treats occupational aspirations as a response variable. Interpret effects.

7.14   Refer to the previous problem. Obtain a more parsimonious logit model by utilizing the ordinal nature of the explanatory variables.

7.15   Refer to Problem 5.18. Using methods of this chapter, analyze the data. (You may wish to compare your results to those of Fowlkes et al. (1988), who provided useful graphics for describing associations and comparing fits of various models.)

7.16   Refer to the independence model for Table 3.4 on income and job satisfaction. Calculate adjusted residuals, and search for any pattern of lack of fit. (Section 8.1 introduces a model that gives a better fit.)

Table 7.17

Gender	Residence	IQ	Socio-economic status	Occupational aspirations	
				High	Low
Male	Rural	High	High	117	47
			Low	54	87
		Low	High	29	78
			Low	31	262
	Small urban	High	High	350	80
			Low	70	85
		Low	High	71	120
			Low	33	265
	Large urban	High	High	151	31
			Low	27	23
		Low	High	30	27
			Low	12	52
Female	Rural	High	High	102	69
			Low	52	119
		Low	High	32	73
			Low	28	349
	Small urban	High	High	338	96
			Low	44	99
		Low	High	76	107
			Low	22	344
	Large urban	High	High	148	35
			Low	17	39
		Low	High	21	47
			Low	6	116

Source: Reprinted with permission from Sewell and Orenstein (1965).

7.17 Refer to Problem 6.4. Conduct a residual analysis with the model of no three-factor interaction to describe the nature of the interaction.

7.18   Refer to Table 5.1. Use the Cochran-Mantel-Haenszel procedure to test independence of death penalty verdict and victim's race, controlling for defendant's race. Give another way of conducting this test, and compare results.

7.19   Table 7.18, given by Chin et al. (1961), classifies 174 poliomyelitis cases in Des Moines, Iowa by age of subject, paralytic status, and by whether the subject had been injected with the Salk vaccine.

Table 7.18

Age	Salk Vaccine	Paralysis No	Paralysis Yes
0–4	Yes	20	14
	No	10	24
5–9	Yes	15	12
	No	3	15
10–14	Yes	3	2
	No	3	2
15–19	Yes	7	4
	No	1	6
20–39	Yes	12	3
	No	7	5
40+	Yes	1	0
	No	3	2

Source: Reprinted with permission, based on data from Chin et al. (1961).

a. Test the hypothesis that severity is independent of whether vaccinated, controlling for age.

b. Use another procedure for testing this hypothesis, and compare results to those obtained in (a).

7.20   Refer to the previous problem.

a. Calculate the Mantel-Haenszel estimator of the common odds ratio between severity and whether vaccinated, and interpret.

b. Calculate the unconditional ML estimator of the common odds ratio.

c. If you have a computer program, compute the conditional ML estimator. Compare results.

7.21 Two treatments were compared for 40 pairs of subjects who were matched on relevant covariates. For each pair, treatments were randomly assigned to the subjects, and a binary response was subsequently measured. Twenty pairs of subjects made the same response. For six pairs, there was a success for the subject receiving treatment A and a failure for the subject receiving treatment B, whereas for the other fourteen pairs there was a success for treatment B and a failure for treatment A. Use the Cochran–Mantel–Haenszel procedure to test independence of response and treatment. (Section 10.1 presents an equivalent test, McNemar's test.)

7.22 Refer to Section 7.6.2.

a. What sample size is needed for the test to have approximate power 0.80?

b. What sample size is needed to obtain approximate power 0.80 for detecting $\pi_A - \pi_B = 0.06$ for the one-sided alternative hypothesis $H_a: \pi_A > \pi_B$?

7.23 In an experiment designed to compare two treatments on a response having three categories, a researcher expects the conditional distributions to be approximately (0.2, 0.2, 0.6) and (0.3, 0.3, 0.4).

a. For significance level $\alpha = 0.05$, find the approximate power for using X^2 to compare the distributions when (1) 25, (2) 50, (3) 100 observations are made for each treatment.

b. What sample size is needed for each treatment for the test in (a) to have approximate power 0.90?

c. Repeat (a) using G^2, and compare results to those for X^2.

7.24 Refer to Table 7.13. An alternative scenario presented by O'Brien has the same joint distribution for the new and standard tests, but has P(Nonroutine Care) values equal to (0.50, 0.45, 0.40, 0.45, 0.25, 0.15) instead of (0.40, 0.32, 0.27, 0.30, 0.22, 0.15). Calculate the noncentrality for the likelihood-ratio model-based test of N–O partial association. Find the approximate powers for sample sizes

400, 600, 800, and 1000, for a 0.05-level test. How large a sample is needed to achieve power 0.90?

7.25 Suppose $n_{11+} = 0$. Do ML estimates of model parameters exist for loglinear model (XY, Z)? Explain.

7.26 Give an example of variables for a contingency table in which certain cells contain structural zeroes.

7.27 Use the first panel of Table 7.15 to illustrate how the degree of sparseness affects whether G^2 gives a conservative or a liberal test of independence.

Theory and Methods

7.28 Suppose loglinear model M_2 is a special case of loglinear model M_1.

   a. Show the fitted values for the two models are identical in the sufficient marginal distributions for M_2.

   b. Use (a) to prove expression (7.2) for $G^2(M_2 | M_1)$ applied to models (XY, Z) and (X, Y, Z).

7.29 For loglinear models, Haberman (1974a) showed that when M_2 is a special case of M_1 and $\{\hat{m}_i\}$ satisfy any model that is a special case of M_2,

$$\sum \hat{m}_{1i} \log(\hat{m}_i) = \sum \hat{m}_{2i} \log(\hat{m}_i) .$$

(An interpretation is that $\hat{\mathbf{m}}_2$ is the orthogonal projection of $\hat{\mathbf{m}}_1$ onto the linear manifold of $\{\log \mathbf{m}\}$ values satisfying model M_2.)

   a. Assuming this result, show the special cases

$$\sum \hat{m}_{1i} \log(\hat{m}_{2i}) = \sum \hat{m}_{2i} \log(\hat{m}_{2i})$$
$$\sum n_i \log(\hat{m}_{ai}) = \sum \hat{m}_{ai} \log(\hat{m}_{ai}), \quad a = 1, 2 .$$

   b. Use (a) to prove $G^2(M_2) - G^2(M_1) = 2 \sum \hat{m}_{1i} \log(\hat{m}_{1i}/\hat{m}_{2i})$.

7.30 For the independence model in a two-way table, show that

$$\sum \sum n_{ij} \log \hat{m}_{ij} = \sum \sum \hat{m}_{ij} \log \hat{m}_{ij} .$$

7.31 Refer to Section 7.1.4. Show that $G^2(M_j \mid M_{j-1})$ is identical to G^2 for testing independence in the 2×2 table in which the first column combines columns 1 through $j - 1$ and the second column is column j of the original table.

7.32 For a cross-classification of T variables $X_1, \ldots, X_T$, show

$$G^2(X_1, X_2, \ldots, X_T) = G^2(X_1, X_2) + G^2(X_1 X_2, X_3)$$
$$+ \cdots + G^2(X_1 X_2 \cdots X_{T-1}, X_T).$$

7.33 For a cross-classification of T variables, show

$$G^2(X_1 \cdots X_{T-1}, X_T) = G^2(X_1, X_T) + G^2(X_1 X_T, X_1 X_2)$$
$$+ \cdots + G^2(X_1 X_2 \cdots X_{T-1}, X_1 X_2 \cdots X_{T-2} X_T).$$

That is, X_T is jointly independent of $X_1, \ldots, X_{T-1}$ if X_T is independent of X_1, X_T is independent of X_2 given $X_1, \ldots$, and X_T is independent of X_{T-1} given $X_1, \ldots, X_{T-2}$.

7.34 Suppose there are s nested models $M_1, \ldots, M_s$, where M_1 is the most complex model. Let ν denote the difference in residual degrees of freedom between M_s and M_1. Suppose model M_k holds, so that M_j also holds when $j < k$.

a. Show that for all $j < k$, $G^2(M_k \mid M_j) \leqslant G^2(M_k \mid M_1)$.

b. Show that for all $j < k$, as $n \to \infty$, $P[G^2(M_k \mid M_j) > \chi_\nu^2(\alpha)] \leqslant \alpha$.

c. Gabriel (1966) suggested a simultaneous testing procedure in which, for each pair of models, the critical value for differences between G^2 values is $\chi_\nu^2(\alpha)$. The final model accepted must be more complex than any model rejected in a pairwise comparison. Since (b) is true for all $j < k$, argue that Gabriel's procedure has Type-I error probability no greater than α.

7.35 Suppose $\{n_i\}$ are independent Poisson random variables having means $\{m_i\}$.

a. Let $z_i = (n_i - m_i)/(m_i)^{1/2}$. Show that the sum of variances of $\{z_i\}$ equals the number of cells in the table.

b. Let $e_i = (n_i - \hat{m}_i)/(\hat{m}_i)^{1/2}$, where $\{\hat{m}_i\}$ are fitted values for a model $\{m_i\}$ satisfy. Give a heuristic argument that the sum of variances of $\{e_i\}$ asymptotically equals df for testing the fit of the model. Hence, the average variance for individual e_i is less than 1.

7.36 Obtain adjusted residuals (7.5) for the independence model as a special case of Haberman's general formula (7.6).

7.37 Derive the likelihood equations for model (7.7), and show that there is a perfect fit for Department A. Generalize the model to permit a common A–G association for Departments B–F, and a different association for Department A. Derive the likelihood equations, and give the residual df.

7.38 Give population probabilities in a $2 \times 2 \times 2$ table for which there is conditional X-Y dependence, but $G^2[(XZ, YZ)\,|\,(XY, XZ, YZ)]$ is not a consistent test. (Hint: Find a table for which $\lambda^{XY} = 0$ but $\lambda^{XYZ} \neq 0$.) Explain why $G^2(XZ, YZ)$ gives a consistent test for this table.

7.39 Show that $G^2[(Y, XZ)\,|\,(XY, XZ)]$ is identical to $G^2[(X, Y)]$ for the X–Y marginal table.

7.40 Show that ML estimates do not exist for Table 7.14. (Hint from Haberman (1973b, 1974a p. 398): If $\hat{m}_{111} = c > 0$, then marginal constraints imposed by the model imply $\hat{m}_{222} = -c$.)

7.41 We want to compare the association between X and Y among K levels of a control variable Z. Let ζ denote a generic measure of association. For independent multinomial samples of sizes $\{n_k\}$, suppose $\sqrt{n_k}(\hat{\zeta}_k - \zeta_k) \xrightarrow{d} N(0, \sigma_k^2)$ as $n_k \to \infty$. A measure of average partial association is

$$\bar{\zeta} = \frac{\sum \left(\dfrac{n_k}{\hat{\sigma}_k^2}\right) \hat{\zeta}_k}{\sum \left(\dfrac{n_k}{\hat{\sigma}_k^2}\right)}.$$

a. Show that $\sum z_k^2 = V + \bar{\zeta}^2 / \hat{\sigma}^2(\bar{\zeta})$, where

$$V = \sum \frac{n_k(\hat{\zeta}_k - \bar{\zeta})^2}{\hat{\sigma}_k^2}, \qquad z_k = \frac{n_k^{1/2} \hat{\zeta}_k}{\hat{\sigma}_k}, \qquad \hat{\sigma}^2(\bar{\zeta}) = \left(\sum \frac{n_k}{\hat{\sigma}_k^2}\right)^{-1}.$$

b. Suppose $n \to \infty$ with $n_k/n \to \rho_k > 0$, $k = 1, \dots, K$. Give the asymptotic chi-squared distribution for each component in this partitioning. Indicate the hypothesis each component tests.

CHAPTER 8

Loglinear-Logit Models For Ordinal Variables

The loglinear models discussed in the previous three chapters have achieved widespread use in recent years. These models have a limitation, however. They treat all classifications as nominal, in the sense that parameter estimates and chi-squared statistics are invariant to orderings of categories. These models ignore important information when at least one variable is ordinal. This limitation is serious, because ordinal data are common. For instance, Moses et al. (1984) reported that ordered categorical data occurred in 32 of 168 articles in volume 36 (1982) of the *New England Journal of Medicine*.

Table 8.1 illustrates the inadequacy of nominal loglinear models for analyzing ordinal data. In Section 3.3 we analyzed this cross-classification of income and job satisfaction using chi-squared tests of independence. The test statistics are $G^2 = 12.03$ and $X^2 = 11.99$, based on df = 9. There is only weak evidence of association, but these tests ignore the category orderings. If we interchange cell counts in the first two rows, for instance, the fitted values also interchange, and G^2 and X^2 are unchanged. These statistics are not powerful for detecting departures from independence that reflect the orderings, such as "Job satisfaction increases as income increases."

Table 8.1 also contains fitted values and adjusted residuals (7.5) for the independence model. Large positive residuals occur in the corners of the table where both variables are at high level and where both are at low level. Large negative residuals occur in the corners where one variable is high and the other is low. These residuals indicate lack of fit in the form of a positive trend.

This chapter adapts loglinear models so that they *do* exploit ordinality. The models have association and interaction terms that reflect ordinal

Table 8.1 Observed and Fitted Frequencies and Adjusted Residuals for Testing Independence of Income and Job Satisfaction

	Job Satisfaction			
Income ($)	Very Dissatisfied	Little Dissatisfied	Moderately Satisfied	Very Satisfied
<6000	20	24	80	82
	(14.2)	(24.7)	(72.9)	(94.2)
	1.83	−0.17	1.17	−1.94
6000–15,000	22	38	104	125
	(19.9)	(34.6)	(102.3)	(132.2)
	0.60	0.74	0.25	−1.02
15,000–25,000	13	28	81	113
	(16.2)	(28.2)	(83.2)	(107.5)
	−0.95	−0.04	−0.35	0.84
>25,000	7	18	54	92
	(11.8)	(20.5)	(60.5)	(78.2)
	−1.60	−0.65	−1.16	2.35

characteristics such as "monotone trend" or "stochastic ordering." Advantages of these models over the nominal-scale models of Chapters 5–7 include the following:

1. Model parameters describe types of trends and are simpler to interpret than those in nominal models.
2. Unsaturated models exist in situations in which nominal models are saturated. For instance, there are unsaturated ordinal models for association in two-way tables and for three-factor interaction in three-way tables. The ordinal models have structured association and interaction terms that contain fewer parameters, hence retaining more residual degrees of freedom than the nominal models.
3. Tests based on ordinal models have improved power for detecting certain types of association and interaction.

We begin by discussing loglinear models for two-way tables. Section 8.1 introduces a model that treats both variables as ordinal. Section 8.2 introduces a more general model that treats one variable as ordinal and one as nominal. Section 8.3 generalizes these models for multidimensional tables.

Section 8.4 presents statistics for testing independence or conditional independence. These tests are analogs of the model-based tests and Cochran-Mantel-Haenszel test of Section 7.4 that utilize orderings of classifications.

Section 8.5 presents models resembling loglinear models of Sections 8.1 and 8.2, except they replace fixed scores in the models by parameters. We discuss a link between these models and *correspondence analysis*, a popular descriptive method for contingency tables.

The final section discusses model selection for ordinal variables. Chapter 9 presents alternative models for ordinal variables, based on various logit transformations.

8.1 LINEAR-BY-LINEAR ASSOCIATION

For two-way tables, we rarely expect the independence model to fit well. For a model to have much scope, it must allow association yet retain some residual degrees of freedom; that is, it nests between the independence model and the saturated model. This section presents a simple model of this type for association between two ordinal variables.

8.1.1 Using Scores to Structure the Association

The model requires assigning scores $\{u_i\}$ and $\{v_j\}$ to the rows and columns. To reflect category orderings, we take $u_1 \le u_2 \le \cdots \le u_I$ and $v_1 \le v_2 \le \cdots \le v_J$. The model is

$$\log m_{ij} = \mu + \lambda_i^X + \lambda_j^Y + \beta u_i v_j . \tag{8.1}$$

The independence model is the special case $\beta = 0$. Since $\{u_i\}$ and $\{v_j\}$ are fixed, model (8.1) has only one more parameter (β) than that model. Its residual degrees of freedom are

$$\mathrm{df} = IJ - [1 + (I-1) + (J-1) + 1] = IJ - I - J .$$

The model is unsaturated for all but 2×2 tables. It is the special case of the saturated model

$$\log m_{ij} = \mu + \lambda_i^X + \lambda_j^Y + \lambda_{ij}^{XY} \tag{8.2}$$

in which λ_{ij}^{XY} takes the structured form $\beta u_i v_j$. Model (8.1) requires only one parameter to describe association, regardless of the numbers of rows and columns, whereas the saturated model requires $(I-1)(J-1)$ parameters.

For identifiability in (8.1), we impose a constraint such as $\Sigma \lambda_i^X = \Sigma \lambda_j^Y = 0$. The definition of the scores affects the interpretation of β, which describes strength of association. In choosing scores, we might attempt to approximate distances between midpoints of categories for an assumed underlying interval scale. Simple interpretations result when scores are equal-interval; that is, when $u_2 - u_1 = u_3 - u_2 = \cdots = u_I - u_{I-1}$ and $v_2 - v_1 = \cdots = v_J - v_{J-1}$. The unit-spaced scores $\{u_i = i\}$ and $\{v_j = j\}$ or their zero-sum equivalents $\{u_i = i - (I+1)/2\}$ and $\{v_j = j - (J+1)/2\}$ are simple equal-interval scores. For the latter scores, $\Sigma_i \beta u_i v_j = \Sigma_j \beta u_i v_j = 0$. It is sometimes useful to linearly transform scores to achieve the standardization

$$\sum u_i \pi_{i+} = \sum v_j \pi_{+j} = 0$$

$$\sum u_i^2 \pi_{i+} = \sum v_j^2 \pi_{+j} = 1 .$$

We can express model (8.1) as

$$\log m_{ij} = \text{independence} + \beta u_i v_j .$$

For the coding $\Sigma u_i = \Sigma v_j = 0$, we can regard $\beta u_i v_j$ as a deviation of $\log m_{ij}$ from independence. If $\beta > 0$, $\beta u_i v_j$ is positive for (small X, small Y) and (large X, large Y) values. For such cells, m_{ij} is greater than if X and Y were independent. The expected number of (large X, small Y) or (small X, large Y) observations is smaller than if X and Y were independent. The deviation from independence increases in the directions of the four corner cells. This model is a simple way of imposing a monotone trend, the direction depending on the sign of β.

The deviation $\beta u_i v_j$ of $\log m_{ij}$ from independence is linear in Y for fixed X and linear in X for fixed Y. In column j, for instance, the deviation is a linear function of X (represented by the scores $\{u_i\}$), with slope βv_j. Because of this property, (8.1) is called the *linear-by-linear association* model. Birch (1965), Haberman (1974b), and Goodman (1979a) introduced special cases of this model.

Formula (8.1) has the same structural form as the log density of the bivariate normal distribution. The model tends to fit well when an underlying continuous distribution is approximately bivariate normal. For standardized scores, β is comparable in value to $\rho/(1 - \rho^2)$, where ρ is the underlying correlation (Problem 8.33). For weak associations, $\beta \doteq \rho$.

8.1.2 Uniform Association

We can use odds ratios to interpret the magnitude of β in the linear-by-linear association model. For an arbitrary pair of rows $h < i$ and an arbitrary pair of columns $j < k$,

$$\log\left(\frac{m_{hj}m_{ik}}{m_{hk}m_{ij}}\right) = \beta(u_i - u_h)(v_k - v_j) . \tag{8.3}$$

The absolute log odds ratio is larger for pairs of rows or columns that are farther apart. We interpret β as the log odds ratio per unit distances $u_i - u_h = v_k - v_j = 1$ on X and Y. The odds ratio equals e^β whenever rows are one unit apart and columns are one unit apart.

To describe properties of models for ordinal variables, it is useful to use the *local odds ratios*

$$\theta_{ij} = \frac{m_{ij}m_{i+1,j+1}}{m_{i,j+1}m_{i+1,j}} . \tag{8.4}$$

For the linear-by-linear association model,

$$\log \theta_{ij} = \beta(u_{i+1} - u_i)(v_{j+1} - v_j) .$$

For equal-interval scores, all local odds ratios are equal. Goodman (1979a) referred to this case as *uniform association*. For unit-spaced scores, all $\log \theta_{ij} = \beta$ and all $\theta_{ij} = e^\beta$. Unless an uneven spacing of scores is natural, we suggest using such scores so β can be interpreted simply as the common local log odds ratio. (see Figure 8.1).

8.1.3 Logit Model for Adjacent Responses

A logit formulation of the model is sensible when Y is a response and X is explanatory. Let $\pi_{j|i}$ be the probability that Y is in column j, for observations in row i of variable X. We utilize the ordinality of Y by constructing logits for adjacent response categories. For the linear-by-linear association model, this gives

$$\log\left(\frac{\pi_{j+1|i}}{\pi_{j|i}}\right) = \log\left(\frac{m_{i,j+1}}{m_{ij}}\right) = (\lambda_{j+1}^Y - \lambda_j^Y) + \beta(v_{j+1} - v_j)u_i .$$

For unit-spaced $\{v_j\}$, this simplifies to

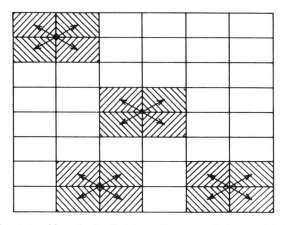

Figure 8.1 Constant odds ratio implied by uniform association model. (*Note:* β = the constant log odds ratio for adjacent rows and adjacent columns.)

$$\log\left(\frac{\pi_{j+1\,|\,i}}{\pi_{j\,|\,i}}\right) = \alpha_j + \beta u_i \tag{8.5}$$

where $\alpha_j = \lambda^Y_{j+1} - \lambda^Y_j$. The same linear logit effect applies simultaneously for all $(J-1)$ pairs of adjacent response categories. There is a simple interpretation: The odds of classification in column $j+1$ instead of column j are multiplied by e^β for each unit change in X. In using equal-interval response scores, we implicitly assume the effect of X is the same on each of the $J-1$ logits for Y.

For this model, the conditional distributions on Y are stochastically ordered (Problem 8.18). They are stochastically increasing in X if $\beta > 0$ and stochastically decreasing in X if $\beta < 0$.

8.1.4 Fitting the Model

For Poisson sampling, the kernel of the log-likelihood is $L(\mathbf{m}) = \Sigma\,\Sigma\,n_{ij}\log(m_{ij}) - \Sigma\,\Sigma\,m_{ij}$. For linear-by-linear association (8.1), this simplifies to

$$L(\mathbf{m}) = n\mu + \sum n_{i+}\lambda^X_i + \sum n_{+j}\lambda^Y_j + \beta \sum\sum u_i v_j n_{ij}$$
$$- \sum\sum \exp[\mu + \lambda^X_i + \lambda^Y_j + \beta u_i v_j].$$

By differentiating $L(\mathbf{m})$ with respect to λ^X_i, λ^Y_j, and β, and setting the three partial derivatives equal to zero, we obtain the likelihood equations

$$\hat{m}_{i+} = n_{i+}, \quad i = 1, \ldots, I$$

$$\hat{m}_{+j} = n_{+j}, \quad j = 1, \ldots, J$$

$$\sum \sum u_i v_j \hat{m}_{ij} = \sum \sum u_i v_j n_{ij}.$$

As happens for the independence model, the fitted table has the same marginal distributions as the observed table. Let $p_{ij} = n_{ij}/n$, and let $\hat{\pi}_{ij} = \hat{m}_{ij}/n$ denote the model ML estimate of π_{ij}. The third likelihood equation is

$$\sum \sum u_i v_j \hat{\pi}_{ij} = \sum \sum u_i v_j p_{ij}.$$

The expected cross-product of scores is the same for the fitted joint distribution as for the observed one. Since marginal distributions are identical for fitted and observed distributions, marginal expectations such as means and variances of the scores are the same. Thus, the third equation implies that the correlation between the scores for X and Y is the same for both distributions.

The likelihood equations do not have a direct solution, but we can obtain ML estimates using the Newton-Raphson method. For description and inference, the important parameter is β. For large samples, $\hat{\beta}$ is approximately normally distributed. We obtain its estimated standard error from the inverse of the estimated information matrix. This is a by-product of the Newton-Raphson routine for fitting the model.

8.1.5 Job Satisfaction Example

For the data on income and job satisfaction (Table 8.1), the independence model does not fit poorly, yet adjusted residuals indicate systematic lack of fit. Table 8.2 contains fitted values for the independence (I) model and the uniform association (U) model. Uniform association provides a much improved fit in the corners of the table, where it predicts the greatest departures from independence. Table 8.3 summarizes goodness of fit. The likelihood-ratio statistic for the independence model equals $G^2(I) = 12.03$, based on df $= 9$, giving a P-value of 0.21. The uniform association model has $G^2(U) = 2.39$, based on df $= 8$.

For unit-spaced scores, the estimate of the association parameter is $\hat{\beta} = 0.112$ (ASE $= 0.036$). The positive value indicates that job satisfaction tends to be greater at higher income levels. The estimated uniform local odds ratio is $\hat{\theta}_{ij} = \exp(0.112) = 1.12$. The estimated odds that job satisfaction is in category $j + 1$ instead of j increase by a multiplicative factor of

Table 8.2 Observed and Fitted Frequencies for Independence and Uniform Association Models

| | Job Satisfaction | | | | |
Income (\$)	Very Dissatisfied	Little Dissatisfied	Moderately Satisfied	Very Satisfied	Total
<6000	20	24	80	82	206
	$(14.2)^a$	(24.7)	(72.9)	(94.2)	
	$(19.3)^b$	(29.4)	(74.9)	(82.3)	
6000–15,000	22	38	104	125	289
	(19.9)	(34.6)	(102.3)	(132.2)	
	(21.4)	(36.4)	(103.7)	(127.4)	
15,000–25,000	13	28	81	113	235
	(16.2)	(28.2)	(83.2)	(107.5)	
	(13.6)	(25.9)	(82.4)	(113.2)	
>25,000	7	18	54	92	171
	(11.8)	(20.5)	(60.5)	(78.2)	
	(7.6)	(16.3)	(58.0)	(89.1)	
Total	62	108	319	412	901

[a] For independence model $\hat{m}_{ij}$.
[b] For uniform association model $\hat{m}_{ij}$.

1.12 for each category change in income. Using the four corner cells alone, the estimated odds of being very satisfied instead of very dissatisfied is $\exp[0.112(4-1)(4-1)] = 2.74$ times higher for the (>\$25,000) income group than for the (<\$6000) income group. A 95% confidence interval for $\theta_{ij} = \exp(\beta)$ is $\exp(0.112 \pm 1.96 \times 0.036)$, or (1.04, 1.20).

For the scores $\{v_j = j\}$, the marginal mean and standard deviation for job satisfaction are 3.20 and 0.90. The standardized scores are $\{(j - 3.20)/0.90\}$, or (−2.44, −1.33, −0.22, 0.89). The standardized equal-interval scores for income are (−1.36, −0.40, 0.57, 1.53). For the

Table 8.3 Goodness-of-Fit of Models for Job Satisfaction Data

Model	G^2	df
Independence	12.03	9
Uniform association	2.39	8
Independence, given uniform association	9.64	1

standardized scores, $\hat{\beta} = 0.105$. By setting $\hat{\beta} = \hat{\rho}/(1 - \hat{\rho}^2)$, we obtain $\hat{\rho} = 0.104$. If there is an underlying bivariate normal distribution, we estimate the correlation to be 0.104. Though the uniform association model provides evidence of a positive association, the strength of the association appears to be rather weak.

8.1.6 Directed Test of Independence

Denote the linear-by-linear association model by $L \times L$ and the independence model by I. Assuming the $L \times L$ model holds, we assess the statistical significance of the association by testing H_0: $\beta = 0$. One test statistic is the reduction in G^2

$$G^2(I \mid L \times L) = G^2(I) - G^2(L \times L) \qquad (8.6)$$

based on df $= 1$. For the job satisfaction data, the reduction $12.03 - 2.39 = 9.64$ results from adding the uniform association term to the independence model. The P-value is 0.002 for testing H_0: $\beta = 0$. This is much stronger evidence of association than given by the test based directly on $G^2(I)$. If a positive association truly exists between job satisfaction and income, we have been more successful at detecting it using this ordinal approach.

The ratio $z = \hat{\beta}/\text{ASE}$ is an alternative statistic for testing H_0:$\beta = 0$. Unlike $G^2(I \mid L \times L)$, it retains information about direction of association, so it is useful for the alternatives H_a: $\beta > 0$ and H_a: $\beta < 0$. Another statistic of this type is the *score* statistic, described in Problem 8.29.

It follows from the likelihood equations that $\beta = 0$ is equivalent to a population correlation of zero, for the given scores. Thus these ordinal tests are analogs for ordered categorical variables of the t-test for a linear regression slope or correlation for continuous variables.

8.2 ROW EFFECTS AND COLUMN EFFECTS MODELS

This section presents a loglinear model that treats X as nominal and Y as ordinal. The model is appropriate for two-way tables with ordered column classifications.

8.2.1 Row Effects Model

We assign ordered scores $v_1 \leqslant v_2 \leqslant \cdots \leqslant v_J$ to reflect the ordering of the columns. The rows are now unordered, so it does not make sense to assign scores to them. We continue to use linear-by-linear structure for the association, replacing the ordered values $\{\beta u_i\}$ in model (8.1) by unordered parameters $\{\mu_i\}$. This gives the model

$$\log m_{ij} = \mu + \lambda_i^X + \lambda_j^Y + \mu_i v_j . \tag{8.7}$$

The $\{v_j\}$ are fixed constants, and the $\{\mu_i\}$ parameters are called *row effects*.

For identifiability, we use constraints $\Sigma \lambda_i^X = \Sigma \lambda_j^Y = \Sigma \mu_i = 0$. Model (8.7) has $I - 1$ more parameters (the $\{\mu_i\}$) than the independence model, and has residual df equal to

$$\text{df} = IJ - [1 + (I - 1) + (J - 1) + (I - 1)] = (I - 1)(J - 2) .$$

It is called the *row effects model*. The linear-by-linear association model is the special case in which $\{\mu_i\}$ have the linear pattern $\{\mu_i = \beta u_i\}$, for a set of ordered row scores $\{u_i\}$. The independence model is the special case in which all μ_i are equal.

Odds ratio interpretations of $\{\mu_i\}$ are simplest for unit-spaced $\{v_j\}$. Then, the log odds ratio for an arbitrary pair of rows h and i and adjacent columns j and $j + 1$ is

$$\log \left(\frac{m_{hj} m_{i,j+1}}{m_{h,j+1} m_{ij}} \right) = \mu_i - \mu_h . \tag{8.8}$$

The odds ratio is identical for all $J - 1$ pairs of adjacent columns. The $\{\mu_i - \mu_h\}$ describe differences among rows with respect to their conditional distributions on Y. When $\mu_i = \mu_h$, rows h and i have identical conditional distributions on Y. If $\mu_i > \mu_h$, Y is stochastically higher in row i than row h (Problem 8.18). As μ_i increases, subjects in row i are more likely to make higher responses on the ordinal scale.

When we permute rows of a sample table, the row effects model has the same form and produces the same G^2 value. It treats the row variable as nominal, but is a valid model whether rows are nominal or ordinal. We might use it for ordered rows when we do not want to assign scores $\{u_i\}$ to them. The estimates $\{\hat{\mu}_i\}$ then indicate that model form (8.1) fits best when $\{u_i\}$ are a linear transformation of $\{\hat{\mu}_i\}$. Or, we might use it when we want to compare rows' conditional distributions on Y, but we do not want to impose a particular pattern (e.g., linear) for the way those distributions change.

A corresponding *column effects model* has association term $u_i v_j$. It treats the row variable as ordinal, represented by ordered scores $\{u_i\}$, and the column variable as nominal, with unknown parameters $\{v_j\}$. The row effects and column effects models were developed by Haberman (1974b), Simon (1974), and Goodman (1979a).

8.2.2 Logit Model for Adjacent Responses

When the ordinal variable Y is a response variable, a logit model for adjacent responses gives a simple way to describe effects of X on Y. For the row effects model,

$$\log\left(\frac{m_{i,j+1}}{m_{ij}}\right) = (\lambda_{j+1}^Y - \lambda_j^Y) + \mu_i(v_{j+1} - v_j) .$$

For unit-spaced column scores, this has form

$$\log\left(\frac{\pi_{j+1|i}}{\pi_{j|i}}\right) = \alpha_j + \mu_i . \tag{8.9}$$

This is a simple logit model for an ordinal response and nominal explanatory variable. The $\{\mu_i\}$ are the row effects of making response $j + 1$ instead of j. The effect in row i is identical for each pair of adjacent responses. When these logits are plotted against i ($i = 1, \ldots, I$), the plots for different j are parallel. Goodman (1983) referred to model (8.9) as the *parallel odds* model.

8.2.3 Fitting the Model

The likelihood equations for the row effects model are

$$\hat{m}_{i+} = n_{i+}, \qquad i = 1, \ldots, I$$

$$\hat{m}_{+j} = n_{+j}, \qquad j = 1, \ldots, J \tag{8.10}$$

$$\sum_j v_j \hat{m}_{ij} = \sum_j v_j n_{ij}, \quad i = 1, \ldots, I .$$

The fitted and observed counts have the same marginal distributions. Let $\hat{\pi}_{j|i} = \hat{m}_{ij}/\hat{m}_{i+}$ and $p_{j|i} = n_{ij}/n_{i+}$. Since $\hat{m}_{i+} = n_{i+}$, the third likelihood equation is $\Sigma_j v_j \hat{\pi}_{j|i} = \Sigma_j v_j p_{j|i}$. For the conditional distribution within each row, the mean of the column scores is the same when the distribution is based on the fitted values as when it is based on the sample data. The likelihood equations can be solved iteratively, using the Newton-Raphson algorithm.

Assuming the model holds, we can test independence by testing H_0: $\mu_1 = \cdots = \mu_I$. The test statistic is

$$G^2(I \mid R) = G^2(I) - G^2(R)$$

where $G^2(R)$ is the likelihood-ratio statistic for testing the fit of the row effects model. This test has $df = (I - 1)(J - 1) - (I - 1)(J - 2) = I - 1$. Given that the row effects model holds, $\mu_1 = \cdots = \mu_I$ is equivalent to equality of the I means of the conditional distributions within the rows. Thus the $G^2(I \mid R)$ test is an analog for an ordered categorical response of the one-way ANOVA F-test for a continuous response. Similarly, the row effects model is to the linear-by-linear association model as the one-way ANOVA model is to the linear regression model for continuous response variables.

Since the linear-by-linear association model is the special case of the row effects model with $\mu_i = \beta u_i$, we can test whether row effect parameters have the pattern given by a fixed set of scores $\{u_i\}$. The relevant test statistic is $G^2(L \times L \mid R) = G^2(L \times L) - G^2(R)$, based on $df = I - 2$, where the $L \times L$ model uses the fixed scores. This test is useful for checking the validity of a given choice of row scores for the $L \times L$ model.

The parameter estimates in the row effects model have an asymptotic normal distribution, with asymptotic covariance matrix equal to the inverse of the information matrix. That matrix also determines standard errors for $\{\hat{\mu}_i - \hat{\mu}_h\}$, which estimate log odds ratios for describing differences among rows.

8.2.4 Political Ideology Example

Table 8.4 is based on data presented by Hedlund (1978). It presents the relationship between an ordinal variable, political ideology, and a nominal variable, political party affiliation, for a sample of voters in the 1976 presidential primary in Wisconsin. This table contains fitted values for the independence model and for the row effects model with scores $\{v_j = j\}$.

The estimated row effects of party affiliation on ideology, with ASE values in parentheses, are $\hat{\mu}_1 = -0.495 \, (0.062)$, $\hat{\mu}_2 = -0.224 \, (0.059)$, and $\hat{\mu}_3 = 0.719 \, (0.080)$. The further $\hat{\mu}_i$ falls in the positive direction, the greater the tendency for the ith party to locate at the conservative end of the ideology scale. In this sample the Democrats (row 1) are the least conservative group, and the Republicans are much more conservative than the other two groups.

The row effects model with equal-interval scores predicts constant odds ratios for adjacent columns of political ideology. To illustrate,

Table 8.4 Observed Frequencies and Estimated Expected Frequencies for Independence Model and for Loglinear Row Effects Model for Political Ideology Data

Party Affiliation	Political ideology			Total
	Liberal	Moderate	Conservative	
Democrat	143	156	100	399
	$(102.0)^a$	(161.4)	(135.6)	
	$(136.6)^b$	(168.7)	(93.6)	
Independent	119	210	141	470
	(120.2)	(190.1)	(159.7)	
	(123.8)	(200.4)	(145.8)	
Republican	15	72	127	214
	(54.7)	(86.6)	(72.7)	
	(16.6)	(68.9)	(128.6)	

[a] For independence model $\hat{m}_{ij}$.
[b] For row effects model $\hat{m}_{ij}$.

$\hat{\mu}_3 - \hat{\mu}_1 = 1.214$ means that the odds of being classified conservative instead of moderate, and the odds of being classified moderate instead of liberal, are estimated to be $\exp(1.214) = 3.37$ times higher for Republicans than for Democrats.

Table 8.5 summarizes goodness of fit of the independence and row effects models. The independence model is completely inadequate, with $G^2(I) = 105.66$ based on df $= 4$. The addition of the two linearly independent row effect parameters produces a much better fit, with $G^2(R) = 2.81$ based on df $= 2$. The improved fit is especially noticeable at the ends of the ordinal scale, where the row effects model predicts the greatest deviation from independence. For these data, $G^2(I \mid R) = 105.66 - 2.81 = 102.85$, based on df $= 4 - 2 = 2$, shows very strong evidence of an association.

To compare a pair of rows using the row effects model, we can form a confidence interval for $\mu_b - \mu_a$ or for its antilog odds ratio. For Table 8.4

Table 8.5 Goodness-of-Fit of Models for Political Ideology Data

Model	G^2	df
Independence	105.66	4
Row effects	2.81	2
Independence, given row effects	102.85	2

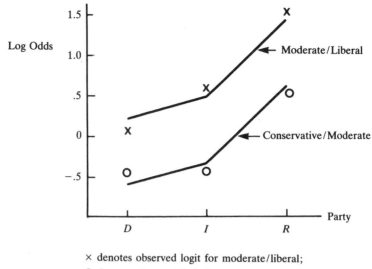

× denotes observed logit for moderate/liberal;
○ denotes observed logit for conservative/moderate.

Figure 8.2 Observed and Predicted Logits for Adjacent Response Categories.

these intervals provide strong evidence of a difference between each pair of party affiliations, with the conclusion $\mu_1 < \mu_2 < \mu_3$. To illustrate, the estimated covariance between $\hat{\mu}_3$ and $\hat{\mu}_1$ is -0.00337. Thus the ASE of $(\hat{\mu}_3 - \hat{\mu}_1)$ is $[(0.062)^2 + (0.080)^2 - 2(-0.00337)]^{1/2} = 0.130$. An approximate 95% confidence interval for $\exp(\mu_3 - \mu_1)$ is $\exp[1.214 \pm 1.96(0.130)]$, or $(2.60, 4.34)$.

For unit-spaced scoring of columns, from (8.9) the predicted logits for adjacent responses have form $\hat{\alpha}_j + \hat{\mu}_i$. For Table 8.4, the predicted logits for the row effects model have the parallelism shown in Figure 8.2.

8.3 MODELS FOR ORDINAL VARIABLES IN MULTIDIMENSIONAL TABLES

Multidimensional tables having ordinal variables can utilize generalizations of loglinear models introduced in Sections 8.1 and 8.2. In three dimensions, the rich collection of models includes (1) partial association models that are more parsimonious and simpler to interpret than the nominal-scale model (XY, XZ, YZ), and (2) models permitting three-factor interaction that, unlike model (XYZ), are unsaturated.

8.3.1 No Three-Factor Interaction

Models that assume a lack of three-factor interaction are special cases of model (XY, XZ, YZ). We achieve parsimony and simplicity of description by replacing λ association terms in this model by structured terms that account for ordinality.

For instance, suppose we want to model the $X-Y$ conditional association. This is described in model (XY, XZ, YZ) by the general term λ_{ij}^{XY}. When X is nominal and Y is ordinal, we might replace this with the type of term used in a row effects model; that is, $\mu_i v_j$, where $\{\mu_i\}$ are parameters for the levels of X and $\{v_j\}$ are fixed scores for the levels of Y. When both X and Y are ordinal, we could replace λ_{ij}^{XY} by a row effects term $\mu_i v_j$, a column effects term $u_i v_j$, or a linear-by-linear term $\beta u_i v_j$, depending on whether we expect a stochastic ordering of conditional distributions within rows, within columns, or both.

Replacing the λ_{ij}^{XY} term in (XY, XZ, YZ) by a linear-by-linear term gives

$$\log m_{ijk} = \mu + \lambda_i^X + \lambda_j^Y + \lambda_k^Z + \beta u_i v_j + \lambda_{ik}^{XZ} + \lambda_{jk}^{YZ}. \tag{8.11}$$

The conditional local odds ratios (5.18) then satisfy

$$\log \theta_{ij(k)} = \beta(u_{i+1} - u_i)(v_{j+1} - v_j) \quad \text{for all } k.$$

Since the association is the same in different partial tables, we say there is *homogeneous linear-by-linear association*.

8.3.2 Dumping Severity Example

Table 8.6 is a $4 \times 4 \times 3$ cross-classification of operation (O), hospital (H), and dumping severity (D) presented by Grizzle et al. (1969). The operations refer to treatments for duodenal ulcer patients. Operation A is

Table 8.6 Cross-Classification of Operation, Hospital, and Dumping Severity[a]

Operation	Hospital 1			Hospital 2			Hospital 3			Hospital 4		
	N	S	M	N	S	M	N	S	M	N	S	M
A	23	7	2	18	6	1	8	6	3	12	9	1
B	23	10	5	18	6	2	12	4	4	15	3	2
C	20	13	5	13	13	2	11	6	2	14	8	3
D	24	10	6	9	15	2	7	7	4	13	6	4

Source: Reprinted with permission from the Biometric Society (Grizzle et al. 1969).
[a] N = none, S = slight, M = moderate dumping severity.

drainage and vagotomy, B is 25% resection and vagotomy, C is 50% resection and vagotomy, and D is 75% resection. Vagotomy is cutting of a nerve to reduce the flow of gastric juice, and resection refers to removal of part of the stomach. The operation categories have a natural ordering, with A the least severe operation and D having the greatest removal of stomach. The dumping severity variable describes a possible undesirable side effect of the operation. It is characterized by sweating, weakness, or collapse of vascular or nervous system responses, soon after eating. Its categories are also ordered, with "none" representing the most desirable response. The data were collected from four hospitals. One purpose of the study was to determine the influence of type of operation on dumping severity.

Since dumping severity is the response variable, we treat as fixed the number of patients at each $O-H$ combination. Models that contain the λ^{OH} term force the fitted values to have these same $O-H$ marginal totals. Table 8.7 summarizes goodness-of-fit tests for standard loglinear models of this type.

The model (OH, D), according to which dumping severity is jointly independent of both operation and hospital, appears to fit adequately. We should be cautious about this tentative conclusion, since many cell counts are small. Somewhat reassuring is the fact that G^2 tends to be liberal (give P-values that are too small) when many expected frequencies are between 1 and 5. To test the hypothesis that D is statistically independent of O, we compare this model to model (OH, OD), obtaining

$$G^2[(OH, D)|(OH, OD)] = 31.64 - 20.76 = 10.88$$

based on $df = 30 - 24 = 6$. This statistic tests the hypothesis that the λ^{OD} terms equal zero in model (OH, OD). It has P-value 0.092 and shows slight evidence of an $O-D$ association. By Section 7.4.2, this result is identical to the test of independence in the $O-D$ marginal table.

Table 8.7 Goodness-of-Fit Tests for Loglinear Models Relating Operation (O), Dumping Severity (D), and Hospital (H)

Model	G^2	df	P-value
(OH, D)	31.64	30	0.38
(OH, OD)	20.76	24	0.65
(OH, HD)	23.54	24	0.49
(OH, OD, HD)	12.50	18	0.82

A disadvantage of this test is that it treats both O and D as nominal. Standardized values of $\{\hat{\lambda}^{OD}\}$ for model (OH, OD), reported in Table 8.8, give evidence of a monotone trend in this association. Positive $\hat{\lambda}^{OD}$ values occur in corners of the table where operation severity and dumping severity both take high values or both take low values. Moderate negative $\hat{\lambda}^{OD}$ values occur where one variable is at the high end of the scale and the other is at the low end. This table suggests replacing λ_{ik}^{OD} by a linear-by-linear term $\beta u_i v_k$. Model

$$\log m_{ijk} = \mu + \lambda_i^O + \lambda_j^H + \lambda_k^D + \lambda_{ij}^{OH} + \beta u_i v_k \qquad (8.12)$$

fitted using equal-interval scores has residual $G^2 = 25.35$ based on df = 29. In this model, dumping severity and operation have a uniform association that is the same for each hospital, and dumping severity is independent of hospital for each operation.

For model (8.12), $\beta = 0$ represents the hypothesis of conditional independence of dumping severity and operation. A test statistic is the difference between $G^2(OH, D) = 31.64$ and $G^2 = 25.35$ for model (8.12), based on df = 1. This statistic value of 6.29 has P-value 0.012. Although model (OH, D) seems to fit well, it gives a poorer fit than model (8.12). We obtain much stronger evidence of O–D association when we utilize their ordinal scalings. Since model (8.12) satisfies collapsibility conditions (5.21) for the O–D association, we obtain the same result by comparing the independence model and the uniform association model for the marginal O–D table.

Table 8.9 summarizes results of fitting several nested loglinear models. Comparing the third and fourth models in that table, we see there is not much evidence of an H–D association. Model (8.12) seems to describe these data adequately.

Since D is a response variable, we now consider equivalent logit models. For unit-spaced $\{v_k\}$, model (8.12) is equivalent to the adjacent-categories linear logit model

Table 8.8   Standardized Values of O–D Association Parameter Estimates, for Model (OD, OH)

Operation	Dumping Severity		
	None	Slight	Moderate
A	1.59	0.53	−1.39
B	1.21	−2.00	0.80
C	−0.93	1.17	−0.23
D	−2.11	0.43	1.18

Table 8.9 Analysis of Loglinear Models for Dumping Severity Data

Association Term					Difference	Difference
$O–H$	$O–D$	$H–D$	G^2	df	in G^2	in df
λ_{ij}^{OH}	–	–	31.64	30		
λ_{ij}^{OH}	$ik\beta^{OD}$	–	25.35	29	6.29	1
λ_{ij}^{OH}	λ_{ik}^{OD}	–	20.76	24	4.59	5
λ_{ij}^{OH}	λ_{ik}^{OD}	λ_{jk}^{HD}	12.50	18	8.26	6

$$\log\left(\frac{m_{ij,k+1}}{m_{ijk}}\right) = \alpha_k + \beta u_i .$$

When $\{u_i\}$ are also unit-spaced, the ML estimate of the $O–D$ association parameter is $\hat{\beta} = 0.163$. The estimated odds of moderate instead of slight dumping, or of slight instead of no dumping, are $\exp(0.163) = 1.18$ times higher for each additional 25% of stomach removal.

8.3.3 Modeling Three-Factor Interaction

When a relationship exhibits interaction, structured terms for ordinal variables often make the interaction simpler to interpret. In interpreting models for three-factor interaction, we refer to conditional odds ratios (5.18), and also to their ratios

$$\theta_{ijk} = \frac{\theta_{ij(k+1)}}{\theta_{ij(k)}} = \frac{\theta_{i(j+1)k}}{\theta_{i(j)k}} = \frac{\theta_{(i+1)jk}}{\theta_{(i)jk}} . \qquad (8.13)$$

The $\{\theta_{ijk}\}$ describe three-factor interaction in $2 \times 2 \times 2$ sub-tables consisting of adjacent rows, adjacent columns, and adjacent layers. There is an absence of three-factor interaction when all $(I-1)(J-1)(K-1)$ of the $\{\theta_{ijk}\}$ equal 1.0.

Suppose the no three-factor interaction model (XY, XZ, YZ) is inadequate. We can exploit category orderings to construct models that are nested between (XY, XZ, YZ) and the saturated model (XYZ). Such models permit three-factor interaction but have positive residual df. For example, suppose all three variables are ordinal and are assigned ordered scores $\{u_i\}$, $\{v_j\}$, and $\{w_k\}$. The model

$$\log m_{ijk} = \mu + \lambda_i^X + \lambda_j^Y + \lambda_k^Z + \lambda_{ij}^{XY} + \lambda_{ik}^{XZ} + \lambda_{jk}^{YZ} + \beta u_i v_j w_k$$

$$(8.14)$$

has residual df $= (I-1)(J-1)(K-1) - 1$, so it is unsaturated whenever I, J, or K exceeds 2. Its interpretation is that a log odds ratio for any two variables changes linearly across the levels of the third variable. When this model holds with unit-spaced scores,

$$\log \theta_{ijk} = \beta \quad \text{for all } i, j, k$$

and it is called the *uniform interaction* model (Goodman 1979a).

More structured models provide patterns for associations as well as interactions, maintaining the principle of a hierarchical model. For instance, suppose we want to describe how the conditional X–Y association changes across levels of Z. If X and Y are ordinal, we might fit the heterogeneous linear-by-linear association model

$$\log m_{ijk} = \mu + \lambda_i^X + \lambda_j^Y + \lambda_k^Z + \lambda_{ik}^{XZ} + \lambda_{jk}^{YZ} + (\beta + \beta_k) u_i v_j$$

$$(8.15)$$

where $\Sigma \beta_k = 0$. For this model applied with unit-spaced scores,

$$\log \theta_{ij(k)} = \beta + \beta_k \quad \text{for all } i \text{ and } j.$$

It implies uniform association between X and Y within each level of Z, but heterogeneity among levels of Z in the degree of that association.

Fitting model (8.15) corresponds to fitting the linear-by-linear association model (8.1) separately at each level of Z. The model has residual df $= K(IJ - I - J)$. The homogeneous linear-by-linear association model (8.11) is the special case in which all β_k are equal. When $I > 2$ or $J > 2$, model (8.15) does not contain (XY, XZ, YZ) as a special case, because of the structured form of the X–Y association. This model treats Z as nominal, and is not comparable to (8.14), neither being a special case of the other. Suppose Z is ordinal, and we replace $\beta + \beta_k$ in (8.15) by βw_k. That model used with unit-spaced scores *is* a special case of the uniform interaction model, one in which the heterogeneity across Z in the X–Y uniform association is linear.

8.3.4 Effect of Smoking Example

Table 8.10, taken from Forthofer and Lehnen (1981, p. 21), displays associations among smoking status (S), breathing test results (B), and age (A) for Caucasians in certain industrial plants in Houston, Texas, during 1974 and 1975. The loglinear model (SA, SB, BA) fits poorly, with

Table 8.10   Cross-Classification of Houston Industrial Workers by Breathing Test Results, Smoking Status, and Age

Age	Smoking Status	Breathing Test Results		
		Normal	Borderline	Abnormal
<40	Never smoked	577	27	7
	Former smoker	192	20	3
	Current smoker	682	46	11
40–59	Never smoked	164	4	0
	Former smoker	145	15	7
	Current smoker	245	47	27

Source: From *Public Program Analysis* by R. N. Forthofer and R. G. Lehnen. Copyright © 1981 by Lifetime Learning Publications, Belmont, CA 94002, a division of Wadsworth, Inc. Reprinted by permission of Van Nostrand Reinhold. All rights reserved.

$G^2 = 25.93$ based on df = 4, so we consider models permitting three-factor interaction.

Breathing test is a response variable, and age and smoking status are explanatory variables. We treat smoking status as ordinal, in terms of how recently one was a smoker. Since age has only two categories, it is equivalent to treat it as nominal or ordinal. Let m_{ijk} denote the expected frequency for level i of smoking status, level j of age, and level k of breathing test result. Let $\{u_i\}$ be scores for smoking status and $\{v_j\}$ scores for age. The adjacent-categories logit model

$$\log\left(\frac{m_{ij,k+1}}{m_{ijk}}\right) = \alpha_k + \beta_1 u_i + \beta_2 v_j + \beta_3 u_i v_j$$

assumes a linear effect of smoking status that differs according to the level of age. For instance, when we set age scores $v_1 = 0$ and $v_2 = 1$, β_1 describes the effect of smoking when age < 40, and $\beta_1 + \beta_3$ describes the effect of smoking for age 40–59.

This logit model is equivalent to loglinear model

$$\log m_{ijk} = \mu + \lambda_i^S + \lambda_j^A + \lambda_k^B + \lambda_{ij}^{SA} + \beta_1 u_i w_k + \beta_2 v_j w_k + \beta_3 u_i v_j w_k$$

$$(8.16)$$

where $\{w_k = k\}$ for levels of breathing test. The loglinear model has a general λ term for the S–A association, so that $\{\hat{m}_{ij+} = n_{ij+}\}$. For equal-interval $\{u_i\}$, it implies a uniform association between S and B that has parameters β_1 and $\beta_1 + \beta_3$ for the two levels of age. Hence, it is a special case of the heterogeneous uniform association model, one in which there is also uniform association between A and B that changes linearly across levels of S.

Table 8.11 contains results of fitting this model. Although it is not directly comparable to (SA, SB, BA), it gives a better fit using fewer parameters. It gives strong evidence of three-factor interaction, since the difference $\hat{\beta}_3 = 0.663$ between the uniform association parameters has ASE = 0.164. The effect of smoking is much stronger for the older age group, and is not particularly significant for the younger age group. The estimated local odds ratio between S and B is $\exp(0.115) = 1.12$ for the < 40 age group and $\exp(0.778) = 2.18$ for the 40-59 age group. For the older age group, the estimated odds of abnormal rather than borderline breathing are 2.18 times higher for current smokers than for former smokers, and $\exp(2 \times 0.778) = 4.74$ times higher for current smokers than for never smokers. We square these values to get the estimated odds of abnormal rather than normal breathing.

Model (8.16) is a special case of the uniform interaction model, which adds the term $\beta u_i v_j w_k$ to model (SA, SB, BA). The uniform interaction model fits very well, with $G^2 = 2.74$ based on df = 3. The ML estimate of $\log \theta_{ijk}$ is $\hat{\beta} = 0.83$, with ASE = 0.19. Any local odds ratio is estimated to be $\hat{\theta}_{ij1} = \exp(0.83) = 2.3$ times greater for the higher age group than for the lower age group. This model gives a marginal improvement in fit over model (8.16), though both models give the same basic conclusion. Effects are a bit simpler to interpret with model (8.16).

The difference of 23.19 in G^2 values between model (SA, SB, BA) and the uniform interaction model gives a single-degree-of-freedom test of the hypothesis of no three-factor interaction. The improvement in fit is dramatic and gives very strong evidence of interaction.

Table 8.11 Results of Model for Breathing Test

Parameter	Estimate	Standard Error	P-value
Smoking	0.115	0.086	0.18
Age	0.311	0.151	0.04
Smoking $\times$ age	0.663	0.164	0.00

$G^2 = 10.82$,   df = 7, $P = 0.15$

8.4 TESTING INDEPENDENCE FOR ORDINAL CLASSIFICATIONS

Sections 8.1 and 8.2 presented tests of independence that utilize the ordinality of classifications. The test statistics have form $G^2(I \mid M) = G^2(I) - G^2(M)$, where M is an ordinal model such as $L \times L =$ linear-by-linear association or $R =$ row effects. These tests yield different results than the standard $G^2(I)$ and $X^2(I)$ tests of independence. This is because $G^2(I \mid M)$ statistics are directed toward narrower alternatives. The $G^2(I \mid L \times L)$ statistic is designed to detect monotonicity, positive or negative associations whereby all local log odds ratios have the same sign. The $G^2(I \mid R)$ statistic focuses on ordered alternatives characterized by a stochastic ordering of rows with respect to conditional distributions on Y. Though $G^2(I)$ and $X^2(I)$ can detect *any* departure from the null hypothesis, they sacrifice power for the directional alternative hypotheses that are usually relevant for ordinal variables.

8.4.1 Improved Power with Ordinal Tests

When the $L \times L$ model holds, the ordinal test based on $G^2(I \mid L \times L)$ is asymptotically more powerful than the test based on $G^2(I)$. This is true for the same reason given in Section 4.4 for the linear logit model. The power of a chi-squared test increases when the degrees of freedom decrease, for fixed noncentrality. The ordinal test focuses the departure from independence on a single degree of freedom. When the $L \times L$ model holds, the noncentrality of $G^2(I \mid L \times L)$ is the same as the noncentrality for $G^2(I)$; thus $G^2(I \mid L \times L)$ is more powerful, since its df is smaller. For the same reason, the statistic $X^2(I \mid L \times L)$, which is (7.3) with $M_1 = L \times L$ and $M_2 = I$, is more powerful than $X^2(I)$.

Table 8.12 illustrates these remarks. It reports power approximations, based on Monte Carlo simulation, for the tests based on $X^2(I)$ and $X^2(I \mid L \times L)$. We compare the Pearson rather than the likelihood-ratio statistics, since the null chi-squared approximation is much better for $X^2(I)$ than $G^2(I)$. Results are given for various nominal α-levels, table sizes, and sample sizes. For these comparisons, the $L \times L$ model holds for a joint distribution corresponding to an underlying normal distribution with correlation 0.2. The cutpoints for the marginal $N(\mu, \sigma)$ distributions were selected at μ when $I = 2$, at $\mu \pm 0.6\sigma$ when $J = 3$, at μ and $\mu \pm 0.8\sigma$ when $I = J = 4$, and at μ, $\mu \pm 0.6\sigma$, $\mu \pm 1.2 \sigma$ when $I = J = 6$.

The power advantage of the ordinal test over the nominal test increases as I and J increase. As I and J increase, for fixed n, $X^2(I \mid L \times L)$ becomes more powerful but $X^2(I)$ loses power. For these tables, the noncentrality increases as I and J increase. As this happens, it remains

Table 8.12   Estimated Powers, When the $L \times L$ Model Holds and There is an Underlying Normal Distribution with Correlation 0.2

Statistic	$I \times J$	$n = 50$			$n = 100$		
	α:	0.01	0.05	0.10	0.01	0.05	0.10
$X^2(I)$	2×3	0.03	0.13	0.22	0.09	0.23	0.35
	4×4	0.02	0.10	0.17	0.05	0.16	0.26
	6×6	0.02	0.07	0.14	0.03	0.12	0.20
$X^2(I \mid L \times L)$	2×3	0.05	0.18	0.28	0.13	0.30	0.42
	4×4	0.09	0.24	0.35	0.19	0.42	0.54
	6×6	0.10	0.26	0.38	0.25	0.47	0.61

focused on a single degree of freedom for $X^2(I \mid L \times L)$. However, df also increases for $X^2(I)$, more than offsetting the increase in non-centrality.

The model-based tests also have the advantage of faster convergence to their asymptotic distribution. From Section 7.7.6, $G^2(I \mid L \times L)$ and $X^2(I \mid L \times L)$ depend on the data only through sufficient statistics for the $L \times L$ model, which are the row totals, columns totals, and the correlation. By contrast $G^2(I)$ and $X^2(I)$ depend also on individual cell counts. For sparse tables, the sufficient statistics for the $L \times L$ model are more nearly normally distributed than are cell counts. Thus, the chi-squared distribution applies better for the model-based goodness-of-fit statistics.

8.4.2   Mantel Score Test of Conditional Independence

We now turn our attention to tests of conditional independence in three-way tables. Section 7.4 introduced the Cochran-Mantel-Haenszel test of conditional independence in $2 \times 2 \times K$ tables. Mantel (1963) introduced generalized statistics of this type designed to detect association between ordinal variables.

Suppose we expect a monotone conditional relation between X and Y, with the same direction at each level of Z. If we assign monotone scores $\{u_i\}$ to levels of X and $\{v_j\}$ to levels of Y, then a correlation or covariance measure is sensitive to such a trend. There is evidence of a positive trend if, within each stratum, the statistic $\Sigma_i \Sigma_j u_i v_j n_{ijk}$ is greater than its expectation under independence. Given the marginal totals in each stratum, under conditional independence of X and Y,

$$E\left(\sum_i \sum_j u_i v_j n_{ijk}\right) = \frac{\left(\sum_i u_i n_{i+k}\right)\left(\sum_j v_j n_{+jk}\right)}{n_{++k}}$$

$$\text{Var}\left(\sum_i \sum_j u_i v_j n_{ijk}\right) = \frac{1}{(n_{++k}-1)}\left[\sum_i u_i^2 n_{i+k} - \frac{\left(\sum_i u_i n_{i+k}\right)^2}{n_{++k}}\right]$$

$$\times \left[\sum_j v_j^2 n_{+jk} - \frac{\left(\sum_j v_j n_{+jk}\right)^2}{n_{++k}}\right].$$

The statistic $[\Sigma \Sigma u_i v_j n_{ijk} - E(\Sigma \Sigma u_i v_j n_{ijk})]/[\text{Var}(\Sigma \Sigma u_i v_j n_{ijk})]^{1/2}$ is the correlation between X and Y in stratum k, multiplied by the square root of the sample size minus one for that stratum.

To summarize the correlation information from the K strata, Mantel (1963) proposed the statistic

$$M^2 = \frac{\left\{\sum_k \left[\sum_i \sum_j u_i v_j n_{ijk} - E\left(\sum_i \sum_j u_i v_j n_{ijk}\right)\right]\right\}^2}{\sum_k \text{Var}\left(\sum_i \sum_j u_i v_j n_{ijk}\right)}. \qquad (8.17)$$

When the total sample size is large, $\Sigma \Sigma u_i v_j n_{ij+}$ has approximately a normal distribution, so M^2 has an approximate chi-squared null distribution with df $= 1$.

8.4.3   Other Ordinal Tests of Conditional Independence

We can also test conditional independence of ordinal classifications by generalizing model-based tests such as $G^2(I \mid L \times L)$. For instance, when both variables are ordinal, we can compare the baseline model (XZ, YZ) of X–Y conditional independence to the homogeneous linear-by-linear association model (8.11). This test, like Mantel's score test, focuses evidence of partial association on a single degree of freedom. It also uses correlation information, since $\Sigma \Sigma u_i v_j n_{ij+}$ is in the minimal sufficient set for model (8.11) but not for (XZ, YZ).

When only Y is ordinal or when we do not wish to utilize the ordinal quality of X, we could compare (XZ, YZ) to model (8.11) with $\beta u_i v_j$ replaced by $\mu_i v_j$ for a set of row effect parameters $\{\mu_i\}$. This comparison has df $= I - 1$. The statistics in the minimal sufficient set for (8.11) but not for (XZ, YZ) are then the row means in the X–Y marginal table.

In principle, it is also possible to conduct exact tests. Under the assumption that model (8.11) holds and conditional on the strata margins, the distribution of the data depends only on β. In fact, that distribution involves the data only through the value of $\Sigma \Sigma u_i v_j n_{ij+}$. Thus, to test conditional independence ($\beta = 0$) against the alternative $\beta > 0$, the P-value is the probability of those tables having the same strata margins as the observed table but having values of $\Sigma \Sigma u_i v_j n_{ij+}$ at least as large as the observed table. For $K = 1$ stratum, Agresti et al. (1990) gave a fast algorithm for exact permutation tests and confidence intervals for β, conditional on row and column totals.

8.4.4 Deaths from Leukemia Example

Table 8.13, analyzed by Sugiura and Otake (1974) and Landis et al. (1978), shows the relationship between deaths from leukemia during 1950–1970 and estimated radiation dosage from atomic bombing at the end of World War II. Subjects are stratified according to their age at time of bombing. We use the midpoint scores (0, 5, 30, 75, 150, 300) for the levels of dosage. Survival status has only two levels, so the choice of scores for it is irrelevant.

Let S = survival status, A = age, and D = dosage. For these data,

Table 8.13 Deaths from Leukemia Observed at Atomic Bomb Casualty Commission (1950–1970)

Age	Survival Status[a]	Dose (rad)					
		Not in City	0–9	10–49	50–99	100–199	200+
0–9	LD	0	7	3	1	4	11
	NLD	5015	10752	2989	694	418	387
10–19	LD	5	4	6	1	3	6
	NLD	5973	11811	2620	771	792	820
20–34	LD	2	8	3	1	3	7
	NLD	5669	10828	2798	797	596	624
35–49	LD	3	19	4	2	1	10
	NLD	6158	12645	3566	972	694	608
50+	LD	3	7	3	2	2	6
	NLD	3695	9053	2415	655	393	289

Source: Reprinted from Sugiura and Otake (1974), by courtesy of Marcel Dekker, Inc.
[a] LD = death from leukemia, NLD = nondeath from leukemia.

Mantel's statistic for testing conditional independence of survival status and dosage is $M^2 = 426.3$, based on df $= 1$. There is very strong evidence of an effect, as would be expected. An alternative approach compares G^2 values for the loglinear model (AS, AD) of S–D conditional independence and the model of homogeneous S–D linear-by-linear association. The difference $203.7 - 32.4 = 171.3$, based on df $= 25 - 24 = 1$, also gives strong evidence of an effect.

It is more illuminating to estimate a parameter that describes the strength of the S–D association. The statistic $G^2[(AD, SD)|(AS, AD, SD)] = 32.5 - 27.8 = 4.7$, based on df $= 24 - 20 = 4$, shows little evidence of association of age with survival status, so we collapse over age in describing the effect of dose on survival status. The S–D linear-by-linear association model is equivalent to the linear logit model with survival status as the response variable. That model fits quite well, with $G^2 = 4.9$ and df $= 4$. The linear logit effect of dose on the probability of death from leukemia has $\hat{\beta} = 0.0102$ (ASE $= 0.00064$). The estimated odds of death from leukemia are $\exp(-7.189 + 0.0102 \times \text{dose})$. This corresponds to a predicted probability of death from leukemia of 0.0008 at zero dosage and 0.0159 at dosage level 300 rad.

8.4.5 Generalized Mantel Tests

Landis et al. (1978) generalized Mantel's score statistic. Let $\mathbf{n}_k$ denote a column vector of the cell counts in stratum k, let $\mathbf{m}_k$ denote their expected values and $\mathbf{S}_k$ denote their covariance matrix under conditional independence. Let $\mathbf{B}_k = \mathbf{U}_k \otimes \mathbf{V}_k$ denote a matrix of constants based on row scores $\mathbf{U}_k$ and column scores $\mathbf{V}_k$ for the kth stratum, where $\otimes$ denotes the Kronecker product. The generalized statistic is

$$L^2 = \left[\sum_k \mathbf{B}_k(\mathbf{n}_k - \mathbf{m}_k)\right]' \left[\sum_k \mathbf{B}_k \mathbf{S}_k \mathbf{B}_k'\right]^{-1} \left[\sum_k \mathbf{B}_k(\mathbf{n}_k - \mathbf{m}_k)\right]. \quad (8.18)$$

When the scores $\mathbf{U}_k = (u_1, \ldots, u_I)$ and $\mathbf{V}_k = (v_1, \ldots, v_J)$ are the same for all strata, $\mathbf{U}_k \otimes \mathbf{V}_k$ is a $1 \times IJ$ vector of cross-product scores, and L^2 simplifies to Mantel's statistic M^2. When both variables are nominal, it makes sense to let $\mathbf{U}_k$ be an $(I-1) \times I$ matrix $(\mathbf{I}, -\mathbf{1})$, where $\mathbf{I}$ is an identity matrix of size $(I-1)$ and $\mathbf{1}$ denotes a column vector of $I-1$ ones, and where $\mathbf{V}_k$ is the analogous matrix of size $(J-1) \times J$. Then L^2 simplifies to (7.9), the df $= (I-1)(J-1)$ generalization of the Cochran-Mantel-Haenszel statistic for testing conditional independence in an $I \times J \times K$ table. When X is nominal and Y is ordinal, it makes sense to use this $\mathbf{U}_k$ together with $\mathbf{V}_k = (v_1, \ldots, v_J)$. Then L^2 sums over the K

strata information about how I row means compare to their null expected values, and it has df $= I - 1$. For $K = 1$, this ANOVA-type measure of variation in row means equals

$$(n-1) \frac{\sum_i n_{i+} \left(\dfrac{\sum_j v_j n_{ij}}{n_{i+}} - \dfrac{\sum_j v_j n_{+j}}{n} \right)^2}{\left[\sum_j v_j^2 n_{+j} - \dfrac{\left(\sum_j v_j n_{+j} \right)^2}{n} \right]}. \tag{8.19}$$

Landis et al. also permitted rank-type scores in these statistics. The rank versions are analogs for ordered categorical responses of strata-adjusted Spearman correlation and Kruskal-Wallis tests.

An advantage that the Mantel and Landis tests have over the model-based tests is that they require large $X-Y$ marginal totals, rather than large $X-Y$, $X-Z$, and $Y-Z$ marginal totals. When the number K of strata is large and there are few observations per strata, the distribution of L^2 is closer to its asymptotic chi-squared distribution. An advantage of the model-based approach is that it leads directly to measures of the strength of partial association.

8.5 OTHER MODELS HAVING PARAMETER SCORES

The linear-by-linear association $(L \times L)$ model is a special case of the row effects (R) model and the column effects (C) model. The R model has parameter row scores, and the C model has parameter column scores. These models are special cases of a more general model in which row *and* column scores are parameters.

8.5.1 Row and Column Effects Model

When we replace the row scores $\{u_i\}$ and column scores $\{v_j\}$ in the $L \times L$ model (8.1) by parameters, we obtain the *row and column effects* (RC) model

$$\log m_{ij} = \mu + \lambda_i^X + \lambda_j^Y + \beta \mu_i v_j. \tag{8.20}$$

The row and column effects determine the nature of the association, as described by the local log odds ratios

$$\log \theta_{ij} = \beta(\mu_{i+1} - \mu_i)(\nu_{j+1} - \nu_j) \, .$$

The residual df for the RC model equals $(I-2)(J-2)$. Unlike the $L \times L$, R, and C models, this model is not loglinear, because the log expected frequency is a multiplicative (rather than linear) function of the model parameters μ_i and ν_j. It is also referred to as the *log-multiplicative* or *log-bilinear* model.

Though replacing fixed scores by parameters is appealing for some applications, the RC model presents complications that do not occur for loglinear models. The likelihood may not be concave, and may have local maxima. Also, though the model of independence is a special case, it is awkward to test independence using this model. Haberman (1981) showed that $G^2(I) - G^2(RC)$ has null distribution that is not chi-squared, but instead is that of the maximum eigenvalue from a Wishart matrix.

When one set of parameter scores is fixed, the RC model simplifies to the R or C model. Goodman (1979a) suggested a simple iterative algorithm for fitting the RC model that uses this fact. A cycle of the algorithm has two steps, each step consisting of the iterative fitting of the R or C model. First, for some initial guess of column parameter scores (such as $\{\nu_j = j\}$), we estimate the row scores, as in the R model. Then, treating the estimated row scores from the first step as fixed, we estimate the column scores, as in the C model. Those estimates serve as fixed column scores in the first step of the next cycle, for re-estimating the row scores in the R model. There is no guarantee that this algorithm converges to ML estimates, but it seems to do so when the model fits well.

The RC model is a special case of the saturated model. Goodman (1985) expressed the association term in the saturated model in a form that is a generalization of the $\beta\mu_i\nu_j$ term in the RC model, namely

$$\lambda_{ij}^{XY} = \sum_{k=1}^{M} \beta_k \mu_{ik} \nu_{jk} \tag{8.21}$$

where $M = \min(I-1, J-1)$. The parameters satisfy constraints such as

$$\sum_i \mu_{ik} \pi_{i+} = \sum_j \nu_{jk} \pi_{+j} = 0 \quad \text{for all } k$$

$$\sum_i \mu_{ik}^2 \pi_{i+} = \sum_j \nu_{jk}^2 \pi_{+j} = 1 \quad \text{for all } k \tag{8.22}$$

$$\sum_i \mu_{ik} \mu_{ih} \pi_{i+} = \sum_j \nu_{jk} \nu_{jh} \pi_{+j} = 0 \quad \text{for all } k \neq h \, .$$

Though this model treats classifications as nominal, parameter interpretation is simplest when at least one variable is ordinal.

When there is some value M^* between 1 and M such that $\beta_k = 0$ for $k > M^*$, model (8.21) is called the $RC(M^*)$ model. The RC model (8.20) is the special case $M^* = 1$. The row effects model is the special case in which $\{\nu_{j1}\}$ are known scores, $\{\mu_i = \beta_1 \mu_{i1}\}$ are unknown parameters, and $\beta_k = 0$ for $k \geq 2$. The linear-by-linear association model is the special case in which $\{\mu_{i1}\}$ and $\{\nu_{j1}\}$ are known scores and $\beta_k = 0$ for $k \geq 2$.

8.5.2 Mental Health Status Example

Table 8.14, taken from Srole et al. (1978, p. 289), describes the relationship between mental impairment and parents' socioeconomic status for a sample of residents of Manhattan. Goodman (1979a, 1985, 1986) used these data to illustrate several analyses of categorical data.

The RC model fits well, with $G^2(RC) = 3.57$ based on df $= 8$. For scaling (8.22), the ML estimates are $(-1.11, -1.12, -0.37, 0.03, 1.01, 1.82)$ for the row scores, $(-1.68, -0.14, 0.14, 1.41)$ for the column scores, and $\hat{\beta} = 0.17$. Nearly all estimated local log odds ratios are positive, indicating a tendency for mental health to be better at higher levels of parents' socioeconomic status.

Simple loglinear models also describe these data well. For equal-interval scores, the $L \times L$ model implies uniform local association. That model fits very well, with $G^2(L \times L) = 9.89$ based on df $= 14$. To test the hypothesis that row and column scores in the RC model are equal-interval, we use the statistic $G^2(L \times L \mid RC) = 6.32$, based on df $= 6$. The

Table 8.14 Cross-Classification of Mental Health Status and Parents' Socioeconomic Status

Parents' Socioeconomic Status	Mental Health Status			
	Well	Mild Symptom Formation	Moderate Symptom Formation	Impaired
A (high)	64	94	58	46
B	57	94	54	40
C	57	105	65	60
D	72	141	77	94
E	36	97	54	78
F (low)	21	71	54	71

Source: Reprinted with permission from Srole (1978).

parameter scores do not give a significantly better fit than equal-interval scores. Thus, it is sufficient to use a fitted uniform local odds ratio to describe the table. For unit-spaced scores, $\hat{\beta} = 0.091$, so the fitted local odds ratio is $\exp(0.091) = 1.09$. The ASE of $\hat{\beta}$ is 0.015. There is strong evidence of positive association, but the degree of association seems rather weak.

8.5.3 Correlation Models

The *correlation model* is a model for two-way tables that has many features in common with the *RC* model. In its simplest form, it is

$$\pi_{ij} = \pi_{i+} \pi_{+j}(1 + \lambda x_i y_j) \tag{8.23}$$

where $\{x_i\}$ and $\{y_j\}$ are score parameters satisfying

$$\sum x_i \pi_{i+} = \sum y_j \pi_{+j} = 0 \quad \text{and} \quad \sum x_i^2 \pi_{i+} = \sum y_j^2 \pi_{+j} = 1 .$$

The parameter λ is the correlation between the scores for joint distribution (8.23).

The correlation model is also called the *canonical correlation model*, because ML estimates of the scores are the values that maximize the correlation. A more general canonical correlation model is

$$\pi_{ij} = \pi_{i+} \pi_{+j}\left(1 + \sum_{k=1}^{M} \lambda_k x_{ik} y_{jk}\right)$$

where $0 \le \lambda_M \le \cdots \le \lambda_1 \le 1$ and where score parameters satisfy constraints such as in (8.22). The parameter λ_k is the correlation between $\{x_{ik}, i = 1, \ldots, I\}$ and $\{y_{jk}, j = 1, \ldots, J\}$. The $\{x_{i1}\}$ and $\{y_{j1}\}$ are standardized scores that maximize the correlation λ_1 for the joint distribution; $\{x_{i2}\}$ and $\{y_{j2}\}$ are standardized scores that maximize the correlation λ_2, subject to $\{x_{i1}\}$ and $\{x_{i2}\}$ being uncorrelated and $\{y_{j1}\}$ and $\{y_{j2}\}$ being uncorrelated, and so forth.

Unsaturated models result from replacing M by $M^* < \min(I - 1, J - 1)$. Goodman (1985) and Gilula and Haberman (1986) discussed ML estimation for such models. When λ is close to zero in correlation model (8.23), Goodman (1981a, 1985, 1986) noted that ML estimates of λ and the score parameters are very similar to those of β and the score parameters in the *RC* model. Correlation models also can be applied with fixed scores in place of parameter scores.

Goodman argued that the RC model and its loglinear special cases have certain advantages over correlation models. The correlation model is not defined for all possible combinations of score values, ML fitted values do not have the same marginal totals as the observed data, and it is awkward to incorporate structural zeroes. In addition, the correlation model is not simply generalizable to multi-way tables. Gilula and Haberman (1988) analyzed multi-way tables with canonical correlation models by treating explanatory variables as a single variable and response variables as a second variable.

8.5.4   Correspondence Analysis

Correspondence analysis is a graphical way of representing associations in two-way contingency tables. The rows and columns are represented by points on a graph, the positions of which indicate associations. Goodman (1985, 1986) noted that coordinates of the points are reparameterizations of the $\{x_{ik}\}$ and $\{y_{jk}\}$ scores in the general canonical correlation model. Correspondence analysis uses adjusted scores

$$x_{ik}^* = \lambda_k x_{ik}, \qquad y_{jk}^* = \lambda_k y_{jk} . \tag{8.24}$$

The adjusted scores are close to zero for dimensions k in which the correlation λ_k is close to zero. A correspondence analysis graph uses the first two dimensions, plotting (x_{i1}^*, x_{i2}^*) for each row and (y_{j1}^*, y_{j2}^*) for each column.

Goodman (1985, 1986) used Table 8.14 to illustrate the similarities of correspondence analysis to analyses using correlation models and RC-type association models. For the general canonical correlation model, $M = \min(I - 1, J - 1) = 3$, and the squared correlations for the three dimensions are (0.0260, 0.0014, and 0.0003). The association is weak. Table 8.15 contains estimated row and column scores for the correspondence analysis of these three dimensions. Both sets of scores in the first dimension fall in a monotone increasing pattern, except for a slight discrepancy between the first two row scores. This indicates an overall positive association between the variables. The scores for the second and third dimension are close to zero, reflecting the relatively small estimates of λ_2 and λ_3.

Figure 8.3 is a graphical display that exhibits the results of correspondence analysis. The horizontal axis refers to estimates for the first dimension and the vertical axis to estimates for the second dimension. Six points (circles) represent the six rows, with point i giving $(\hat{x}_{i1}^*, \hat{x}_{i2}^*)$. Similarly, four points (squares) display the estimates of the first two

Table 8.15   Scores from Correspondence Analysis Applied to Table 8.14

Column Score	Dimension			Row Score	Dimension		
	1	2	3		1	2	3
1	0.260	0.012	0.023	1	0.181	−0.018	0.028
2	0.030	0.024	−0.019	2	0.185	−0.011	−0.026
3	−0.013	−0.069	−0.002	3	0.059	−0.021	−0.010
4	−0.236	0.019	0.016	4	−0.008	0.042	0.011
				5	−0.164	0.044	−0.009
				6	−0.287	−0.061	0.005

Source: Reprinted with permission of the Institute of Mathematical Statistics, based on Goodman (1985).

dimensions of adjusted column scores ($\hat{y}^{*}_{j1}$, $\hat{y}^{*}_{j2}$). Both sets of points lie close to the horizontal axis, since the first dimension is much more important than the second one.

Row points that are close together represent rows in which conditional distributions across the columns are similar; column points that are close represent columns in which conditional distributions across the rows are similar. Row points that are close to column points represent combinations that are more likely than would be expected under independence. Figure 8.3 shows a tendency for subjects at the high end of one scale to be at the high end of the other, and for subjects at the low end of one scale to be at the low end of the other.

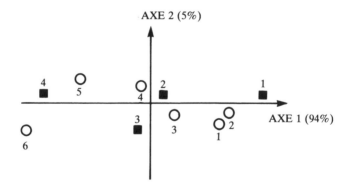

Note: Circles refer to ($\hat{x}^{*}_{i1}$, $\hat{x}^{*}_{i2}$), squares refer to ($\hat{y}^{*}_{j1}$, $\hat{y}^{*}_{j2}$).

Figure 8.3   Graphical Display of Scores from First Two Dimensions of Correspondence Analysis. *Source*: Reprinted with permission, based on Escoufier (1982).

Correspondence analysis has been used mainly as a descriptive tool. Goodman (1986) showed that inferential methods can also be applied to correspondence analyses. For Table 8.14, inferential analyses reveal that the first dimension, accounting for 94% of the total squared correlation, is adequate for describing whatever association exists. Goodman argued that scores for these data should be estimated for the unsaturated version of the model employing only one dimension, and that graphics need display fitted scores for that dimension alone. Then the correspondence analysis is equivalent to a ML analysis using the correlation model (8.23). The estimated scores for that model are $(-1.09, -1.17, -0.37, 0.05, 1.01, 1.80)$ for the rows and $(-1.60, -0.19, 0.09, 1.48)$ for the columns. The model fits well, with $G^2 = 2.75$ based on df $= 8$. The quality of fit and the estimated scores are very similar to those given previously for the RC model. More parsimonious correlation models also fit these data well, such as versions of (8.23) using fixed equal-interval scores.

8.6 MODEL SELECTION FOR ORDINAL VARIABLES

This chapter has introduced ways of using category orderings in model building. We next discuss some issues relating to model selection.

8.6.1 Strategy for Model Building

When we allow linear-by-linear or row or column effects for associations and analogous effects for interactions, the variety of potential models is much greater than the standard models of Chapters 5–7. This variety can provide an embarrassment of riches, and it is necessary to develop a strategy for choosing among the models.

One approach uses standard loglinear models for guidance. Once we find a standard model that fits well, we simplify by replacing some parameters with structured terms for ordinal classifications. For instance, consider a three-way table in which Z is an ordinal response variable. Suppose the model (XY, XZ, YZ) fits well, but no model permitting conditional independence of two variables fits well. Then we could try to simplify the model by replacing the λ^{XZ} and λ^{YZ} terms by structured terms. Suppose we expect the conditional distributions on Z to be stochastically ordered across the levels of X, but perhaps not monotonically. Then we could try replacing λ_{ik}^{XZ} by a row effects term $\mu_i w_k$ in which parameters refer to levels of X and ordered scores refer to levels of Z.

For two-way tables, Chuang et al. (1985) proposed a detailed set of diagnostic rules for selecting a model from the general class of $RC(M)$ models. Their approach uses a singular value decomposition of a transformation of the cell counts.

8.6.2 Assignment of Scores

The loglinear models studied in this chapter require assigning scores to levels of ordinal variables. Parameter interpretations are simplest for equally-spaced scores. Often it is unclear how to assign scores, and we may not want to assume equal spacings. Cochran (1954) noted that "any set of scores gives a *valid* test, provided that they are constructed without consulting the results of the experiment. If the set of scores is poor, in that it badly distorts a numerical scale that really does underlie the ordered classification, the test will not be sensitive. The scores should therefore embody the best insight available about the way in which the classification was constructed and used." Ideally, the scale is chosen by a consensus of "experts," and subsequent interpretations use that same scale.

How sensitive are ordinal analyses to the choice of scores? There is no simple answer to this question, but there are examples (e.g., Graubard and Korn 1987) in which different scoring systems give quite different results. Gross (1981) noted that for linear logit model (4.14), the local asymptotic relative efficiency for testing independence using a statistic based on an incorrect set of scores is given by the square of the Pearson correlation between the true and incorrect scores. Simon (1978) and Tarone and Gart (1980) gave related asymptotic results. When there is no natural set of scores, it is wise to conduct a sensitivity analysis. Assign scores a variety of ways that seem reasonable, to check whether substantive conclusions depend on the choice.

An alternative approach treats the scores as parameters to estimate from the data, rather than numbers to preassign. The resulting ML estimates of scores need not be monotone, though we can force monotonicity by maximizing the likelihood subject to order restrictions (Agresti et al. 1987). There are disadvantages, however, in treating scores as parameters. The model becomes less parsimonious. In fact, it is no longer loglinear when both row and column scores are parameters (RC model). Also, for detecting association, there may be a reduction in power compared to using a model with preassigned scores that describes the association nearly as well. The test of independence $G^2(I \mid L \times L)$ has df = 1, whereas the comparable test $G^2(I \mid R)$ for the row effects model has df = $I - 1$. Even though preassigned row scores for the $L \times L$ model need not be as "good" as parameter scores estimated with the R model, the test $G^2(I \mid L \times L)$ is likely to be more powerful than $G^2(I \mid R)$ because it focuses the effect on fewer degrees of freedom.

Even when a model with preassigned scores does not fit well, it may be more powerful for detecting associations than methods that ignore the ordering. A model such as linear-by-linear association can help detect an important component of an association. As Mantel (1963) argued in a similar context, "that a linear regression is being tested does not mean that an assumption of linearity is being made. Rather it is that that test of a linear component of regression provides power for detecting any progressive association which may exist."

The necessity of selecting scores for ordinal classifications need not be a drawback. It is unnecessary to treat the scores as reasonable scalings of ordinal variables in order for the models to be valid. For example, if the linear-by-linear association model fits well with equally-spaced row and column scores, we have a simple description of the nature of the association (uniform local odds ratios), regardless of whether the scores are sensible indexes of "true" distances between ordered categories.

Graubard and Korn (1987) argued that the perceived advantage of procedures not requiring preassigned scores is illusory. They noted that procedures that automatically generate scores, such as nonparametric methods that utilize midranks (Problem 8.30), can produce inappropriate scores. For instance, when there are relatively few observations in two adjacent categories, rank-type scores are necessarily close.

Section 9.4 introduces a class of logit models for ordinal response variables that does not require preassigned or parameter scores.

CHAPTER NOTES

Section 8.1: Linear-by-Linear Association

8.1 Goodman's (1979a) article was highly influential and largely responsible for the explosion of research in the past decade on loglinear models for ordinal data. His work builds on ideas presented by Haberman (1974b), who expressed the general λ^{XY} association term as an expansion involving orthogonal polynomials. The model with linear-by-linear term resembles Tukey's approach of using a single degree of freedom for nonadditivity in two-way ANOVA. Graubard and Korn (1987) listed 14 tests for $2 \times J$ tables that utilize the correlation-type statistic $\Sigma \Sigma u_i v_j n_{ij}$ that occurs in the minimal sufficient set for the linear-by-linear association model.

Alternative methods for fitting ordinal loglinear models include a unidimensional version of the Newton-Raphson method (Goodman 1979a), and a generalization of iterative scaling (Darroch and Ratcliff 1972). These methods can be much slower than Newton-Raphson, but neither requires matrix inversion at each step.

8.2 Karl Pearson often analyzed contingency tables by assuming there was an underlying normal distribution. There is a large literature on estimating the correlation for an assumed underlying bivariate normal distribution. See, for instance, Becker (1989b), Goodman (1981b), Kendall and Stuart (1979, Chaps. 26 and 33), Lancaster and Hamdan (1964), Lancaster (1969, Chap. X), and Pearson (1904).

Section 8.3: Models for Ordinal Variables in Multidimensional Tables

8.3 For more general models for partial association and for interaction in multi-way tables with ordinal variables, see Agresti and Kezouh (1983), Becker (1989a), Becker and Clogg (1989), Clogg (1982b), and Goodman (1986).

Section 8.4: Testing Independence for Ordinal Classifications

8.4 Agresti and Yang (1987) gave numerical illustrations of the superiority of $G^2(I \mid M)$ and $X^2(I \mid M)$ over $G^2(I)$ and $X^2(I)$ in convergence to their null chi-squared distributions.

For $K = 1$ stratum, Yates (1948) and Bhapkar (1968) proposed statistics similar to the Mantel M^2 statistic (8.17) for testing independence in a two-way table, and also proposed ANOVA-type statistics similar to (8.19) for singly-ordered two-way tables. See also Nair (1986, 1987) and Williams (1952). For three-way tables, Birch (1965) discussed the theory of exact tests based on $\Sigma \Sigma u_i v_j n_{ij+}$.

An alternative test of independence for ordinal variables directly uses sample proportions of concordant and discordant pairs, denoted by $\hat{\Pi}_c$ and $\hat{\Pi}_d$. By the delta method, the asymptotic standard error of $\hat{\Pi}_c - \hat{\Pi}_d$ is $\sigma / \sqrt{n}$, where $\sigma^2 = \Sigma \Sigma \pi_{ij} \phi_{ij}^2$ with $\phi_{ij} = 2(\pi_{ij}^{(c)} - \pi_{ij}^{(d)})$, using notation defined in Section 3.4.5. The test statistic is the ratio of $(\hat{\Pi}_c - \hat{\Pi}_d)$ to its ASE. Another test statistic is $\hat{\gamma}$ divided by its ASE. Simon (1978) showed that all tests using measures for which numerators are the difference between the proportions of concordant and discordant pairs have the same local power for testing independence.

Section 8.5: Models Having Parameter Scores

8.5 Several authors have studied the *RC* model, including Goodman (1979a, 1981a, 1981b), Andersen (1980, p. 211), and Clogg (1982a). Goodman (1985, 1986), Becker and Clogg (1989), and Becker (1989a, b) discussed the RC(M) model and its generalizations for multi-way tables. Anderson (1984) discussed a related model. Kendall and Stuart (1979, Chap. 33) surveyed the literature on canonical correlation methods for categorical data, prior to Goodman's work.

The correspondence analysis approach has gained widespread popularity in recent years in Europe (particularly France and the Netherlands), under the influence of a series of articles by Benzecri (see, e.g., 1973) and more recently Escoufier (1982) and van der Heijden and de Leeuw (1985). Greenacre (1981, 1984) reviewed its development and application, and showed its relation to the singular value decomposition of a matrix. Greenacre (1984, p. 8) attributed the mathematical origins of correspondence analysis to an article by H. O. Hartley, published under his original German name (Hirschfeld, 1935). For other discussion, see Greenacre and Hastie (1987), Hill

(1974), Lebart et al. (1984), van der Heijden and de Leeuw (1985) and van der Heijden et al. (1989). For related work, see Gabriel (1971), Aitkin et al. (1987), and Worsley (1987). Greenacre's book (pp. 317–325) lists several data sets to which correspondence analysis has been applied.

PROBLEMS

Applications

8.1 Refer to Table 8.14.
  a. Fit the independence and uniform association models. Comment.
  b. Interpret the value of $\hat{\beta}$ for the uniform association model.
  c. Test independence, using the category orderings.

8.2 Refer to the previous problem.
  a. Using standardized scores, find $\hat{\beta}$ for the uniform association model. Comment on the strength of association.
  b. Fit a model in which scores for mental health status are parameters. Interpret the estimated scores, and compare the fit to the uniform association model.

8.3 Table 8.16, taken from Holmes and Williams (1954), classifies 1398 children on tonsil size and on whether they are carriers of the virus *Streptococcus pyogenes*. Find an unsaturated model that fits these data well. Interpret parameters, and test for a difference in tonsil size between carriers and noncarriers of the virus.

Table 8.16

	Tonsil Size		
	Not Enlarged	Enlarged	Greatly Enlarged
Noncarriers	497	560	269
Carriers	19	29	24

Source: Reprinted with permission from Cambridge University Press (Holmes and Williams 1954).

8.4 Refer to Table 8.1. Fit the column effects model. Compare estimated column scores to the scores used in the uniform association model. Test the hypothesis that the column scores are equal-interval, given that the column effects model holds.

8.5 Refer to Table 8.4.

  a. Use a statistical computer package to fit the row effects model and obtain the results in Section 8.2 for these data.

  b. For each pair of rows a and b, construct a 95% confidence interval for $\exp(\mu_b - \mu_a)$, and interpret.

8.6 Use models discussed in this chapter to analyze Table 2.10.

8.7 Fit model (8.16) for the breathing test data. Report the fitted values, and construct a 95% confidence interval for the local S-B odds ratio for the older age group. Interpret.

8.8 Refer to Problem 6.4. Using the ordinal nature of education, find an unsaturated model that fits these data adequately. Is your model equivalent to a logit model? Describe how the effect of education on following politics regularly varies by nationality.

Table 8.17

Age	Alzheimer's Disease	Severe	Moderate	Mild	Borderline	Unaffected
		Cognitive Impairment				
65–69	Highly probable	1	1	0	0	0
	Probable	0	4	5	0	0
	Possible	0	4	11	9	0
	Unaffected	0	0	2	1	45
70–74	Highly probable	1	0	0	0	0
	Probable	1	8	3	0	0
	Possible	1	6	16	11	0
	Unaffected	0	1	3	3	40
75–79	Highly probable	1	4	0	0	0
	Probable	5	17	8	0	0
	Possible	1	5	17	14	0
	Unaffected	0	0	2	2	30
80–84	Highly probable	4	7	0	0	0
	Probable	2	15	9	0	0
	Possible	1	7	24	12	0
	Unaffected	0	0	0	3	28
85+	Highly probable	9	8	1	0	0
	Probable	17	16	8	0	0
	Possible	0	13	22	9	0
	Unaffected	0	0	2	2	11

Source: Dr. Laurel Smith, Department of Biostatistics, Harvard University.

8.9 Table 8.17 shows the relationship between Alzheimer's disease and cognitive impairment for a sample of elderly people, stratified by age. Using methods of this chapter, analyze these data.

8.10 Clogg and Shockey (1988) reported Table 8.18, taken from the 1982 General Social Survey. Analyze these data using models described in this chapter.

Table 8.18

		Job Satisfaction			
Race	Degree Attained	Very Satisfied	Moderately Satisfied	A Little Dissatisfied	Very Dissatisfied
White	≤HS	400	319	81	52
	>HS	112	81	16	10
Black	≤HS	48	46	21	7
	>HS	8	14	1	0

Source: 1982 General Social Survey; see Clogg and Shockey (1988).

8.11 Refer to Table 4.12. Treating political views as ordinal, use an ordinal model to describe the association structure among these variables.

8.12 Refer to Table 9.9.
  a. Test independence using $X^2(I)$ or $G^2(I)$.
  b. Test independence with a method that uses the ordering of categories. Compare results in (a) and (b).

8.13 Use the Mantel score statistic to test conditional independence of dumping severity and operation, given hospital, for Table 8.6. Compare results with those obtained using loglinear models.

8.14 Refer to Table 8.17. Test conditional independence of Alzheimer's disease and cognitive impairment, controlling for age. Interpret.

8.15 Fit the RC model to Table 8.1. Interpret the estimated scores, and test whether this model gives a better fit than the uniform association model.

Theory and Methods

8.16 For $I \times 2$ contingency tables, show that linear logit model (4.14) is equivalent to the linear-by-linear association model.

8.17 Consider the linear-by-linear association model.

   a. When $\beta < 0$, explain which corner cells are expected to have more observations than when X and Y are independent.

   b. Show (8.3).

   c. Give a formula for the sample Pearson correlation between the two sets of scores, and use the likelihood equations to prove that this equals the fitted correlation.

   d. We decide to replace $\{v_j = j - (J+1)/2\}$ by $\{2v_j\}$. Show the estimate of β is halved, but $\{\hat{m}_{ij}\}$, $\{\hat{\theta}_{ij}\}$, and G^2 are unchanged.

8.18 Lehmann (1966) defined two random variables (X, Y) to be *positively likelihood-ratio dependent* if their joint density satisfies $f(x_1, y_1)f(x_2, y_2) \geq f(x_1, y_2)f(x_2, y_1)$ whenever $x_1 < x_2$ and $y_1 < y_2$. For such a joint distribution, the conditional distribution of Y is stochastically increasing as X increases, and the conditional distribution of X is stochastically increasing as Y increases (Alam and Wallenius 1976, Goodman 1981a).

   a. For the linear-by-linear association model, infer that the conditional distributions of Y are stochastically ordered and the conditional distributions of X are stochastically ordered. What is the nature of this ordering if $\beta > 0$?

   b. In row effects model (8.7), if $\mu_i - \mu_h > 0$, show that the conditional distribution of Y is stochastically higher in row i than in row h.

8.19 Consider the row effects model.

   a. Show there is no loss of generality in assuming the parameters satisfy $\Sigma \lambda_i^X = \Sigma \lambda_j^Y = \Sigma \mu_i = 0$.

   b. Show an alternative permissible scaling is $\lambda_1^X = \lambda_1^Y = \mu_1 = 0$. Interpret $\{\mu_i\}$ in that case.

   c. When $I = 2$, show it is equivalent to the linear-by-linear association model.

   d. Show that minimal sufficient statistics are $\{n_{i+}\}$, $\{n_{+j}\}$, and $\{\Sigma_j v_j n_{ij}, i = 1, \ldots, I\}$, and show that the likelihood equations are (8.10).

8.20 Refer to the previous problem.

   a. Show that if the model holds and the rows are permuted, then the model still holds with the corresponding permutation of $\{\mu_i\}$.

b. Suppose the model holds and the columns are permuted but the scores are still assigned in a monotone manner. Does the model still apply?

c. Show there is a strict monotone relationship between the $\{\hat{\mu}_i\}$ and the sample means of the conditional distributions within the rows.

8.21 Yule (1906) defined a table to be *isotropic* if the rows and columns can be ordered such that the local log odds ratios are all non-negative.

a. Show that a table is isotropic if it satisfies (i) the linear-by-linear association model, (ii) the row effects model, (iii) the *RC* model.

b. Show that a table that is isotropic for a certain ordering is still isotropic when rows or columns are combined.

8.22 A table has positive marginal counts for every row and every column, but in one row only the first cell count is positive. Explain why ML estimates of parameters do not exist for the row effects model. (Hint: Consider the third likelihood equation.) More generally, note that ML estimates do not exist for the R or $L \times L$ model when a sufficient statistic for an association parameter takes its maximum or minimum possible value for the given marginal counts.

8.23 Refer to the homogeneous linear-by-linear association model (8.11).

a. Show that the likelihood equations are

$$\hat{m}_{i+k} = n_{i+k}, \qquad \hat{m}_{+jk} = n_{+jk} \quad \text{for all } i, j, \text{ and } k$$

$$\sum \sum u_i v_j \hat{m}_{ij+} = \sum \sum u_i v_j n_{ij+}.$$

b. Show that residual df $= K(I-1)(J-1) - 1$.

c. When $I = J = 2$, show that this model is equivalent to (XY, XZ, YZ).

d. Show how the likelihood equations change when we permit heterogeneous linear-by-linear X–Y association (See (8.15)). Show the equations imply that, in each stratum, the fitted X–Y correlation equals the observed correlation.

8.24 Construct a model having general X-Z and Y-Z associations, but row effects for the X–Y association that are (a) homogeneous across levels of Z, (b) heterogeneous across levels of Z. Give the likelihood equations and residual df for each case, and explain how to interpret the row effect parameters.

8.25 Suppose Z is an ordinal response variable, X is nominal, and Y is ordinal. For fixed scores $\{v_j\}$ and $\{w_k\}$ for the levels of Y and Z, consider the loglinear model

$$\log m_{ijk} = \mu + \lambda_i^X + \lambda_j^Y + \lambda_k^Z + \lambda_{ij}^{XY} + \mu_i w_k + \beta v_j w_k .$$

a. Interpret the parameters for the conditional X-Z and Y-Z associations.
b. Using unit-spaced scores for Z, construct an equivalent logit model. Explain how this resembles an analysis of covariance model.
c. Derive the likelihood equations, and interpret.
d. Find the residual df.

8.26 Refer to the previous problem, but suppose only Z is ordinal.
a. Formulate a loglinear model that assumes no three-factor interaction and fully utilizes the ordinality of Z. Interpret the parameters.
b. Using unit-spaced scores for Z, construct an equivalent logit model, and explain how it resembles a two-way ANOVA model.
c. Give the likelihood equations, and interpret.
d. Find the residual df.
e. Show that the model in the previous problem is a special case of this one, and that both models are special cases of (XY, XZ, YZ).

8.27 Refer to Problem 8.25. To allow for interaction, we add the term $\gamma_i v_j w_k$. Using unit-spaced $\{w_k\}$, show this model is equivalent to logit model

$$\log(m_{ij,k+1}/m_{ijk}) = \alpha_k + \mu_i + \beta v_j + \gamma_i v_j .$$

Explain how this logit model resembles an analysis of covariance model, allowing interaction. Interpret parameters.

8.28 Consider model (8.14).

    a. Show that log odds ratios for any two variables change linearly across the levels of the third variable.

    b. Show the likelihood equations are those for model (XY, XZ, YZ) plus

$$\sum \sum \sum u_i v_j w_k \hat{m}_{ijk} = \sum \sum \sum u_i v_j w_k n_{ijk} .$$

    c. Suppose we want a test of homogeneity of log odds ratios in K strata of a $2 \times 2 \times K$ table that is sensitive to the alternative of linear change across the strata. Explain why the difference in G^2 values between model (XY, XZ, YZ) and this model is an appropriate statistic.

8.29 For testing a null hypothesis about a parameter β, the *score* (Note 4.4) is obtained by differentiating the log likelihood function with respect to β, and evaluating it at null hypothesis estimates of parameters.

    a. For the linear-by-linear association model, show that the score for testing independence is

$$\sum \sum u_i v_j (p_{ij} - p_{i+} p_{+j}) .$$

    b. Use the delta method to show that its null ASE is

$$\left\{ \left[\sum u_i^2 p_{i+} - \left(\sum u_i p_{i+} \right)^2 \right] \left[\sum v_j^2 p_{+j} - \left(\sum v_j p_{+j} \right)^2 \right] \Big/ n \right\}^{1/2} .$$

    c. Construct a statistic for testing independence using the score. Show it does not require fitting the model, and is essentially Mantel's correlation test (8.17) for $K = 1$. (Under H_0 and for local alternatives, it follows from Cox and Hinkley (1974, pp. 322-324) that the test is asymptotically equivalent to ones based on $G^2(I \mid L \times L)$ and on $\hat{\beta}/\text{ASE}$. Hirotsu (1982) gave a family of score tests for testing hypotheses with ordered alternatives.)

8.30 *Ridits*, defined by Bross (1958), are scores for response categories based on cumulative proportions for the sample marginal distribution $\{p_{+j}, j = 1, \ldots, J\}$ of Y. The jth sample ridit is the average cumulative proportion within category j of the ordinal response,

$$\hat{r}_j = \sum_{k=1}^{j-1} p_{+k} + (\tfrac{1}{2})p_{+j}.$$

The sample mean ridit in the ith row is

$$\hat{R}_i = \sum_j \hat{r}_j p_{j|i}.$$

A statistic related to the variance of $\{\hat{R}_i\}$, namely

$$W = \frac{\dfrac{12n}{(n+1)} \sum_i n_{i+}(\hat{R}_i - 0.50)^2}{\left[1 - \dfrac{\sum_j (n_{+j}^3 - n_{+j})}{n^3 - n}\right]}$$

is a rank analog of statistic (8.19), having df $= I - 1$ for testing independence in the nominal − ordinal table.

a. Show that $\Sigma_j\, p_{+j}\hat{r}_j = 0.50$.
b. Show that $\Sigma_i\, p_{i+}\hat{R}_i = 0.50$.
c. Use W to test independence for the political ideology data, and compare results to those obtained in Section 8.2.4.

For further details on use and interpretion of ridits, see Bross (1958), Fleiss (1981, Section 9.4), Landis et al. (1978), Semenya et al. (1983), and Agresti (1984, Sections 9.3 and 10.2).

8.31 For the RC model, show that residual df $= (I - 2)(J - 2)$. (Note each set of scores satisfies two constraints.) Find and interpret the likelihood equations, and explain why the fit of the model is invariant to category orderings.

8.32 For three dimensions, give a generalization of the RC model that is a special case of model (XY, XZ, YZ).

8.33 Express the RC model as a probability function for cell probabilities $\{\pi_{ij}\}$, and demonstrate the similarity of this function to the bivariate normal density having unit standard deviations. Show that β in the RC model corresponds to $\rho/(1 - \rho^2)$ for the bivariate normal density, where ρ is the correlation coefficient. See Goodman (1981a, b, 1985) and Becker (1989b).

8.34 Refer to the previous problem. Suppose we fit the linear-by-linear association model, scaling scores to have a standard deviation of 1 in each dimension. If the model fits well but the association is weak, explain why $\hat{\beta}$ takes value similar to the sample Pearson correlation between the scores.

8.35 Refer to correlation model (8.23).
   a. Show that λ is the correlation between the scores.
   b. If this model holds, show that $\Sigma_i x_i(\pi_{ij}/\pi_{+j}) = \lambda y_j$ and $\Sigma_j y_j(\pi_{ij}/\pi_{i+}) = \lambda x_i$. Interpret.
   c. For the correlation model with λ close to zero, show that $\log(\pi_{ij})$ has form $\gamma_i + \delta_j + \lambda x_i y_j + o(\lambda)$, where $o(\lambda)/\lambda \to 0$ as $\lambda \to 0$. Thus, when the association is weak, the correlation model is similar to the linear-by-linear association model with $\beta = \lambda$ and scores $\{u_i = x_i\}$ and $\{v_j = y_j\}$.

8.36 For the general canonical correlation model, show that $\Sigma \lambda_k^2 = \Sigma \Sigma(\pi_{ij} - \pi_{i+}\pi_{+j})^2/\pi_{i+}\pi_{+j}$. Thus, the squared correlations give a partitioning of a dependence measure which is the noncentrality (7.11) of X^2 for the independence model with $n = 1$. See Goodman (1986) for other partitionings.

CHAPTER 9

Multinomial Response Models

The loglinear models discussed in Chapters 5–8 do not distinguish between response and explanatory variables. This chapter presents models that do make this distinction. The models, designed for categorical response variables having more than two categories, are generalizations of logit models for binary responses.

The first two sections of this chapter present generalized logit models for *nominal* response variables. Section 9.1 introduces models having categorical explanatory variables. Like binary logit models for categorical variables, they are equivalent to loglinear models for multi-way contingency tables. Section 9.2 discusses models that incorporate both continuous and categorical explanatory variables. That section also introduces a general *multinomial logit* model used for *discrete-choice* modeling of a subject's choice from one of several response options.

Chapter 8 noted that loglinear models for *ordinal* variables have representations as logit models using adjacent response categories. Section 9.3 shows other ways of forming logits for ordinal response variables. A popular ordinal logit model using *cumulative logits* is studied in greater detail in Section 9.4.

Other models for ordinal variables use different transformations of response probabilities. Section 9.5 presents a family of models using transformations of the cumulative distribution function of the response. Section 9.6 presents a model that mimics regression models for continuous response variables, describing how a response mean depends on explanatory variables.

Because models in this chapter explicitly treat one variable as a response, they use a sampling model that fixes counts at combinations of levels of the explanatory variables. We assume the response counts at each combination have a multinomial distribution, and that multinomial counts at different combinations are independent. Because of this sampling model and the identification of a response variable, we refer to the models as *multinomial response* models.

9.1 GENERALIZED LOGIT MODELS AND LOGLINEAR MODELS

If Y is a categorical response with J categories, there are $\binom{J}{2}$ pairs of responses for which we can construct logits. Given a certain choice of $J - 1$ of these, the rest are redundant.

9.1.1 Baseline-Category Logits

To illustrate logit modeling for multi-category responses, we begin by constructing a model for a three-way contingency table. Suppose A and B are explanatory factors. Let $\pi_{j \mid hi}$ denote the probability of response j, when A is at level h and B is at level i. We construct a logit model equivalent to loglinear model (AB, AY, BY). One way to form logits pairs each response category with a baseline category. When the final category is the baseline, the jth logit for model (AB, AY, BY) is

$$\log\left(\frac{\pi_{j \mid hi}}{\pi_{J \mid hi}}\right) = \log\left(\frac{m_{hij}}{m_{hiJ}}\right) = \alpha_j + \beta_{hj}^A + \beta_{ij}^B \tag{9.1}$$

for $j = 1, \ldots, J - 1$, where $\alpha_j = (\lambda_j^Y - \lambda_J^Y)$, $\beta_{hj}^A = (\lambda_{hj}^{AY} - \lambda_{hJ}^{AY})$, and $\beta_{ij}^B = (\lambda_{ij}^{BY} - \lambda_{iJ}^{BY})$. The intercept, effects $\{\beta_{hj}^A\}$ of A, and effects $\{\beta_{ij}^B\}$ of B depend on j; that is, they vary according to the response paired with the baseline. For given j, effects of A and B are additive.

Parameters in this model determine parameters for logits using other pairs of response categories, since

$$\log\left(\frac{\pi_{a \mid hi}}{\pi_{b \mid hi}}\right) = \log\left(\frac{\pi_{a \mid hi}}{\pi_{J \mid hi}}\right) - \log\left(\frac{\pi_{b \mid hi}}{\pi_{J \mid hi}}\right).$$

For instance, the effect of category h of A on the log odds of classification in category a of Y instead of category b is $\beta_{ha}^A - \beta_{hb}^A$. Identifiability requires that, for fixed j, the parameters in (9.1) satisfy a linear constraint. When the loglinear parameters sum to zero over the levels of A and B, $\Sigma_h \beta_{hj}^A = \Sigma_i \beta_{ij}^B = 0$.

9.1.2 Alligator Food Choice Example

Table 9.1 was taken from an investigation by Delany and Moore (1987) of factors influencing the primary food choice of alligators. The sample

Table 9.1 Primary Food Choice of Alligators, by Lake, Gender, and Size

| Lake | Gender | Size | Primary Food Choice | | | | |
			Fish	Invertebrate	Reptile	Bird	Other
Hancock	Male	≤2.3	7	1	0	0	5
		>2.3	4	0	0	1	2
	Female	≤2.3	16	3	2	2	3
		>2.3	3	0	1	2	3
Oklawaha	Male	≤2.3	2	2	0	0	1
		>2.3	13	7	6	0	0
	Female	≤2.3	3	9	1	0	2
		>2.3	0	1	0	1	0
Trafford	Male	≤2.3	3	7	1	0	1
		>2.3	8	6	6	3	5
	Female	≤2.3	2	4	1	1	4
		>2.3	0	1	0	0	0
George	Male	≤2.3	13	10	0	2	2
		>2.3	9	0	0	1	2
	Female	≤2.3	3	9	1	0	1
		>2.3	8	1	0	0	1

Source: Clint Moore, Wildlife Research Laboratory, Florida Game and Fresh Water Fish Commission.

consisted of 219 alligators captured in four Florida lakes, during September 1985. The response variable is the primary food type, in volume, found in an alligator's stomach. This variable, which we symbolize by F, had five categories: Fish, Invertebrate, Reptile, Bird, Other. The invertebrates found in the stomachs were primarily apple snails, aquatic insects, and crayfish. The reptiles were primarily turtles (though one stomach contained tags of 23 baby alligators that had been released in the lake during the previous year!). The Other category consisted of amphibian, mammal, plant material, stones or other debris, or no food or dominant type.

Table 9.1 also classifies the alligators according to explanatory variables L = lake (Hancock, Oklawaha, Trafford, George), G = gender (male, female), and S = size (≤2.3 meters long, >2.3 meters long). We use generalized logit models to investigate the effects of L, G, and S on the nominal response variable, F. These logit models correspond to loglinear models containing the LGS term for interaction among the explanatory variables.

Table 9.2 contains G^2 and X^2 values for several loglinear models. The data are sparse, 219 observations scattered among 80 cells. Though G^2 is

Table 9.2   Goodness-of-Fit of Models for Table 9.1

Model	G^2	X^2	df
(F, LGS)	116.8	106.2	60
(FG, LGS)	114.7	101.0	56
(FS, LGS)	101.6	86.6	56
(FL, LGS)	73.6	79.4	48
(FL, FS, LGS)	52.5	57.9	44
(FG, FL, FS, LGS)	50.3	52.5	40
Collapsed Table			
(F, LS)	81.4	73.0	28
(FS, LS)	66.2	54.3	24
(FL, LS)	38.2	32.7	16
(FS, FL, LS)	17.1	15.0	12

Table 9.3   Observed and Fitted Values for Model (FS, FL, LS)

Lake	Size	Primary Food Choice				
		Fish	Invertebrate	Reptile	Bird	Other
Hancock	≤2.3	23	4	2	2	8
		(20.9)	(3.6)	(1.9)	(2.7)	(9.9)
	>2.3	7	0	1	3	5
		(9.1)	(0.4)	(1.1)	(2.3)	(3.1)
Oklawaha	≤2.3	5	11	1	0	3
		(5.2)	(12.0)	(1.5)	(0.2)	(1.1)
	>2.3	13	8	6	1	0
		(12.8)	(7.0)	(5.5)	(0.8)	(1.9)
Trafford	≤2.3	5	11	2	1	5
		(4.4)	(12.4)	(2.1)	(0.9)	(4.2)
	>2.3	8	7	6	3	5
		(8.6)	(5.6)	(5.9)	(3.1)	(5.8)
George	≤2.3	16	19	1	2	3
		(18.5)	(16.9)	(0.5)	(1.2)	(3.8)
	>2.3	17	1	0	1	3
		(14.5)	(3.1)	(0.5)	(1.8)	(2.2)

unreliable for testing goodness of fit for such sparse data, we can use it to compare models. The statistics $G^2[(F, LGS)|(FG, LGS)] = 2.1$ and $G^2[(FL, FS, LGS)|(FG, FL, FS, LGS)] = 2.2$, each based on df = 4, suggest simplifying the analysis by collapsing the table over gender. G^2 and X^2 values for the collapsed table indicate that both lake and size have effects on primary food choice. Table 9.3 contains fitted values for model (FL, FS, LS) for the collapsed table. The fit appears to be adequate.

We study effects by expressing loglinear model (FL, FS, LS) as a generalized logit model. Fish was the most common primary food choice, and we estimate the effects of L and S on the odds that alligators select other primary food types instead of fish. Thus, we identify fish (level 1 of F) as the baseline category, and use model

$$\log\left(\frac{\pi_{j|hi}}{\pi_{1|hi}}\right) = \alpha_j + \beta_{hj}^L + \beta_{ij}^S$$

for $h = 1, 2, 3, 4$, $i = 1, 2$, and $j = 2, 3, 4, 5$ for levels of F.

Table 9.4 contains ML estimates of effect parameters, using the scaling $\Sigma_h \hat{\beta}_{hj}^L = \Sigma_i \hat{\beta}_{ij}^S = 0$. Size of alligator has a noticeable effect on the odds of selecting invertebrates instead of fish. For a given lake, the estimated odds that primary food choice was invertebrates instead of fish is $\exp(2 \times 0.73) = 4.3$ times higher for the smaller alligators than the larger ones. The lake effects indicate that the estimated odds that primary food choice was invertebrates instead of fish is relatively higher at Lakes Trafford and Oklawaha and relatively lower at Lake Hancock; the estimated odds that primary food choice was reptiles instead of fish is higher in Lake Trafford and lower in Lake George.

9.1.3 Simultaneous vs. Separate Fitting

We fit generalized logit models by maximizing the likelihood subject to *simultaneously* satisfying the $J - 1$ equations that specify the model. We can do this by applying iterative procedures (such as Newton–Raphson) to the generalized logit model, or by fitting an equivalent loglinear model and constructing estimates from those for that model. An alternative approach fits logit models *separately* for the $J - 1$ pairings of responses. We could fit model (9.1) for $j = 1$ alone, using only observations in categories 1 or J of the response variable, to obtain estimates of α_1, $\{\beta_{h1}^A\}$, and $\{\beta_{i1}^B\}$; we could fit (9.1) using only categories 2 and J and obtain estimates of α_2, $\{\beta_{h2}^A\}$, and $\{\beta_{i2}^B\}$; in this manner, we obtain $J - 1$ separate fits of logit models.

Parameter estimates obtained in separate fitting of logit models differ

Table 9.4 Estimated Parameters in Logit Model for Alligator Food Choice

	Size		Lake			
Logit[a]	<2.3	>2.3	Hancock	Oklawaha	Trafford	George
$\log(\pi_I/\pi_F)$	0.73(0.20)	−0.73(0.20)	−1.76(0.44)	0.84(0.33)	1.02(0.34)	−0.10(0.29)
$\log(\pi_R/\pi_F)$	−0.18(0.29)	0.18(0.29)	−0.42(0.56)	0.80(0.47)	1.28(0.47)	−1.66(0.79)
$\log(\pi_B/\pi_F)$	−0.32(0.32)	0.32(0.32)	0.41(0.51)	−0.94(0.82)	0.80(0.54)	−0.28(0.56)
$\log(\pi_O/\pi_F)$	0.17(0.22)	−0.17(0.22)	0.24(0.35)	−0.58(0.51)	0.93(0.38)	−0.59(0.39)

[a]I = invertebrate, R = reptile, B = bird, O = other, F = fish; ASE values in parentheses.

from those obtained by fitting the logit model simultaneously to the $J - \overset{.}{1}$ logits. The estimators in the separate-fitting approach are less efficient. However, Begg and Gray (1984) showed that those estimators are not very inefficient unless the probability of classification in the baseline category is small. Unless there is a natural baseline category, it is best to use the response category having highest prevalence as the baseline in the separate fitting approach.

To illustrate the separate-fitting approach, we fit model

$$\log\left(\frac{\pi_{I \mid hi}}{\pi_{F \mid hi}}\right) = \alpha + \beta_h^L + \beta_i^S$$

using the data for categories Invertebrate and Fish alone. The model fits adequately: $G^2 = 3.8$ with df = 3. The estimated size effects are $\hat{\beta}_1^S = 0.83$ (ASE = 0.21) and $\hat{\beta}_2^S = -0.83$, and the estimated lake effects are $\hat{\beta}_1^L = -1.90$ (ASE = 0.45), $\hat{\beta}_2^L = 0.93$ (0.34), $\hat{\beta}_3^L = 1.09$ (0.36), and $\hat{\beta}_4^L = -0.12$ (0.30). The effects are similar to those obtained using simultaneous fitting with all five response categories (see the first row of Table 9.4). The estimated standard errors are only slightly larger, since 155 of the 219 observations were in the Fish or Invertebrate categories of food type.

A logit model fitted using data from only two response categories is the same as a regular logit model fitted *conditional* on classification into one of those two categories. For instance, the jth generalized logit is a logit of conditional probabilities

$$\log\left[\frac{\pi_{j \mid hi}/(\pi_{j \mid hi} + \pi_{J \mid hi})}{\pi_{J \mid hi}/(\pi_{j \mid hi} + \pi_{J \mid hi})}\right] = \log\left(\frac{\pi_{j \mid hi}}{\pi_{J \mid hi}}\right).$$

9.1.4 Ordinal Explanatory Variables

Suppose an explanatory variable is ordinal. Then useful generalized logit models are equivalent to ordinal loglinear models discussed in Chapter 8. To illustrate, suppose Y is a nominal response and X is an ordinal explanatory variable. The loglinear column effects model

$$\log m_{ij} = \mu + \lambda_i^X + \lambda_j^Y + u_i \nu_j$$

applies to ordinal-nominal tables, where $\{u_i\}$ are monotone scores for X and $\{\nu_j\}$ are parameters for Y. It corresponds to generalized logit model

$$\log(m_{ij}/m_{iJ}) = (\lambda_j^Y - \lambda_J^Y) + u_i(\nu_j - \nu_J)$$

for all i and for $j = 1, \ldots, J - 1$. This model has form

$$\log(m_{ij}/m_{iJ}) = \alpha_j + u_i\tau_j, \quad i = 1, \ldots, I, \ j = 1, \ldots, J-1. \quad (9.2)$$

This is a linear logit model in the scores for X, with slope τ_j varying according to the response categories paired for the logit.

9.2 MULTINOMIAL LOGIT MODELS

The previous section introduced logit models for nominal response variables. This section studies more general models that can handle explanatory variables that are continuous or take different values for different categories of the response.

9.2.1 Continuous Explanatory Variables

Generalized logit models, like ordinary regression models, can contain continuous or discrete explanatory variables. Let $\pi_j(\mathbf{x}_i)$ denote the probability of response j, $j = 1, \ldots, J$, at the ith setting of values of k explanatory variables $\mathbf{x}_i = (1, x_{i1}, \ldots, x_{ik})'$. In terms of the response probabilities, the generalized logit model is

$$\pi_j(\mathbf{x}_i) = \frac{\exp(\boldsymbol{\beta}_j'\mathbf{x}_i)}{\displaystyle\sum_{h=1}^{J} \exp(\boldsymbol{\beta}_h'\mathbf{x}_i)} . \quad (9.3)$$

For identifiability, we take $\boldsymbol{\beta}_J = \mathbf{0}$, in which case

$$\log[\pi_j(\mathbf{x}_i)/\pi_J(\mathbf{x}_i)] = \boldsymbol{\beta}_j'\mathbf{x}_i, \quad j = 1, \ldots, J-1.$$

A separate parameter vector applies for each of the $J-1$ nonredundant logit equations.

To obtain ML estimates, we maximize the independent multinomial likelihood subject to constraint (9.3). The parameter estimates have large-sample normal distributions with asymptotic standard errors obtained from the inverse of the information matrix. The log likelihood is concave, and an iterative procedure such as Newton–Raphson yields the estimates. Though simple in principle, the process can be computationally time-consuming when some variables are continuous, because of data sparseness. This sparseness also adversely affects behavior of chi-squared goodness-of-fit statistics. Such statistics are inappropriate for testing fit of models, but can be used to compare models (Haberman 1974a, pp. 372–373).

9.2.2 Occupational Attainment Example

Schmidt and Strauss (1975) modeled occupational attainment in the United States, using explanatory variables S = years of schooling, E = labor market experience (calculated as age − years of schooling − 5), R = race (1 = white, 0 = black), and G = gender (1 = male, 0 = female). The categories of occupational attainment are professional (P), white collar (W), blue collar (B), craft (C), and menial (M).

For a sample of 1000 full-time workers drawn at random from the 1967 Public Use Samples, they obtained the fitted logit equations summarized in Table 9.5. The menial category is the baseline for each equation. For instance, the final equation is

$$\log(\hat{\pi}_P/\hat{\pi}_M) = -5.959 + 0.429S + 0.008E + 0.976R + 0.656G .$$

The estimated odds of holding a professional (instead of menial) job are $\exp(0.429) = 1.54$ times higher for each additional year of schooling, 1.01 times higher for each additional year of experience, 2.65 times higher for whites than blacks, and 1.93 times higher for men than women, controlling for the other factors. Schmidt and Strauss showed that the addition of interaction terms does not improve the fit.

Using Table 9.5, we can obtain parameter estimates for logits using any pair of occupational groups. However, we need the entire estimated covariance matrix of the estimates in Table 9.5 in order to obtain standard errors of parameter estimates for other logits.

Table 9.5 Estimated Parameters in Logit Model for Occupational Attainment, Using Menial as Baseline Category[a]

Logit	Intercept	Schooling	Experience	Race	Gender
$\log(\pi_B/\pi_M)$	1.056	−0.124	−0.015	0.700	1.252
	(0.677)	(0.049)	(0.009)	(0.302)	(0.243)
$\log(\pi_C/\pi_M)$	−3.769	−0.001	−0.008	1.458	3.112
	(0.902)	(0.047)	(0.010)	(0.435)	(0.474)
$\log(\pi_W/\pi_M)$	−3.305	0.225	0.003	1.762	−0.523
	(0.783)	(0.053)	(0.009)	(0.413)	(0.236)
$\log(\pi_P/\pi_M)$	−5.959	0.429	0.008	0.976	0.656
	(0.793)	(0.054)	(0.009)	(0.372)	(0.247)

Source: Reprinted with permission from Schmidt and Strauss (1975).
[a]ASE values in parentheses.

Table 9.6 Occupation Probabilities, Given Average Schooling and Experience

		Occupation				
Race	Gender	Menial	Blue Collar	Craft	White Collar	Professional
Black	Female	0.396	0.188	0.011	0.219	0.187
	Male	0.222	0.368	0.136	0.073	0.202
White	Female	0.153	0.146	0.018	0.492	0.192
	Male	0.089	0.296	0.232	0.169	0.214

Source: Reprinted with permission from Schmidt and Strauss (1975).

The estimates in the Race column of Table 9.5 indicate that occupational groups are ordered (W, C, P, B, M) in terms of relative number of white workers; that is, controlling for the other factors, the odds a worker is white is highest for white-collar occupations. The occupational groups are ordered (C, B, P, M, W) in terms of relative number of male workers. Table 9.6 illustrates the effects of race and gender by listing predicted probabilities for occupation, evaluated at the sample means for schooling and experience.

9.2.3 Multinomial Logit Models

An important application of logit models is determining effects of explanatory variables on a subject's choice of one of a discrete set of options—for instance, the choice of transportation system to take to work (drive, bus, subway, walk, bicycle), housing (buy house, buy condominium, rent), primary shopping location (downtown, mall A, mall B, other), brand of toothpaste, political party preference, or occupation. Models for response variables consisting of a discrete set of choices are called *discrete-choice* models. Such models are important research tools for economists and geographers.

In many discrete-choice applications, for a given subject, an explanatory variable takes different values for different response choices. As predictors of choice of transportation system, "cost" and "transit time to reach destination" take different values for each choice. As predictors of choice of brand of product, "price" and "amount spent on advertising" take different values for each choice. Explanatory variables of this type are "characteristics of the choices." They differ from the usual explanatory variables, for which values remain constant across the choice set. Such variables, "characteristics of the chooser," include income, education, and other demographic characteristics.

McFadden (1974) proposed a discrete-choice model for explanatory variables that are characteristics of the choices. His model permits the set of choices to vary by subject. For instance, some subjects may not have the subway as an option for travel to work. McFadden's model is related to models proposed by Bradley and Terry (1952) and Luce (1959). For subject i and response choice j, let $\mathbf{x}_{ij} = (x_{ij1}, \ldots, x_{ijk})'$ denote the values of the k explanatory variables. Conditional on the set of response choices C_i for subject i, the model is

$$\pi_j(\mathbf{x}_{ij}) = \frac{\exp(\boldsymbol{\beta}'\mathbf{x}_{ij})}{\sum_{h \in C_i} \exp(\boldsymbol{\beta}'\mathbf{x}_{ih})} . \tag{9.4}$$

For each pair of alternatives a and b, this model has the logit form

$$\log[\pi_a(\mathbf{x}_{ia})/\pi_b(\mathbf{x}_{ib})] = \boldsymbol{\beta}'(\mathbf{x}_{ia} - \mathbf{x}_{ib}) . \tag{9.5}$$

Conditional on a subject's choice being a or b, the influence of a particular explanatory variable depends on the distance between the subject's values of that variable for those alternatives. If the values are the same, the model asserts that that variable has no influence on the choice between alternatives a and b.

From (9.5), the odds of choosing a over b do not depend on the other alternatives in the choice set or on values of the explanatory variables for those alternatives. Luce (1959) called this property *independence from irrelevant alternatives*. It is not a realistic assumption in some applications. For instance, for travel alternatives auto and red bus, suppose 80% choose auto. Now suppose the options are auto, red bus, and blue bus. According to (9.5), the odds are still 4.0 of choosing auto instead of red bus, but intuitively we expect them to be about 8.0 (10% choosing each bus option). McFadden (1974) suggested that "application of the model should be limited to situations where the alternatives can plausibly be assumed to be distinct and weighed independently in the eyes of each decision-maker."

Model (9.4) can also incorporate explanatory variables that are characteristics of the chooser. This may seem surprising, since (9.4) has a single parameter for each explanatory variable; that is, the parameter vector is the same for each pair of choices. However, we can express (9.3) in form (9.4) by replacing such an explanatory variable by J artificial variables. The jth variable is the product of the explanatory variable with a dummy variable that equals 1 when the response choice is j. For instance, suppose there is a single explanatory variable, and let x_i denote its value for the *i*th subject. For $j = 1, \ldots, J$, let δ_{jt} equal 1 when $t = j$ and 0 otherwise, and let

$$\mathbf{z}_{ij} = (\delta_{j1}, \ldots, \delta_{jJ}, \delta_{j1}x_i, \ldots, \delta_{jJ}x_i)'$$

Let $\boldsymbol{\beta} = (\alpha_1, \ldots, \alpha_J, \beta_1, \ldots, \beta_J)'$. Then $\boldsymbol{\beta}'\mathbf{z}_{ij} = \alpha_j + \beta_j x_i$, and (9.3) is

$$\pi_j(x_i) = \frac{\exp(\alpha_j + \beta_j x_i)}{\exp(\alpha_1 + \beta_1 x_i) + \cdots + \exp(\alpha_J + \beta_J x_i)}$$

$$= \frac{\exp(\boldsymbol{\beta}'\mathbf{z}_{ij})}{\exp(\boldsymbol{\beta}'\mathbf{z}_{i1}) + \cdots + \exp(\boldsymbol{\beta}'\mathbf{z}_{iJ})}$$

which has form (9.4). One pair of parameters in $\boldsymbol{\beta}$ is redundant, so we can take $\alpha_J = \beta_J = 0$.

Using this approach, we can formulate *mixed* logit models, containing both characteristics of the chooser and characteristics of the choices. Thus, model (9.4) is very general. Originally referred to by McFadden as a *conditional logit* model, it is now usually called the *multinomial logit* model. All models studied so far in this chapter are special cases of the multinomial logit model.

9.2.4 Shopping Choice Example

McFadden (1974) and Domencich and McFadden (1975, pp. 170–173) used a multinomial logit model to describe how a sample of residents of Pittsburgh, Pennsylvania chose a shopping destination. There were five possible destinations, corresponding to different city zones. One explanatory variable was a measure of shopping opportunities, defined to be the retail employment in the zone as a percentage of total retail employment in the region. The other explanatory variable was price of the trip, defined from a separate analysis in terms of auto in-vehicle time and auto operating cost.

The ML estimates of model parameters were -1.06 (ASE $= 0.28$) for price of trip and 0.84 (ASE $= 0.23$) for shopping opportunity. From (9.5),

$$\log(\hat{\pi}_a/\hat{\pi}_b) = -1.06(P_a - P_b) + 0.84(S_a - S_b)$$

where $P =$ price and $S =$ shopping opportunity. A destination is more attractive as the price of getting there is relatively less and as the shopping opportunity is relatively higher. Given values of P and S for each destination, we could use the sample analog of (9.4) to obtain predicted probabilities of choosing each destination.

9.3 LOGITS FOR ORDINAL RESPONSES

When response categories have a natural ordering, logit models should utilize that ordering. We can incorporate the ordering directly in the way we construct logits. This section discusses three types of logits for ordered response categories.

9.3.1 Adjacent-Categories Logits

Let $\{\pi_1(\mathbf{x}), \ldots, \pi_J(\mathbf{x})\}$ denote response probabilities at value $\mathbf{x}$ for a set of explanatory variables. Chapter 8 showed that loglinear models for ordinal variables have simple representations as logit models for adjacent response categories. The adjacent-categories logits are

$$L_j = \log\left(\frac{\pi_j(\mathbf{x})}{\pi_{j+1}(\mathbf{x})}\right), \quad j = 1, \ldots, J-1. \tag{9.6}$$

These logits are a basic set equivalent to the baseline-category logits

$$L_j^* = \log\left(\frac{\pi_j(\mathbf{x})}{\pi_J(\mathbf{x})}\right), \quad j = 1, \ldots, J-1. \tag{9.7}$$

Either set determines logits for all $\binom{J}{2}$ pairs of response categories.

We can express adjacent-categories logit models as baseline-category logit models, and fit them using methods and computer software for the latter models. For instance, suppose we wanted to fit an adjacent-categories logit model

$$L_j = \alpha_j + \boldsymbol{\beta}'\mathbf{x}, \quad j = 1, \ldots, J-1.$$

We could use the relationship $L_j^* = L_j + L_{j+1} + \cdots + L_{J-1}$ to obtain the equivalent baseline-category logit model

$$L_j^* = \sum_{k=j}^{J-1} \alpha_k + \boldsymbol{\beta}'(J-j)\mathbf{x}, \quad j = 1, \ldots, J-1$$

$$= \alpha_j^* + \boldsymbol{\beta}'\mathbf{u}_j, \quad j = 1, \ldots, J-1$$

with $\mathbf{u}_j = (J-j)\mathbf{x}$. The adjacent-categories logit model corresponds to a baseline-category logit model with adjusted model matrix.

9.3.2 Continuation-Ratio Logits

Continuation-ratio logits are defined as

$$L_j = \log\left(\frac{\pi_j(\mathbf{x})}{\pi_{j+1}(\mathbf{x}) + \cdots + \pi_J(\mathbf{x})}\right), \quad j = 1, \ldots, J-1 \qquad (9.8)$$

or as

$$L_j^* = \log\left(\frac{\pi_{j+1}(\mathbf{x})}{\pi_1(\mathbf{x}) + \cdots + \pi_j(\mathbf{x})}\right), \quad j = 1, \ldots, J-1.$$

Let $\rho_j(\mathbf{x})$ denote the probability of response j, given response j or higher; that is,

$$\rho_j(\mathbf{x}) = \frac{\pi_j(\mathbf{x})}{\pi_j(\mathbf{x}) + \cdots + \pi_J(\mathbf{x})}, \quad j = 1, \ldots, J-1.$$

The continuation-ratio logits $\{L_j\}$ are ordinary logits of these conditional probabilities, namely

$$L_j = \log\left(\frac{\rho_j(\mathbf{x})}{1 - \rho_j(\mathbf{x})}\right), \quad j = 1, \ldots, J-1.$$

At a particular setting of $\mathbf{x}$, let $\{n_j(\mathbf{x}), j = 1, \ldots, J\}$ denote the response counts and let $n(\mathbf{x}) = \Sigma_j \, n_j(\mathbf{x})$. One can show that the multinomial mass function for $\{n_j(\mathbf{x}), j = 1, \ldots, J\}$ has factorization

$$b[n(\mathbf{x}), n_1(\mathbf{x}); \rho_1(\mathbf{x})]b[n(\mathbf{x}) - n_1(\mathbf{x}), n_2(\mathbf{x}); \rho_2(\mathbf{x})] \cdots$$
$$b[n(\mathbf{x}) - n_1(\mathbf{x}) - \cdots - n_{J-2}(\mathbf{x}), n_{J-1}(\mathbf{x}); \rho_{J-1}(\mathbf{x})] \qquad (9.9)$$

where $b(n, y; \rho)$ denote the binomial probability of y "successes" in n trials, when the success probability is ρ on each trial. The full likelihood is the product of multinomial mass functions from the different $\mathbf{x}$ values. Thus, the log likelihood is a sum of terms such that different ρ_j enter into different terms. When parameters in the model specification for L_j are distinct from those for L_k, whenever $j \neq k$, we can maximize the log likelihood by maximizing each term separately. It follows that separate fitting of models for different continuation-ratio logits gives the same results as simultaneous fitting. We can sum the $J-1$ separate G^2 statistics to obtain an overall goodness-of-fit statistic pertaining to the simultaneous fitting of $J-1$ models, one for each $\text{logit}(\rho_j)$, $j = 1, \ldots, J-1$.

Because these logits refer to a binary response in which one category combines levels of the original scale, we can perform separate fitting using methods for binary logit models or corresponding loglinear models. Similar remarks apply to continuation-ratio logits $\{L_j^*\}$, though those logits and the subsequent analysis do not give equivalent results. Neither set gives models equivalent to those for baseline-category logits or adjacent-category logits.

9.3.3 Developmental Toxicity Study with Pregnant Mice

We illustrate continuation-ratio logits using Table 9.7, based on a developmental toxicity study presented by Price et al. (1987). Developmental toxicity experiments with rodents are commonly used to test and regulate substances posing potential danger to developing fetuses. Diethylene glycol dimethyl ether (diEGdiME), one such substance, is an industrial solvent used in the manufacture of protective coatings such as lacquers and metal coatings.

In their study, Price et al. administered diEGdiME in distilled water to pregnant mice. Each mouse was exposed to one of five concentration levels, for ten days early in the pregnancy. (The mice exposed to level 0 formed a control group.) Two days later, the uterine contents of the pregnant mice were examined for defects. Each fetus had three possible outcomes: Non-live; Malformation; Normal. The outcomes are ordered, with a non-live fetus the least desirable result. We use continuation-ratio logits to model (1) the probability π_1 of a non-live fetus, and (2) the conditional probability $\pi_2/(\pi_2 + \pi_3)$ of a malformed fetus, given that the fetus was live.

We fitted the continuation-ratio logit models

Table 9.7 Outcomes for Pregnant Mice in Developmental Toxicity Study[a]

Concentration (mg/kg per day)	Response		
	Non-live	Malformation	Normal
0 (controls)	15	1	281
62.5	17	0	225
125	22	7	283
250	38	59	202
500	144	132	9

[a]Based on results in Price et al. (1987). I thank Dr. Louise Ryan for showing me these data.

$$L_j = \alpha_j + \beta_j x_i, \quad j = 1, 2$$

using scores $\{x_1 = 0, \ x_2 = 62.5, \ x_3 = 125, \ x_4 = 250, \ x_5 = 500\}$ for concentration level. The two models are linear logit models in which the responses are column 1 and columns 2–3 combined in Table 9.7 for $j = 1$, and columns 2 and 3 for $j = 2$. The parameter estimates are $\hat{\beta}_1 = 0.0064$ (ASE $= 0.0004$), and $\hat{\beta}_2 = 0.0174$ (ASE $= 0.0012$). In each case, the less desirable outcome is more likely as the concentration increases. For instance, given that a fetus was live, the estimated odds that it was malformed rather than normal changes by a multiplicative factor of $\exp(1.74) = 5.7$ for every 100-unit increase in the concentration of diEGdeME. The likelihood-ratio statistics are $G^2 = 5.78$ for $j = 1$, and $G^2 = 6.06$ for $j = 2$, each based on df $= 3$. We summarize the fit by their sum, $G^2 = 11.84$, based on df $= 6$.

This analysis treats pregnancy outcomes for different fetuses as independent observations. In fact, each pregnant mouse had a litter of fetuses, and there may be statistical dependence among different fetuses in the same litter. At a fixed concentration level, there may also be heterogeneity among the pregnant mice in their response probabilities. The total G^2 gives some evidence of lack of fit, but may reflect overdispersion caused by these factors, rather than an inappropriate choice of response curve.

9.3.4 Cumulative Logits

Another way to use ordered response categories is by forming logits of cumulative probabilities,

$$F_j(\mathbf{x}) = \pi_1(\mathbf{x}) + \cdots + \pi_j(\mathbf{x}), \quad j = 1, \ldots, J.$$

The *cumulative logits* are defined as

$$L_j = \text{logit}[F_j(\mathbf{x})] = \log\left(\frac{F_j(\mathbf{x})}{1 - F_j(\mathbf{x})}\right)$$

$$= \log\left(\frac{\pi_1(\mathbf{x}) + \cdots + \pi_j(\mathbf{x})}{\pi_{j+1}(\mathbf{x}) + \cdots + \pi_J(\mathbf{x})}\right), \quad j = 1, \ldots, J - 1. \tag{9.10}$$

Each cumulative logit uses all J response categories.

A model for cumulative logit L_j is an ordinary logit model for a binary response in which categories 1 to j form a single category, and categories $j + 1$ to J form the second category. More general models simultaneously provide a structure for all $J - 1$ cumulative logits. Unlike models for other logits studied in this section, some cumulative logit models are not equivalent to binary logit or multinomial logit or loglinear models. We discuss such models in the next section.

9.4 CUMULATIVE LOGIT MODELS

We now incorporate all $J - 1$ cumulative logits for a J-category response into a single, parsimonious model. Let $\{L_j(\mathbf{x}) = \text{logit}[F_j(\mathbf{x})], \ j = 1, \ldots, J - 1\}$, where $F_j(\mathbf{x}) = P(Y \leq j \mid \mathbf{x})$ is the cumulative probability for response category j, when the explanatory variables take value $\mathbf{x}$.

The simplest cumulative logit model

$$L_j(\mathbf{x}) = \alpha_j, \quad j = 1, \ldots, J - 1$$

implies the response variable is simultaneously independent of all explanatory variables. The $\{\alpha_j\}$ are called *cutpoint* parameters. They are nondecreasing in j, since the cumulative logit is an increasing function of $F_j(\mathbf{x})$, which is itself increasing in j for fixed $\mathbf{x}$.

9.4.1 Proportional Odds Model

To include effects of explanatory variables, we use model

$$L_j(\mathbf{x}) = \alpha_j + \boldsymbol{\beta}'\mathbf{x}, \quad j = 1, \ldots, J - 1. \tag{9.11}$$

This model assumes a variable's effect on the odds of response below category j is the same for all j. It satisfies

$$L_j(\mathbf{x}_1) - L_j(\mathbf{x}_2) = \log\left[\frac{P(Y \leq j \mid \mathbf{x}_1)/P(Y > j \mid \mathbf{x}_1)}{P(Y \leq j \mid \mathbf{x}_2)/P(Y > j \mid \mathbf{x}_2)}\right] = \boldsymbol{\beta}'(\mathbf{x}_1 - \mathbf{x}_2).$$

The odds ratio of cumulative probabilities in this expression is called a *cumulative odds ratio*. The log of the cumulative odds ratio is proportional to the distance between the values of the explanatory variables, with the same proportionality constant applying to each cutpoint. Because of this property, model (9.11) is called a *proportional odds model*. Its interpretation is that the odds of making response $\leq j$ are $\exp[\boldsymbol{\beta}'(\mathbf{x}_1 - \mathbf{x}_2)]$ times higher at $\mathbf{x} = \mathbf{x}_1$ than at $\mathbf{x} = \mathbf{x}_2$.

Figure 9.1 depicts model (9.11) for a single explanatory variable x and an ordinal response variable having four categories. For fixed j, the response curve looks like a logistic regression curve for a binary response with outcomes $Y \leq j$ and $Y > j$. The response curves for $j = 1$, 2, and 3 have the same shape. The curve for $P(Y \leq k)$ is the curve for $P(Y \leq j)$ translated by $(\alpha_j - \alpha_k)/\beta$ units in the x direction; that is,

$$F_j(x) = F_k[x + (\alpha_j - \alpha_k)/\beta] .$$

When $\beta_i > 0$ in model (9.11), each cumulative logit increases as x_i increases, so each cumulative probability increases. This means that relatively more probability mass falls at the low end of the Y scale; that is, Y tends to be smaller at higher values of x_i. To make $\beta_i > 0$ have the more usual meaning of Y tending to be *larger* at higher values of x_i, we replace β in (9.11) by $-\beta$. For this parameterization,

$$L_j(\mathbf{x}) = \alpha_j - \beta'\mathbf{x}, \quad j = 1, \ldots, J - 1 . \tag{9.12}$$

We can motivate the common effect β for different j in the proportional odds model by assuming that a regression model holds when we measure the response more finely (Anderson and Philips 1981). Let Y^* denote an underlying continuous response variable having cdf $G(y^* - \eta)$, where η is a location parameter dependent on $\mathbf{x}$ through $\eta(\mathbf{x}) = \beta'\mathbf{x}$. Suppose $-\infty = \alpha_0 < \alpha_1 < \cdots < \alpha_J = \infty$ are such that the observed ordinal response Y satisfies

$$Y = j \quad \text{if } \alpha_{j-1} < Y^* \leq \alpha_j , \tag{9.13}$$

as depicted in Figure 9.2. That is, we observe a response in category j

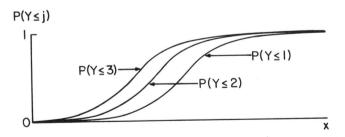

Figure 9.1 Depiction of cumulative logit model with effect independent of cutpoint.

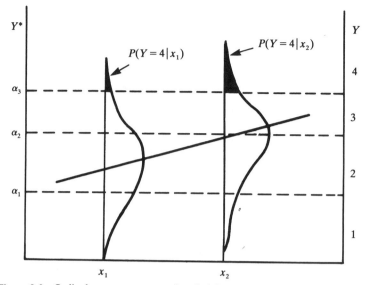

Figure 9.2   Ordinal measurement, and underlying continuous regression model.

when the underlying continuous response falls in the jth interval of values. Then,

$$F_j(\mathbf{x}) = P(Y \le j \mid \mathbf{x}) = P(Y^* \le \alpha_j \mid \mathbf{x}) = G(\alpha_j - \boldsymbol{\beta}'\mathbf{x}) \, .$$

The appropriate model for Y applies link G^{-1} to $F_j(\mathbf{x})$. If $Y^* = \boldsymbol{\beta}'\mathbf{x} + \epsilon$ where the cdf G of ϵ is the logistic distribution (Section 4.2.5), then G^{-1} is the logit and we obtain model (9.12).

Explanatory variables in a cumulative logit model can be continuous or categorical. Because the model constrains the $J - 1$ response curves to have the same shape, we cannot fit it by fitting separate logit models for each cutpoint. We must maximize the multinomial likelihood subject to constraint (9.12). Unlike other logit models, cumulative logit models making the proportional odds assumption are not equivalent to loglinear models. They cannot be fitted using simple procedures for binary responses. Walker and Duncan (1967) and McCullagh (1980) gave Fisher scoring algorithms for iterative calculation of ML estimates of parameters. The algorithms resemble the Newton-Raphson iterative method, except that expected (rather than observed) values are used in the second derivative matrix.

9.4.2 Mental Impairment Example

To illustrate cumulative logit models, we analyze Table 9.8, relating mental impairment to two explanatory variables. Mental impairment is an ordinal response, with categories (well, mild symptom formation, moderate symptom formation, impaired). The life events index X_1 is a composite measure of both the number and severity of important life events (such as birth of child, new job, divorce, or death in family) that occurred to the subject within the past three years. The other explanatory variable is a binary measurement ($X_2 = 1$, high; $X_2 = 0$, low) of socioeconomic status (SES).

For the main effects model

$$L_j(\mathbf{x}) = \alpha_j - \beta_1 x_1 - \beta_2 x_2 ,$$

the parameter estimates are $\hat{\beta}_1 = 0.319$ (ASE = 0.121) and $\hat{\beta}_2 = -1.111$ (ASE = 0.611). The chance of higher levels of mental impairment increases as the life events score increases, and it decreases at the higher

Table 9.8 Mental Impairment by SES and Life Events

Subject	Mental Impairment	SES	Life Events	Subject	Mental Impairment	SES	Life Events
1	Well	1	1	21	Mild	1	9
2	Well	1	9	22	Mild	0	3
3	Well	1	4	23	Mild	1	3
4	Well	1	3	24	Mild	1	1
5	Well	0	2	25	Moderate	0	0
6	Well	1	0	26	Moderate	1	4
7	Well	0	1	27	Moderate	0	3
8	Well	1	3	28	Moderate	0	9
9	Well	1	3	29	Moderate	1	6
10	Well	1	7	30	Moderate	0	4
11	Well	0	1	31	Moderate	0	3
12	Well	0	2	32	Impaired	1	8
13	Mild	1	5	33	Impaired	1	2
14	Mild	0	6	34	Impaired	1	7
15	Mild	1	3	35	Impaired	0	5
16	Mild	0	1	36	Impaired	0	4
17	Mild	1	8	37	Impaired	0	4
18	Mild	1	2	38	Impaired	1	8
19	Mild	0	5	39	Impaired	0	8
20	Mild	1	5	40	Impaired	0	9

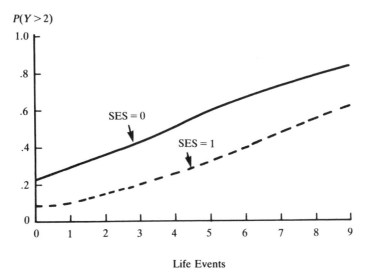

Figure 9.3   Predicted values of $P(Y > 2)$ for Table 9.8.

level of SES. At a given life events score, the odds that mental impairment is below any fixed level is estimated to be $e^{1.111} = 3.04$ times as great at the high SES level as at the low level. The cutpoint parameter estimates are $\hat{\alpha}_1 = -0.28$, $\hat{\alpha}_2 = 1.21$, and $\hat{\alpha}_3 = 2.21$. Using these and $\{\hat{\beta}_i\}$, we can calculate predicted logits, and hence predicted values of $P(Y \leq j)$, $P(Y > j)$, or $P(Y = j)$. Figure 9.3 plots predicted values of $P(Y > 2)$ as a function of the life events index, at the two levels of SES.

Adding an interaction term $\beta_3 x_1 x_2$, we obtain $\hat{\beta}_1 = 0.420$ (ASE = 0.186), $\hat{\beta}_2 = -0.371$ (1.136), and $\hat{\beta}_3 = -0.181$ (0.238). The estimated effect of life events on the cumulative logit is 0.420 for the low SES group and 0.239 for the high SES group. The impact of life events seems more severe for the low SES group, though the difference in effects is not significant for this small sample.

9.4.3   A Uniform Association Model for Cumulative Odds Ratios

We now present a uniform association model for cumulative odds ratios. The model applies to two-way tables having ordinal response Y and explanatory variable X. Let $L_{j|i}$ denote the jth cumulative logit within row i; that is,

$$L_{j|i} = \log\left(\frac{\pi_{1|i} + \cdots + \pi_{j|i}}{\pi_{j+1|i} + \cdots + \pi_{J|i}}\right) = \log\left(\frac{m_{i1} + \cdots + m_{ij}}{m_{i,j+1} + \cdots + m_{iJ}}\right).$$

A proportional odds model of form (9.12) is

$$L_{j|i} = \alpha_j - \beta u_i, \quad i = 1, \ldots, I, \quad j = 1, \ldots, J - 1 \qquad (9.14)$$

where $\{u_i\}$ are preassigned scores for the rows. Each of the $J - 1$ logits is linearly related to the explanatory variable, with common slope β for all logits. As u_i changes, the cumulative response probability changes in the same direction at each j. Thus response distributions at different levels of X are stochastically ordered. When $\beta > 0$, the conditional Y distributions are stochastically higher at higher levels of X.

For unit-spaced scores, model (9.14) implies that for adjacent rows of X,

$$L_{j|i} - L_{j|i+1} = \log\frac{(m_{i1} + \cdots + m_{ij})(m_{i+1,j+1} + \cdots + m_{i+1,J})}{(m_{i,j+1} + \cdots + m_{iJ})(m_{i+1,1} + \cdots + m_{i+1,j})} = \beta.$$

This difference in logits is the log cumulative odds ratio for the 2×2 table consisting of rows i and $i + 1$ and the binary response having cutpoint following category j. The odds ratio is "local" in the rows but "global" in the columns, since it uses all J column categories. The log odds ratio equals β uniformly for all $i = 1, \ldots, I - 1$ and $j = 1, \ldots, J - 1$. In other words, $\exp(\beta)$ is the common cumulative odds ratio for the $(I - 1)(J - 1)$ 2×2 tables for all pairs of adjacent rows and all binary collapsings of the response. We refer to model (9.14) with equal-interval scores as the cumulative logit *uniform association* model. Figure 9.4 illustrates the constant cumulative odds ratio implied by this model.

Model (9.14) has $J - 1$ logits in each of I rows, a single association parameter (β), and $J - 1$ cutpoint parameters. The residual df for fitting the model is df $= I(J - 1) - J = IJ - I - J$. This is the same as the

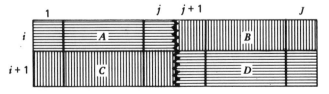

Figure 9.4 Uniform odds ratios AD/BC for all pairs of adjacent rows and all response cutpoints, for cumulative logit uniform association model.

residual df for a loglinear model for ordinal-ordinal tables, the linear-by-linear association model (8.1). The models are not equivalent unless $J = 2$.

Given that model (9.14) holds, the null hypothesis of independence is $H_0: \beta = 0$. We can test H_0 using the difference in G^2 values between the independence model and this model, based on df = 1. The cumulative logit independence model is equivalent to the loglinear independence model (5.2), and has fitted values $\{\hat{m}_{ij} = n_{i+}n_{+j}/n\}$.

9.4.4 A Row Effects Model for Nominal Explanatory Variable

We now generalize model (9.14) by replacing $\{\beta u_i\}$ by a set of unordered parameters $\{\mu_i\}$. The resulting model

$$L_{j|i} = \alpha_j - \mu_i, \quad i = 1, \ldots, I, \ j = 1, \ldots, J - 1 \tag{9.15}$$

treats levels of X as nominal. The row effects $\{\mu_i\}$ satisfy a linear constraint such as $\Sigma \ \mu_i = 0$. We refer to this model as the cumulative logit *row effects* model.

The $\{\mu_i\}$ describe the association. For each pair of rows a and b,

$$L_{j|b} - L_{j|a} = \mu_a - \mu_b$$

is constant for all $J - 1$ cutpoints j for the logit. Thus the cumulative odds ratio for the 2×2 table formed by taking rows a and b of the table and collapsing the response is assumed to be identical for all $J - 1$ collapsings. If $\mu_a > \mu_b$, the conditional distribution of Y is stochastically higher in row a than in row b.

This model has residual df = $(I - 1)(J - 2)$, the same as the loglinear row effects model (8.7) for nominal-ordinal tables. The models are not equivalent unless $J = 2$. The independence model is the special case $\mu_1 = \mu_2 = \cdots = \mu_I$. We can test independence using the difference in likelihood-ratio statistics for the independence and row effects models. This chi-squared statistic has df = $I - 1$.

9.4.5 Dumping Severity Example

To illustrate cumulative logit models for ordinal and nominal explanatory variables, we analyze the dumping severity data introduced in Table 8.6. Let H = hospital, O = operation, and D = dumping severity. Let $L_{j|hi}$ represent, for hospital h and operation i, the cumulative logit when the cutpoint for D follows category j. We assume independent multinomial sampling on D at each O–H combination.

Since D has three response categories, there are 32 cumulative logits, two at each of the 16 $O-H$ combinations. The model

$$L_{j \mid hi} = \alpha_j, \quad j = 1, 2$$

states that D is jointly independent of O and H. It is equivalent to loglinear model (D, OH). The G^2 value is 31.64, and since the model has only two parameters, its residual df $= 30$.

The ordinality of operation is recognized in the model

$$L_{j \mid hi} = \alpha_j - \beta u_i . \tag{9.16}$$

This model assumes a linear effect of O on the logit for D, the effect being the same for both logits ($j = 1, 2$) and the same for each hospital. It also assumes D is conditionally independent of H for each operation. It has only one more parameter than the joint independence model, and it yields $G^2 = 25.03$ based on df $= 29$ when fitted with scores $\{u_i = i\}$. The reduction in G^2 of 6.61, based on df $= 1$, gives strong evidence of an effect of operation on dumping, evidence also provided by the ML estimate $\hat{\beta} = 0.225$ and its ASE of 0.088. The odds that dumping severity is below a certain point rather than above it are estimated to be $\exp(0.225) = 1.25$ times higher for operation i than operation $i + 1$. We obtain the same result by fitting the cumulative logit model to the O-D marginal table, Table 9.9.

We represent potential effects of the nominal variable H by adding row-effect-type parameters to the model, giving

$$L_{j \mid hi} = \alpha_j - \mu_h - \beta u_i .$$

For this model, each effect parameter is homogeneous for the two ways of forming the cumulative logits, and there is an absence of three-factor interaction. Since $G^2 = 22.48$ with residual df $= 26$, it does not fit much better than the previous model. For each operation there is little evidence that the distribution of dumping severity differs among the four hospitals.

In fitting the cumulative logit uniform association model to the $O-D$ marginal table, we obtain $\hat{\alpha}_1 = 0.883$, $\hat{\alpha}_2 = 2.636$, and $\hat{\beta} = 0.225$. The first predicted logit in row 1 is $0.883 - 0.225(1) = 0.658$, for which the predicted value of $P(Y \leqslant 1)$ is $\exp(0.658)/[1 + \exp(0.658)] = 0.659$. It follows that $\hat{m}_{11} = n_{1+} \hat{\pi}_{1 \mid 1} = 96(0.659) = 63.2$. Table 9.9 contains ML fitted values for the $O-D$ marginal table, both for the independence model and the cumulative logit model. The cumulative logit model fits quite a bit better than the independence model in the corners of the table.

Table 9.9   Observed and Fitted Frequencies for Independence Model and Cumulative Logit Uniform Association Model

		Dumping Severity		
Operation	None	Slight	Moderate	Total
A	61	28	7	96
	(55.3)[a]	(29.7)	(11.0)	
	(63.2)[b]	(24.9)	(7.9)	
B	68	23	13	104
	(59.9)	(32.2)	(12.0)	
	(63.1)	(30.4)	(10.5)	
C	58	40	12	110
	(63.3)	(34.0)	(12.7)	
	(60.7)	(35.7)	(13.6)	
D	53	38	16	107
	(61.6)	(33.1)	(12.3)	
	(53.1)	(37.9)	(16.0)	

[a]For independence model $\hat{m}_{ij}$.
[b]For uniform association model $\hat{m}_{ij}$.

9.4.6   More Complex Models

Compared to loglinear models, logit models are simple to formulate because of the similarity of their structure to regression models. For instance, suppose we want to permit interaction in the effects of a nominal factor A and an ordinal factor X (having scores $\{u_i\}$) on the cumulative logit of an ordinal variable Y. For model

$$L_{j \mid hi} = \alpha_j - \mu_h - \beta u_i - \tau_h u_i \, ,$$

the cumulative logit is linearly related to X, but the slope $\beta + \tau_h$ differs among levels of A. This model is analogous to the analysis of covariance model (allowing interaction) for a continuous response variable and nominal and interval explanatory variables.

Models in this section all make the proportional odds assumption that each effect is the same for the different cutpoints for forming cumulative logits. An advantage of such models is that effects are simple to summarize and interpret. The models can be generalized to include nonhomogeneous logit effects. However, such models violate the proper order among the $J - 1$ cumulative probabilities for some x values. If cumulative logit models assuming proportional odds fail to fit satisfactorily, models incorporating dispersion effects or using other transformations of the cumulative probabilities (or other probabilities, such as adjacent response probabilities) may permit homogeneous logit effects. The next section presents a generalization of the cumulative logit model that permits a variety of transformations.

9.5 CUMULATIVE LINK MODELS

The previous section used the logit to link the cumulative distribution function (cdf) of the response to the explanatory variables. Other monotone transformations provide alternative links for this purpose.

Let G denote the cdf of a continuous random variable having positive density over the entire real line. Then G^{-1} is a strictly monotone function from $(0, 1)$ onto the real line. The *cumulative link* model

$$G^{-1}[F_j(\mathbf{x})] = \alpha_j - \boldsymbol{\beta}'\mathbf{x} ,$$

or, equivalently,

$$F_j(\mathbf{x}) = G(\alpha_j - \boldsymbol{\beta}'\mathbf{x}) \tag{9.17}$$

links the cumulative probabilities $\{F_j(\mathbf{x}), j = 1, \ldots, J\}$ to the real line, using link function G^{-1}. This model assumes that effects of $\mathbf{x}$ are the same for each cutpoint, $j = 1, \ldots, J - 1$. Section 9.4.1 showed that this assumption holds if there is a linear regression for an underlying continuous response having standardized cdf G.

9.5.1 Types of Cumulative Links

McCullagh (1980) discussed several cumulative link models. The logit link, $G^{-1}(u) = \log[u/(1 - u)]$, gives cumulative logit models. The logit link function is the inverse of the logistic cdf. Use of the standard normal cdf Φ for G gives the *cumulative probit* model, a generalization of the binary probit model to ordered response categories. This model is appropriate when the underlying response distribution is normal. Cumulative probit and cumulative logit models provide similar fits, because of the similarity of logistic and normal distributions. Parameters in logit models are simpler to interpret.

The link $G^{-1}(u) = \log[-\log(1 - u)]$ is appropriate when the underlying distribution is exponential or of a type used in survival analysis (Cox 1972). This *complementary log-log* link, introduced in Section 4.5, is the inverse cdf for the extreme-value distribution. The ordinal model using this link is called a *proportional hazards* model. Its interpretation uses the property

$$1 - F_k(\mathbf{x}_1) = [1 - F_k(\mathbf{x}_2)]^{\exp[\boldsymbol{\beta}'(\mathbf{x}_2 - \mathbf{x}_1)]} .$$

The $\log[-\log(1 - u)]$ link is similar to the logit or probit for small u, but it tends to ∞ much more slowly for large values. This link is sensible when $P(Y \leq j)$ converges to 1.0 at a faster rate than it converges to 0.0. The related transformation $\log[-\log(u)]$ is another possible link, one that is appropriate when the $\log[-\log(1 - u)]$ link holds for the categories listed in reverse order.

9.5.2 ML Estimation for Cumulative Link Models

McCullagh (1980) and Thompson and Baker (1981) treated cumulative link models as multivariate generalized linear models. McCullagh presented an iterative routine for calculating ML estimates. The routine expresses response probabilities in the likelihood in terms of cumulative probabilities, and applies a Fisher scoring algorithm. McCullagh showed that a unique maximum of the likelihood is guaranteed for sufficiently large sample sizes, though infinite parameter values can arise with sparse data sets containing certain patterns of zeroes. Burridge (1981) and Pratt (1981) showed the log likelihood is concave for many cumulative link models, including the logit, probit, and complementary log-log. Iterative routines such as McCullagh's usually give rapid convergence to the ML estimates.

9.5.3 Life Table Example

Table 9.10 is taken from a recent *Statistical Abstract of the United States* (U.S. Bureau of the Census, 1984, p. 69). It gives the life length distribution for residents of the United States in 1981, by race (white, black) and gender (male, female). Life length is measured with five ordered categories. We would expect the underlying continuous cdf of life length to increase slowly at small to moderate ages, but then increase quite sharply at older ages. This suggests the complementary log-log link. This link can also be motivated by assuming the hazard rate increases exponentially with age, which happens for an extreme value distribution (the Gompertz; see Cox and Oakes, 1984).

Let $F_{j \mid hi}$ denote the cumulative probability at life length category j, for gender h and race i. Table 9.10 contains fitted distributions for the proportional hazards model

Table 9.10 Life Length Distribution, in Percentages, of US Residents in 1981[a]

Lifelength	Males		Females	
	White	Black	White	Black
0–20	2.4 (2.4)	3.6 (4.4)	1.6 (1.2)	2.7 (2.3)
20–40	3.4 (3.5)	7.5 (6.4)	1.4 (1.9)	2.9 (3.4)
40–50	3.8 (4.4)	8.3 (7.7)	2.2 (2.4)	4.4 (4.3)
50–65	17.5 (16.7)	25.0 (26.1)	9.9 (9.6)	16.3 (16.3)
Over 65	72.9 (73.0)	55.6 (55.4)	84.9 (84.9)	73.7 (73.7)

[a]Values in parentheses give fit of proportional hazards model.

$$\log[-\log(1 - F_{j \mid hi})] = \alpha_j - \beta_h^G - \beta_i^R .$$

Goodness-of-fit statistics are irrelevant, since the table contains population distributions. The model does a very good job describing the four distributions. The parameter values for the fit are $\beta_1^G = -\beta_2^G = -0.329$ and $\beta_1^R = -\beta_2^R = 0.313$. Since $\beta_2^G - \beta_1^G = 0.658$, the fitted cdfs satisfy

$$1 - F_{j \mid 1i} = (1 - F_{j \mid 2i})^{\exp(0.658)} .$$

Given race, the proportion of men living longer than a fixed time equals the proportion of women living longer than that time taken to the $\exp(0.658) = 1.93$ power. Given gender, the proportion of blacks living longer than a fixed time equals the proportion for whites taken to the $\exp(0.626) = 1.87$ power. The parameter values indicate that white men and black women have similar life length distributions, that white women tend to have longest lives, and that black men tend to have shortest lives. If the probability of living longer than some fixed time equals ρ for white women, then that probability is about ρ^2 for white men and black women, and ρ^4 for black men.

9.6 MEAN RESPONSE MODELS

The cumulative link models discussed in the previous section have form

$$F_j(\mathbf{x}) = G(\alpha_j - \boldsymbol{\beta}'\mathbf{x}) .$$

We visualized the ordinal response as a discrete measurement of an underlying continuous variable from a location-parameter family having cdf $G(y - \boldsymbol{\beta}'\mathbf{x})$. The parameters $\{\alpha_j\}$ are category cutpoints on the continuous scale. When the density corresponding to G is symmetric with mean 0, the conditional mean of the underlying response is $\beta'\mathbf{x}$. In this sense, cumulative logit and probit models are regression models, describing a linear relationship between explanatory variables and the conditional expectation of this underlying response.

9.6.1 Regression Model for Ordered Response

We now present a regression-type model that applies directly to an observed ordinal response, rather than to an assumed underlying scale. For scores $v_1 \leq v_2 \leq \cdots \leq v_J$ and fixed value $\mathbf{x}$ for explanatory variables, let

$$M(\mathbf{x}) = \sum_j v_j \pi_j(\mathbf{x})$$

denote the mean response. The model

$$M(\mathbf{x}) = \alpha + \beta'\mathbf{x} \tag{9.18}$$

assumes a linear relationship between the conditional mean and the explanatory variables. When there are two response categories, this is the linear probability model (Section 4.2). When there are more than two response categories, this model does not specify a structural form for response probabilities, but simply describes how the conditional mean depends on $\mathbf{x}$.

Assuming independent multinomial sampling at different levels of $\mathbf{x}$, Bhapkar (1968), Grizzle et al. (1969), and Williams and Grizzle (1972) presented weighted least squares (WLS) solutions for mean response models. The WLS solution applies when all explanatory variables are categorical. The residual df for the model equals the number of observed means (i.e., the number of levels of $\mathbf{x}$ at which observations occur) minus the number of parameters in the model.

ML fitting of this model maximizes the multinomial likelihood subject to (9.18). The ML approach applies for categorical or continuous explanatory variables, but it has received scant attention in the literature. Haber (1985) presented an algorithm for obtaining ML fits for a family including mean response models.

9.6.2 Olive Preference Example

Table 9.11 is taken from a book by Bock and Jones (1968) that was one of the first to present sophisticated models for categorical data. Using a nine-point ordinal scale, subjects indicated their preference for black olives. Bock and Jones reported results for a condensed six-point scale,

Table 9.11 Preference for Black Olives, by Urbanization and Location[a]

Urbanization	Location	Preference						Means	
		A	B	C	D	E	F	Sample	Predicted
Urban	MW	20	15	12	17	16	28	5.22	4.87
	NE	18	17	18	18	6	25	4.94	5.00
	SW	12	9	23	21	19	30	5.72	5.92
Rural	MW	30	22	21	17	8	12	4.00	4.28
	NE	23	18	20	18	10	15	4.46	4.41
	SW	11	9	26	19	17	24	5.54	5.33

Source: Reprinted with permission from Holden-Day (Bock and Jones 1968, p. 244).
[a]*Key*: A, Dislike extremely; B, dislike very much or dislike moderately; C, dislike slightly or neither like nor dislike; D, Like slightly; E, Like moderately; F, Like very much or like extremely.

combining categories 2 and 3, 4 and 5, and 8 and 9 of the original scale. The sample consists of independent samples of Armed Forces personnel selected from six combinations of urbanization (urban, rural) and location (NE, MW, SW).

We analyzed these data using mean response models, assuming equal-interval scores for the original nine-point scale. For the reported six-point scale, we assigned scores (1, 2.5, 4.5, 6, 7, 8.5) to levels of preference. Let M_{ij} be the mean preference at urbanization level i and location j. The model

$$M_{ij} = \alpha + \beta_i^U + \beta_j^L$$

resembles a two-way ANOVA model for a 2×3 table of cell means. We fitted the model assuming multinomial response, rather than normal response with constant variance. For constraints $\Sigma \beta_i^U = \Sigma \beta_j^L = 0$, the WLS estimates are:

$$\hat{\beta}_1^U = -\hat{\beta}_2^U = 0.297 \quad (\text{ASE} = 0.101)$$

$$\hat{\beta}_1^L = -0.393, \qquad \hat{\beta}_2^L = -0.266, \qquad \hat{\beta}_3^L = 0.659$$

and $\hat{\alpha} = 4.97$. Given location, there is strong evidence that preference for black olives averages higher for subjects in urban than rural areas, the predicted difference in means being $2(0.297) = 0.59$ categories. A Wald chi-squared statistic for testing $\beta_1^L = \beta_2^L = \beta_3^L$ equals 23.4, with df = 2. In substantive terms, preference is similar in the NE and MW, and subjects in the SW exhibit a more positive preference for black olives.

The residual chi-squared for the WLS solution (discussed in Section

13.3) is 4.81. Since there are six (urbanization, location) combinations and four parameters in the model, residual df = 2. Testing the fit of the model corresponds to testing the hypothesis of no interaction between urbanization and location in the effects on preference. Though we do not reject the model, inspection of the sample means in Table 9.11 shows a slightly higher mean for NE than MW subjects in rural areas, but the reverse in urban areas.

9.6.3 Advantages and Disadvantages of Mean Response Models

We have treated ordinal variables in a quantitative manner in models presented in this book. Because of the quantitative aspects of ordinal responses, some statisticians argue that models for them should resemble regression models for continuous variables more than loglinear and logit models for nominal variables. Mean response models have this advantage. Fitting them is justified if the categorical nature of the response reflects crude measurement of an inherently continuous variable.

For $J = 2$ responses, Section 4.2 noted that linear probability models have a structural difficulty resulting from the restriction of probabilities to the interval [0, 1]. A similar difficulty occurs here, since a linear model can give predicted means outside the range of assigned scores. This happens less frequently when the number of response categories is large and there is reasonable dispersion of responses throughout the domain of interest for the explanatory variables. The notion of an underlying continuous distribution makes more sense for an ordinal variable than for a strictly binary response, so this difficulty has less relevance here.

Unlike logit and loglinear models, mean response models use a summary response measure that does not uniquely determine cell probabilities. Thus we cannot easily use mean response models to make conclusions about structural aspects, such as stochastic response orderings. These models do not represent the categorical structure of the data as well as do models for expected cell frequencies, and conditions such as independence do not occur as special cases. For this reason, mean response models are primarily appropriate when we want to make inferences about underlying continuous variables.

CHAPTER NOTES

Section 9.2: Multinomial Logit Models

9.1 For further discussion of generalized logit models, see Amemiya (1981), Bock (1970), Haberman (1974a, pp. 352–373; 1979, Chap. 6), Nerlove and Press (1973), and Theil

(1969, 1970). Lesaffre and Albert (1989) presented regression diagnostics. Amemiya (1981) and Wrigley (1985) gave several applications of discrete-choice models.

R-squared measures for multinomial response models were presented by Amemiya (1981), Haberman (1982a), and Theil (1970). McFadden (1974) discussed a link between multinomial logit models and utility maximization. Amemiya discussed generalizations that do not assume independence from irrelevant alternatives. Brownstone and Small (1989) and McFadden (1981, 1982) discussed hierarchical logit models, in which there is a nesting of choices in a tree-like structure. For further details on multinomial logit modeling, see Amemiya (1981), Ben-Akiva and Lerman (1985), Borsch-Supan (1987), Maddala (1983), McFadden (1974, 1981, 1982, 1984), Small (1988), and Train (1986).

Section 9.3: Logits for Ordinal Responses

9.2 The ratio of a probability density function to the complement of the cdf is the *hazard* function (Section 6.6). For discrete variables, this is the ratio found in continuation-ratio logits. Hence, the continuation-ratio logit is sometimes interpreted as a log hazard. Thompson (1977) used these logits in models for the analysis of discrete survival-time data. He showed that when lengths of the time intervals approach zero, his model converges to Cox's (1972) proportional hazards model for survival data. Fienberg and Mason (1979) used continuation-ratio logits in age-period-cohort models for the analysis of discrete archival data.

Kupper et al. (1986) and Ryan (1989) discussed modeling overdispersion caused by litter effects in developmental toxicity studies. See Follman and Lambert (1989), Kupper and Haseman (1978), Lefkopoulou et al. (1989), and Paul (1985) for related material.

Section 9.4: Cumulative Logit Models

9.3 Models for cumulative logits have been presented by Anderson and Philips (1981), Aranda-Ordaz (1983), Bennett (1983), Bock and Jones (1968), Clayton (1974), Hastie and Tibshirani (1987), Landis et al. (1987), McCullagh (1984), Pettitt (1984), Simon (1974), Snapinn and Small (1986), Snell (1964), Stram et al. (1988), Walker and Duncan (1967), and Williams and Grizzle (1972). McCullagh's (1980) article has been highly influential in advancing their use.

G. Taguchi proposed an alternative way of testing independence using cumulative probabilities, called *accumulation analysis*. Nair (1986) showed difficulties with the Taguchi approach.

Section 9.5: Cumulative Link Models

9.4 Aitchison and Silvey (1957), Bock and Jones (1968, Chap. 8), and Gurland et al. (1960) have used cumulative probit models. Gurland et al. (1960) assumed an underlying normal tolerance distribution in modeling how the dosage of an insecticide affects whether an insect is alive, moribund, or dead. Prentice and Gloeckler (1978) used the complementary log-log model to analyze grouped survival data. Farewell (1982) generalized it to allow for variation among the sample in the values regarded as category boundaries for the underlying scale. Genter and Farewell (1985) introduced a generalized link function that permits comparison of fits provided by probit, complementary log-log, and other links.

When levels of explanatory variables are not stochastically ordered on the response, it is often because the dispersion also varies across those levels. A cumulative link model that incorporates dispersion effects is

$$F_j(\mathbf{x}) = G\left(\frac{\alpha_j - \boldsymbol{\beta}'\mathbf{x}}{\tau_\mathbf{x}}\right)$$

where τ is a scale parameter. This model is a special case of a nonlinear model proposed by McCullagh (1980). Nair (1987) and the subsequent discussion and Hamada and Wu (1990) presented alternative ways of detecting dispersion effects.

Section 9.6: Mean Response Models

9.5 In (9.18), response scores can be data-generated, rather than preassigned. Semenya et al. (1983) gave WLS analyses for models for the mean ridit of the marginal response distribution.

PROBLEMS

Applications

9.1 For the study described in Section 9.1.2, Table 9.12 contains data for the 63 alligators caught in Lake George. This table classifies primary food choice as Fish, Invertebrate, or Other, and it gives length in meters. Alligators are classified as subadults if length is less than 1.83 meters (6 feet), and adults if length exceeds 1.83 meters. Measuring length with the binary classification (adult, subadult), find a generalized logit model that adequately describes effects of gender and length on primary food choice. Use parameter estimates to interpret effects.

9.2 Refer to the previous problem. Find a logit model that adequately describes effects of gender and binary length on whether primary food choice is Fish or Invertebrate (i.e., use only observations for which primary food choice was F or I). Compare parameter estimates and standard errors for this separate-fitting approach to those obtained using simultaneous fitting in Problem 9.1.

9.3 Refer to Problem 9.1. Since length is continuous, information is lost by measuring it as binary. Treating length as continuous, find a generalized logit model that adequately describes effects of gender and length on primary food choice. Compare results to those obtained in Problem 9.1.

Table 9.12

Males				Females			
Length	Choice	Length	Choice	Length	Choice	Length	Choice
1.30	I	1.80	F	1.24	I	2.56	O
1.32	F	1.85	F	1.30	I	2.67	F
1.32	F	1.93	I	1.45	I	2.72	I
1.40	F	1.93	F	1.45	O	2.79	F
1.42	I	1.98	I	1.55	I	2.84	F
1.42	F	2.03	F	1.60	I		
1.47	I	2.03	F	1.60	I		
1.47	F	2.31	F	1.65	F		
1.50	I	2.36	F	1.78	I		
1.52	I	2.46	F	1.78	O		
1.63	I	3.25	O	1.80	I		
1.65	O	3.28	O	1.88	I		
1.65	O	3.33	F	2.16	F		
1.65	I	3.56	F	2.26	F		
1.65	F	3.58	F	2.31	F		
1.68	F	3.66	F	2.36	F		
1.70	I	3.68	O	2.39	F		
1.73	O	3.71	F	2.41	F		
1.78	F	3.89	F	2.44	F		
1.78	O						

9.4 Table 9.13 describes the effect on political party identification of gender and race. Find a generalized logit model that fits these data well. Obtain and interpret estimates of the effects of race and gender on the log odds of party identification being Democrat instead of Republican.

Table 9.13

		Party Identification		
Gender	Race	Democrat	Republican	Independent
Male	White	132	176	127
	Black	42	6	12
Female	White	172	129	130
	Black	56	4	15

9.5 Refer to Table 9.5. Construct the prediction equation for $\log(\pi_W/\pi_B)$, and interpret the parameter estimates.

9.6 Refer to Table 8.1, treating job satisfaction as the response.

    a. Find a cumulative logit model that gives a good fit, and interpret the estimated effect.

    b. Analyze the data using adjacent-categories logits. Test the fit, and interpret parameter estimates. Give the equivalent loglinear model.

    c. Analyze the data using one other type of multinomial response model. Interpret results, and summarize your conclusions from fitting the models.

9.7 Refer to the previous problem. If you have access only to a computer package that fits baseline-category logit models, specify its model matrix so that the effect parameter is the same as β for adjacent-categories logit model (8.5) for uniform local association.

9.8 Refer to Table 8.14. Find a multinomial response model that fits these data well, and interpret the parameter estimates.

9.9 Refer to Table 8.4. Analyze these data using cumulative logits, treating political ideology as the response. Test the model fit, interpret parameter estimates, and describe the influence of party affiliation on ideology. Compare results to those obtained in Section 8.2 using loglinear models.

9.10 Table 9.14, based on data reported by Jennings (1987), describes a sample of subjects who graduated from high school in 1965. Subjects were classified as protestors if they had taken part in at least one demonstration, protest march, or sit-in. They were classified also according to their party identification in 1982.

Table 9.14

Party Identification	Non-Protestors	Protestors
Strong Democrat	10	18
Weak Democrat	59	38
Leaning Democrat	41	22
Independent	26	7
Leaning Republican	44	10
Weak Republican	47	7
Strong Republican	29	2

Source: Reprinted with permission, based on data from Jennings (1987).

a. Find a multinomial response model that describes well the difference between protestors and nonprotestors in their party identifications. Interpret.

b. Treat whether a protestor as the response variable, and find a logit model that describes the table well. Compare the interpretation to the one in (a).

9.11 Table 9.15 is based on a sample described by Madsen (1976) of 1681 residents of twelve areas in Copenhagen. The variables are type of housing (H), degree of contact with other residents (C), feeling of influence on apartment management (I), and satisfaction with housing conditions (S). Analyze these data.

Table 9.15

Housing/ Influence		Low contact			High contact		
	Satisfaction:	Low	Medium	High	Low	Medium	High
Tower blocks	Low	21	21	28	14	19	37
	Medium	34	22	36	17	23	40
	High	10	11	36	3	5	23
Apartments	Low	61	23	17	78	46	43
	Medium	43	35	40	48	45	86
	High	26	18	54	15	25	62
Atrium houses	Low	13	9	10	20	23	20
	Medium	8	8	12	10	22	24
	High	6	7	9	7	10	21
Terraced houses	Low	18	6	7	57	23	13
	Medium	15	13	13	31	21	13
	High	7	5	11	5	6	13

Source: Reprinted with permission from Madsen (1976).

9.12 Refer to the previous problem.

a. Treating S as the response variable, find a cumulative logit model that describes the data adequately. Interpret the estimated effects.

b. Find an adjacent-categories logit model that fits adequately. Interpret the estimated effects, and compare results to those in (a). Give an equivalent loglinear model.

c. Use continuation-ratio logits to analyze the data.

d. Find a mean response model that adequately describes the data.

e. Compare interpretations obtained with the various models.

9.13 Analyze Table 8.17 using methods described in this chapter,
    a. Treating Alzheimer's disease as the response variable,
    b. treating cognitive impairment as the response.

9.14 Use a multinomial response model to analyze Table 8.18.

9.15 Refer to Problem 9.28. Fit this model to Table 9.16, taken from McCullagh (1980). Interpret the difference between the income distributions.

Table 9.16   Family Income Distributions (Percent) in the US Northeast[a]

Year	0–3	3–5	5–7	7–10	10–12	12–15	15+
1960	6.5	8.2	11.3	23.5	15.6	12.7	22.2
1970	4.3	6.0	7.7	13.2	10.5	16.3	42.1

Source: Reproduced with permission from the Royal Statistical Society, London (McCullagh 1980).

9.16 Find a mean response model that fits Table 9.9 well. Interpret results.

9.17 Why is it improper to fit mean response models for categorical responses using ordinary least squares, as is done for normal regression models? Under what circumstances do you feel that this simpler approach is justified?

9.18 Bock and Jones (1968) used cumulative logit models to analyze Table 9.11. Analyze the data using that form of model.

Theory and Methods

9.19 Suppose Y is a nominal response variable, A is nominal and X is ordinal.

    a. Give the generalized logit model that corresponds to loglinear model

$$\log m_{hij} = \mu + \lambda_h^A + \lambda_i^X + \lambda_j^Y + \lambda_{hi}^{AX} + \lambda_{hj}^{AY} + u_i \nu_j$$

where $\{u_i\}$ are fixed scores for X and $\{v_j\}$ are parameters for Y. How does this simplify when the response is binary?

b. Explain how to interpret parameters in the generalized logit model.

9.20 Give a model of form (9.4) for two explanatory variables, one a characteristic of the chooser and the other a characteristic of the choices. Interpret the model parameters.

9.21 Give a multiparameter generalization of the natural exponential family (4.1). Show that the multinomial distribution for sample size n and parameters $\{\pi_j, j = 1, \ldots, J\}$ is in the $(J-1)$-parameter exponential family, with baseline-category logits as natural parameters.

9.22 Let $\mathbf{y}'_i = (y_{i1}, \ldots, y_{iJ})$ represent the response on a nominal variable for subject i, $i = 1, \ldots, n$, where each $y_{ij} = 0$ or 1, and $\Sigma_j \, y_{ij} = 1$. Let $\mathbf{x}'_i = (x_{i1}, \ldots, x_{iK})$ denote values of K explanatory variables for subject i. Suppose

$$f(y_{i1}, \ldots, y_{iJ}) = \prod_{j=1}^{J} \pi_j(\mathbf{x}_i)^{y_{ij}}, \quad i = 1, \ldots, n$$

with

$$\pi_j(\mathbf{x}_i) = \frac{\exp\left(\alpha_j + \sum_{k=1}^{K} \beta_{jk} x_{ik}\right)}{\sum_{h=1}^{J} \exp\left(\alpha_h + \sum_{k=1}^{K} \beta_{hk} x_{ik}\right)}.$$

a. If the n multinomial trials are independent, show sufficient statistics are

$$np_j = \sum_i y_{ij}, \quad j = 1, \ldots, J$$

$$S_{jk} = \sum_i x_{ik} y_{ij}, \quad j = 1, \ldots, J, \quad k = 1, \ldots, K.$$

(There is some further reduction in sufficiency, for instance by using the parameterization $\alpha_J = \beta_{J1} = \cdots = \beta_{JK} = 0$.)

b. Condition on $\Sigma_i \, y_{ij}$, $j = 1, \ldots, J$. Under the null hypothesis that explanatory variables have no effect on the response, show

$$E(S_{jk}) = np_j \bar{x}_k, \quad \text{where } \bar{x}_k = \left(\sum_i x_{ik} \right) \Big/ n .$$

$$\text{Var}(S_{jk}) = np_j(1 - p_j)s_{kk}^2$$

$$\text{Cov}(S_{jk}, S_{rt}) = -np_j p_r s_{kt}^2$$

where

$$s_{kt}^2 = \left[\sum_i (x_{ik} - \bar{x}_k)(x_{it} - \bar{x}_t) \right] \Big/ (n - 1) .$$

c. Show how expressions in (b) simplify for a binary response and a single binary explanatory variable, taking values 0 and 1.

d. Let $\mathbf{S} = (S_{11}, \ldots, S_{1K}, \ldots, S_{J1}, \ldots, S_{JK})'$. Show that

$$E(\mathbf{S}) = n(\mathbf{p} \otimes \mathbf{m}), \quad \text{Var}(\mathbf{S}) = n(\mathbf{V} \otimes \mathbf{\Sigma})$$

where $\mathbf{p} = (p_1, \ldots, p_J)'$, $\mathbf{m} = (\bar{x}_1, \ldots, \bar{x}_K)'$, $\mathbf{\Sigma}$ has elements (s_{kt}^2), $\mathbf{V}$ has elements $v_{ii} = p_i(1 - p_i)$ and $v_{ij} = -p_i p_j$, and $\otimes$ denotes Kronecker product. (Thanks to Prof. M. Zelen for showing me this representation.)

9.23 Prove factorization (9.9) for the multinomial distribution.

9.24 For a two-way table, $L_{j|i}$ is the jth cumulative logit in row i.

a. Show that statistical independence is equivalent to

$$L_{j|i} = \alpha_j, \quad i = 1, \ldots, I, \, j = 1, \ldots, J - 1 .$$

b. Suppose the model

$$L_{j|i} = \alpha_j - \mu_i$$

holds. Show that if $\mu_a > \mu_b$, the conditional response distribution is stochastically higher in row a than in row b.

9.25 Consider a three-way table in which A is nominal, X is ordinal, and Y is an ordinal response.

a. Construct a cumulative logit model that has main effects but assumes no three-factor interaction.

b. Construct an adjacent-categories logit model for the same situation, and give the equivalent loglinear model.

    c. Compare the residual df values for the models in (a) and (b). Explain the differences in parameter interpretations.

9.26 Suppose model (9.14) holds for a $2 \times J$ table, and let $u_2 - u_1 = 1$. Let $F_j = \pi_{+1} + \cdots + \pi_{+j}$, $j = 0, \ldots, J$, with $F_0 = 0$.

    a. Show that all $J - 1$ cumulative odds ratios equal $\exp(\beta)$.

    b. Show that local odds ratios $\{\theta_{1j}\}$ are related to the cumulative log odds ratio β by

$$\log \theta_{1j} = \beta(F_{j+1} - F_{j-1}) + o(\beta), \quad j = 1, \ldots, J - 1$$

    where $o(\beta)/\beta \to 0$ as $\beta \to 0$ (McCullagh and Nelder 1983, p. 122).

    c. Explain why, when model (9.14) holds and $J > 2$, we expect local log odds ratios to be smaller in absolute value than cumulative log odds ratios.

    d. Show that if (9.14) holds and $|\beta|$ is small, the linear-by-linear association model should fit well if we use the ridit scores $\{v_j = (F_{j-1} + F_j)/2\}$, and then its β parameter should be about twice the value of β for the cumulative logit model.

9.27 Consider the cumulative link model, $F_j(\mathbf{x}) = G(\alpha_j - \boldsymbol{\beta}'\mathbf{x})$. Show that for j and k between 1 and $J - 1$, $F_j(\mathbf{x}) = F_k(\mathbf{x}^*)$, where $\mathbf{x}^*$ is obtained by increasing the ith component of $\mathbf{x}$ by $(\alpha_j - \alpha_k)/\beta_i$ units. Interpret.

9.28 Consider the row effects version of the proportional hazards model

$$\log[-\log(1 - F_{j|i})] = \alpha_j - \mu_i, \quad i = 1, \ldots, I, \, j = 1, \ldots, J - 1$$

    where $\Sigma \, \mu_i = 0$.

    a. Show the residual df for testing goodness of fit equals $(I - 1)(J - 2)$, as is also true for the loglinear and cumulative logit row effects models (see (8.7) and (9.15)).

    b. When this model holds, show that independence corresponds to equality of $\{\mu_i\}$.

    c. When this model holds, show the rows are stochastically ordered on the column variable.

9.29 Let F_1 be the cdf of an exponential random variable with parameter λ; that is, $F_1(y) = 1 - \exp(-\lambda y)$ for $y > 0$. Let F_2 be an exponential cdf with parameter μ. Show that the difference between the cdfs on a complementary log-log scale is identical for all y. Give implications for categorical data analysis.

9.30 Consider the model $\text{Link}[\rho_j(\mathbf{x})] = \alpha_j - \boldsymbol{\beta}_j'\mathbf{x}$, where $\rho_j(\mathbf{x}) = \pi_j(\mathbf{x})/[\pi_j(\mathbf{x}) + \cdots + \pi_J(\mathbf{x})]$.

a. Explain why this model can be fitted separately for $j = 1, \ldots, J - 1$.

b. For the complementary log-log link, show this model is equivalent to one using the same link for cumulative probabilities (Läärä and Matthews 1985).

CHAPTER 10

Models For Matched Pairs

The next two chapters introduce methods for analyzing categorical data from *dependent* samples. Dependent samples result from repeated observations on the response variable for a set of subjects. For example, suppose the response is a subject's opinion about the President's performance in office, measured as approval or disapproval. The Gallup company measures the opinions of a sample of residents in the United States each month. If they use the same sample each month, the samples are dependent. After 6 months of observations, there are six dependent samples, the ith sample referring to subjects' opinions at month i. Such data are often called *repeated measures* or *longitudinal* data.

Dependent samples also occur in *matched-pair* designs. In occupational mobility tables, each observation is a pairing of parent's occupation with child's occupation. For a sample of matched pairs, responses are summarized by a two-way table in which both classifications have the same categories. Thus, the table is *square*, having $I = J$. This chapter presents analyses for dependent samples classified in square tables.

Two matters have special importance in the analysis of square tables. First, cell probabilities or associations may exhibit a symmetric pattern about the main diagonal of the table. Section 10.2 presents models of this type. Second, the two marginal distributions may differ in some systematic way. Section 10.3 presents ways of comparing marginal distributions. Section 10.4 introduces models for ordinal classifications, which have more parsimonious symmetry patterns and ways of comparing marginal distributions. For 2×2 tables, the study of symmetry is equivalent to the study of marginal distributions. We treat this case separately in Section 10.1.

Sections 10.5 and 10.6 describe two important applications having square contingency tables. Section 10.5 describes measurement and modeling of rater agreement when two observers rate a sample of subjects on

347

the same categorical scale. Square contingency tables also occur for pairwise comparison of a set of items, such as pairwise comparison of quality of different wines or pairwise comparison of different athletic teams. Section 10.6 presents the Bradley–Terry model for pairwise comparisons.

10.1 COMPARING DEPENDENT PROPORTIONS

We refer to levels of the variable determining the dependent samples (e.g., month, generation) as *occasions*. This chapter studies the two-occasion case, and Chapter 11 studies the several-occasion case. Let π_{ij} denote the probability of response i at occasion 1 and response j at occasion 2, and let $p_{ij} = n_{ij}/n$ denote the sample proportion, $i = 1, \ldots, I$, $j = 1, \ldots, I$. Then p_{i+} is the relative frequency of the ith response at occasion 1, and p_{+i} is its relative frequency at occasion 2. This section studies the binary response case, $I = 2$. We compare samples by comparing p_{1+} with p_{+1}. Since the samples are dependent, these proportions are *correlated*, and methods of Section 3.4.2 for independent proportions are inappropriate.

For binary responses, when $\pi_{1+} = \pi_{+1}$, then $\pi_{2+} = \pi_{+2}$ also, and there is *marginal homogeneity*. For 2×2 tables,

$$\pi_{1+} - \pi_{+1} = \pi_{12} - \pi_{21}$$

and marginal homogeneity is equivalent to *symmetry* of probabilities across the main diagonal—that is, $\pi_{12} = \pi_{21}$.

10.1.1 Increased Precision with Dependent Samples

We now treat $\{n_{ij}\}$ in the 2×2 table as a sample from a multinomial $(n; \{\pi_{ij}\})$ distribution. Let

$$d = p_{1+} - p_{+1} = p_{+2} - p_{2+} .$$

From formulas (3.19), (3.20) for covariances of multinomial proportions, $\text{Cov}(p_{1+}, p_{+1}) = \text{Cov}(p_{11} + p_{12}, p_{11} + p_{21})$ simplifies to $(\pi_{11}\pi_{22} - \pi_{12}\pi_{21})/n$, so that

$$\text{Var}(\sqrt{n}d) = \pi_{1+}(1 - \pi_{1+}) + \pi_{+1}(1 - \pi_{+1}) - 2(\pi_{11}\pi_{22} - \pi_{12}\pi_{21}) .$$

$$(10.1)$$

By contrast, suppose we use *independent* samples of size n each to estimate binomial probabilities π_1 and π_2 and their difference. The covariance for the sample proportions is zero, and

$$\text{Var}(\sqrt{n} \times \text{difference in sample proportions}) = \pi_1(1 - \pi_1) + \pi_2(1 - \pi_2) .$$

In (10.1), the dependence in the sample marginal proportions asserts itself in the final term, involving the covariance of p_{1+} and p_{+1}. Dependent samples usually exhibit a positive dependence between responses; that is, $\log \theta = \log[\pi_{11}\pi_{22}/\pi_{12}\pi_{21}] > 0$, or $\pi_{11}\pi_{22} > \pi_{12}\pi_{21}$. From (10.1), positive dependence implies the variance of the difference in sample proportions is smaller than when the samples are independent. A study design having dependent samples can help improve the precision of statistical inferences, and the improvement is substantial when samples are highly correlated.

10.1.2 Inference for Dependent Proportions

For large samples, $d = p_{1+} - p_{+1}$ has approximately a normal sampling distribution. A large-sample $100(1 - \alpha)$ percent confidence interval for $\pi_{1+} - \pi_{+1}$ is

$$(p_{1+} - p_{+1}) \pm z_{\alpha/2}\hat{\sigma}(d) ,$$

where

$$\hat{\sigma}^2(d) = [p_{1+}(1 - p_{1+}) + p_{+1}(1 - p_{+1}) - 2(p_{11}p_{22} - p_{12}p_{21})]/n .$$

$$(10.2)$$

The ratio $z = d/\hat{\sigma}(d)$ is a test statistic for the hypothesis $H_0: \pi_{1+} = \pi_{+1}$. Under H_0, an alternative estimated variance is

$$\hat{\sigma}_0^2(d) = \frac{p_{1+} + p_{+1} - 2p_{11}}{n} = \frac{n_{12} + n_{21}}{n^2} . \qquad (10.3)$$

The statistic $z_0 = d/\hat{\sigma}_0(d)$ simplifies to

$$z_0 = \frac{n_{12} - n_{21}}{(n_{12} + n_{21})^{1/2}} . \qquad (10.4)$$

This depends only on the observations classified in *different* categories at the two occasions. The $n_{11} + n_{22}$ observations on the main diagonal are irrelevant to inference about whether π_{1+} and π_{+1} differ. This may seem surprising, but notice that *all* observations are relevant to inference about *how much* π_{1+} and π_{+1} differ. For instance, we use all observations to estimate $\pi_{1+} - \pi_{+1}$ and the standard error of that estimate.

For small samples, an exact test of H_0: $\pi_{1+} = \pi_{+1}$ conditions on the off-main-diagonal total $n^* = n_{12} + n_{21}$. The null hypothesis is equivalent to $\pi_{12} = \pi_{21}$, or $\pi_{12}/(\pi_{12} + \pi_{21}) = 0.5$. Under H_0, n_{12} has a binomial $(n^*, \frac{1}{2})$ distribution, for which $E(n_{12}) = (\frac{1}{2})n^*$ (Mosteller 1952). A P-value is the sum of binomial probabilities for n_{12} values at least as far from $(\frac{1}{2})n^*$ as observed. For large samples, the normal approximation to the binomial suggests the statistic

$$\frac{n_{12} - (\frac{1}{2})n^*}{[n^*(\frac{1}{2})(\frac{1}{2})]^{1/2}} = \frac{n_{12} - n_{21}}{(n_{12} + n_{21})^{1/2}} .$$

This is identical to the statistic z_0 derived by treating the main-diagonal counts as part of the data set. The square of z_0 is a single-degree-of-freedom chi-squared statistic, for which the significance test is called *McNemar's test* (McNemar 1947).

10.1.3 Presidential Approval Rating Example

For a random sample of 1600 voting-age Americans, 944 people indicate approval of the President's performance in office. A month later, 880 of these same 1600 people indicate approval. Table 10.1 shows the results. Relatively few people changed their opinion. There is strong association between opinions at the two occasions, the sample odds ratio being $(794 \times 570)/(150 \times 86) = 35.1$.

The sample proportions indicating approval of the President's performance are $p_{1+} = 944/1600 = 0.59$ for the first survey and $p_{+1} = 880/1600 = 0.55$ for the second survey. Using (10.2), a 90% confidence interval for the true change $\pi_{+1} - \pi_{1+}$ is $(0.55 - 0.59) \pm 1.645(0.0095)$, or

Table 10.1　Performance of President

First Survey	Second Survey		Total
	Approve	Disapprove	
Approve	794	150	944
Disapprove	86	570	656
Total	880	720	1600

$(-0.056, -0.024)$. The approval rating appears to have dropped between 2% and 6%. Incidentally, *independent* samples of size 1600 each have a standard error of 0.0175 for the difference between sample proportions 0.59 and 0.55, nearly double the standard error achieved with these dependent samples.

From (10.3), the null estimated standard error of $p_{+1} - p_{1+}$ is $\hat{\sigma}_0(d) = 0.0096$. The statistic $z_0 = -0.04/0.0096 = -4.17$ gives strong evidence of a negative change in the approval rating.

10.1.4 A Logistic Model for Binary Matched Pairs

Cox (1958b, 1970) presented a logistic model that provides justification for comparing marginal distributions using only the $n^* = n_{12} + n_{21}$ pairings having observations in different categories at the two occasions. Let (Y_{1h}, Y_{2h}) denote the hth pair of observations, $h = 1, \ldots, n$, where a "1" response denotes category 1 (success) and "0" denotes category 2. Consider the model

$$\text{Logit}[P(Y_{1h} = 1)] = \alpha_h, \qquad \text{Logit}[P(Y_{2h} = 1)] = \alpha_h + \beta$$

which permits separate response distributions for each pair, but assumes a common effect. For pair h, the probability of success equals $\exp(\alpha_h)/[1 + \exp(\alpha_h)]$ at occasion 1 and $\exp(\alpha_h + \beta)/[1 + \exp(\alpha_h + \beta)]$ at occasion 2. The $\{\alpha_h\}$ are nuisance parameters, and β is the parameter of interest for comparing marginal distributions. From the success probabilities, we see that $\exp(\beta)$ is an odds ratio: For each of the n matched pairs, the odds of success are $\exp(\beta)$ times higher at occasion 2 than at occasion 1. The value $\beta = 0$ implies marginal homogeneity.

For this model, assuming independence of responses for different matched pairs and for the two responses in the same pair, the joint mass function for $\{(y_{11}, y_{21}), \ldots, (y_{1n}, y_{2n})\}$ is

$$\prod_{h=1}^{n} \left(\frac{\exp(\alpha_h)}{1 + \exp(\alpha_h)} \right)^{y_{1h}} \left(\frac{1}{1 + \exp(\alpha_h)} \right)^{1-y_{1h}}$$

$$\times \left(\frac{\exp(\alpha_h + \beta)}{1 + \exp(\alpha_h + \beta)} \right)^{y_{2h}} \left(\frac{1}{1 + \exp(\alpha_h + \beta)} \right)^{1-y_{2h}}.$$

In terms of the data, this is proportional to

$$\exp\left[\sum \alpha_h(y_{1h} + y_{2h}) + \beta\left(\sum y_{2h} \right) \right].$$

We eliminate the nuisance parameters $\{\alpha_h\}$ by conditioning on the pairwise success totals $\{S_h = y_{1h} + y_{2h}\}$. Given $S_h = 0$, $P(Y_{1h} = Y_{2h} = 0) = 1$, and given $S_h = 2$, $P(Y_{1h} = Y_{2h} = 1) = 1$. The distribution of (Y_{1h}, Y_{2h}) depends on β only when $S_h = 1$; that is, only when responses differ at the two occasions. Given $y_{1h} + y_{2h} = 1$, the conditional distribution is

$$P(Y_{1h} = y_{1h}, Y_{2h} = y_{2h} \mid S_h = 1)$$

$$= P(Y_{1h} = y_{1h}, Y_{2h} = y_{2h}) / [P(Y_{1h} = 1, Y_{2h} = 0) + P(Y_{1h} = 0, Y_{2h} = 1)]$$

$$= \frac{\left(\dfrac{\exp(\alpha_h)}{1 + \exp(\alpha_h)}\right)^{y_{1h}} \left(\dfrac{1}{1 + \exp(\alpha_h)}\right)^{1-y_{1h}} \left(\dfrac{\exp(\alpha_h + \beta)}{1 + \exp(\alpha_h + \beta)}\right)^{y_{2h}} \left(\dfrac{1}{1 + \exp(\alpha_h + \beta)}\right)^{1-y_{2h}}}{\left(\dfrac{\exp(\alpha_h)}{1 + \exp(\alpha_h)}\right)\left(\dfrac{1}{1 + \exp(\alpha_h + \beta)}\right) + \left(\dfrac{1}{1 + \exp(\alpha_h)}\right)\left(\dfrac{\exp(\alpha_h + \beta)}{1 + \exp(\alpha_h + \beta)}\right)}$$

$$= \exp(\beta)/[1 + \exp(\beta)], \qquad y_{1h} = 0, \qquad y_{2h} = 1$$

$$= 1/[1 + \exp(\beta)], \qquad y_{1h} = 1, \qquad y_{2h} = 0.$$

Each pair having $S_h = 1$ has the same probabilities for these two outcomes. For these pairs, $\Sigma\, y_{1h} = n_{12}, \Sigma\, y_{2h} = n_{21}$, and $\Sigma\, S_h = n^* = n_{12} + n_{21}$. Since n_{21} is the sum of n^* independent, identical Bernoulli random variables, its conditional distribution is binomial with parameter $\exp(\beta)/[1 + \exp(\beta)]$. For testing marginal homogeneity ($\beta = 0$), the parameter equals $\frac{1}{2}$. In summary, the conditional analysis for the logistic model implies that pairs in which the response is identical at the two occasions are irrelevant to inference about the pairwise odds ratio of success.

Conditional on $S_h = 1$, the joint distribution of the matched pairs is

$$\prod_{S_h = 1} \left(\frac{1}{1 + \exp(\beta)}\right)^{y_{1h}} \left(\frac{\exp(\beta)}{1 + \exp(\beta)}\right)^{y_{2h}} = [\exp(\beta)]^{\Sigma\, y_{2h}} [1 + \exp(\beta)]^{-n^*}$$

where the product refers to all pairs having $S_h = 1$. Differentiating the log of this conditional likelihood yields the result that the ML estimator of the odds ratio $\exp(\beta)$ in the logistic model is n_{21}/n_{12}. Liang and Zeger (1988) gave an alternative estimator of the odds ratio that also uses the main-diagonal observations. It is a compromise between this conditional ML estimator and one (Problem 10.20) that ignores the information on the pairings.

10.2 SYMMETRY MODELS

The definitions of symmetry and marginal homogeneity extend directly when there are $I > 2$ categories. For an $I \times I$ joint distribution $\{\pi_{ij}\}$, there is *marginal homogeneity* if

$$\pi_{i+} = \pi_{+i}, \quad i = 1, \ldots, I. \tag{10.5}$$

There is *symmetry* if

$$\pi_{ij} = \pi_{ji} \quad \text{whenever } i \neq j. \tag{10.6}$$

When there is symmetry, $\pi_{i+} = \Sigma_j \pi_{ij} = \Sigma_j \pi_{ji} = \pi_{+i}$ for all i, so there is also marginal homogeneity. For $I = 2$, symmetry is equivalent to marginal homogeneity, but for $I > 2$, symmetry is not implied by marginal homogeneity.

10.2.1 Symmetry as a Loglinear Model

When all $\pi_{ij} > 0$, symmetry has a loglinear model representation. For expected frequencies $\{m_{ij} = n\pi_{ij}\}$, the symmetry model is

$$\log m_{ij} = \mu + \lambda_i + \lambda_j + \lambda_{ij} \tag{10.7}$$

where all $\lambda_{ij} = \lambda_{ji}$. Both classifications have the same single-factor parameters $\{\lambda_i\}$. Identifiability requires constraints such as $\Sigma \lambda_i = 0$ and $\Sigma_i \lambda_{ij} = 0, j = 1, \ldots, I$.

The symmetry model treats classifications as nominal. The likelihood equations are

$$\hat{m}_{ij} + \hat{m}_{ji} = n_{ij} + n_{ji} \quad \text{for all } i \text{ and } j. \tag{10.8}$$

The solution of (10.8) that satisfies the symmetry condition $m_{ij} = m_{ji}$ is

$$\hat{m}_{ij} = \frac{n_{ij} + n_{ji}}{2} \quad \text{for all } i \text{ and } j \tag{10.9}$$

for which $\hat{m}_{ii} = n_{ii}$.

In the symmetry model, the $\binom{I}{2}$ values $\{\lambda_{ij} \text{ for } i < j\}$ determine $\{\lambda_{ij} \text{ for } i > j\}$ by the symmetry condition and determine $\{\lambda_{jj}\}$ by the constraint $\Sigma_i \lambda_{ij} = 0$. Therefore, the residual df are

$$\text{df} = I^2 - [1 + (I - 1) + I(I - 1)/2] = I(I - 1)/2 .$$

The Pearson statistic for testing the fit of the symmetry model simplifies to

$$X^2 = \sum\sum_{i<j} \frac{(n_{ij} - n_{ji})^2}{n_{ij} + n_{ji}} \tag{10.10}$$

presented by Bowker (1948). For $I = 2$ this is McNemar's statistic, the square of (10.4). The adjusted residuals are $\{r_{ij} = (n_{ij} - n_{ji})/(n_{ij} + n_{ji})^{1/2}\}$.

10.2.2 Quasi Symmetry

The symmetry model rarely fits well, because it is highly structured and has limited scope. For instance, whenever there is marginal heterogeneity, the symmetry model cannot hold. One way to broaden the scope of (10.7) is by deleting the equality constraint on the single-factor parameters. The model

$$\log m_{ij} = \mu + \lambda_i^X + \lambda_j^Y + \lambda_{ij}^{XY} \tag{10.11}$$

where $\lambda_{ij}^{XY} = \lambda_{ji}^{XY}$ for $i \neq j$, is called the *quasi-symmetry* model. It was introduced by Caussinus (1965). The symmetry model is the special case $\lambda_i^X = \lambda_i^Y$ for $i = 1, \ldots, I$. Like the symmetry model, the quasi-symmetry model treats classifications as nominal.

Unlike the symmetry model, the quasi-symmetry model does not imply marginal homogeneity. Its likelihood equations are

$$\hat{m}_{i+} = n_{i+}, \quad i = 1, \ldots, I \tag{10.12a}$$

$$\hat{m}_{+j} = n_{+j}, \quad j = 1, \ldots, I \tag{10.12b}$$

$$\hat{m}_{ij} + \hat{m}_{ji} = n_{ij} + n_{ji} \quad \text{for } i \neq j . \tag{10.12c}$$

Only one of the first two sets of equations is needed. The other is redundant, given the third set. The residual degrees of freedom equal $(I - 1)(I - 2)/2$.

From (10.12c), $\hat{m}_{ii} = n_{ii}$ for $i = 1, \ldots, I$. Otherwise, the likelihood equations do not have a direct solution, but can be solved using Newton–Raphson or other iterative methods. Caussinus (1965) obtained $\{\hat{m}_{ij}$ for $i \neq j\}$ by applying iterative proportional fitting to an amended table

$$n_{ij}^* = n_{ij}, \quad i \neq j$$
$$= 0, \quad for \ i = j.$$

Let $\{m_{ij}^{(0)} = 1$ for $i \neq j$ and 0 for $i = j\}$. Caussinus scaled these initial estimates to match successively the amended row margin, amended column margin, and pair totals $\{n_{ij} + n_{ji}\}$. Each cycle has form

$$m_{ij}^{(t)} = m_{ij}^{(t-1)} \left(\frac{n_{i+}^*}{m_{i+}^{(t-1)}} \right), \qquad m_{ij}^{(t+1)} = m_{ij}^{(t)} \left(\frac{n_{+j}^*}{m_{+j}^{(t)}} \right) \qquad (10.13a)$$

$$m_{ij}^{(t+2)} = m_{ij}^{(t+1)} \left(\frac{n_{ij} + n_{ji}}{m_{ij}^{(t+1)} + m_{ji}^{(t+1)}} \right). \qquad (10.13b)$$

This sequence converges to the ML fitted values for the off-main-diagonal cells. Problem 10.27 outlines another way of fitting this model, using any iterative procedure for loglinear model (XY, XZ, YZ).

The quasi-symmetry model has multiplicative form

$$\pi_{ij} = \alpha_i \beta_j \gamma_{ij}, \quad \text{where } \gamma_{ij} = \gamma_{ji} \text{ all } i \text{ and } j \qquad (10.14)$$

and all parameters are positive. The symmetry model is (10.14) with $\alpha_i = \beta_i$ for all i. Equation (10.14) indicates that a table satisfying quasi symmetry is the cell-wise product of a table satisfying independence with a table satisfying symmetry. Quasi symmetry also has characterizations in terms of symmetry of odds ratios. It is the model defined by the property

$$\frac{m_{ij} m_{II}}{m_{iI} m_{Ij}} = \frac{m_{ji} m_{II}}{m_{jI} m_{Ii}} \quad \text{for all } i \text{ and } j \qquad (10.15)$$

or the property $\theta_{ij} = \theta_{ji}$ for local odds ratios. Because of this, Goodman (1979a) referred to it as the *symmetric association* model. Many useful models are special cases of the quasi-symmetry model.

10.2.3 Quasi Independence

An effect of the dependence between matched pairs is that square tables usually have larger counts on the main diagonal than the independence model predicts. Conditional on the event that a matched pair falls off the main diagonal, though, there may be a simple structure for the relationship.

A square contingency table satisfies *quasi independence* when the variables are independent, given that the row response differs from the column response. This means that there exist constants $\{\alpha_i\}$ and $\{\beta_j\}$ such that

$$\pi_{ij} = \alpha_i \beta_j \quad \text{for } i \neq j. \tag{10.16}$$

Quasi independence is the special case of the quasi-symmetry model (10.14) in which $\{\gamma_{ij} \text{ for } i \neq j\}$ are identical. Caussinus (1965, p. 146) showed the models are equivalent when $I = 3$.

The quasi-independence model has loglinear form

$$\log m_{ij} = \mu + \lambda_i^X + \lambda_j^Y + \delta_i I(i = j),$$

where $I(\cdot)$ is the indicator function

$$I(i = j) = 1, \quad i = j$$
$$= 0, \quad i \neq j.$$

The first three terms in the model specify independence, and the $\{\delta_i\}$ parameters permit $\{m_{ii}\}$ to depart from this pattern and have arbitrary positive values. Odds ratios equal 1.0 for all rectangular sets of four cells falling off the main diagonal. The likelihood equations are

$$\hat{m}_{i+} = n_{i+}, \quad i = 1, \ldots, I$$
$$\hat{m}_{+j} = n_{+j}, \quad j = 1, \ldots, I$$
$$\hat{m}_{ii} = n_{ii}, \quad i = 1, \ldots, I.$$

The $\{\hat{m}_{ij} \text{ for } i \neq j\}$ result using the Newton–Raphson method, or from iterative use of equations (10.13a) for the quasi-symmetry model. The scaling process successively fits row and column totals of the amended table having zeroes on the main diagonal. The model has residual $\mathrm{df} = (I-1)^2 - I$ and applies to tables of size 3×3 and larger. It treats the classifications as nominal.

10.2.4 Migration Example

Table 10.2 compares region of residence in 1985 with 1980, for a sample selected by the U.S. Bureau of the Census. Relatively few people changed region, 95% of the observations falling on the main diagonal. Not surprisingly, the independence model fits terribly, with $G^2 = 125,923$ and $X^2 = 146,929$ (X^2 is not much less than its maximum possible value for a sample of this size; see Problem 3.24). The symmetry model is also

Table 10.2 Migration from 1980 to 1985

Residence in 1980	Residence in 1985				Total
	Northeast	Midwest	South	West	
Northeast	11,607	100 (126.6)[a] (95.8)[b]	366 (312.9) (370.4)	124 (150.5) (123.8)	12,197
Midwest	87 (117.4) (91.2)	13,677	515 (531.1) (501.7)	302 (255.5) (311.1)	14,581
South	172 (133.2) (167.6)	225 (243.8) (238.3)	17,819	270 (290.0) (261.1)	18,486
West	63 (71.4) (63.2)	176 (130.6) (166.9)	286 (323.0) (294.9)	10,192	10,717
Total	11,929	14,178	18,986	10,888	55,981

Source: Data based on Table 12 of U.S. Bureau of the Census, Current Population Reports, Series P-20, No. 420, *Geographical Mobility: 1985*, U.S. Government Printing Office, Washington, D.C., 1987. Subjects living abroad in 1980 deleted from sample.

[a] Quasi-independence fit;
[b] Quasi-symmetry fit, both models giving perfect fit on main diagonal.

unpromising. For instance, 124 people moved from Northeast to West, but only 63 people made the reverse move. The G^2 statistic for testing symmetry equals 243.55.

Next we consider whether, for people who moved, residence in 1985 is independent of region in 1980. Table 10.2 contains fitted values for the quasi-independence model, for which $G^2 = 69.51$ based on df = 5. This model fits much better than the independence model, primarily because it forces a perfect fit on the main diagonal, where most observations occur. However, lack of fit is apparent off that diagonal. For instance, many more people moved from the Northeast to the South and many fewer moved from the West to the South than would be expected under quasi independence.

Finally, we consider a more general model, quasi symmetry. It has residual $G^2 = 2.99$, based on df = 3. Table 10.2 displays the fit, which is much better than with quasi independence. There is a symmetric association pattern, and the lack of symmetry in cell probabilities seems to reflect slight marginal heterogeneity.

10.3   MARGINAL HOMOGENEITY

Next we study methods for comparing marginal distributions of a square table. Marginal homogeneity is not equivalent to a loglinear model. However, some loglinear models imply marginal homogeneity when their parameters take certain values. We can test the hypothesis of marginal homogeneity by comparing the fit of such a model to the fit of the special case implying marginal homogeneity.

10.3.1   Test Using Quasi-Symmetry Model

Quasi symmetry is a useful model for studying marginal homogeneity. Caussinus (1965) showed that symmetry is equivalent to quasi symmetry and marginal homogeneity simultaneously holding. We have seen previously that symmetry implies both quasi symmetry and marginal homogeneity. Now we give Caussinus's argument for the converse, that the joint occurrence of quasi symmetry and marginal homogeneity implies symmetry.

From (10.14), if quasi symmetry holds, then $\pi_{ij} = \alpha_i \beta_j \gamma_{ij}$, where $\gamma_{ij} = \gamma_{ji} > 0$ for all i and j. Equivalently,

$$\pi_{ij} = \rho_i \delta_{ij} \, ,$$

where $\rho_i = \alpha_i / \beta_i$ and $\delta_{ij} = \beta_i \beta_j \gamma_{ij}$ also satisfies $\delta_{ij} = \delta_{ji} > 0$ for all i and j. If there is also marginal homogeneity, then

$$\pi_{k+} = \rho_k \sum_j \delta_{kj} = \sum_i \rho_i \delta_{ik} = \pi_{+k} \, ,$$

or

$$\rho_k = \left(\sum_i \rho_i \delta_{ik} \right) \Big/ \left(\sum_j \delta_{kj} \right)$$

$$= \left(\sum_i \rho_i \delta_{ik} \right) \Big/ \left(\sum_j \delta_{jk} \right), \quad k = 1, \ldots, I . \qquad (10.17)$$

Each ρ_k is a weighted average of $\{\rho_i\}$, with weights $\{\delta_{ik} / \sum_j \delta_{jk} > 0$, $i = 1, \ldots, I\}$. Any set $\{\rho_i\}$ satisfying (10.17) must be identical. Otherwise there would be a ρ_k that is no greater than any ρ_i but smaller than at least one, and hence it could not be a positive weighted average of all of them. But, since $\{\rho_i\}$ are identical, $\pi_{ij} = \rho_i \delta_{ij} = \rho_j \delta_{ij} = \rho_j \delta_{ji} = \pi_{ji}$, so there is symmetry. Thus a table that satisfies both quasi symmetry and marginal homogeneity also satisfies symmetry. Since the converse holds, we have

$$\text{Quasi symmetry} + \text{Marginal homogeneity} = \text{Symmetry}.$$
$$(10.18)$$

It follows that when quasi symmetry (QS) holds, marginal homogeneity (MH) is equivalent to symmetry (S), which is $\{\lambda_i^X = \lambda_i^Y, \; i = 1, \ldots, I\}$ in the QS model. Thus, conditional on quasi symmetry, testing the null hypothesis of marginal homogeneity is equivalent to testing symmetry. We can test marginal homogeneity by comparing goodness-of-fit statistics for the symmetry and quasi-symmetry models,

$$G^2(S \,|\, QS) = G^2(S) - G^2(QS) \qquad (10.19)$$

based on df $= I - 1$.

10.3.2 Tests Using Restricted and Unrestricted ML Estimates

It is possible to test marginal homogeneity without fitting loglinear models. Madansky (1963) and Lipsitz (1988) maximized the multinomial or Poisson likelihood subject to the constraints $m_{i+} = m_{+i}$, $i = 1, \ldots, I$. Iterative methods for this maximization produce ML estimates $\{\hat{m}_{ij}\}$ satisfying this constraint. Evaluated at these fitted values, G^2 or X^2 test marginal homogeneity, with df $= I - 1$.

Bhapkar (1966) and Stuart (1955) derived direct statistics for testing marginal homogeneity by exploiting the asymptotic normality of unrestricted ML estimates of marginal probabilities. Let $d_i = p_{i+} - p_{+i}$, and let $\mathbf{d}' = (d_1, \ldots, d_{I-1})$. It is redundant to include d_I in this vector, since the condition $\Sigma \, d_i = 0$ determines it. Marginal homogeneity corresponds to $E(\mathbf{d}) = \mathbf{0}$. The sample covariance matrix $\hat{\mathbf{V}}$ of $\sqrt{n}\mathbf{d}$ has elements

$$\hat{v}_{ij} = -(p_{ij} + p_{ji}) - (p_{i+} - p_{+i})(p_{j+} - p_{+j}) \text{ for } i \neq j,$$

$$\hat{v}_{ii} = p_{i+} + p_{+i} - 2p_{ii} - (p_{i+} - p_{+i})^2.$$

Now $\sqrt{n}[\mathbf{d} - E(\mathbf{d})]$ has an asymptotic multivariate normal distribution with covariance matrix $\mathbf{V}$. When $E(\mathbf{d}) = \mathbf{0}$, the Wald statistic

$$W = n\mathbf{d}'\hat{\mathbf{V}}^{-1}\mathbf{d} \qquad (10.20)$$

has a large-sample chi-squared distribution with df $= I - 1$.

Bhapkar used W to test marginal homogeneity. Stuart proposed a similar statistic W_0, replacing $\hat{\mathbf{V}}$ by the sample *null* covariance matrix $\hat{V}_0$, which has

$$\hat{v}_{ij0} = - (p_{ij} + p_{ji}) \quad \text{for } i \neq j ,$$
$$\hat{v}_{ii0} = p_{i+} + p_{+i} - 2p_{ii} .$$

Ireland et al. (1969) noted that $W = W_0/(1 - W_0/n)$. For 2×2 tables, W_0 is identical to McNemar's statistic, the square of (10.4).

10.3.3 Migration Example

Table 10.2 on migration from 1980 to 1985 has $G^2(S) = 243.55$ and $G^2(QS) = 2.99$. The difference $G^2(S \mid QS) = 240.56$, based on df $= I - 1 = 3$, gives extremely strong evidence of marginal heterogeneity. Madansky's ML test of marginal homogeneity gives $G^2 = 240.75$, based on df $= 3$. Table 10.3 shows the fitted values obtained in maximizing the likelihood subject to this hypothesis. Statistics based on the vector **d** of differences in sample marginal proportions give similar results. For instance, Bhapkar's statistic $W = 236.49$, based on df $= 3$.

The sample marginal proportions for the four regions were (0.218, 0.260, 0.330, 0.191) in 1980 and (0.213, 0.253, 0.339, 0.194) in 1985. There is little change over such a short time period, and the large test statistics reflect the huge sample size. We can estimate the change for a given region by applying (10.2) to the collapsed 2×2 table that combines the other regions. A 95% confidence interval for $\pi_{+1} - \pi_{1+}$ is (0.2131 −

Table 10.3 Fit of Marginal Homogeneity Model to Table 10.2[a]

Residence in 1980	Residence in 1985				Total
	Northeast	Midwest	South	West	
Northeast	11,607	100	366	124	12,197
	(11,607)	(98.1)	(265.7)	(94.0)	(12,064.7)
Midwest	87	13,677	515	302	14,581
	(88.7)	(13,677)	(379.1)	(232.3)	(14,377.1)
South	172	225	17,819	270	18,486
	(276.5)	(350.8)	(17,819)	(287.3)	(18,733.5)
West	63	176	286	10,192	10,717
	(92.5)	(251.3)	(269.8)	(10,192)	(10,805.6)
Total	11,929	14,178	18,986	10,888	55,981
	(12,064.7)	(14,377.1)	(18,733.5)	(10,805.6)	

[a] Fit obtained using method described by Lipsitz (1988). See Section 14.2.

0.2179) ± 1.96(0.00054), or −0.0048 ± 0.0011. Similarly, a 95% confidence interval for $\pi_{+2} - \pi_{2+}$ is −0.0072 ± 0.0013, for $\pi_{+3} - \pi_{3+}$ is 0.0089 ± 0.0015, and for $\pi_{+4} - \pi_{4+}$ is 0.0030 ± 0.0012. There is strong evidence of change for all four regions.

It is not surprising that the symmetry model fits poorly. Whenever there is marginal heterogeneity, the symmetry model is doomed.

10.4 SQUARE TABLES WITH ORDERED CATEGORIES

Models presented so far in this chapter treat classifications as nominal. When categories are ordered, more parsimonious models often hold.

10.4.1 Conditional Symmetry

For ordered classifications, when symmetry does not hold, often either $\pi_{ij} > \pi_{ji}$ for all $i < j$, or $\pi_{ij} < \pi_{ji}$ for all $i < j$. A generalization of symmetry that has this property is

$$\log m_{ij} = \mu + \lambda_i + \lambda_j + \lambda_{ij} + \tau I(i < j) \qquad (10.21)$$

where all $\lambda_{ij} = \lambda_{ji}$ and where $I(\cdot)$ is the indicator function. The corresponding logit model

$$\log(m_{ij}/m_{ji}) = \tau \quad \text{for } i < j \qquad (10.22)$$

contains only an intercept term. Thus,

$$m_{ij} = m_{ji} e^{\tau} \quad \text{for all } i < j .$$

The symmetry model is the special case $\tau = 0$.

Let X denote the row number and Y the column number of an observation made according to distribution $\{\pi_{ij}\}$. Model (10.21) implies that for all $i < j$,

$$P(X = i, Y = j \mid X < Y) = P(X = j, Y = i \mid X > Y) . \qquad (10.23)$$

The pattern of probabilities for cells above the main diagonal is a mirror image of the pattern for cells below it. Because of this property, the model is called the *conditional symmetry* model (McCullagh 1978).

The likelihood equations for conditional symmetry are

$$\hat{m}_{ij} + \hat{m}_{ji} = n_{ij} + n_{ji} \quad \text{for all } i \text{ and } j \,,$$

and

$$\sum_{i<j}\sum \hat{m}_{ij} = \sum_{i<j}\sum n_{ij}$$

or, equivalently,

$$\sum_{i>j}\sum \hat{m}_{ij} = \sum_{i>j}\sum n_{ij}$$

The solution of these equations that satisfies the model is

$$\hat{\tau} = \log\left(\frac{\displaystyle\sum_{i<j}\sum n_{ij}}{\displaystyle\sum_{i>j}\sum n_{ij}}\right)$$

$$\hat{m}_{ij} = \frac{\exp(\hat{\tau})(n_{ij} + n_{ji})}{\exp(\hat{\tau}) + 1}\,, \quad i < j \tag{10.24a}$$

$$\hat{m}_{ij} = \frac{n_{ij} + n_{ji}}{\exp(\hat{\tau}) + 1}\,, \quad i > j \tag{10.24b}$$

$$\hat{m}_{ii} = n_{ii}\,, \quad i = 1,\dots.I\,. \tag{10.24c}$$

The residual df equal $(I + 1)(I - 2)/2$, one less than symmetry has. We can also fit the model by fitting logit model (10.22), treating $\{n_{ij} \text{ for } i < j\}$ as independent binomial random variables with sample sizes $\{n_{ij} + n_{ji}\}$.

Conditional symmetry is a special case of the *diagonals-parameter symmetry* model

$$m_{ij} = m_{ji}\exp(\tau_{j-i})\,, \quad i < j\,,$$

proposed by Goodman (1972, 1979b). The parameter τ_k is the log odds that an observation falls in a cell (i, j) satisfying $j - i = k$ instead of a cell (i, j) satisfying $j - i = -k$, $k = 1,\dots,I - 1$. The conditional symmetry model satisfies $\tau_1 = \tau_2 = \cdots = \tau_{I-1}$.

10.4.2 Quasi Association

The quasi-independence model rarely fits well for tables with ordered categories. There is often a monotone pattern to the association that remains even after conditioning on the event that responses differ.

Let $\{u_i\}$ be a fixed set of ordered scores. The model

$$\log m_{ij} = \mu + \lambda_i^X + \lambda_j^Y + \beta u_i u_j + \delta_i I(i=j) \qquad (10.25)$$

permits linear-by-linear association off the main diagonal. It is nested between quasi symmetry and quasi independence. It is a parsimonious quasi-symmetry model that has quasi-independence as the special case $\beta = 0$. For equal-interval scores, it implies uniform local association, given that responses differ. Goodman (1979a) called it the *quasi-uniform association* model.

10.4.3 Testing Marginal Homogeneity for Ordinal Classifications

Tests of marginal homogeneity given in Section 10.3 use only nominal information. They use all $I - 1$ degrees of freedom available for comparisons of I pairs of marginal proportions. For ordinal classifications, alternative single-degree-of-freedom statistics are more sensitive to certain departures from the null hypothesis. When the number of categories is large and the dependence between classifications is strong, ordinal tests can be much more powerful than standard tests (Agresti 1983b).

We use conditional symmetry to illustrate a model-based ordinal test of marginal homogeneity. Conditional symmetry implies the marginal distributions are stochastically ordered, the sign of τ determining the direction of order and $\tau = 0$ implying marginal homogeneity. When conditional symmetry (CS) holds, $\tau = 0$ is also equivalent to symmetry (S), so

Conditional symmetry + Marginal homogeneity = Symmetry .

A statistic for testing marginal homogeneity is

$$G^2(S \,|\, CS) = G^2(S) - G^2(CS) ,$$

which has df $= 1$.

By the delta method, the estimated asymptotic variance of $\hat\tau$ is

$$\hat\sigma^2(\hat\tau) = \left(\sum_{i<j}\sum n_{ij}\right)^{-1} + \left(\sum_{i>j}\sum n_{ij}\right)^{-1} .$$

Another large-sample statistic for testing marginal homogeneity is $z = \hat\tau / \hat\sigma(\hat\tau)$. When the conditional symmetry model holds, tests of marginal homogeneity based on it are more powerful asymptotically than tests such as $G^2(S \,|\, QS)$, which ignore category orderings.

The cumulative logit model is also useful for comparing marginal distributions. We discuss it in Section 11.4, for a more general repeated measures setting.

10.4.4 Esophageal Cancer Example

Table 10.4, taken from Breslow (1982), compares 80 esophageal cancer patients with 80 matched control subjects. The response is the number of beverages reported drunk at "burning hot" temperatures. We analyze whether cases tended to drink more beverages burning hot than did controls. Since most cell counts are small, we report Pearson's X^2 for testing fit of models.

The cell counts show little hope for the symmetry model, since $n_{ij} \leq n_{ji}$ whenever $i < j$. Table 10.4 contains fitted values for that model, which has $X^2 = 15.1$ based on df $= 6$. The quasi-symmetry model has $X^2 = 2.5$, based on df $= 3$. To test marginal homogeneity, we form the Pearson statistic $X^2(S \mid QS) = 14.0$, based on df $= 3$. Though G^2 is an unreliable index of fit for sparse data, the difference statistic $G^2(S \mid QS)$ has greater

Table 10.4 Number of Beverages Drunk at Burning Hot Temperatures, for Esophageal Cancer Case-Control Pairs

Case	Control			
	0	1	2	3
0	31	5	5	0
	$(31)^a$	(8.5)	(9.5)	(3.0)
	$(31)^b$	(4.1)	(4.5)	(1.4)
1	12	1	0	0
	(8.5)	(1)	(0.5)	(0.5)
	(12.9)	(1)	(0.2)	(0.2)
2	14	1	2	1
	(9.5)	(0.5)	(2)	(1.0)
	(14.5)	(0.8)	(2)	(0.5)
3	6	1	1	0
	(3.0)	(0.5)	(1.0)	(0)
	(4.6)	(0.8)	(1.5)	(0)

Source: Reprinted with permission from the Biometric Society; data from Breslow (1982).

[a] Symmetry model and [b] conditional symmetry model.

reliability (Section 7.7.6). It equals 15.3, also giving strong evidence of marginal heterogeneity. These statistics treat the classifications as nominal, and are not helpful for telling us *how* marginal distributions depart from homogeneity.

The model-building process can exploit the ordering of response categories. For the symmetry model, uniformly nonnegative residuals below the main diagonal indicate systematic lack of fit remedied by slightly more complex models. The conditional symmetry model gives a much better fit, with $X^2 = 3.6$ based on df = 5. The ML estimate of $\exp(\tau)$ is $11/35 = 0.314$. The probability that a control drank k more beverages burning hot than did the case is estimated to be 0.314 times the probability that the case drank k more beverages burning hot than the control, for $k = 1,2,3$.

The sample cumulative marginal distribution is $(0.51, 0.68, 0.90, 1.0)$ for cases; it is $(0.79, 0.89, 0.99, 1.0)$ for controls, stochastically lower than for cases. The estimate $\hat{\tau} = \log(0.314) = -1.16$, having $\hat{\sigma}(\hat{\tau}) = 0.346$, gives strong evidence that cancer patients tended to drink more beverages burning hot than did the controls. Also, we can use $G^2(S\,|\,CS) = 13.2$ and $X^2(S\,|\,CS) = 12.5$ to test marginal homogeneity, both based on df = 1. They have values nearly as large as $G^2(S\,|\,QS)$ and $X^2(S\,|\,QS)$, but use the ordinality and are based on 1 instead of 3 degrees of freedom.

Conditional symmetry is not a special case of quasi symmetry. The related model

$$m_{ij} = m_{ji}e^{\tau(j-i)}$$

is a special case. It also fits these data well and has simple interpretations (Problem 10.35).

10.5 MEASURING AGREEMENT

Suppose two observers separately classify a sample of subjects using the same categorical scale. For instance, two clinical psychologists might classify the subjects' mental health, using the scale (well, mild symptom formation, moderate symptom formation, mentally impaired). The joint ratings of the observers can be displayed in a square table, with the same categories in each dimension. The main diagonal represents observer agreement for the ratings.

Many categorical scales are subjective, and reliability can be assessed by evaluating agreement between observers. This section studies ways of describing strength of agreement. We distinguish between measuring *agreement* and measuring *association*, because there can be strong association without strong agreement. For an ordinal scale, if observer A consistently rates subjects one level higher than observer B, then the strength of agreement is weak even though the association is strong.

10.5.1 Kappa

Suppose the observers and the categorical scale are fixed, and the subjects rated are a sample. For a randomly selected subject, let π_{ij} denote the probability of classification in the ith category by the first observer and the jth category by the second observer. Then

$$\Pi_o = \sum \pi_{ii}$$

is the probability the two observers agree. Perfect agreement corresponds to $\Pi_o = 1$. If the observers' ratings are statistically independent, $\pi_{ii} = \pi_{i+}\pi_{+i}$, and the probability of agreement equals

$$\Pi_e = \sum \pi_{i+}\pi_{+i} \, .$$

Thus, $\Pi_o - \Pi_e$ is the excess of the observer agreement over that expected purely by chance (i.e., if ratings were statistically independent).

Kappa, introduced by Cohen (1960), is

$$\kappa = \frac{\sum \pi_{ii} - \sum \pi_{i+}\pi_{+i}}{1 - \sum \pi_{i+}\pi_{+i}} = \frac{\Pi_o - \Pi_e}{1 - \Pi_e} \, . \qquad (10.26)$$

The denominator replaces Π_o by its maximum possible value of 1, corresponding to perfect agreement. Kappa equals 0 when the agreement equals that expected by chance, and it equals 1.0 when there is perfect agreement. The stronger the agreement, the higher the value, for a given pair of marginal distributions. Negative values occur when agreement is weaker than expected by chance, but this rarely happens.

For multinomial sampling, the sample measure $\hat{\kappa}$ has a large-sample normal distribution. Its asymptotic variance can be estimated (Fleiss et al. 1969) by

$$\hat{\sigma}^2(\hat{\kappa}) = \frac{1}{n}\left\{ \frac{P_o(1 - P_o)}{(1 - P_e)^2} + \frac{2(1 - P_o)\left[2P_oP_e - \sum p_{ii}(p_{i+} + p_{+i})\right]}{(1 - P_e)^3} \right.$$
$$\left. + \frac{(1 - P_o)^2\left[\sum\sum p_{ij}(p_{j+} + p_{+i})^2 - 4P_e^2\right]}{(1 - P_e)^4} \right\}, \qquad (10.27)$$

where $P_o = \Sigma\, p_{ii}$ and $P_e = \Sigma\, p_{i+} p_{+i}$. It is rarely plausible that agreement is no better than expected by chance. Thus, rather than testing H_0: $\kappa = 0$, it is more important to estimate strength of agreement, by constructing a confidence interval for κ.

10.5.2 Generalized Kappa Measures

Kappa is designed for nominal classifications. When categories are ordered, the seriousness of a disagreement depends on the difference between the ratings. For nominal classifications as well, some disagreements may be considered more substantial than others. The measure *weighted kappa* (Spitzer et al. 1967) uses weights $\{w_{ij}\}$ to describe closeness of agreement. For weights satisfying $0 \leqslant w_{ij} \leqslant 1$, with all $w_{ij} = w_{ji}$ and $w_{ii} = 1$, the weighted agreement is $\Sigma\,\Sigma\, w_{ij}\pi_{ij}$, and weighted kappa is

$$\kappa_w = \frac{\Sigma\,\Sigma\, w_{ij}\pi_{ij} - \Sigma\,\Sigma\, w_{ij}\pi_{i+}\pi_{+j}}{1 - \Sigma\,\Sigma\, w_{ij}\pi_{i+}\pi_{+j}}.$$

For weights $\{w_{ij} = 1 - (i-j)^2/(I-1)^2\}$, suggested by Fleiss and Cohen (1973), agreement is greater for cells nearer the main diagonal.

Whenever we summarize a contingency table by a single number, the reduction in information can be severe. Models are useful for more detailed investigations of the structure of agreement and disagreement. For nominal classifications, the quasi-independence model (10.16) analyzes whether classifications are independent, conditional on the event that the observers disagree. For ordinal classifications, models that take ordering into account are more informative for describing the structure.

10.5.3 Agreement Between Pathologists Example

Table 10.5 is based on data presented by Landis and Koch (1977b) and originally reported by Holmquist et al. (1967). Two pathologists classified each of 118 slides in terms of carcinoma in situ of the uterine cervix, based on the most involved lesion, using the ordered categories (1) Negative; (2) Atypical Squamous Hyperplasia; (3) Carcinoma in Situ; (4) Squamous Carcinoma with Early Stromal Invasion; (5) Invasive Carcinoma. The 5×5 cross-classification of their ratings contains 12 empty cells, indicative of the sparseness that often occurs off the main diagonal for agreement data.

For Table 10.5, the estimates of Π_o and Π_e are $P_o = 0.636$ and $P_e = 0.273$. Sample kappa is $\hat{\kappa} = 0.498$, with estimated standard error

Table 10.5   Pathologist Ratings of Carcinoma, with Fitted Values for Model of Uniform Association Plus Extra Agreement

Pathologist A	Pathologist B				
	1	2	3	4	5
1	22	2	2	0	0
	(22.1)	(1.7)	(2.2)	(0.0)	(0.0)
2	5	7	14	0	0
	(4.4)	(8.9)	(12.5)	(0.1)	(0.0)
3	0	2	36	0	0
	(0.4)	(1.0)	(36.1)	(0.5)	(0.0)
4	0	1	14	7	0
	(0.1)	(0.4)	(15.6)	(5.4)	(0.5)
5	0	0	3	0	3
	(0.0)	(0.0)	(2.6)	(1.0)	(2.4)

Source: Reprinted with permission from Holmquist et al. (1967), copyright 1967, American Medical Association.

0.057. The difference between observed and chance agreement is about 50% of the maximum possible difference. There is strong evidence that agreement is better than if the pathologists' ratings were statistically independent.

To investigate the agreement structure more fully, we fit some models. The symmetry model has $G^2 = 39.2$ and $X^2 = 30.3$, with df = 10. The fit is extremely poor, considering that the model forces a perfect fit on the main diagonal, where most observations fall. The conditional symmetry model does not fit better, having $G^2 = 38.0$ and $X^2 = 29.9$. This is not surprising, since $\log(n_{ij}/n_{ji})$ is not consistently of one sign, and since sample marginal distributions are not stochastically ordered. The marginal distributions do display heterogeneity, however. The quasi-symmetry model fits well, and the statistic $G^2(S \mid QS) = 39.2 - 1.0 = 38.2$, with df = $10 - 6 = 4$, provides strong evidence against marginal homogeneity. The heterogeneity is one reason the agreement is not stronger.

As expected, the ratings *are* associated—the independence model gives $G^2 = 131.2$ and $X^2 = 174.9$, based on df = 16. The quasi-independence model fits substantially better, with $G^2 = 13.6$ and $X^2 = 11.6$, based on df = 11. Like the symmetry model, however, this model fits more poorly than the quasi-symmetry model. Conditional on the event that the pathologists disagree, some association remains between the ratings.

For ratings on ordinal scales, there is almost always a positive association. Conditional on rater disagreement, there usually remains a tendency for high (low) ratings by observer A to occur with relatively high (low) ratings by observer B. Hence, we partition the beyond-chance agreement into two parts: Agreement due to a baseline linear-by-linear association between the ratings, and an increment that reflects agreement in excess of that occurring simply by chance or from the baseline association. The model is

$$\log m_{ij} = \mu + \lambda_i^A + \lambda_j^B + \beta u_i u_j + \delta I(i = j) \qquad (10.28)$$

where $\{u_i\}$ are ordered scores assigned to levels of the ordinal scale.

This is a parsimonious quasi-symmetry model that is also a special case of quasi-association model (10.25), with $\delta_i = \delta$ for all i. Its likelihood equations are

$$\hat{m}_{i+} = n_{i+}, \qquad \hat{m}_{+i} = n_{+i}, \qquad i = 1, \ldots, I,$$

$$\sum \sum u_i u_j \hat{m}_{ij} = \sum \sum u_i u_j n_{ij}, \qquad \sum \hat{m}_{ii} = \sum n_{ii}.$$

The fitted and observed distributions share the same marginal distributions, correlation between ratings, and prevalence of exact agreement. We can fit the model using standard iterative routines for loglinear models.

A useful set of odds ratios for studying agreement is

$$\tau_{ij} = (m_{ii} m_{jj})/(m_{ij} m_{ji}) \quad \text{for all } i \text{ and } j.$$

For model (10.28),

$$\log \tau_{ij} = (u_j - u_i)^2 \beta + 2\delta.$$

The level of agreement increases as β and δ increase, and as the distance between levels increases. The model can also be described using local odds ratios. For unit-spaced scores, $\log \theta_{ii} = \log \tau_{i,i+1} = \beta + 2\delta$ for $i = 1, \ldots, I-1$ and $\log \theta_{ij} = \beta$ whenever none of the four cells is on the main diagonal. For this scoring, model (10.28) assumes "uniform association plus extra agreement."

When fitted to Table 10.5 using unit-spaced scores, model (10.28) has residual $G^2 = 8.4$ based on df $= 14$. Table 10.5 reports fitted values. Though the data are sparse, the model fits well and gives simpler description than does the quasi-symmetry model. The ML parameter estimates are $\hat{\delta} = 1.067$ (ASE = 0.404) and $\hat{\beta} = 1.150$ (ASE = 0.342). There is strong evidence of extra agreement beyond that due to the baseline association, and there is strong evidence of extra association beyond that due to the exact agreement. We summarize beyond-chance agreement as follows: For $i = 1, 2, 3$, the odds that the diagnosis of pathologist A is $i + 1$ rather than i is estimated to be $\exp(\hat{\beta} + 2\hat{\delta}) = 26.7$ times higher when the diagnosis of pathologist B is $i + 1$ than when it is i. We summarize the baseline association similarly: For $|i - j| > 1$, the odds that the diagnosis of pathologist A is $i + 1$ rather than i is estimated to be $\exp(\hat{\beta}) = 2.9$ times higher when the diagnosis of pathologist B is $j + 1$ than when it is j.

10.6 BRADLEY–TERRY MODEL FOR PAIRED COMPARISONS

Suppose I items are compared in a pairwise manner. For instance, a wine critic might rate I chardonnay wines. It might be difficult for a rater to establish an outright ranking, especially if I is large. However, for any given pair of wine brands, a rater could probably state a preference after tasting them at the same occasion. Thus we might obtain results of several pairwise comparisons, and use them to establish an overall ranking of the wines.

10.6.1 The Bradley–Terry Model

Bradley and Terry (1952) proposed a logit model for paired comparison experiments. Let Π_{ij} denote the probability that brand i is preferred to brand j. Suppose $\Pi_{ij} + \Pi_{ji} = 1$ for all pairs; that is, there is no chance of a tie. The Bradley–Terry model assumes nonnegative parameters $\{\pi_i\}$ exist such that

$$\Pi_{ij} = \frac{\pi_i}{\pi_i + \pi_j}. \tag{10.29}$$

Alternatively, letting $\pi_i = \exp(\phi_i)$, we have $\Pi_{ij} = \exp(\phi_i)/[\exp(\phi_i) + \exp(\phi_j)]$, or

$$\log\left(\frac{\Pi_{ij}}{\Pi_{ji}}\right) = \phi_i - \phi_j. \tag{10.30}$$

Thus, Π_{ij} is a monotone function of $\phi_i - \phi_j$, with $\Pi_{ij} = 0.5$ when $\phi_i = \phi_j$.

For each pair of brands $i < j$, let N_{ij} denote the sample number of comparisons. Suppose i is preferred n_{ij} times and j is preferred $n_{ji} = N_{ij} - n_{ij}$ times. We summarize results in a square contingency table, in which cells on the main diagonal are empty. When the N_{ij} comparisons are independent, with the same probability Π_{ij} applying to each, n_{ij} has a binomial (N_{ij}, Π_{ij}) distribution. If we regard comparisons for different pairs of brands as also independent, we can fit the model using ordinary methods for logit models.

10.6.2 Bradley–Terry Model and Quasi Symmetry

Fienberg and Larntz (1976) showed that the Bradley–Terry model is a logit formulation of the quasi-symmetry model. For quasi symmetry, given that an observation is in cell (i, j) or (j, i), the logit of the conditional probability it is in cell (i, j) equals

$$\log\left(\frac{m_{ij}}{m_{ji}}\right) = (\mu + \lambda_i^X + \lambda_j^Y + \lambda_{ij}^{XY}) - (\mu + \lambda_j^X + \lambda_i^Y + \lambda_{ji}^{XY})$$

$$= (\lambda_i^X - \lambda_i^Y) - (\lambda_j^X - \lambda_j^Y) = \phi_i - \phi_j,$$

where $\phi_i = \lambda_i^X - \lambda_i^Y$. Estimates of $\{\lambda_i^X\}$ and $\{\lambda_j^Y\}$ for quasi symmetry yield estimates of $\{\phi_i\}$ for the Bradley–Terry model, and hence of $\{\pi_i = \exp(\phi_i)\}$ and $\{\Pi_{ij}\}$. The usual constraints on $\{\lambda_i^X\}$ and $\{\lambda_j^Y\}$ imply a similar constraint on $\{\phi_i\}$, such as $\Sigma \phi_i = 0$. Any constant multiple of the $\{\pi_i\}$ estimates also satisfy (10.29), so we can scale them to satisfy a constraint such as $\Sigma \pi_i = 1$.

10.6.3 Home Team Advantage in Baseball

Table 10.6 contains results of the 1987 season for professional baseball teams in the Eastern Division of the American League. For instance, for the games between Boston and New York, Boston won 7 and New York won 6. Each team played each other team 13 times. Table 10.6 represents the population of regular-season games played between these teams, but we could regard it as a sample estimate of a conceptual distribution representing "long-run" performance of the teams as constituted in 1987.

When teams i and j play, let Π_{ij} denote the probability that team i beats team j. For Table 10.6, the quasi-symmetry model has $G^2 = 15.74$ and $X^2 = 14.61$, based on df $= 15$. Table 10.6 contains the fitted values. From the likelihood equation (10.12c), $\hat{m}_{ij} + \hat{m}_{ji} = 13$ for all $i \neq j$. Table

Table 10.6   Results of 1987 Season for American League Baseball Teams, with Fit of Bradley–Terry Model

Winning Team	Losing Team						
	Milwaukee	Detroit	Toronto	New York	Boston	Cleveland	Baltimore
Milwaukee	–	7(7.0)	9(7.4)	7(7.6)	7(8.0)	9(9.2)	11(10.8)
Detroit	6(6.0)	–	7(7.0)	5(7.1)	11(7.6)	9(8.8)	9(10.5)
Toronto	4(5.6)	6(6.0)	–	7(6.7)	7(7.1)	8(8.4)	12(10.2)
New York	6(5.4)	8(5.9)	6(6.3)	–	6(7.0)	7(8.3)	10(10.1)
Boston	6(5.0)	2(5.4)	6(5.9)	7(6.0)	–	7(7.9)	12 (9.8)
Cleveland	4(3.8)	4(4.2)	5(4.6)	6(4.7)	6(5.1)	–	6 (8.6)
Baltimore	2(2.2)	4(2.5)	1(2.8)	3(2.9)	1(3.2)	7(4.4)	–

Source: American League Red Book—1988, Sporting News Publishing Co.

10.7 exhibits the sample proportion of games won for each team, and $\{\hat{\pi}_i\}$ for the Bradley–Terry model. When Boston plays New York, the estimated probability of a Boston victory is

$$\hat{\Pi}_{54} = 1 - \hat{\Pi}_{45} = \hat{\pi}_5/(\hat{\pi}_5 + \hat{\pi}_4)$$

$$= \hat{m}_{54}/(\hat{m}_{54} + \hat{m}_{45}) = 0.46 .$$

This model does not take into account which team is the home team. Many sports tend to have a "home-field" advantage, whereby the chance a team wins is higher when it plays at home. Table 10.8 contains results for the 1987 season according to the (home team, away team) classification. For instance, when Boston was the home team, they beat New York 4 times and lost 2 times; when New York was the home team, they beat Boston 4 times and lost 3 times.

Table 10.7   Results of Fitting Bradley–Terry Models to Baseball Data

Team	Winning Percentage	$\hat{\pi}_i$ (10.30)	$\hat{\pi}_i$ (10.31)
Milwaukee	64.1	0.218	0.220
Detroit	60.2	0.189	0.190
Toronto	56.4	0.164	0.164
New York	55.1	0.158	0.157
Boston	51.3	0.136	0.137
Cleveland	39.7	0.089	0.088
Baltimore	23.1	0.045	0.044

Now for all $i \neq j$, let Π_{ij}^* denote the probability that team i beats team j, when team i is the home team. Consider logit model

$$\log\left(\frac{\Pi_{ij}^*}{1 - \Pi_{ij}^*}\right) = \alpha + (\phi_i - \phi_j) \qquad (10.31)$$

for which, letting $\pi_i = \exp(\phi_i)$,

$$\Pi_{ij}^* = \frac{\exp[\alpha + (\phi_i - \phi_j)]}{1 + \exp[\alpha + (\phi_i - \phi_j)]} = \frac{\exp(\alpha)\pi_i}{\pi_j + \exp(\alpha)\pi_i}.$$

When $\alpha > 0$, there is a home field advantage, whereby the home team of two evenly-matched teams has probability $\exp(\alpha)/[1 + \exp(\alpha)]$ of winning.

For Table 10.8, model (10.31) has 42 logits and 7 parameters. We can fit the model as a logit model, regarding the sample as 42 independent binomial samples, or by fitting a corresponding loglinear model (Problem 10.41). The goodness-of-fit statistics are $G^2 = 38.64$ and $X^2 = 34.94$, based on df = 35. Table 10.7 contains $\{\hat{\pi}_i\}$, which are similar to those obtained previously. The estimate of the home-field parameter is $\hat{\alpha} = 0.302$. For two evenly-matched teams, the home team has estimated probability 0.575 of winning. When Boston plays New York, the estimated probability that Boston wins is 0.54 when they play at Boston and 0.39 when they play at New York.

Model (10.31) can be a useful generalization of the Bradley–Terry model whenever we suspect an *order effect*. For instance, in pairwise taste comparisons, there may be slight advantage to the product tasted first.

Table 10.8 Wins/Losses by Home and Away Team, 1987

Home Team	Milwaukee	Detroit	Toronto	New York	Boston	Cleveland	Baltimore
				Away Team			
Milwaukee	–	4-3	4-2	4-3	6-1	4-2	6-0
Detroit	3-3	–	4-2	4-3	6-0	6-1	4-3
Toronto	2-5	4-3	–	2-4	4-3	4-2	6-0
New York	3-3	5-1	2-5	–	4-3	4-2	6-1
Boston	5-1	2-5	3-3	4-2	–	5-2	6-0
Cleveland	2-5	3-3	3-4	4-3	4-2	–	2-4
Baltimore	2-5	1-5	1-6	2-4	1-6	3-4	–

Source: American League Red Book—1988.

CHAPTER NOTES

Section 10.1: Comparing Dependent Proportions

10.1 Altham (1971) gave a Bayesian analysis for dependent proportions. Copas (1973), Gart (1969), Holford (1978), Kenward and Jones (1987), Liang and Zeger (1988), and Miettinen (1969) studied various generalizations of matched-pairs designs. Gart generalized the logistic model to allow for an order effect. Miettinen generalized the McNemar test for case-control pairings having several controls for every case.

Section 10.2: Symmetry Models

10.2 For additional discussion of quasi symmetry, see Caussinus (1965), Darroch (1981, 1986), McCullagh (1982), and Darroch and McCloud (1986). Caussinus's article, besides being influential in introducing new concepts, nicely summarized the 1965 state-of-the-art for categorical data analyses.

A more general definition of quasi independence is $\pi_{ij} = \alpha_i \beta_j$ for some fixed set of cells. See Altham (1975), Caussinus (1965), Goodman (1968), and Fienberg (1972a). The term quasi independence apparently originated in Goodman's article. Caussinus used the concept extensively in analyzing "truncated" tables that deleted a certain set of cells from consideration, and Goodman used it in earlier papers on social mobility analysis.

Section 10.4: Square Tables with Ordered Categories

10.3 For discussion of conditional symmetry and other diagonals-parameter symmetry models, see Bishop et al. (1975, pp. 285–286), McCullagh (1978), Goodman (1979b, 1985), and Hout et al. (1987). For $I > 2$, McCullagh noted that quasi symmetry + conditional symmetry = symmetry.

Section 10.5: Measuring Agreement

10.4 Fleiss (1975) and Fleiss and Cohen (1973) showed the relationship of kappa and weighted kappa to the intraclass correlation coefficient, a measure of inter-rater reliability for interval scales. Kraemer (1979) discussed related properties of kappa. Landis and Koch (1977a,b) used alternative weighting schemes to describe various aspects of observer agreement. Fleiss (1981, Chap. 13) and Kraemer (1983) reviewed the literature on kappa and its generalizations. Goodman (1979b), Tanner and Young (1985), and Agresti (1988) used loglinear models to describe agreement. Darroch and McCloud (1986) showed that quasi symmetry has an important role in agreement modeling, and proposed alternative measures of agreement based on odds ratios $\{\tau_{ij}\}$.

Section 10.6: Bradley–Terry Model for Paired Comparisons

10.5 There have been numerous extensions of the Bradley–Terry model, many discussed by Bradley (1976). Imrey et al. (1976) and Fienberg and Larntz (1976) showed its relation to the quasi-independence model. See David (1988) for further discussion of paired comparisons methods.

PROBLEMS

Applications

10.1 A sample of married couples indicate their candidate preference in a presidential election. Table 10.9 reports the results.

Table 10.9

Husband's Preference	Wife's Preference	
	Democrat	Republican
Democrat	200	25
Republican	75	200

a. Find a 90% confidence interval for the difference between the proportions of wives and husbands prefering the Democrat.

b. Find the *P*-value for testing marginal homogeneity.

10.2 Refer to the previous problem. Calculate and interpret the estimate of the odds ratio for Cox's logistic model for matched pairs.

10.3 Give an example of a square table illustrating that, for $I > 2$, marginal homogeneity does not imply symmetry.

10.4 Refer to Table 2.6.

a. Fit the symmetry model. Use a residual analysis to describe the lack of fit. Would the symmetry model fit well if the "none or other" category were deleted?

b. Test whether the marginal distributions of religious identification differ.

c. Fit the quasi-independence model, and interpret.

10.5 Refer to Table 2.11.

a. Fit the independence model, and use a residual analysis to describe the dependence between student and parent party identification.

b. Fit the quasi-symmetry model, and interpret the result (recall quasi symmetry = quasi independence for 3×3 tables).

c. Test whether the party identification distribution is identical for parents and students. Interpret.

10.6   Use a statistical computer package to fit the symmetry, quasi-symmetry, quasi-independence, and conditional symmetry models to Table 10.4.

10.7   Table 10.10, from Stuart (1955), describes unaided distance vision for a sample of women. Analyze these data.

Table 10.10

	Left Eye Grade			
Right Eye Grade	Best	Second	Third	Worst
Best	1520	266	124	66
Second	234	1512	432	78
Third	117	362	1772	205
Worst	36	82	179	492

Source: Reprinted with permission from the Biometrika Trustees (Stuart 1955).

10.8   Table 10.11, from Glass (1954, p. 183), relates father's and son's occupational status for a British sample.

Table 10.11   Occupational Status for British Father-Son Pairs

Father's Status	Son's Status					Total
	1	2	3	4	5	
1	50	45	8	18	8	129
2	28	174	84	154	55	495
3	11	78	110	223	96	518
4	14	150	185	714	447	1510
5	3	42	72	320	411	848
Total	106	489	459	1429	1017	3500

Source: Reprinted with permission from Glass (1954).

a. Fit the symmetry model, and describe lack of fit.

b. Fit the conditional symmetry model, and use $\hat{\tau}$ to describe the relationship.

c. Test marginal homogeneity, and describe the marginal heterogeneity.

d. Fit the quasi-independence model and the quasi-uniform association model. Does there seem to be an association, given that fathers and sons have different status categories? Interpret.

10.9 Refer to Table 2.10. Using models, describe the relationship between husband's and wife's sexual fun.

10.10 Table 10.12, from Mullins and Sites (1984), relates mother's education to father's education for a sample of eminent black Americans (defined as persons having biographical sketch in the publication, *Who's Who Among Black Americans*). Analyze and interpret these data.

Table 10.12

Mother's Education	Father's Education			
	8th Grade or less	Part High School	High School	College
8th Grade or less	81	3	9	11
Part High School	14	8	9	6
High School	43	7	43	18
College	21	6	24	87

Source: Reprinted with permission from Mullins and Sites (1984).

10.11 Treating the response in Table 10.4 as continuous, use a normal paired-difference procedure to compare cases and controls on the mean number of beverages drunk burning hot. Compare the results to an ordinal test of marginal homogeneity. List assumptions on which each procedure is based.

10.12 Refer to Table 2.6. Use kappa to describe agreement between current religious identification and identification at age 16. Interpret.

10.13 Table 10.13, from Landis and Koch (1977a), displays diagnoses of multiple sclerosis for two neurologists who classified patients in two sites, Winnipeg and New Orleans. The diagnostic classes are (1). Certain multiple sclerosis; (2). Probable multiple sclerosis; (3). Possible multiple sclerosis; (4). Doubtful, unlikely, or definitely not multiple sclerosis.

Table 10.13

New Orleans Neurologist	Winnipeg Neurologist							
	Winnipeg Patients				New Orleans Patients			
	1	2	3	4	1	2	3	4
1	38	5	0	1	5	3	0	0
2	33	11	3	0	3	11	4	0
3	10	14	5	6	2	13	3	4
4	3	7	3	10	1	2	4	14

Source: Reprinted with permission from the Biometric Society (Landis and Koch 1977a).

    a. Test marginal homogeneity for the New Orleans patients (i) by comparing symmetry and quasi-symmetry models, (ii) with a test that uses the ordering of response categories. Interpret.

    b. Use kappa to describe strength of agreement between the neurologists for the New Orleans patients. Interpret.

10.14 Refer to the previous problem. Analyze the data for the Winnipeg patients.

10.15 Refer to the previous two problems. Construct a model that simultaneously describes agreement between neurologists for the two sites.

10.16 Compute kappa for Table 10.14, and explain why there can be strong association without strong agreement.

Table 10.14

	A	B	C	D	E
A	4	16	0	0	0
B	0	4	16	0	0
C	0	0	4	16	0
D	0	0	0	4	16
E	16	0	0	0	4

10.17 Use the Bradley–Terry model to analyze Table 10.15, based on matches for five women tennis players during 1988. Give a ranking of the five players and a predicted probability that Graf beats Navratilova for a match between them in 1988.

Table 10.15

			Loser		
Winner	Graf	Navratilova	Sabatini	Evert	Shriver
Graf	–	1	3	2	2
Navratilova	0	–	3	3	2
Sabatini	2	0	–	2	1
Evert	0	2	1	–	0
Shriver	1	0	1	1	–

Source: Reprinted with permission from World Tennis magazine.

10.18 Refer to model (10.31).

    a. Construct a more general model having home-team parameters $\{\pi_{Hi}\}$ and away-team parameters $\{\pi_{Ai}\}$, such that the probability team i beats team j when i is the home team is $\pi_{Hi}/(\pi_{Hi} + \pi_{Aj})$, where $\Sigma \pi_{Ai} = 1$ but $\Sigma \pi_{Hi}$ is unrestricted.

    b. For the model in (a), interpret the special case $\{\pi_{Hi} = c\pi_{Ai}\}$, when (i) $c = 1$, (ii) $c > 1$, (iii) $c < 1$.

    c. Fit the model in (a) to Table 10.8. Compare the fit to model (10.31).

    d. Compare estimates of $\{\pi_{Hi}\}$ and $\{\pi_{Ai}\}$ to describe how teams play at home and away.

Theory and Methods

10.19 For a 2×2 table, let $d = p_{1+} - p_{+1}$.

    a. Calculate $\text{Cov}(p_{1+}, p_{+1})$, and show that $\text{Var}(\sqrt{n}d)$ equals (10.1).

    b. If $\pi_{1+} = \pi_{+1}$, show that $\text{Var}(\sqrt{n}d) = \pi_{1+} + \pi_{+1} - 2\pi_{11}$.

10.20 Refer to Cox's logistic model for matched pairs.

    a. Show that $\exp(\beta)$ is an odds ratio between response and occasion.

    b. Show that $\pi_{1+}\pi_{+2}/\pi_{2+}\pi_{+1}$ is also an occasion × response odds ratio. Explain the distinction between this "population-averaged" odds ratio and the "subject-specific" odds ratio $\exp(\beta)$ for the logistic model.

    c. Using the conditional distribution, show that the ML estimator of $\exp(\beta)$ is n_{12}/n_{21}. (The estimator $\log(n_{12}/n_{21})$ of β has estimated standard error $[n_{12}^{-1} + n_{21}^{-1}]^{1/2}$.)

10.21   A case-control study has n matched pairs measured on a binary response Y and an explanatory variable X. For the ith matched pair, $Y = 1$ for the case and $Y = 0$ for the control. Show that one can fit the logistic regression model of Y on X for the $2n$ observations by fitting a model with zero intercept to the n pairwise difference scores on Y and X (Breslow and Day, 1980, Section 7.2).

10.22   A wildlife biologist wants to estimate the number of alligators in Lake Lochloosa, Florida. She catches n_{1+} alligators, tags them, and releases them back into the lake. Two weeks later, she catches a second sample of n_{+1} alligators, of which n_{11} were also in the first sample (they are tagged). In Table 10.16, n_{22} is not observed, and the population size $n = n_{11} + n_{12} + n_{21} + n_{22}$ is unknown. Suppose whether an alligator is captured in the second sample is independent of whether it was captured in the first sample. Argue that a reasonable estimator of the population size is $n_{1+}n_{+1}/n_{11}$. (Sekar and Deming (1949) obtained an estimated variance for this estimator of $n_{1+}n_{+1}n_{12}n_{21}/(n_{11})^3$. For discussion of capture − recapture methods, see Bishop et al. (1975, Chap. 6), Cormack (1989), El-Khorazaty et al. (1977) and Fienberg (1972b).

Table 10.16

First Sample	Second Sample		Total
	Present	Absent	
Present	n_{11}	n_{12}	n_{1+}
Absent	n_{21}	−	
Total	n_{+1}		

10.23   We observe a binary response at two times. Data are available for some subjects at both times, for other subjects only at the first time, and for others only at the second time. So, the overall samples for the two times are partly dependent and partly independent. Of n subjects measured both times (the dependent samples), let p_{ij} denote the proportion making response i at time 1 and response j at time 2. Of n_g subjects observed only at time g, let q_g denote the proportion making the first response, $g = 1, 2$. Treat n, n_1, and n_2 as fixed, and let $a = n/(n + n_1)$, $b = n/(n + n_2)$, and $\mathbf{p}' = (p_{11}, p_{12}, p_{21}, q_1, q_2)$.

a. Treating $\{np_{ij}\}$ as a multinomial sample for the 2×2 table for the dependent samples, and treating $n_1 q_1$ and $n_2 q_2$ as independent binomials, show the estimated covariance matrix of **p** is the matrix **S** in Table 10.17.

Table 10.17

$$
S = \begin{bmatrix}
\dfrac{p_{11}(1-p_{11})}{n} & \dfrac{-p_{11}p_{12}}{n} & \dfrac{-p_{11}p_{21}}{n} & 0 & 0 \\[3mm]
\dfrac{-p_{11}p_{12}}{n} & \dfrac{p_{12}(1-p_{12})}{n} & \dfrac{-p_{12}p_{21}}{n} & 0 & 0 \\[3mm]
\dfrac{-p_{11}p_{21}}{n} & \dfrac{-p_{12}p_{21}}{n} & \dfrac{p_{21}(1-p_{21})}{n} & 0 & 0 \\[3mm]
0 & 0 & 0 & \dfrac{q_1(1-q_1)}{n_1} & 0 \\[3mm]
0 & 0 & 0 & 0 & \dfrac{q_2(1-q_2)}{n_2}
\end{bmatrix}
$$

b. Of all subjects observed at time 1, let P_1 denote the proportion making the first response. Of all subjects observed at time 2, let P_2 denote the proportion making the first response. Show that $P_1 = \mathbf{d}_1'\mathbf{p}$ and $P_2 = \mathbf{d}_2'\mathbf{p}$, with $\mathbf{d}_1' = (a, a, 0, 1-a, 0)$ and $\mathbf{d}_2' = (b, 0, b, 0, 1-b)$. Thus the estimated variance of P_i is $\mathbf{d}_i'S\mathbf{d}_i$, and the estimated variance of $P_1 - P_2$ is $(\mathbf{d}_1 - \mathbf{d}_2)'S(\mathbf{d}_1 - \mathbf{d}_2)$.

c. Table 10.18 is taken from a study conducted in the College of Pharmacy at the University of Florida concerning drug use in an elderly population. Subjects were asked whether they took

Table 10.18

Take Drug	1985		
	Yes	No	Not Sampled
1979			
Yes	175	190	230
No	139	1518	982
Not Sampled	64	595	

Source: Mary Moore, Division of Biostatistics, University of Florida.

tranquillizers. Some subjects were interviewed in 1979, some in 1985, and others both times. Calculate P_1 and P_2. Assuming $E(p_{1+}) = E(q_1) = \pi_1$ and $E(p_{+1}) = E(q_2) = \pi_2$, construct a 95% confidence interval for $\pi_1 - \pi_2$.

(Other analyses for *partially classified* data were given by Chen and Fienberg (1976), Dempster et al. (1977), Fay (1986), Fuchs (1982), Gimotty and Brown (1987), Hocking and Oxspring (1971, 1974), Kadane (1985), Koch et al. (1972), Little and Rubin (1987, Chap. 9), Shih (1987), and Woolson and Clarke (1984). Such analyses make assumptions about the missing data. The approach given here is reasonable when data are *missing completely at random* (Rubin 1976).)

10.24 For the symmetry model, derive the likelihood equations and show that $\hat{m}_{ij} = (n_{ij} + n_{ji})/2$.

10.25 For the quasi-symmetry model, derive the likelihood equations and the residual df.

10.26 Show that quasi symmetry is equivalent to

$$(\pi_{ij}\pi_{jk}\pi_{ki})/(\pi_{ji}\pi_{kj}\pi_{ik}) = 1 \quad \text{all } i, j, \text{ and } k$$

(Caussinus 1965). Use this result to prove (10.15).

10.27 For an $I \times I$ table $\{n_{ij}\}$, construct the $I \times I \times 2$ tables $\{n_{ij1} = n_{ij}, n_{ij2} = n_{ji}\}$ and $\{m_{ij1} = m_{ij}, m_{ij2} = m_{ji}\}$.

a. If quasi symmetry holds for $\{m_{ij}\}$, show $\theta_{ij(1)}/\theta_{ij(2)} = 1$ for $\{m_{ijk}\}$, for all i and j.

b. Show that likelihood equations for the quasi-symmetry model for $\{m_{ij}\}$ correspond to likelihood equations for loglinear model (XY, XZ, YZ) for $\{m_{ijk}\}$.

c. Show that $\{\hat{m}_{ij}\}$ for the quasi-symmetry model are identical to $\{\hat{m}_{ij1}\}$ for model (XY, XZ, YZ) fitted to $\{n_{ijk}\}$ (Bishop et al. 1975, pp. 289–290).

d. Show that model $\log m_{ijk} = \mu + \lambda_i^X + \lambda_j^Y + \lambda_{ij}^{XY}$ for $\{m_{ijk}\}$ corresponds to symmetry for $\{m_{ij}\}$.

(Parts (c) and (d) show that computer packages for standard loglinear models can also fit the symmetry and quasi-symmetry models. Since the data comprise one of the two mirror-image layers in the three-way table, true goodness-of-fit statistics equal half those obtained by fitting the indicated models to the three-way table.)

10.28 Derive the likelihood equations and the residual df for the quasi-independence model. Show that the independence model is a special case.

10.29 Suppose there is quasi independence, and let $E_{ij} = 1$ for $i \neq j$ and $E_{ij} = \epsilon > 0$ for $i = j$. Show the amended cell counts $\{n_{ij}^*\}$ with zeroes on the main diagonal have expected values $m_{ij} = E_{ij} \alpha_i \beta_j$, or $m_{ij}/E_{ij} = \alpha_i \beta_j$ for all i and j, as $\epsilon \downarrow 0$. (Thus, fitting a model with structural zeroes corresponds to fitting a general loglinear model of form (6.22), and giving zero weight to certain cells.)

10.30 Derive the covariance matrix for the vector **d** of $\{d_i = p_{i+} - p_{+i}\}$ used to test marginal homogeneity in Section 10.3.2.

10.31 Construct the loglinear model for which there is both marginal homogeneity and statistical independence. Find the residual df for the model, and show $\hat{\pi}_{ij} = (p_{i+} + p_{+i})(p_{j+} + p_{+j})/4$.

10.32 Consider the conditional symmetry model.
  a. Show the model satisfies (10.23).
  b. Derive the likelihood equations.
  c. Show that (10.24) satisfy the model and its likelihood equations, and hence are the ML fitted values.
  d. Derive the residual df.
  e. Derive the asymptotic variance of $\hat{\tau}$.

10.33 Show that the quasi-association model (10.25) is a special case of the quasi-symmetry model.

10.34 Identify loglinear models that correspond to the logit models, for $i < j$, $\log(m_{ij}/m_{ji}) =$
  a. 0.
  b. τ.
  c. $\alpha_i - \alpha_j$.
  d. $\tau(j - i)$.

10.35 Consider the model $\log(m_{ij}/m_{ji}) = \tau(j - i)$.
  a. Show it is a special case of the diagonals-parameter symmetry model. Explain how to interpret τ.
  b. Give the corresponding loglinear model, and show it satisfies quasi symmetry.

    c. Show that this model + marginal homogeneity = symmetry.

    d. Derive sufficient statistics, and explain why tests of marginal homogeneity using this model are sensitive to differences in means of marginal distributions.

    e. Fit the model to Table 10.4, and interpret the results.

10.36 A nonmodel-based ordinal measure of marginal heterogeneity is

$$\hat{\Delta} = \sum\sum_{i<j} p_{i+}p_{+j} - \sum\sum_{i>j} p_{i+}p_{+j}.$$

    a. Show $\hat{\Delta}$ estimates $\Delta = P(Y > X) - P(X > Y)$, where X is selected at random from $\{\pi_{i+}\}$, and Y is selected independently at random from $\{\pi_{+j}\}$.

    b. Show that when Y is stochastically higher than X, $\Delta > 0$, and that marginal homogeneity implies $\Delta = 0$.

    c. Show the estimated asymptotic variance of $\hat{\Delta}$ is

$$\hat{\sigma}^2(\hat{\Delta}) = \left[\sum\sum \hat{\phi}_{ij}^2 p_{ij} - \left(\sum\sum \hat{\phi}_{ij}p_{ij}\right)^2\right] / n ,$$

where

$$\hat{\phi}_{ij} = \hat{F}_{j1} + \hat{F}_{j-1,1} - \hat{F}_{i2} - \hat{F}_{i-1,2} ,$$

and where $\{\hat{F}_{i1}\}$ and $\{\hat{F}_{i2}\}$ are the sample row and column marginal cdfs. (For large n, $z = \hat{\Delta}/\hat{\sigma}(\hat{\Delta})$ tests marginal homogeneity. When we display the marginal distributions in a $2 \times I$ table, $\hat{\Delta}$ equals the difference between discrete analogs of Mann-Whitney statistics.)

10.37 For ordered scores $\{u_i\}$ assigned to categories of an ordinal scale, let $\bar{X} = \Sigma_i u_i p_{i+}$ and $\bar{Y} = \Sigma_i u_i p_{+i}$ denote sample marginal means.

    a. Show that marginal homogeneity implies $E(\bar{X}) = E(\bar{Y})$.

    b. Show the variance of $(\bar{X} - \bar{Y})$ can be estimated by

$$\hat{\sigma}^2(\bar{X} - \bar{Y}) = \frac{1}{n}\left\{\sum_i\sum_j (u_i - u_j)^2 p_{ij} - (\bar{X} - \bar{Y})^2\right\}.$$

    c. Use the sample means to construct a test of marginal homogeneity. Against what alternatives is this test sensitive? (See Bhapkar 1970, Fleiss and Everitt 1971, Koch and Reinfurt 1971, and Meeks and D'Agostino 1983.)

10.38 Consider the multiplicative model

$$\pi_{ij} = \alpha_i \alpha_j (1 - \beta), \quad i \neq j$$
$$= \alpha_i^2 + \beta \alpha_i (1 - \alpha_i), \quad i = j.$$

a. Show the model satisfies (1) symmetry, (2) marginal homogeneity, (3) quasi symmetry, (4) quasi independence.

b. Show that $\alpha_i = \pi_{i+} = \pi_{+i}, i = 1, \ldots, I$.

c. Show that $\beta = $ Cohen's kappa, and interpret $\kappa = 0$ and $\kappa = 1$ for this model.

10.39 Refer to the Bradley–Terry model.

a. Show that $\log(\Pi_{ik}/\Pi_{ki}) = \log(\Pi_{ij}/\Pi_{ji}) + \log(\Pi_{jk}/\Pi_{kj})$.

b. With this model, is it possible that i could be preferred to j (i.e., $\Pi_{ij} > \Pi_{ji}$) and j could be preferred to k, yet k could be preferred to i?

10.40 Give the log likelihood function for the Bradley–Terry model. From the kernel, show that (given $\{N_{ij}\}$) the minimal sufficient statistics are $\{n_{i+}\}$.

10.41 Refer to Table 10.8. Let m_{ijk} denote the expected frequency for outcome k ($k = 1$, win; $k = 2$, lose) for home team i when it plays away team j. Find the loglinear model for $\{m_{ijk}\}$ that is equivalent to logit model (10.31). Check that residual df $= I(I - 2)$, as in the logit model. Show that $\lambda_1^Z - \lambda_2^Z$ in the loglinear model represents the home-team advantage.

CHAPTER 11

Analyzing Repeated Categorical Response Data

Many studies observe the response variable for each subject at *several* occasions—for instance, at several time points or under several conditions. Repeated categorical response data occur commonly in health-related applications. For example, a clinician might evaluate patients at weekly intervals regarding whether a new drug treatment is successful. When the response is observed for each subject at T occasions, a T-dimensional contingency table cross-classifies the T responses. The "occasions" for a repeated response need not refer to different times. For instance, a biomedical response might be measured at T locations on a subject's body, or the dependent responses may refer to a matched set of T subjects. When the occasions refer to times, repeated measurement data are often called *longitudinal* data.

This chapter generalizes methods of the previous chapter, which apply when $T = 2$. Section 11.1 describes generalizations of symmetry, quasi symmetry, and marginal homogeneity for repeated categorical responses. Section 11.2 presents ways of testing marginal homogeneity in T-way tables.

Many longitudinal studies involve comparing different groups over time. In comparing a new drug and a standard drug for treating some chronic disease, a study might use two randomly selected groups of subjects, one group to receive each drug. Responses on a categorical measure of effectiveness might then be compared each week, for a several week period. Sections 11.3 and 11.4 describe models for comparing marginal distributions of T-dimensional tables for different groups. Section 11.3 treats the nominal-response case and Section 11.4 the ordinal-response case. The final section introduces Markov chain models, and applies categorical-data methods to obtain statistical inferences for them.

11.1 SYMMETRY

The definition of symmetry for square tables extends readily to $T > 2$ dependent responses. When the response has I categories, a contingency table with I^T cells summarizes frequencies of the possible responses at the T occasions. Let $\mathbf{i} = (i_1, \ldots, i_T)$ denote the cell in this table corresponding to response i_g at occasion g, $g = 1, \ldots, T$. Let $\pi_{\mathbf{i}}$ denote the probability of classification in this cell, and let $m_{\mathbf{i}} = n\pi_{\mathbf{i}}$ denote the expected frequency. Let $\phi_h(g)$ denote the probability of response h at occasion g. Then

$$\phi_h(g) = \pi_{+ \cdots + h + \cdots +}$$

where the h subscript is in position g, and $\{\phi_h(g), h = 1, \ldots, I\}$ form the gth marginal distribution of the table.

11.1.1 Complete Symmetry and Marginal Homogeneity

There is *complete symmetry* in the T-way table if

$$\pi_{\mathbf{i}} = \pi_{\mathbf{j}} \tag{11.1}$$

for any permutation $\mathbf{j} = (j_1, \ldots, j_T)$ of $\mathbf{i} = (i_1, \ldots, i_T)$. There is *marginal homogeneity* if

$$\phi_h(1) = \phi_h(2) = \cdots = \phi_h(T), \quad \text{for } h = 1, \ldots, I. \tag{11.2}$$

Complete symmetry implies marginal homogeneity, but the converse does not hold except when $T = I = 2$.

For $T = 3$ occasions, for instance, there is complete symmetry if

$$\pi_{ijk} = \pi_{ikj} = \pi_{jik} = \pi_{jki} = \pi_{kij} = \pi_{kji}$$

for all i, j, and k. This corresponds to the loglinear model

$$\log m_{ijk} = \mu + \lambda_i + \lambda_j + \lambda_k + \lambda_{ij} + \lambda_{ik} + \lambda_{jk} + \lambda_{ijk}$$

where the λ terms are identical for any permutations of the subscripts. The ML fitted values are

$$\hat{m}_{ijk} = (n_{ijk} + n_{ikj} + n_{jik} + n_{jki} + n_{kij} + n_{kji})/6.$$

In particular, $\hat{m}_{iii} = n_{iii}$, $i = 1, \ldots, I$. The residual degrees of freedom equal df $= I(I-1)(5I+2)/6$ (see Note 11.1). The marginal probabilities are $\phi_h(1) = \Sigma_j \Sigma_k \pi_{hjk} = \pi_{h++}$, and similarly $\phi_h(2) = \pi_{+h+}$ and $\phi_h(3) = \pi_{++h}$, $h = 1, \ldots, I$. There is marginal homogeneity if

$$\pi_{h++} = \pi_{+h+} = \pi_{++h} \quad \text{for } h = 1, \ldots, I.$$

11.1.2 Quasi Symmetry

An I^T table satisfies *quasi symmetry* if there are T sets $\{\alpha_{gj}, j = 1, \ldots, I\}$, and parameters $\{\gamma_i\}$ that are identical for all permutations of $\mathbf{i} = (i_1, \ldots, i_T)$, such that

$$\pi_{\mathbf{i}} = \alpha_{1i_1} \alpha_{2i_2} \cdots \alpha_{Ti_T} \gamma_{\mathbf{i}}. \tag{11.3}$$

The quasi-symmetry model has symmetric association and higher-order interaction terms, but permits each single-factor marginal distribution to have its own parameters. For identifiability, we impose constraints such as $\Pi_j \alpha_{gj} = 1$ (i.e., $\Sigma_j \log(\alpha_{gj}) = 0$). This model has $(I-1)(T-1)$ more parameters than the complete symmetry model, which is the special case $\{\alpha_{1j} = \cdots = \alpha_{Tj}, j = 1, \ldots, I-1\}$. Equation (11.3) generalizes (10.14) for square tables.

For $T = 3$, the quasi-symmetry model has loglinear form

$$\log m_{ijk} = \mu + \lambda_i^X + \lambda_j^Y + \lambda_k^Z + \lambda_{ij} + \lambda_{ik} + \lambda_{jk} + \lambda_{ijk}$$

where the two-factor and three-factor terms are identical for all permutations of indices. Its likelihood equations are

$$\hat{m}_{i++} = n_{i++}, \qquad \hat{m}_{+j+} = n_{+j+}, \qquad \hat{m}_{++k} = n_{++k}$$

$$\hat{m}_{ijk} + \hat{m}_{ikj} + \hat{m}_{jik} + \hat{m}_{jki} + \hat{m}_{kij} + \hat{m}_{kji}$$

$$= n_{ijk} + n_{ikj} + n_{jik} + n_{jki} + n_{kij} + n_{kji}$$

for all i, j, and k. The model can be fitted using IPF or Newton–Raphson iterative methods.

11.1.3 Attitudes Toward Abortion Example

Table 11.1, reported by Clogg and Shockey (1988), is taken from the 1982 General Social Survey. The response pertains to whether abortion should be legal. Each subject responded to this under four conditions—if there is a strong chance of a defect in the baby (D), if the woman is

Table 11.1 Attitude toward Legalized Abortion

Defect	Rape	Unmarried	Poor	Count
Yes	Yes	Yes	Yes	605
Yes	Yes	Yes	No	68
Yes	Yes	No	Yes	91
Yes	Yes	No	No	320
Yes	No	Yes	Yes	1
Yes	No	Yes	No	0
Yes	No	No	Yes	3
Yes	No	No	No	45
No	Yes	Yes	Yes	7
No	Yes	Yes	No	3
No	Yes	No	Yes	7
No	Yes	No	No	54
No	No	Yes	Yes	2
No	No	Yes	No	0
No	No	No	Yes	3
No	No	No	No	125

Source: 1982 General Social Survey; see Clogg and Shockey (1988).

pregnant as a result of rape (R), if the woman is not married (M), or if the woman is poor (P). The four responses on the same classification results in a 2^4 contingency table.

The complete symmetry model for Table 11.1 has $G^2 = 1334.9$ and $X^2 = 1735.0$, based on df = 11. The table contains a mixture of very large and very small counts, and goodness-of-fit statistics give only crude indices of lack of fit. Nevertheless, this model is clearly inapproriate. For instance, 320 subjects believe abortion should be legal in cases D and R but not M or P, whereas two subjects believe it should be legal in cases M or P but not D or R. The complete symmetry model requires these response patterns to be equally likely.

The quasi-symmetry model has $G^2 = 33.2$ and $X^2 = 63.2$, based on df = 8. Though much better than complete symmetry, it does not fit well. A quick look at the association structure of the table reveals why. The no-three-factor interaction model (DR, DM, DP, RM, RP, MP) has $G^2 = 6.6$ and $X^2 = 8.8$, based on df = 5. The estimated conditional odds ratios for this model are $\hat{\theta}_{DR} = 17.2$, $\hat{\theta}_{DM} = 4.6$, $\hat{\theta}_{DP} = 2.5$, $\hat{\theta}_{RM} = 14.0$, $\hat{\theta}_{RP} = 3.8$, $\hat{\theta}_{MP} = 31.7$. There is considerable variety in the strength of association. The quasi-symmetry model forces these associations to be identical.

11.1.4 Types of Marginal Symmetry

A more general type of symmetry for an I^T table has marginal homogeneity and complete symmetry as special cases. Let

$$\phi_{h_1 h_2 \cdots h_k}(g_1, g_2, \ldots, g_k)$$

denote the probability of response h_j for occasion g_j, $j = 1, \ldots, k$, where k is between 1 and T. This is a k-dimensional marginal probability, the case $k = 1$ giving the single-variable marginal probabilities previously studied. There is kth-order marginal symmetry if for all k-tuple choices of $\mathbf{h} = (h_1, \ldots, h_k)$, ϕ is the same for every permutation of $\mathbf{h}$ and for all combinations $\mathbf{g} = (g_1, \ldots, g_k)$ of k of the T occasions.

For $k = 1$, first-order marginal symmetry is simply marginal homogeneity: For each h, $\phi_h(g)$ is the same for all g. For $k = 2$, $\phi_{hi}(a, b)$ denotes the probability of response h at occasion a and response i at occasion b. The $\{\phi_{hi}(a, b)$ for $h = 1, \ldots, I$ and $i = 1, \ldots, I\}$ are a two-dimensional marginal table of the full T-way table. There is second-order marginal symmetry if, for all h and i, $\phi_{hi}(a, b) = \phi_{ih}(a, b)$ and is the same for all pairs of occasions (a, b). In other words, the two-way marginal tables exhibit symmetry, and they are identical. For $k = T$, Tth order marginal symmetry in an I^T table is complete symmetry.

If there is kth-order marginal symmetry, then there is jth-order marginal symmetry for any $j < k$. For instance, complete symmetry implies second-order marginal symmetry, which itself implies marginal homogeneity. Though this hierarchy is mathematically attractive, the higher-order symmetries are too restrictive to hold very often in practice.

11.2 MARGINAL HOMOGENEITY

In many applications, modeling the full multivariate dependence among the T repeated responses is of less interest than comparing the T first-order marginal distributions of the response. For instance, in treating a chronic disease with some drug, the primary goal might be to study whether the probability of successful treatment changes over the T weeks of a treatment period. The T success probabilities refer to the T binary first-order marginal distributions. The hypothesis that they are identical corresponds to marginal homogeneity of these T distributions.

For square tables with nominal scales, we can test marginal homogeneity in a variety of ways—by comparing symmetry and quasi-symmetry models, by comparing maximized log likelihoods with and without the constraint of marginal homogeneity, or by constructing a quadratic form involving differences in matched marginal proportions (see Section 10.3). This section generalizes these tests to $T > 2$ occasions.

A test proposed by Cochran (1950) is often used to compare one-dimensional margins of a 2^T contingency table. We begin by studying this test, which applies to T matched samples on a binary response.

11.2.1 Cochran's Q Test

For the gth occasion, let $f_h(g)$ denote the sample proportion of responses in category h; that is, the sample version of $\phi_h(g)$. When $I = 2$, $f_1(g) + f_2(g) = 1$ for $g = 1, \ldots, T$. Denote the average proportion of responses in the first category, for the T occasions, by $\bar{f}_1 = \Sigma_g f_1(g)/T$. For the ith subject in the sample, let q_i denote the proportion of occasions for which that subject makes the first response, $i = 1, \ldots, n$. Cochran's statistic is

$$Q = \frac{n^2(T-1)}{T} \frac{\displaystyle\sum_{g=1}^{T} [f_1(g) - \bar{f}_1]^2}{\displaystyle\sum_{i=1}^{n} q_i(1 - q_i)}. \tag{11.4}$$

Problem 11.22 shows that $\{q_i\}$ satisfy $\Sigma_i q_i/n = \bar{f}_1$, and outlines Cochran's derivation of this statistic.

For testing marginal homogeneity, Cochran's statistic is an approximate chi-squared statistic with df $= T - 1$. When $T = 2$ it simplifies to McNemar's statistic $(n_{12} - n_{21})^2/(n_{12} + n_{21})$ for comparing dependent proportions. For $T > 2$, Bhapkar (1970a) showed the asymptotic chi-squared distribution applies if and only if

$$\phi_{11}(a, b) \text{ is the same for all pairs } a \text{ and } b. \tag{11.5}$$

Given (11.5), marginal homogeneity is equivalent to second-order marginal symmetry. Thus, Cochran's statistic actually tests the null hypothesis of second-order marginal symmetry against the alternative that (11.5) holds without marginal homogeneity. When used to test marginal homogeneity, Cochran's test assumes condition (11.5). In this sense it is analogous to model-based tests, such as the test of marginal homogeneity under the assumption of quasi symmetry.

11.2.2 Testing Homogeneity for a Nominal Classification

When the quasi-symmetry model (11.3) holds, marginal homogeneity is equivalent to complete symmetry. There is marginal heterogeneity if quasi symmetry (QS) holds but complete symmetry (S) does not. When the quasi-symmetry model fits well, we can test marginal homogeneity by

$$G^2(S \mid QS) = G^2(S) - G^2(QS) . \qquad (11.6)$$

When there is complete symmetry, this statistic has an asymptotic chi-squared distribution with df $= (I-1)(T-1)$.

Let $L(\hat{\boldsymbol{\pi}})$ denote the multinomial log likelihood evaluated at the unconditional ML estimate of $\boldsymbol{\pi}$, and let $L(\hat{\boldsymbol{\pi}}_{MH})$ denote the maximized log likelihood subject to the constraint of marginal homogeneity. Madansky (1963) and Lipsitz (1988) tested marginal homogeneity using

$$-2[L(\hat{\boldsymbol{\pi}}_{\mathrm{MH}}) - L(\hat{\boldsymbol{\pi}})] .$$

This statistic has the asymptotic chi-squared distribution whenever there is marginal homogeneity, regardless of whether there is quasi symmetry or second-order marginal symmetry.

A similar test, discussed by Bhapkar (1970b, 1973) and Darroch (1981), uses the asymptotic normality of the unconditional ML estimators of marginal probabilities. Let

$$\bar{f}_h = \sum_g f_h(g)/T , \qquad d_h(g) = f_h(g) - \bar{f}_h ,$$

and let $\mathbf{d}$ denote the vector of $\{d_h(g), g = 1, \ldots, T-1, h = 1, \ldots, I-1\}$. Let $\hat{\mathbf{V}}$ denote the unconditional ML estimator of the covariance matrix of $\sqrt{n}\mathbf{d}$. Darroch gave the elements of $\hat{\mathbf{V}}$. The test statistic, a generalization of Bhapkar's statistic (10.20), is the Wald statistic

$$W = n\mathbf{d}'\hat{\mathbf{V}}^{-1}\mathbf{d} . \qquad (11.7)$$

Under the null hypothesis of marginal homogeneity, W also has an asymptotic chi-squared distribution with df $= (I-1)(T-1)$.

These tests treat the response classification as nominal. Section 11.4 presents statistics that are sensitive to stochastic ordering alternatives for ordered categories.

11.2.3 Single Population Drug Comparison Example

We illustrate tests of marginal homogeneity using Table 11.2, taken from Grizzle et al. (1969) and analyzed in several articles, including Bhapkar (1979) and Imrey et al. (1982). Each subject in the sample is classified as having favorable or unfavorable reaction after symptomatic treatment of

Table 11.2 Responses to Three Drugs

	Drug A Favorable		Drug A Unfavorable	
	B Favorable	B Unfavorable	B Favorable	B Unfavorable
C Favorable	6	2	2	6
C Unfavorable	16	4	4	6

Source: Reprinted with permission from the Biometric Society (Grizzle et al. 1969).

a chronic condition by each of three drugs. The table has size 2^3, giving the (favorable, unfavorable) classification for reaction to drug A in the first dimension, drug B in the second dimension, and drug C in the third dimension. We assume the drugs have no carry-over effects, and that the severity of the chronic condition remains stable for each subject over the course of the experiment. These assumptions are reasonable for a chronic condition such as migraine headache.

The sample proportion of favorable responses was 0.61 for drug A, 0.61 for drug B, and 0.35 for drug C. Madansky's likelihood-ratio approach gives test statistic 5.95 for testing marginal homogeneity, based on df = 2. The W statistic (11.7) equals 6.58. These statistics have P-value about 0.05. There is some evidence that the proportion of favorable responses is lower for drug C. From formula (10.2), the estimate 0.26 of the difference between drugs A and C in the proportion of favorable responses has an estimated standard error of 0.108. A 95% confidence interval for the true difference is (0.05, 0.47). The same interval holds for comparison of drugs B and C.

For these data, Cochran's Q statistic equals 8.5, and $G^2(S \mid QS) = 18.9 - 10.4 = 8.5$ based on df $= 4 - 2 = 2$. The use of these statistics is somewhat dubious, since assumptions on which the null distribution is chi-squared do not seem to be fulfilled. For instance, the quasi-symmetry model fits poorly.

11.2.4 Marginal Homogeneity and Mantel–Haenszel Tests

Some statistics for testing marginal homogeneity are special cases of the generalized Cochran–Mantel–Haenszel (CMH) statistic (Section 7.4.6). The CMH statistic applies to n strata of $T \times I$ tables. The kth stratum gives the T responses for subject k. The gth row in that stratum contains a 1 in the column that is the response at occasion g, and 0 in all other columns. For instance, Table 11.3 shows a stratum for a subject classified

Table 11.3   Stratum for Subject Classified in (A favorable, B favorable, C unfavorable) Cell of Table 11.2

	Response	
Drug	Favorable	Unfavorable
A	1	0
B	1	0
C	0	1

favorable for drugs A and B and unfavorable for drug C. The set of strata for the n subjects forms a $T \times I \times n$ table. Observations in different strata are statistically independent.

In formulating probability distributions for this setup, we might allow the response distribution to vary by subject and occasion, such as in Cox's logistic model for $T = 2$. Suppose, for each subject, that the T response probability distributions are identical; that is, there is conditional independence in the $T \times I \times n$ table between occasion and response, given subject. This implies there is marginal homogeneity for the original I^T table. Thus, we can test marginal homogeneity using the CMH statistic for testing conditional independence in the $T \times I \times n$ table.

For the CMH approach, the row totals in each stratum equal 1, and we condition also on the column totals. These totals give the number of responses of each type observed over the T occasions. Under the null hypothesis of independence within each stratum, we assume the indicator random variables in each stratum follow a multiple hypergeometric distribution.

When the response scale has $I = 2$ categories and there are $T = 2$ occasions, this approach gives the Cochran–Mantel–Haenszel statistic (7.8) without the continuity correction, for a $2 \times 2 \times n$ table. That statistic is identical to the McNemar statistic. That is, McNemar's test is a special case of the Cochran–Mantel–Haenszel test applied to n matched pairs. In addition, the Cochran–Mantel–Haenszel estimator of a common odds ratio (Section 7.5.1) for the $2 \times 2 \times n$ table simplifies to the ML estimator n_{12}/n_{21} derived in Section 10.1.4 for the logistic model for matched pairs. When $I = 2$ and $T \geq 2$, the generalized CMH statistic is identical to Cochran's Q statistic. For further details, see Mantel and Byar (1978), Darroch (1981), and White et al. (1982).

11.3 MODELING A REPEATED CATEGORICAL RESPONSE

The previous section presented tests for differences among marginal distributions of a repeated response. Next we introduce methods for *modeling* how the response changes. We expand our scope to incorporate comparisons of groups, such as different treatments.

Modeling the marginal distributions provides answers to questions such as "Does the response tend to improve with time?" or "At a particular occasion, are there differences among the response distributions for groups A and B?" or "Is the difference in response distributions for A and B the same at each occasion?" Models presented in this section describe how covariates affect these marginal distributions. The discussion is partly based on an article by Koch et al. (1977).

At occasion g, let $\phi_h(g; \mathbf{x})$ denote the probability of response h, when a set of covariates takes value $\mathbf{x}$. When $\mathbf{x}$ is categorical, different levels of $\mathbf{x}$ usually represent subpopulations whose response distributions we want to compare. The probabilities $\{\phi_1(g; \mathbf{x}), \ldots, \phi_I(g; \mathbf{x})\}$ form the gth marginal distribution for the I^T table for subpopulation $\mathbf{x}$.

A primary goal in some studies is to analyze how the marginal probabilities depend on $\mathbf{x}$, for fixed g. At each occasion g, there are no differences among the subpopulations if for all h, $\phi_h(g; \mathbf{x})$ is the same for all $\mathbf{x}$. In other cases a primary goal is to analyze how the marginal distribution changes across the occasions $g = 1, \ldots, T$, for fixed $\mathbf{x}$. For each subpopulation $\mathbf{x}$, there are no differences across occasions if

$$\phi_h(1; \mathbf{x}) = \phi_h(2; \mathbf{x}) = \cdots = \phi_h(T; \mathbf{x}) \quad \text{for } h = 1, \ldots, I.$$

This section formulates models for which these two types of homogeneity occur as special cases. Interpretations depend on whether the difference between marginal distributions for two covariate values is the same for all occasions; that is, on whether there is occasion × covariate interaction.

11.3.1 Logit Models for Repeated Binary Response

Suppose the response variable has $I = 2$ categories, and suppose subpopulations are indexed by $i = 1, \ldots, s$. Denote the logit for the response distribution of the ith subpopulation at the gth occasion by

$$L(g; i) = \log[\phi_1(g; i)/\phi_2(g; i)].$$

The model

$$L(g; i) = \alpha + \beta_i^S + \beta_g^O \qquad (11.8)$$

permits the response distribution to vary by S = subpopulation and by O = occasion, but assumes a lack of interaction. Identifiability requires constraints such as $\Sigma \, \beta_i^S = 0$ and $\Sigma \, \beta_g^O = 0$, or $\beta_s^S = \beta_T^O = 0$.
For this model, there are no differences among subpopulations, for each occasion, when $\beta_1^S = \cdots = \beta_s^S$. There is marginal homogeneity across occasions, for each subpopulation, if $\beta_1^O = \cdots = \beta_T^O$. We can readily adapt model (11.8) to permit alternative structure for the relationship. For instance, if occasions are ordered (such as times or dosages), we might use linear occasion effects $\{\beta_g^O = \beta v_g\}$ for fixed scores $\{v_g\}$.

Models for marginal distributions of a repeated binary response generalize readily to responses having more than two categories. For instance, generalized logit models describe the odds of making each response, relative to a baseline response. We could construct a model that describes how generalized logits of marginal distributions depend on occasions and covariates.

It is not simple to use ML estimation to fit models such as (11.8). The model utilizes marginal probabilities rather than the cell probabilities to which the multinomial likelihood refers. The marginal counts do not have independent multinomial distributions. ML estimation requires an iterative routine for maximizing the full-table likelihood, subject to the constraint that the marginal distributions satisfy the model. Aitchison and Silvey (1958) described such routines. Koch et al. (1977) used weighted least squares (WLS) to fit models of this type. Section 13.3 describes the WLS approach to model-fitting for categorical data. We used WLS to analyze the following data set, taken from the article by Koch et al.

11.3.2 Longitudinal Growth Curve Example

This longitudinal study compares a new drug and a standard drug for treatment of subjects suffering mental depression. Subjects were placed into two diagnosis groups, according to whether degree of depression was mild or severe. In each diagnosis group, subjects were randomly assigned to one of two treatments. The first treatment used the standard drug and the second used the new drug. Following 1 week, 2 weeks, and 4 weeks of treatment, each subject's suffering from mental depression was classified as normal or abnormal. Table 11.4 exhibits the data. There are four subpopulations, the combinations of treatment and prior diagnosis. Since the binary response (normal, abnormal) is observed at $T = 3$ occasions, the contingency table has size $2 \times 2 \times 2^3 = 4 \times 2^3$.

Let $\phi_1(g; hi)$ denote the probability the response is normal at occasion

Table 11.4 Cross-Classification of Responses at Three Times (N = Normal, A = Abnormal) by Diagnosis and Treatment

		Response at Three Times							
Diagnosis	Treatment	NNN	NNA	NAN	NAA	ANN	ANA	AAN	AAA
Mild	Standard	16	13	9	3	14	4	15	6
Mild	New Drug	31	0	6	0	22	2	9	0
Severe	Standard	2	2	8	9	9	15	27	28
Severe	New Drug	7	2	5	2	31	5	32	6

Source: Reprinted with permission from the Biometric Society (Koch et al. 1977).

g ($g = 1$, Week 1; $g = 2$, Week 2; $g = 3$, Week 4), for diagnosis h ($h = 1$, Mild; $h = 2$, Severe) and for treatment i ($i = 1$, Standard; $i = 2$, New Drug). Let $L(g; hi)$ denote its logit. Koch et al. noted that if the occasion (time) metric reflects cumulative dosage of the drugs, then a logit scale often has a linear effect for the logarithm of time. Hence they used scores $\{v_1 = 0, v_2 = 1, v_3 = 2\}$, the logs to base 2 of week number, for the three occasions.

There are 12 marginal distributions, since observations occur at three occasions for each of four subpopulations. Table 11.5 contains sample values $\{f_1(g; hi)\}$ of $\{\phi_1(g; hi)\}$ for these 12 cases. The sample proportion of normal responses (1) increased over time for each subpopulation, (2) increased over time at a faster rate for the new drug than for the standard drug, for each fixed diagnosis, and (3) was higher for the mild diagnosis than for the severe diagnosis, for each treatment at each occasion.

First, consider the model

$$L(g; hi) = \alpha + \beta_h^D + \beta_i^T + \beta v_g$$

where $\{\beta_h^D\}$ and $\{\beta_i^T\}$ denote diagnosis and treatment main effects. This model assumes a linear occasion effect β that is the same for each subpopulation, and it assumes a lack of interaction between diagnosis and treatment. The WLS chi-squared statistic for testing its fit is 31.1. Since there are 12 logits (one for each marginal distribution) and 4 parameters, residual df = 8. This model does not fit well, reflecting the sample's faster rate of improvement for the new drug.

The more complex model

$$L(g; hi) = \alpha + \beta_h^D + \beta_i^T + \beta v_g + \gamma_i v_g$$

Table 11.5 Sample Proportions and Predicted Probabilities of Normal Response[a]

Diagnosis	Treatment	Sample Proportion			Predicted Probability		
		Week 1	Week 2	Week 4	Week 1	Week 2	Week 4
Mild	Standard	0.51(0.06)	0.59(0.06)	0.68(0.05)	0.49(0.04)	0.61(0.03)	0.71(0.03)
	New Drug	0.53(0.06)	0.79(0.05)	0.97(0.02)	0.48(0.05)	0.80(0.03)	0.95(0.01)
Severe	Standard	0.21(0.04)	0.28(0.04)	0.46(0.05)	0.21(0.03)	0.30(0.02)	0.41(0.04)
	New Drug	0.18(0.04)	0.50(0.05)	0.83(0.04)	0.21(0.03)	0.53(0.03)	0.83(0.03)

[a]Values in parentheses are ASEs of estimated proportions.

Table 11.6   Parameter Estimates for Model Fitted to Table 11.4

Parameter	Estimate	ASE
α	-1.382	0.185
β_1^D	1.282	0.146
β_1^T	0.052	0.228
β (Occ)	1.474	0.154
γ_1 (Occ $\times$ T)	-0.994	0.192

permits the linear occasion effect to differ for the two treatments. This model fits well, with goodness-of-fit statistic of 4.2 based on df $= 7$. Table 11.6 contains WLS parameter estimates, using constraints $\beta_2^D = \beta_2^T = \gamma_2 = 0$.

The estimated odds of normal response are $\exp(1.282) = 3.6$ times higher when the diagnosis is mild than when it is severe, for each treatment $\times$ occasion combination. The estimated occasion effects are $\hat{\beta} + \hat{\gamma}_1 = 0.480$ (ASE $= 0.115$) for the standard treatment, and $\hat{\beta} = 1.474$ (ASE $= 0.154$) for the new treatment. The estimated difference in slopes of -0.994 (ASE $= 0.192$) shows strong evidence that the true rate of improvement is faster for the new drug. The estimate 0.052 of β_1^T (ASE $= 0.228$) indicates an insignificant difference between the treatments after 1 week (for which $v_1 = 0$). At occasion g, the estimated odds of normal response are $\exp(-0.052 + 0.994 v_g)$ times higher for the new drug than for the standard one, for each prior diagnosis.

The parameter estimates yield predicted logits, from which we can predict probabilities of a normal response. Table 11.5 lists these, and shows the predicted values are quite close to the sample values. The ASEs of the model-based probability estimates are smaller, due to the smoothing provided by the model (Section 6.4.4).

11.3.3   Rasch Model

A more complex way of modeling repeated categorical data still focuses on marginal distributions but incorporates subject effects. Let Y_{gh} denote the response at occasion g for subject h. We permit a separate response distribution for each subject at each occasion.

Rasch (1961) proposed a logistic model for a binary response that uses subject effects in addition to occasion effects. A common application of the Rasch model is for responses of a sample of subjects to a battery of T questions on an exam. The Rasch model describes how the probability of correct response depends on the subject's overall ability and the difficulty of the question.

For question g and subject h, let $Y_{gh} = 1$ for a correct response and 0 for an incorrect response, $g = 1, \ldots, T$ and $h = 1, \ldots, n$. Suppose we display the data in a $T \times 2 \times n$ table, as in the generalized CMH approach discussed in Section 11.2.4. There is a single observation at each subject $\times$ question combination. Letting $\phi_1(g; h) = P(Y_{gh} = 1)$, the Rasch model is

$$\text{Logit}[\phi_1(g; h)] = \alpha + \lambda_h + \beta_g ,$$

where λ_h represents the ability of the hth subject and β_g represents the easiness of the gth question. The probability of correct response increases as λ_h or β_g increases. When there are covariates, we add a $\boldsymbol{\beta}'\mathbf{x}$ term. This model assumes a lack of subject $\times$ question interaction.

The occasion (question) effects in the Rasch model differ from those in models discussed earlier in this section. For instance, ignoring subpopulations, model (11.8) has form

$$\text{Logit} = \alpha + \beta_g^O ,$$

which refers to the occasion $\times$ response marginal table, collapsed over subjects. The $\{\beta_g^O\}$ in (11.8) differ from $\{\beta_g\}$ in the Rasch model. For $T = 2$, we interpret

$$\beta_1 - \beta_2 = \text{logit}[P(Y_{1h} = 1)] - \text{logit}[P(Y_{2h} = 1)] ,$$

whereas in model (11.8),

$$\beta_1^O - \beta_2^O = \text{logit}[P(Y_{1h} = 1)] - \text{logit}[P(Y_{2i} = 1)] ,$$

where subject h is randomly selected at occasion 1 and subject i is randomly selected at occasion 2 (i.e., h and i are *independent* observations). Effects in the Rasch model are *subject-specific* effects, rather than *population-averaged* effects.

Unfortunately, ML estimates are not consistent for the Rasch model, since the number of parameters has the same order as the number of subjects (Andersen 1980, p.244). The difficulty is similar to that of estimating a common odds ratio in several 2×2 tables when the number of strata is large and there are few observations per strata (Section 7.5). If we treat $\{\lambda_h\}$ as nuisance parameters and concentrate on estimation of $\{\beta_g\}$, we can obtain consistent estimates using a conditional ML approach. Let $S_h = \Sigma_g Y_{gh}$ denote the number of correct responses for subject h. Conditional on $\{S_h, h = 1, \ldots, n\}$, the distribution of the responses is independent of $\{\lambda_h\}$. The conditional ML estimates of $\{\beta_g\}$ are ones that maximize the conditional likelihood. The analysis is a generalization of the one given in Section 10.1.4 for Cox's logistic model for matched pairs. For further discussion, see Andersen (1980).

11.4 MODELING A REPEATED ORDINAL RESPONSE

For ordinal responses, Section 9.4 showed that cumulative logit models have simple interpretations, using cumulative odds ratios. This section formulates cumulative logit models for repeated ordinal responses.

11.4.1 Cumulative Logits for a Repeated Ordinal Response

At occasion g and covariate value $\mathbf{x}$, the marginal response distribution $\{\phi_j(g; \mathbf{x}), j = 1, \ldots, I\}$ has $I - 1$ cumulative logits,

$$L_j(g; \mathbf{x}) = \log\{[(\phi_1(g; \mathbf{x}) + \cdots + \phi_j(g; \mathbf{x})]/[\phi_{j+1}(g; \mathbf{x}) + \cdots + \phi_I(g; \mathbf{x})]\},$$

$$j = 1, \ldots, I - 1,$$

one for each response cutpoint. To begin, suppose there are no covariates. A proportional odds model is

$$L_j(g) = \alpha_j - \beta_g, \quad j = 1, \ldots, I - 1, \, g = 1, \ldots, T. \quad (11.9)$$

The cutpoint parameters $\{\alpha_j\}$ are usually nuisance parameters. The odds of making response above category j are $\exp(\beta_a - \beta_b)$ times greater at occasion a than at occasion b, for all j. The model implies a stochastic ordering of the marginal distributions at the T occasions. Since the model has $(I - 1)T$ cumulative logits and $(I - 1) + (T - 1)$ parameters, its residual df equal $(T - 1)(I - 2)$, unsaturated whenever $I > 2$.

Marginal homogeneity is the special case of (11.9) in which $\beta_1 = \cdots = \beta_T$. Likelihood-ratio or Wald tests of this hypothesis utilize the ordering of response categories and have an asymptotic chi-squared distribution with df $= T - 1$. When the model holds, these tests are asymptotically more powerful than the ones described in Section 11.2.2, which treat the classification as nominal and have df $= (I - 1)(T - 1)$. For $T = 2$, these tests of marginal homogeneity are alternatives to the ordinal tests described in Section 10.4.3.

We could include covariates in the usual manner. For instance, to compare s subpopulations across T occasions, the model

$$L_j(g; i) = \alpha_j - \beta_g^O - \beta_i^S \qquad (11.10)$$

has main effects for occasions and subpopulations that are the same for each cutpoint, $j = 1, \ldots, I - 1$. More general models permit interaction between subpopulations and occasions.

11.4.2 Insomnia Example

Table 11.7, taken from Francom et al. (1989), shows results of a randomized, double-blind clinical trial comparing an active hypnotic drug with a placebo in patients with insomnia. The outcome variable is patient response to the question, "How quickly did you fall asleep after going to bed?", measured using categories (<20 minutes, 20–30 minutes, 30–60 minutes, and >60 minutes). Patients were asked this question before and following a two-week treatment period. The two treatments, Active and Placebo, are levels of a binary explanatory variable. The subjects receiving the two treatments were independent samples.

Table 11.8 contains sample marginal distributions for the four combinations of treatment and occasion. From the initial to follow-up occasion, the distribution of time to falling asleep seems to shift downwards both for active and placebo treatments. The degree of shift seems greater for the active treatment, indicating possible interaction. We consider a proportional odds model

Table 11.7 Time to Falling Asleep, by Treatment and Occasion

		Time to Falling Asleep			
	Initial	Follow-up			
Treatment		<20	20–30	30–60	>60
Active	<20	7	4	1	0
	20–30	11	5	2	2
	30–60	13	23	3	1
	>60	9	17	13	8
Placebo	<20	7	4	2	1
	20–30	14	5	1	0
	30–60	6	9	18	2
	>60	4	11	14	22

Source: Reprinted with permission from John Wiley & Sons, Ltd. (Francom et al. 1989).

Table 11.8   Observed and Fitted Marginal Distributions for Cumulative Logit Model

		Response			
Treatment	Occasion	<20	20–30	30–60	>60
Active	Initial	0.101	0.168	0.336	0.395
		(0.102)	(0.184)	(0.303)	(0.411)
	Follow-up	0.336	0.412	0.160	0.092
		(0.385)	(0.303)	(0.200)	(0.111)
Placebo	Initial	0.117	0.167	0.292	0.425
		(0.098)	(0.179)	(0.301)	(0.421)
	Follow-up	0.258	0.242	0.292	0.208
		(0.239)	(0.286)	(0.273)	(0.202)

$$L_j(g; i) = \alpha_j - \beta_g^O - \beta_i^T - \gamma_{gi}. \qquad (11.11)$$

It permits interaction between occasion (initial, follow-up) and treatment (active, placebo), but assumes effects are the same for each response cutpoint.

For WLS fitting of the model, the residual chi-squared statistic is 7.4, based on df = 6. For the coding $\beta_2^C = \beta_2^T = \gamma_{12} = \gamma_{21} = \gamma_{22} = 0$, the WLS estimates are $\hat{\alpha}_1 = -1.16$, $\hat{\alpha}_2 = 0.10$, $\hat{\alpha}_3 = 1.37$, $\hat{\beta}_1^O = 1.05$ (ASE = 0.16), $\hat{\beta}_1^T = -0.69$ (ASE = 0.23), and $\hat{\gamma}_{11} = 0.65$ (ASE = 0.25). There is substantial evidence of interaction. At the initial observation, the estimated odds that time to falling asleep is below any fixed level is $\exp(0.69 - 0.65) = 1.04$ times as high for the active group as for the placebo; at the follow-up observation, the effect is $\exp(0.69) = 2.0$. In other words, the placebo and active groups had similar distributions of time to falling asleep at the initial observation, but at the follow-up observation the active group tended to fall asleep more quickly.

We can use fitted models to construct fitted marginal logits, and hence fitted marginal probabilities. Table 11.8 also contains the fitted marginal probabilities for model (11.11).

11.4.3   Alternative Approaches for Modeling Repeated Ordinal Responses

We can generalize the models discussed in this section by using an alternative link function for the cumulative probabilities, such as probit or complementary log-log. We could also use alternative logit transformations, such as continuation-ratio logits and adjacent-categories logits. Or, to achieve greater similarity with regression models for continuous variables, we could model the mean response for some fixed set of response scores, as in Section 9.6.

To illustrate, for Table 11.7, let $M(g; i)$ denote the mean time to falling asleep for occasion g and treatment i, $g = 1,2$, $i = 1, 2$. We fitted the model

$$M(g; i) = \alpha + \beta_g^O + \beta_i^T$$

using WLS, with response scores {10, 25, 45, 75} for time to falling asleep. There are four response means and three parameters, so the residual chi-squared of 9.3 is based on df = 1. Adding an interaction term gives a saturated model, whereby fitted means equal the observed ones. The initial means were 50.0 for the active group and 50.3 for the placebo, and the difference in means between the initial response and the follow-up was 22.2 for the active group and 13.0 for the placebo. The difference between these differences of means equals 9.2, with ASE = 3.0, indicating that the change was significantly greater for the active group.

A more complex way of modeling repeated ordered categorical data generalizes the Rasch model, permitting a separate response distribution for each subject at each occasion. Let Y_{gh} denote the response for subject h at occasion g, and let $L_j(gh)$ be the cumulative logit at cutpoint j for its distribution. The model

$$L_j(gh) = \alpha_j - \lambda_h - \nu_g \tag{11.12}$$

refers to the $T \times I \times n$ table of marginal distributions discussed in Section 11.2.4. When there are covariates, we add a $-\beta'x$ term.

Model (11.12) differs from model (11.9),

$$L_j(g) = \alpha_j - \beta_g$$

which refers to the occasion $\times$ response marginal table, collapsed over subjects. For $T = 2$, we interpret

$$\nu_2 - \nu_1 = \text{logit}[P(Y_{1h} \leq j)] - \text{logit}[P(Y_{2h} \leq j)] \,,$$

whereas

$$\beta_2 - \beta_1 = \text{logit}[P(Y_{1h} \leq j)] - \text{logit}[P(Y_{2i} \leq j)]$$

where subject h is randomly selected at occasion 1 and subject i is randomly selected at occasion 2. Effects in (11.12) are subject-specific, rather than population-averaged. The parameters do agree when there is marginal homogeneity. If all $\nu_g = 0$, then all $\beta_g = 0$, because of the marginal occasion × subject independence.

The distinction between testing marginal homogeneity by testing all $\beta_g = 0$ in (11.9) and testing all $\nu_g = 0$ in (11.12) parallels the distinction between population-averaged and subject-specific types of tests of marginal homogeneity for nominal variables. The first case is similar to Bhapkar's approach (11.7) of describing heterogeneity using the summary sample marginal distributions. That population-averaged approach uses subject-specific change only to estimate the covariance structure of differences in marginal proportions. The second case approaches the data in the subject-specific form used by Cochran–Mantel–Haenszel statistics, each stratum being a subject's matched set of observations.

Likelihood techniques are inappropriate for model (11.12), since the number of parameters has the same order as the number of subjects. For $T = 2$ and no covariates, McCullagh (1977) gave a conditional approach that eliminates the nuisance parameters $\{\lambda_h\}$.

11.4.4 Comparisons that Control for Initial Response

At each initial response, suppose the distribution for follow-up response is identical for treatment and placebo groups. If the marginal distributions for initial response are identical for active and placebo groups, then the follow-up marginal distributions are also identical. If the marginal distributions for initial response are not identical, however, the difference between follow-up and initial marginal distributions may differ for treatment and placebo groups, even though their conditional distributions for follow-up response are identical. Though models for marginal distributions can be useful for describing longitudinal effects when initial marginal distributions differ, they usually do not tell the whole story. It is useful also to construct models that enable us to control for the initial response in making comparisons.

Such models are simple to construct when there are only $T = 2$ occasions—for instance, when we want to model a follow-up response in terms of effects of explanatory variables, adjusting for an initial response. Let $L_j(hi)$ denote the jth cumulative logit for the follow-up response, for group i with baseline response h, and let $\{x_h\}$ be fixed scores for the baseline levels. The model

$$L_j(hi) = \alpha_j - \beta_i - \beta x_h \tag{11.13}$$

uses $\{\beta_i\}$ to compare the follow-up distributions for the groups, controlling for baseline observation. This model is an analog of an analysis of covariance model, in which the response and covariate are ordinal rather than continuous. It is a special case of the type of cumulative logit model presented in Section 9.4, since it refers to interior cells of the table rather than marginal distributions.

11.5 MARKOV CHAIN MODELS

For a given subject, let X_t denote the response on a categorical variable at time t, $t = 0, 1, \ldots, T$. The sequence $(X_0, X_1, X_2, \ldots)$ is an example of a *stochastic process*, an indexed family of random variables. The *state space* of the process is the set of possible values for X_t. The value X_0 is the *initial state*. When the state space is categorical and observations occur at a discrete set of times, the process $\{X_t, t = 0, 1, 2, \ldots\}$ has *discrete state space* and *discrete time*.

This section introduces discrete-time *Markov chains*, a simple type of stochastic process having discrete state space. Statistical inferences for Markov chains utilize basic methods of categorical data analysis. These inferences apply when we observe the stochastic process for a large number of subjects at the same set of times.

11.5.1 First-Order Markov Chains

A *Markov chain* is a stochastic process for which, for all t, the conditional distribution of X_{t+1}, given $X_0, \ldots, X_t$, is identical to the conditional distribution of X_{t+1}, given X_t. Thus, given X_t, X_{t+1} is conditionally independent of $X_0, \ldots, X_{t-1}$. When we know the present state of a Markov chain, information about past states does not help us make predictions about the future.

Let $f(x_0, \ldots, x_T)$ denote the joint probability mass function of $(X_0, \ldots, X_T)$. The factorization using conditional mass functions,

$$f(x_0, \ldots, x_T) = f(x_0)f(x_1 \mid x_0)f(x_2 \mid x_0 x_1) \ldots f(x_T \mid x_0 x_1 \ldots x_{T-1})$$

simplifies for Markov chains to

$$f(x_0)f(x_1 \mid x_0)f(x_2 \mid x_1) \ldots f(x_T \mid x_{T-1}). \tag{11.14}$$

From notation introduced in Section 11.1.4, the two-way marginal probabilities $\{\phi_{i,j}(t-1, t), i = 1, \ldots, I, j = 1, \ldots, I\}$ specify the joint distribution of X_{t-1} and X_t. The conditional probability the process is in state j at time t, given it is in state i at time $t-1$, is

$$\pi_{j|i}(t) = P[X_t = j \mid X_{t-1} = i] = \phi_{i,j}(t-1, t) / \left[\sum_k \phi_{i,k}(t-1, t) \right].$$

The $\{\pi_{j|i}(t), j = 1, \ldots, I\}$, which satisfy $\Sigma_j \pi_{j|i}(t) = 1$, are called *transition probabilities*. The $I \times I$ matrix $\{\pi_{j|i}(t), i = 1, \ldots, I, j = 1, \ldots, I\}$ is a transition probability matrix. It is called a *one-step* matrix, to distinguish it from the matrix of probabilities for k-step transitions from time $t - k$ to time t.

From (11.14), the joint distribution for a Markov chain depends only on one-step transition probabilities and the marginal distribution for the initial state. It also follows that the joint distribution satisfies loglinear model

$$(X_0 X_1, X_1 X_2, \ldots, X_{T-1} X_T).$$

When we have data on a large number of realizations of a stochastic process, we can form a contingency table displaying counts of the possible sequences. We can test whether the process satisfies the Markov property by testing the fit of this loglinear model.

Statistical inference for Markov chains uses standard methods of categorical data analysis. To illustrate, we derive ML estimators of transition probabilities. Let $n_{ij}(t)$ denote the observed number of transitions from state i at time $t - 1$ to state j at time t. For fixed t, $\{n_{ij}(t)\}$ form the two-way marginal table for dimensions $t - 1$ and t of the full I^{T+1} contingency table. For the $n_{i+}(t)$ subjects classified in category i at time $t - 1$, suppose the counts $\{n_{ij}(t), j = 1, \ldots, I\}$ have a multinomial distribution with parameters $\{\pi_{j|i}(t)\}$. Let $\{n_{i0} = n_{i+}(1)\}$ denote the initial counts, and suppose they also have a multinomial distribution, with parameters $\{\pi_{i0}\}$. If subjects behave independently, it follows from (11.14) that the likelihood function is proportional to

$$\left\{ \prod_{i=1}^{I} \pi_{i0}^{n_{i0}} \right\} \left\{ \prod_{t=1}^{T} \prod_{i=1}^{I} \left[\prod_{j=1}^{I} \pi_{j|i}(t)^{n_{ij}(t)} \right] \right\}. \tag{11.15}$$

The transition probabilities are parameters of IT independent multinomial distributions. It follows that the ML estimates are

$$\hat{\pi}_{j|i}(t) = n_{ij}(t) / n_{i+}(t)$$

(Anderson and Goodman 1957).

A stochastic process is a *kth-order Markov chain* if, for all t, the conditional distribution of X_{t+1}, given $X_0, \ldots, X_t$, is identical to the conditional distribution of X_{t+1}, given $(X_t, \ldots, X_{t-k+1})$. Given the states at the previous k times, the future behavior of the chain is independent of past behavior before those k times. The Markov chains studied above are *first-order* Markov chains.

11.5.2 Respiratory Illness Example

Table 11.9 refers to respiratory illness in children. The data are taken from a longitudinal study of health effects of air pollution, described by Ware et al. (1988). The children were examined annually at ages 9–12, and classified each time according to presence or absence of wheeze.

Denote the binary variable (wheeze, no wheeze) by X_i at age i, $i = 9$, 10, 11, 12. The loglinear model $(X_9X_{10}, X_{10}X_{11}, X_{11}X_{12})$ represents a first-order Markov chain. This model fits poorly, with $G^2 = 122.90$ based on df $= 8$. Given the state at time t, classification at time $t + 1$ depends on states at times previous to time t. The model $(X_9X_{10}X_{11}, X_{10}X_{11}X_{12})$ represents a second-order Markov chain, whereby there is conditional independence at ages 9 and 12, given states at ages 10 and 11. This model also fits poorly, with $G^2 = 23.86$ based on df $= 4$. The poor fits may partly reflect subject heterogeneity, since these analyses ignore several relevant covariates, such as parental smoking behavior.

The model $(X_9X_{10}, X_9X_{11}, X_9X_{12}, X_{10}X_{11}, X_{10}X_{12}, X_{11}X_{12})$ that permits association at each pair of ages fits well, with $G^2 = 1.46$ based on df $= 5$. Table 11.10 shows the ML estimates of pairwise conditional log odds ratios. The association seems similar for pairs of ages 1 year apart, and somewhat weaker for pairs of ages more than 1 year apart. The simpler model in which

Table 11.9 Results of Breath Test at Four Ages[a]

X_9	X_{10}	X_{11}	X_{12}	Count	X_9	X_{10}	X_{11}	X_{12}	Count
1	1	1	1	94	2	1	1	1	19
1	1	1	2	30	2	1	1	2	15
1	1	2	1	15	2	1	2	1	10
1	1	2	2	28	2	1	2	2	44
1	2	1	1	14	2	2	1	1	17
1	2	1	2	9	2	2	1	2	42
1	2	2	1	12	2	2	2	1	35
1	2	2	2	63	2	2	2	2	572

Source: Dr. James Ware and Dr. Stuart Lipsitz.

[a] "1" = wheeze and "2" = no wheeze.

Table 11.10   Estimated Conditional Log Odds Ratios for Table 11.9

Association	Estimate	Simpler Structure
$X_9–X_{10}$	1.81	1.75
$X_{10}–X_{11}$	1.65	1.75
$X_{11}–X_{12}$	1.85	1.75
$X_9–X_{11}$	0.95	1.04
$X_9–X_{12}$	1.05	1.04
$X_{10}–X_{12}$	1.07	1.04

$$\lambda_{ij}^{X_9 X_{10}} = \lambda_{ij}^{X_{10} X_{11}} = \lambda_{ij}^{X_{11} X_{12}}$$

and

$$\lambda_{ij}^{X_9 X_{11}} = \lambda_{ij}^{X_9 X_{12}} = \lambda_{ij}^{X_{10} X_{12}}$$

fits very well, with $G^2 = 2.27$ based on df $= 9$. The estimated log odds ratios are 1.75 in the first case, and 1.04 in the second.

11.5.3   Stationary Transition Probabilities

A first-order Markov chain has *stationary* transition probabilities if the one-step transition probability matrices are identical; that is, if for all i and j,

$$\pi_{j \mid i}(1) = \pi_{j \mid i}(2) = \cdots = \pi_{j \mid i}(T) = \pi_{j \mid i} .$$

Let $\mathbf{P} = (\pi_{j \mid i})$ denote the one-step transition probability matrix of a Markov chain having stationary transition probabilities. Then, the k-step transition probability matrix is $\mathbf{P}^k$. As k increases, if each entry in $\mathbf{P}$ is positive (and even under much weaker conditions), $\mathbf{P}^k$ converges to a matrix in which each row is identical. The limiting distribution within each row is an *equilibrium* distribution. It gives the "long-run" probability of being in a state, and that probability is the same for each initial state.

Let X, Y, and Z denote the classifications for the $I \times I \times T$ table consisting of $\{n_{ij}(t), \ i = 1, \ldots, I, \ j = 1, \ldots, I, \ t = 1, \ldots, T\}$. Each stratum of the table gives frequencies of the various transitions at a particular time. The transition probabilities are stationary if expected frequencies for this table satisfy loglinear model (XY, XZ); that is, for each originating state, the conditional distribution for one-step transitions from that state is identical at each of the T times for transitions. The likelihood-ratio statistic for testing stationarity of transition probabilities in a first-order Markov chain is simply the G^2 statistic for testing model (XY, XZ) for $\{n_{ij}(t)\}$. We can check stationarity of probabilities for transitions from a fixed state i by fitting the independence model to the $T \times I$ table consisting of transitions from that state at the T different transition times. The goodness-of-fit statistic for (XY, XZ) is the sum of the I statistics for testing independence for each fixed originating state.

Let $n_{ij} = \Sigma_t \, n_{ij}(t)$. The $\{n_{ij}\}$ are the $X-Y$ marginal table for the cross-classification just described. Under the assumption that the chain has stationary transition probabilities, the ML estimators of the entries in **P** are

$$\hat{\pi}_{j\,|\,i} = n_{ij}/n_{i+} \; .$$

Under the stationarity assumption, Anderson and Goodman (1957) proposed a simple test that the chain has the first-order Markov property (Problem 11.31). Unfortunately, in practice transition probabilities seem to be stationary only in rare, idealized situations.

CHAPTER NOTES

Section 11.1: Symmetry

11.1 For $T = 3$, the complete symmetry model has form $\log m_{ijk} = \lambda_{abc}$, where a is the minimum of (i, j, k) and c is the maximum. Haberman (1978, p. 518) noted that the number of $\{\lambda_{ijk}\}$ parameters is the number of ways of selecting 3 out of I items with replacement, which is $\binom{I+2}{3} = (I+2)(I+1)I/6$. Thus residual df $= I^3 - (I+2)(I+1)I/6$.

Section 11.2: Marginal Homogeneity

11.2 Somes (1982) reviewed Cochran's Q statistic. Bhapkar and Somes (1977) derived its distribution when there is marginal homogeneity but condition (11.5) is not satisfied. For small samples, Cochran (1950) gave an exact test for Q. Darroch (1981) gave a very informative survey of relationships among various statistics for testing marginal homogeneity.

The Q statistic was only one of William Cochran's many contributions to categorical data analysis. His 1954 article is a classic mixture of new methodology and advice for applied statisticians. In that article, as in several earlier articles, he dealt with the distribution of the Pearson statistic when expected frequencies are small, giving guidelines for when X^2 is acceptable (see Section 7.7.3). That article also stressed the importance of directing inferences toward narrow (e.g., single-degree-of-freedom) alternatives. One instance of this was his proposed test of conditional independence, which was practically the same as one proposed by Mantel and Haenszel 5 years later (Section 7.4.3). Another was a test for a linear trend in proportions (Section 4.4.3). He also presented several ways of partitioning chi-squared statistics into components. Fienberg (1984) reviewed Cochran's contributions to categorical data analysis.

Section 11.3: Modeling a Repeated Categorical Response

11.3 For further discussion of the Rasch model and alternative ways of estimating parameters, see Andersen (1980, Section 6.4) and Lindsay et al. (1989). Haberman (1977b) showed that one can achieve consistency with ML estimators when both n and T approach infinity at suitable rates. Conaway (1989), Lipsitz (1988), and Ware et al. (1988; see also accompanying commentary by Zeger and others) discussed other ways of analyzing longitudinal categorical data.

Liang and Zeger (1986) discussed generalized linear models for longitudinal data analysis. They showed ways of obtaining consistent estimates of model parameters without specifying a joint distribution for the repeated responses. Estimates based on treating repeated observations as independent are consistent, and they showed how to obtain the asymptotic covariance matrix of those estimates when the observations are truly dependent. They presented efficiency comparisons for estimating parameters under an assumed dependence structure relative to estimation treating repeated observations as independent. See also Zeger et al. (1985, 1988).

A common complication with longitudinal data is missing observations for some subjects at certain occasions. Stanish et al. (1978), Woolson and Clarke (1984), and Lipsitz (1988) showed how to handle missing data for WLS analyses. For ML analyses, see Little and Rubin (1987, Chap. 9) and Stram et al. (1988).

Section 11.4: Modeling a Repeated Ordinal Response

11.4 The WLS approach does not work if some covariates are continuous, or if there are so many covariates that the marginal distributions are sparse. Stram et al. (1988) presented an alternative, semi-parametric way of using cumulative logits to model marginal distributions. Their approach is valid even in cases where WLS fails. Stram et al. used ML to fit models separately at each occasion. They empirically generated the covariance matrix of the separate estimates, and used Wald tests to compare effects from different occasions. Their approach easily handles missing data and time-dependent covariates.

Section 11.5: Markov Chain Models

11.5 For further discussion of statistical inference for Markov chains, see Andersen (1980, Section 7.7), Anderson and Goodman (1957), Billingsley (1961), Bishop et al. (1975, Chap. 7), and Goodman (1962). Bonney (1987), Conaway (1989), Stiratelli et al. (1984), Ware et al. (1988), and Zeger and Qaqish (1988) presented other analyses of repeated categorical data that focused on the conditional dependence structure.

PROBLEMS

Applications

11.1 Refer to Table 11.1. Collapse the data over the first response, leaving a 2^3 table.

  a. Fit the complete symmetry model.

  b. Fit the quasi-symmetry model.

  c. Interpret the association structure for these data.

11.2 Refer to Table 11.1. Test whether the proportion who believe abortion should be legal is the same under the four conditions. Describe the marginal heterogeneity. In which situation is support for abortion highest?

11.3 Refer to Table 11.9.

  a. Use Cochran's Q to test marginal homogeneity for the proportion of wheeze at the four times.

  b. Does this statistic take into account the quantitative nature of the time scale? Construct a logit model for the marginal probabilities that assumes a linear effect of time. Show how to test marginal homogeneity using this model. How does the result compare to Cochran's test?

11.4 Using a statistical computer package, analyze Table 11.4.

11.5 Table 11.11 is based on data presented by Woolson and Clarke (1984), taken from the Muscatine Coronary Risk Factor Study, a longitudinal study of coronary risk factors in school children. A sample of children aged 11–13 in 1977 were classified by gender and by relative weight (obese, not obese) in 1977, 1979, and 1981. Using the methods of this chapter, analyze these data.

Table 11.11 Children Classified by Gender and Relative Weight

Gender	Responses[a]							
	NNN	NNO	NON	NOO	ONN	ONO	OON	OOO
Male	119	7	8	3	13	4	11	16
Female	129	8	7	9	6	2	7	14

Source: Reproduced with permission from the Royal Statistical Society, London (Woolson and Clarke 1984).

[a]NNN indicates not obese in 1977, 1979, and 1981, NNO indicates not obese in 1977 and 1979, but obese in 1981, and so forth.

11.6 Use the cumulative logit model for marginal distributions to test marginal homogeneity for the esophageal cancer data in Table 10.4. Compare results to those obtained in Section 10.4.4.

11.7 Refer to Table 11.7.

    a. To compare effects while controlling for initial response, fit model (11.13), using scores {10, 25, 45, 75} for time to falling asleep. Also fit the interaction model, and describe the lack of fit. (Note that for the first two baseline levels, the active and placebo treatments have similar sample response distributions at the follow-up; at higher baseline levels, the active treatment seems more successful than the placebo.)

    b. Fit the interaction model

$$L_j(hi) = \alpha_j - \beta_h - \delta_i - \gamma_{hi}$$

    that constrains effects $\{\beta_h + \delta_i + \gamma_{hi}\}$ to follow the pattern $(\tau, \tau, \lambda + \sigma, \lambda)$ for the active group and $(\tau, \tau, \sigma, 0)$ for the placebo group. Interpret the results.

11.8 Formulate a model using adjacent-categories logits that is analogous to model (11.9) for cumulative logits. Interpret the parameters.

11.9 Refer to the previous problem. Find an adjacent-categories logit model that fits Table 11.7 well. Interpret parameter estimates, and compare substantive conclusions to those obtained with cumulative logit modeling.

11.10 Refer to Table 11.11. Combine the data for the two genders.

    a. Assuming the classification on relative weight over time is a first-order Markov chain, test whether transition probabilities are stationary.

    b. Test whether the first-order Markov assumption is valid.

11.11 Table 11.12 is taken from the Harvard Study of Air Pollution and Health. Using methods of this chapter, analyze these data. (Thanks to Dr. J. Ware for supplying these data.)

11.12 Two gamblers, A and B, have a total of I dollars. They play a sequence of games. For each game, they each bet 1 dollar, and the winner takes the other's dollar. Suppose the outcomes of the

Table 11.12

Child's Respiratory Illness			No Maternal Smoking		Maternal Smoking	
Age 7	Age 8	Age 9	Age 10		Age 10	
			No	Yes	No	Yes
No	No	No	237	10	118	6
		Yes	15	4	8	2
	Yes	No	16	2	11	1
		Yes	7	3	6	4
Yes	No	No	24	3	7	3
		Yes	3	2	3	1
	Yes	No	6	2	4	2
		Yes	5	11	4	7

Source: Dr. James Ware.

games are statistically independent, with A having probability π of winning any game, and B having probability $1 - \pi$ of winning. Play stops when one player has all the money. Let X_t denote A's monetary total after t games.

a. Show that $\{X_t\}$ is a first-order Markov chain.

b. Show that $\{X_t\}$ has stationary transition probabilities, and give the transition probability matrix for the $I + 1$ states. (For this "gambler's ruin" problem, 0 and I are called *absorbing* states, and the other states are called *transient*.)

Theory and Methods

11.13 For an I^3 table, derive likelihood equations and the ML estimates of $\{m_{ijk}\}$ for the complete symmetry model.

11.14 Refer to Note 11.1. Explain how to fit the complete symmetry model in T dimensions, and give the residual df.

11.15 Show that the ML fit of model (W, XYZ) for the $6 \times I^3$ table with entries $\{n_{1ijk}^* = n_{ijk},\ n_{2ijk}^* = n_{ikj},\ n_{3ijk}^* = n_{jik},\ n_{4ijk}^* = n_{jki},\ n_{5ijk}^* = n_{kij},\ n_{6ijk}^* = n_{kji}\}$ has entries $\{\hat{m}_{hijk}^*\}$ related to the ML fit $\{\hat{m}_{ijk}\}$ for the complete symmetry model by $\{\hat{m}_{ijk} = \hat{m}_{1ijk}^*\}$.

11.16 Devise a way to fit the quasi-symmetry model to an I^3 table by fitting a standard loglinear model to a $6 \times I^3$ table. (Hint: Find a model having likelihood equations equivalent to those for the quasi-symmetry model).

11.17 Construct a loglinear model for an I^3 table having the following quasi-independence interpretation: Conditional on the event that the three responses are completely different, the responses are mutually independent. Find the residual df.

11.18 Prove that if there is kth-order marginal symmetry, then there is jth-order marginal symmetry for any $j < k$.

11.19 Show that under condition (11.5) for a 2^T table, the null hypothesis of marginal homogeneity is equivalent to second-order marginal symmetry.

11.20 Show that when $T = 2$, Cochran's Q simplifies to McNemar's statistic $(n_{12} - n_{21})^2/(n_{12} + n_{21})$.

11.21 Show that the value of Cochran's Q is unaffected by deleting all observations in which the subject has the same response at each occasion.

11.22 Suppose we observe a binary response for n subjects at T occasions. Let b_{ig} indicate the result for the ith subject at the gth occasion: $b_{ig} = 1$ for response in the first category, and $b_{ig} = 0$ for response in the second category.

    a. For notation used in Cochran's Q, show $b_{i.} = q_i$ and $b_{.g} = f_1(g)$, so that $b_{..} = (\Sigma_i q_i)/n = [\Sigma_g f_1(g)]/T$. (A . subscript denotes the average with respect to that index.)

    b. Regard $\{b_{i+}\}$ as fixed, and consider the hypothesis that each way of allocating the b_{i+} "successes" to b_{i+} of the occasions is equally likely. Show that $E(b_{ig}) = q_i$, $\mathrm{Var}(b_{ig}) = q_i(1 - q_i)$, and $\mathrm{Cov}(b_{ig}, b_{ik}) = -q_i(1 - q_i)/(T - 1)$ for $g \neq k$. (Hint: The covariance is the same for any pair of cells in the same row, and $\mathrm{Var}(\Sigma_g b_{ig}) = 0$ since b_{i+} is fixed.)

    c. Refer to (b). For a large number n of independent subjects, argue that $\{b_{.g}, g = 1, \ldots, T\}$ have an approximate multivariate normal distribution, with pairwise correlation $\rho = -1/(T - 1)$. Conclude that

$$\frac{n^2(T-1) \sum\limits_{g=1}^{T} (b_{.g} - b_{..})^2}{T \sum\limits_{i=1}^{n} b_{i.}(1 - b_{i.})}$$

has an approximate chi-squared distribution with $df = (T - 1)$. (One way is to note that if $(Y_1, \ldots, Y_T)$ has a multivariate normal distribution with common mean and common variance σ^2 for $\{Y_g\}$, and common correlation ρ for pairs (Y_g, Y_k), then $\Sigma(Y_g - \bar{Y})^2/\sigma^2(1 - \rho)$ has a chi-squared distribution with $df = (T - 1)$).

d. Show that the statistic in (c) is Cochran's Q.

11.23 For quasi-symmetry model (11.3), prove that marginal homogeneity is equivalent to complete symmetry.

11.24 Refer to Section 11.2.4. Show the Mantel–Haenszel estimator of a common odds ratio in this $2 \times 2 \times n$ table simplifies to the conditional ML estimator for the odds ratio for Cox's logistic model for matched pairs.

11.25 Refer to the Rasch model.

a. Assuming independence of responses for different subjects and for different responses by the same subject, show that the log likelihood is

$$\alpha \left(\sum \sum y_{gh} \right) + \sum \sum \lambda_h y_{gh} + \sum \sum \beta_g y_{gh}$$
$$- \sum \sum \log[1 + \exp(\alpha + \lambda_h + \beta_g)] .$$

b. Show that the likelihood equations equate y_{g+} to $\Sigma_h \phi_1(g; h)$ and y_{+h} to $\Sigma_g \phi_1(g; h)$ for all g and h.

c. To obtain a distribution that does not depend on $\{\lambda_h\}$, explain why it makes sense to condition on $\{y_{+h}\}$.

11.26 Refer to the mean response model in Section 11.4.3. When $I > 2$, why is marginal homogeneity across occasions not a special case of this model?

11.27 Refer to models (11.9) and (11.12). Explain in detail why all $\nu_g = 0$ implies all $\beta_g = 0$.

11.28 Explain what is wrong with the following statement: "For a first-order Markov chain, X_t is independent of X_{t-2}."

11.29 Specify the loglinear model corresponding to a kth order Markov chain.

11.30 Suppose a first-order Markov chain has stationary transition probabilities. Show how the likelihood in (11.15) simplifies, and derive ML estimates of the transition probabilities.

11.31 Suppose a Markov chain has stationary transition probabilities. Let n_{ijk} denote the number of times there are transitions from i to j to k over two successive steps. For the table $\{n_{ijk}\}$, argue that the goodness-of-fit of loglinear model (XY, YZ) gives a test of the null hypothesis that the chain is first-order, against the alternative hypothesis that it is second-order.

CHAPTER 12

Asymptotic Theory For Parametric Models

The final two chapters have a more theoretical flavor than previous chapters. This chapter presents asymptotic theory for parametric models for categorical data. The emphasis is on deriving large-sample properties of model parameter estimators and goodness-of-fit statistics. Chapter 13 studies fixed-sample-size estimation theory, for maximum likelihood and other methods of estimation.

Section 12.1 presents the delta method, which implies large-sample normality of many statistics. Section 12.2 uses the delta method to derive asymptotic normal distributions for ML estimators of parameters in models for contingency tables. Section 12.3 derives asymptotic distributions of cell residuals and the X^2 and G^2 goodness-of-fit statistics.

The results in this chapter have quite a long history. Karl Pearson (1900) derived the asymptotic chi-squared distribution of X^2 for testing goodness-of-fit for a specified multinomial distribution. Fisher (1924) showed the necessary adjustment in degrees of freedom when multinomial probabilities are functions of unknown parameters. Cramér (1946, pp. 424–434) provided a formal proof of this result, under the assumption that ML estimators of the parameters are consistent. Rao (1957, 1958) proved consistency of the ML estimators under general conditions. He also gave the asymptotic distribution of the ML estimators, though the primary emphasis of his articles was on proving consistency. Birch (1964a) proved these results under weaker conditions. Haberman (1974a), Bishop et al. (1975), Andersen (1980), and Cox (1984) provided other proofs.

Our presentation borrows from each source, especially Rao's work as summarized in his 1973 book. As in Cramér's and Rao's proofs, the derivation regards the ML estimator as a point in the parameter space where the derivative of the log likelihood function is zero. Birch regarded it as a point at which the likelihood takes value arbitrarily near its supremum. Though his approach is more powerful, the proofs are also much more complex.

We avoid a formal "theorem-proof" style of exposition. Instead, we intend for the reader to observe that powerful results follow from quite simple mathematical ideas.

12.1 DELTA METHOD

Suppose a statistic has a large-sample normal distribution. This section shows that many functions of that statistic are also asymptotically normal. For instance, for multinomial sampling, cell proportions in a contingency table have a large-sample normal distribution. So do many functions of those proportions, such as logits, odds ratios, and other measures of association.

12.1.1 O, o Rates of Convergence

"Little o" and "big O" notation is convenient for describing limiting behavior of random variables. For a sequence of real numbers $\{z_n\}$, the notation $o(z_n)$ represents a term that has *smaller* order than z_n as $n \to \infty$, in the sense that $o(z_n)/z_n \to 0$ as $n \to \infty$. For instance, $\sqrt{n}$ is $o(n)$ as $n \to \infty$, since $\sqrt{n}/n \to 0$ as $n \to \infty$. A sequence that is $o(1)$ satisfies $o(1)/1 = o(1) \to 0$; for instance $n^{-1/2} = o(1)$ as $n \to \infty$.

The "big O" notation $O(z_n)$ represents terms that have the *same* order of magnitude as z_n, in the sense that $|O(z_n)/z_n|$ is bounded as $n \to \infty$. For instance, $(3/n) + (28/n^2) = O(n^{-1})$ as $n \to \infty$.

Similar notation applies to sequences of random variables. The symbol $o_p(z_n)$ denotes a random variable of *smaller* order than z_n for large n, in the sense that $o_p(z_n)/z_n$ *converges in probability* to 0; that is, for any fixed $\epsilon > 0$, $P(|o_p(z_n)/z_n| \le \epsilon) \to 1$ as $n \to \infty$. The notation $O_p(z_n)$ represents a random variable such that for every $\epsilon > 0$, there is a constant K and an integer n_0 such that $P[|O_p(z_n)/z_n| < K] > 1 - \epsilon$ for all $n > n_0$.

To illustrate, let $\bar{Y}_n$ denote the sample mean of n independent observations $Y_1, \ldots, Y_n$ from a distribution having $E(Y_i) = \mu$. Then $(\bar{Y}_n - \mu) = o_p(1)$, since $(\bar{Y}_n - \mu)/1$ converges in probability to zero as $n \to \infty$, by the law of large numbers. By Tchebychev's Inequality, the difference between a random variable and its mean has the same order of magnitude as the standard deviation of that random variable. Since $\bar{Y}_n - \mu$ has standard deviation $\sigma/\sqrt{n}$, $(\bar{Y}_n - \mu) = O_p(n^{-1/2})$.

A random variable that is $O_p(n^{-1/2})$ is also $o_p(1)$. An example is $(\bar{Y}_n - \mu)$. Note that $\sqrt{n}(\bar{Y}_n - \mu) = n^{1/2}O_p(n^{-1/2}) = O_p(n^{1/2}n^{-1/2}) = O_p(1)$. If the difference between two random variables is $o_p(1)$ as $n \to \infty$, Slutzky's Theorem states that those random variables have the same limiting distribution.

12.1.2 Delta Method for Function of Random Variable

Let T_n denote a statistic, the subscript expressing its dependence on the sample size n. For large samples, suppose T_n is approximately normally distributed about θ, with approximate standard error $\sigma/\sqrt{n}$. More precisely, as $n \to \infty$, suppose the cdf of $\sqrt{n}(T_n - \theta)$ converges to the cdf of a normal random variable with mean 0 and variance σ^2. This limiting behavior is an example of *convergence in distribution*, denoted here by

$$\sqrt{n}(T_n - \theta) \xrightarrow{d} N(0, \sigma^2) . \tag{12.1}$$

We now derive the limiting distribution of a statistic $g(T_n)$.

Suppose g is at least twice differentiable at θ. We use the Taylor series expansion for g, evaluated in a neighborhood of θ. For some θ^* between t and θ,

$$g(t) = g(\theta) + (t - \theta)g'(\theta) + (t - \theta)^2 g''(\theta^*)/2$$
$$= g(\theta) + (t - \theta)g'(\theta) + O(|t - \theta|^2) .$$

Substituting the random variable T_n for t, we have

$$\sqrt{n}[g(T_n) - g(\theta)] = \sqrt{n}(T_n - \theta)g'(\theta) + \sqrt{n}O(|T_n - \theta|^2)$$
$$= \sqrt{n}(T_n - \theta)g'(\theta) + O_p(n^{-1/2}) \tag{12.2}$$

since

$$\sqrt{n}O(|T_n - \theta|^2) = \sqrt{n}O[O_p(n^{-1})] = O_p(n^{-1/2}) .$$

Thus, $\sqrt{n}[g(T_n) - g(\theta)]$ has the same limiting distribution as $\sqrt{n}(T_n - \theta)g'(\theta)$; that is, $g(T_n) - g(\theta)$ behaves like the constant multiple $g'(\theta)$ of the difference $(T_n - \theta)$. Since $(T_n - \theta)$ is approximately normal with variance σ^2/n, $g(T_n) - g(\theta)$ is approximately normal with variance $\sigma^2[g'(\theta)]^2/n$. More precisely,

$$\sqrt{n}[g(T_n) - g(\theta)] \xrightarrow{d} N(0, \sigma^2[g'(\theta)]^2) . \tag{12.3}$$

Figure 12.1 illustrates this result. The dispersion of $g(T_n)$ values about $g(\theta)$ equals about $|\,g'(\theta)\,|$ times the dispersion of T_n values about θ. If the slope of g at θ is $\frac{1}{2}$, then g maps a region of T_n values into a region of $g(T_n)$ values that is only about half as wide.

Result (12.3) is called the *delta method* for obtaining asymptotic distributions. Since $\sigma^2 = \sigma^2(\theta)$ and $g'(\theta)$ generally depend on the unknown parameter θ, the asymptotic variance is also unknown. When $g'(\cdot)$ and $\sigma = \sigma(\cdot)$ are continuous at θ, $\sigma(T_n)g'(T_n)$ is a consistent estimator of $\sigma(\theta)g'(\theta)$; thus, confidence intervals and tests use the result that $\sqrt{n}\,[\,g(T_n) - g(\theta)]\,/\sigma(T_n)|\,g'(T_n)\,|$ is asymptotically standard normal. For instance,

$$g(T_n) \pm 1.96\sigma(T_n)|\,g'(T_n)\,|\,/\sqrt{n}$$

is a large-sample 95% confidence interval for $g(\theta)$.

When $g'(\theta) = 0$, (12.3) is uninformative because the limiting variance equals zero. In that case, $\sqrt{n}\,[\,g(T_n) - g(\theta)] = o_p(1)$, and higher-order terms in the Taylor series expansion yield the asymptotic distribution (see Note 12.1).

12.1.3 Delta Method Applied to Sample Logit

To illustrate the delta method, we let $T_n = p$ denote the sample proportion of successes in n independent Bernoulli trials, each having parameter π. Then $p = (\Sigma Y_i)/n$, where $P(Y_i = 1) = \pi$ and $P(Y_i = 0) = 1 - \pi$. Also,

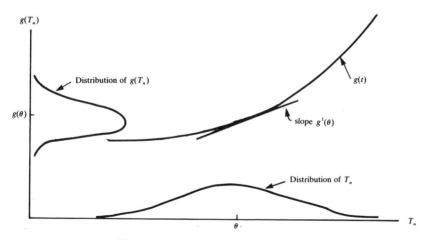

Figure 12.1 Depiction of Delta Method.

$$E(Y_i) = 1(\pi) + 0(1 - \pi) = \pi \quad \text{and} \quad E(Y_i^2) = 1^2(\pi) + 0^2(1 - \pi) = \pi$$

so

$$\text{Var}(Y_i) = \pi(1 - \pi) \quad \text{and} \quad \text{Var}(p) = \text{Var}(\bar{Y}) = \pi(1 - \pi)/n .$$

Now consider the sample logit transformation

$$g(p) = \log[p/(1 - p)] .$$

Evaluated at π, its derivative equals $1/\pi(1 - \pi)$. The delta method implies the asymptotic variance of the sample logit is $\pi(1 - \pi)/n$ (the variance of p) multiplied by the square of $[1/\pi(1 - \pi)]$. That is

$$\sqrt{n}\left[\log\left(\frac{p}{1 - p}\right) - \log\left(\frac{\pi}{1 - \pi}\right)\right] \overset{d}{\to} N\left[0, \frac{1}{\pi(1 - \pi)}\right] .$$

For $0 < \pi < 1$, the asymptotic variance of the sample logit is finite. By contrast, the true exact variance does not exist: Since $p = 0$ or 1 with positive probability, the logit can equal $-\infty$ or ∞ with positive probability. The probability of an infinite logit converges to zero rapidly as n increases. For large n, the distribution of the sample logit looks very similar to normality with mean $\log[\pi/(1 - \pi)]$ and standard deviation $[n\pi(1 - \pi)]^{-1/2}$. Thus, for the logit, the asymptotic variance actually has greater use than the exact variance.

12.1.4 Delta Method for Function of Random Vector

The delta method generalizes to functions of random *vectors*. Suppose $\mathbf{T}_n = (T_{n1}, \ldots, T_{nN})'$ is asymptotically multivariate normal with mean $\boldsymbol{\theta} = (\theta_1, \ldots, \theta_N)'$ and covariance matrix $\boldsymbol{\Sigma}/n$. Suppose the function $g(t_1, \ldots, t_N)$ has a nonzero differential $\boldsymbol{\phi} = (\phi_1, \ldots, \phi_N)'$ at $\boldsymbol{\theta}$, where

$$\phi_i = \frac{\partial g}{\partial t_i}\bigg|_{\mathbf{t}=\boldsymbol{\theta}}$$

Then

$$\sqrt{n}[g(\mathbf{T}_n) - g(\boldsymbol{\theta})] \overset{d}{\to} N(0, \boldsymbol{\phi}'\boldsymbol{\Sigma}\boldsymbol{\phi}) . \tag{12.4}$$

For large samples, $g(\mathbf{T}_n)$ has distribution similar to the normal with mean $g(\boldsymbol{\theta})$ and variance $\boldsymbol{\phi}'\boldsymbol{\Sigma}\boldsymbol{\phi}/n$.

The proof of (12.4) follows from the expansion

$$g(\mathbf{T}_n) - g(\boldsymbol{\theta}) = (\mathbf{T}_n - \boldsymbol{\theta})'\boldsymbol{\phi} + o(\|\mathbf{T}_n - \boldsymbol{\theta}\|)$$

where $\| \mathbf{z} \| = (\Sigma \, z_i^2)^{1/2}$ denotes the length of vector $\mathbf{z}$. For large n, $g(\mathbf{T}_n) - g(\boldsymbol{\theta})$ behaves like a linear function of the approximately normal variate $(\mathbf{T}_n - \boldsymbol{\theta})$.

12.1.5 Asymptotic Normality of Functions of Multinomial Counts

The delta method for random vectors implies asymptotic normality of many functions of cell counts in contingency tables. Suppose cell counts $(n_1, \ldots, n_N)$ have a multinomial distribution with cell probabilities $\boldsymbol{\pi} = (\pi_1, \ldots, \pi_N)'$. Let $n = n_1 + \cdots + n_N$, and let $\mathbf{p} = (p_1, \ldots, p_N)'$ denote the sample proportions, where $p_i = n_i/n$.

Denote the ith of the n observations by $\mathbf{Y}_i = (Y_{i1}, \ldots, Y_{iN})$, where $Y_{ij} = 1$ if it falls in cell j, and $Y_{ij} = 0$ otherwise, $i = 1, \ldots, n$. For instance, $\mathbf{Y}_6 = (0, 0, 1, 0, 0, \ldots, 0)$ means that the sixth observation fell in the third cell of the table. Then $\Sigma_j Y_{ij} = 1$, $Y_{ij}Y_{ik} = 0$ when $j \neq k$, $p_j = \Sigma_i Y_{ij}/n$, and

$$E(Y_{ij}) = P(Y_{ij} = 1) = \pi_j = E(Y_{ij}^2)$$
$$E(Y_{ij}Y_{ik}) = 0 \quad \text{if} \quad j \neq k .$$

It follows that

$$E(\mathbf{Y}_i) = \boldsymbol{\pi} \quad \text{and} \quad \text{Cov}(\mathbf{Y}_i) = \boldsymbol{\Sigma}, \quad i = 1, \ldots, n$$

where $\boldsymbol{\Sigma} = (\sigma_{jk})$ with

$$\sigma_{jj} = \text{Var}(Y_{ij}) = E(Y_{ij}^2) - [E(Y_{ij})]^2 = \pi_j(1 - \pi_j)$$
$$\sigma_{jk} = \text{Cov}(Y_{ij}, Y_{ik}) = E(Y_{ij}Y_{ik}) - E(Y_{ij})E(Y_{ik}) = -\pi_j\pi_k \quad \text{for } j \neq k .$$

The matrix $\boldsymbol{\Sigma}$ has form

$$\boldsymbol{\Sigma} = \text{Diag}(\boldsymbol{\pi}) - \boldsymbol{\pi}\boldsymbol{\pi}'$$

where $\text{Diag}(\boldsymbol{\pi})$ is the diagonal matrix with the elements of $\boldsymbol{\pi}$ on the main diagonal.

Since $\mathbf{p}$ is a sample mean of independent observations,

$$\mathbf{p} = \frac{\sum\limits_{i=1}^{n} \mathbf{Y}_i}{n} ,$$

we have the result

$$\text{Cov}(\mathbf{p}) = [\textbf{Diag}(\boldsymbol{\pi}) - \boldsymbol{\pi}\boldsymbol{\pi}']/n \ . \tag{12.5}$$

This covariance matrix is singular, because of the linear dependence $\Sigma p_i = 1$. The multivariate Central Limit Theorem (Rao 1973, p. 128) implies

$$\sqrt{n}(\mathbf{p} - \boldsymbol{\pi}) \overset{d}{\to} N[\mathbf{0}, \textbf{Diag}(\boldsymbol{\pi}) - \boldsymbol{\pi}\boldsymbol{\pi}'] \ . \tag{12.6}$$

By the delta method, functions of $\mathbf{p}$ having nonzero differential at $\boldsymbol{\pi}$ are also asymptotically normally distributed. To illustrate, let $g(t_1, \ldots, t_N)$ be a differentiable function, and let

$$\phi_i = \partial g / \partial \pi_i, \quad i = 1, \ldots, N$$

denote $\partial g / \partial t_i$ evaluated at $\mathbf{t} = \boldsymbol{\pi}$. By the delta method,

$$\sqrt{n}[g(\mathbf{p}) - g(\boldsymbol{\pi})] \overset{d}{\to} N(0, \boldsymbol{\phi}'[\textbf{Diag}(\boldsymbol{\pi}) - \boldsymbol{\pi}\boldsymbol{\pi}']\boldsymbol{\phi}) \tag{12.7}$$

where $\boldsymbol{\phi} = (\phi_1, \ldots, \phi_N)'$. The asymptotic covariance matrix equals

$$\boldsymbol{\phi}'\textbf{Diag}(\boldsymbol{\pi})\boldsymbol{\phi} - (\boldsymbol{\phi}'\boldsymbol{\pi})^2 = \sum \pi_i \phi_i^2 - \left(\sum \pi_i \phi_i \right)^2 \ .$$

We used this formula in Section 3.4 to derive large-sample variances of sample measures of association.

12.1.6 Delta Method for Vector Functions of Random Vectors

The delta method generalizes further to a *vector* of functions of an asymptotically normal random vector. Let $\mathbf{g}(\mathbf{t}) = (g_1(\mathbf{t}), \ldots, g_q(\mathbf{t}))'$ and let $(\partial \mathbf{g}/\partial \boldsymbol{\theta})$ denote the $q \times N$ matrix for which the entry in row i and column j is $\partial g_i(\mathbf{t}) / \partial t_j$ evaluated at $\mathbf{t} = \boldsymbol{\theta}$. Then

$$\sqrt{n}[g(\mathbf{T}_n) - \mathbf{g}(\boldsymbol{\theta})] \overset{d}{\to} N[\mathbf{0}, (\partial \mathbf{g}/\partial \boldsymbol{\theta})\boldsymbol{\Sigma}(\partial \mathbf{g}/\partial \boldsymbol{\theta})'] \ . \tag{12.8}$$

The rank of the limiting normal distribution equals the rank of the asymptotic covariance matrix.

Expression (12.8) is useful for finding large-sample joint distributions. For instance, from (12.6) and (12.8), the asymptotic distribution of several functions of multinomial proportions has covariance matrix of the form

$$\text{Asymp. Cov}\{\sqrt{n}[\mathbf{g}(\mathbf{p}) - \mathbf{g}(\boldsymbol{\pi})]\} = \boldsymbol{\Phi}[\textbf{Diag}(\boldsymbol{\pi}) - \boldsymbol{\pi}\boldsymbol{\pi}']\boldsymbol{\Phi}'$$

where $\boldsymbol{\Phi}$ is the Jacobian evaluated at $\boldsymbol{\pi}$.

12.1.7   Joint Asymptotic Normality of Log Odds Ratios

We illustrate formula (12.8) by finding the joint asymptotic distribution of a set of odds ratios in a contingency table. As noted in Section 3.4, convergence to normality is more rapid on the log scale.

Let $\mathbf{g}(\boldsymbol{\pi}) = \log(\boldsymbol{\pi})$ denote the vector of natural logs of cell probabilities, for which

$$\partial \mathbf{g}/\partial \boldsymbol{\pi} = \mathbf{Diag}(\boldsymbol{\pi})^{-1}\ .$$

The covariance of the asymptotic distribution of $\sqrt{n}[\log(\mathbf{p}) - \log(\boldsymbol{\pi})]$ is

$$\mathbf{Diag}(\boldsymbol{\pi})^{-1}[\mathbf{Diag}(\boldsymbol{\pi}) - \boldsymbol{\pi}\boldsymbol{\pi}']\mathbf{Diag}(\boldsymbol{\pi})^{-1} = \mathbf{Diag}(\boldsymbol{\pi})^{-1} - \mathbf{11}'$$

where $\mathbf{1}$ is an $N \times 1$ vector of 1 elements.

For a $q \times N$ matrix of constants $\mathbf{C}$, it follows that

$$\sqrt{n}\mathbf{C}[\log(\mathbf{p}) - \log(\boldsymbol{\pi})] \overset{d}{\to} N[\mathbf{0}, \mathbf{C}\,\mathbf{Diag}(\boldsymbol{\pi})^{-1}\mathbf{C}' - \mathbf{C11}'\mathbf{C}']\ . \quad (12.9)$$

We can apply this formula to a set of log odds ratios by letting each row of $\mathbf{C}$ contain zeroes except for two $+1$ elements and two -1 elements. The second term in the covariance matrix in (12.9) is then zero. If a particular odds ratio uses the cells numbered h, i, j, and k, then the variance of the asymptotic distribution is

$$\text{Asymp. Var}[\sqrt{n}(\text{sample log odds ratio})] = \pi_h^{-1} + \pi_i^{-1} + \pi_j^{-1} + \pi_k^{-1}\ .$$

When two log odds ratios have no cells in common, their asymptotic covariance in the limiting normal distribution equals zero.

12.2   ASYMPTOTIC DISTRIBUTIONS OF ESTIMATORS OF MODEL PARAMETERS AND CELL PROBABILITIES

We now derive fundamental results of large-sample model-based inference for categorical data. The key tool is the delta method.

The data are counts $(n_1, \ldots, n_N)$ in N cells of a contingency table. The asymptotics regard N as fixed and let $n = \Sigma\, n_i \to \infty$. We assume the cell counts have a multinomial distribution with cell probabilities $\boldsymbol{\pi} = (\pi_1, \ldots, \pi_N)'$. Let $\mathbf{p} = (p_1, \ldots, p_N)'$ with $p_i = n_i/n$. The model relates $\boldsymbol{\pi}$ to a smaller number of parameters $\boldsymbol{\theta}' = (\theta_1, \ldots, \theta_t)$. We express it as

$$\boldsymbol{\pi} = \boldsymbol{\pi}(\boldsymbol{\theta})\ .$$

In other words, $\pi_i(\boldsymbol{\theta})$ denotes the function that relates the parameters to π_i, $i = 1, \ldots, N$.

As $\boldsymbol{\theta}$ ranges over its parameter space, $\boldsymbol{\pi}(\boldsymbol{\theta})$ ranges over a subset of the space of probabilities for N categories. When we add components to $\boldsymbol{\theta}$, the model becomes more complex, and the space of $\boldsymbol{\pi}$ that satisfy the model is larger. We use $\boldsymbol{\theta}$ and $\boldsymbol{\pi}$ to denote generic parameter and probability values, and $\boldsymbol{\theta}_0 = (\theta_{10}, \ldots, \theta_{t0})'$ and $\boldsymbol{\pi}_0 = (\pi_{10}, \ldots, \pi_{N0})' = \boldsymbol{\pi}(\boldsymbol{\theta}_0)$ to denote true values for a particular application. When the model does not hold, there is no $\boldsymbol{\theta}_0$ for which $\boldsymbol{\pi}(\boldsymbol{\theta}_0)$ gives the true probabilities $\boldsymbol{\pi}_0$; that is, $\boldsymbol{\pi}_0$ falls outside the subset of $\boldsymbol{\pi}$ values that is the range of $\boldsymbol{\pi}(\boldsymbol{\theta})$ for the space of possible $\boldsymbol{\theta}$. We defer discussion of this case to Section 12.3.5.

We first derive the asymptotic distribution of the ML estimator $\hat{\boldsymbol{\theta}}$ of $\boldsymbol{\theta}$. We use that to derive the asymptotic distribution of the model-based ML estimator $\hat{\boldsymbol{\pi}} = \boldsymbol{\pi}(\hat{\boldsymbol{\theta}})$ of $\boldsymbol{\pi}$. The approach follows Rao (1973, Section 5e) and Bishop et al. (1975, Section 14.7–8). We assume the regularity conditions:

1. $\boldsymbol{\theta}_0$ is not on the boundary of the parameter space,
2. all $\pi_{i0} > 0$,
3. $\boldsymbol{\pi}(\boldsymbol{\theta})$ has continuous first-order partial derivatives in a neighborhood of $\boldsymbol{\theta}_0$, and
4. the Jacobian matrix $(\partial \boldsymbol{\pi} / \partial \boldsymbol{\theta})$ has full rank t at $\boldsymbol{\theta}_0$.

These conditions ensure that $\boldsymbol{\pi}(\boldsymbol{\theta})$ is locally smooth and one-to-one at $\boldsymbol{\theta}_0$, and that Taylor series expansions exist in neighborhoods around $\boldsymbol{\theta}_0$ and $\boldsymbol{\pi}_0$. When the Jacobian does not have full rank, we can often satisfy this condition by reformulating the model using fewer parameters.

12.2.1 Distribution of Model Parameter Estimator

The key to deriving the asymptotic distribution and covariance of $\hat{\boldsymbol{\theta}}$ is to express $\hat{\boldsymbol{\theta}}$ as a linearized function of $\mathbf{p}$. We can then apply the delta method, using the asymptotic normality of $\mathbf{p}$. The linearization has two steps, first relating $\mathbf{p}$ to $\hat{\boldsymbol{\pi}}$, and then $\hat{\boldsymbol{\pi}}$ to $\hat{\boldsymbol{\theta}}$.

The kernel of the multinomial log likelihood is

$$L(\boldsymbol{\theta}) = \log \prod_{i=1}^{N} \pi_i(\boldsymbol{\theta})^{n_i} = n \sum_{i=1}^{N} p_i \log \pi_i(\boldsymbol{\theta}) .$$

The likelihood equations are

$$\frac{\partial L(\boldsymbol{\theta})}{\partial \theta_j} = n \sum_i \frac{p_i}{\pi_i(\boldsymbol{\theta})} \left(\frac{\partial \pi_i(\boldsymbol{\theta})}{\partial \theta_j} \right) = 0 , \quad j = 1, \ldots, t \qquad (12.10)$$

which depend on the functional form $\boldsymbol{\pi}(\boldsymbol{\theta})$ used in the model. We regard the ML estimator $\hat{\boldsymbol{\theta}}$ as a solution to these equations. We note that

$$\sum_i \frac{\partial \pi_i(\boldsymbol{\theta})}{\partial \theta_j} = \frac{\partial}{\partial \theta_j} \left[\sum_i \pi_i(\boldsymbol{\theta}) \right] = \frac{\partial}{\partial \theta_j} (1) = 0 . \qquad (12.11)$$

Let $\partial \pi_i / \partial \hat{\theta}_j$ represent $\partial \pi_i(\boldsymbol{\theta}) / \partial \theta_j$ evaluated at $\hat{\boldsymbol{\theta}}$. Subtracting a common term from both sides of the jth likelihood equation, we obtain

$$\sum_i \frac{n(p_i - \pi_{i0})}{\hat{\pi}_i} \left(\frac{\partial \pi_i}{\partial \hat{\theta}_j} \right) = \sum_i \frac{n(\hat{\pi}_i - \pi_{i0})}{\hat{\pi}_i} \left(\frac{\partial \pi_i}{\partial \hat{\theta}_j} \right) \qquad (12.12)$$

since the first sum on the right-hand side equals zero.

Next we express $\hat{\boldsymbol{\pi}}$ in terms of $\hat{\boldsymbol{\theta}}$ using

$$\hat{\pi}_i - \pi_{i0} = \sum_k (\hat{\theta}_k - \theta_{k0}) \left(\frac{\partial \pi_i}{\partial \bar{\theta}_k} \right)$$

where $\partial \pi_i / \partial \bar{\theta}_k$ represents $\partial \pi_i / \partial \theta_k$ evaluated at some point $\bar{\boldsymbol{\theta}}$ falling between $\hat{\boldsymbol{\theta}}$ and $\boldsymbol{\theta}_0$. Substituting this into the right-hand side of (12.12), and dividing both sides by $\sqrt{n}$, we obtain for each j

$$\sum_i \frac{\sqrt{n}(p_i - \pi_{i0})}{\hat{\pi}_i} \left(\frac{\partial \pi_i}{\partial \hat{\theta}_j} \right) = \sum_k \sqrt{n}(\hat{\theta}_k - \theta_{k0}) \left\{ \sum_i \frac{1}{\hat{\pi}_i} \left(\frac{\partial \pi_i}{\partial \hat{\theta}_j} \right) \left(\frac{\partial \pi_i}{\partial \bar{\theta}_k} \right) \right\} .$$

$$(12.13)$$

We next introduce notation that lets us express more simply the dependence of $\hat{\boldsymbol{\theta}}$ on $\mathbf{p}$. Let $\mathbf{A}$ denote the matrix having elements

$$a_{ij} = \pi_{i0}^{-1/2} \left(\frac{\partial \pi_i(\boldsymbol{\theta})}{\partial \theta_{j0}} \right) .$$

That is, $\mathbf{A}$ is the $N \times t$ matrix

$$\mathbf{A} = \mathbf{Diag}(\boldsymbol{\pi}_0)^{-1/2} (\partial \boldsymbol{\pi} / \partial \boldsymbol{\theta}_0) \qquad (12.14)$$

where $(\partial \boldsymbol{\pi} / \partial \boldsymbol{\theta}_0)$ denotes the Jacobian $(\partial \boldsymbol{\pi} / \partial \boldsymbol{\theta})$ evaluated at $\boldsymbol{\theta}_0$. As $\hat{\boldsymbol{\theta}}$ converges to $\boldsymbol{\theta}_0$, the term in brackets on the right-hand side of (12.13) converges to the element in row j and column k of $\mathbf{A}'\mathbf{A}$. As $\hat{\boldsymbol{\theta}} \rightarrow \boldsymbol{\theta}_0$, the set of equations (12.13) has form

$$\mathbf{A}' \, \mathbf{Diag}(\boldsymbol{\pi}_0)^{-1/2}\sqrt{n}(\mathbf{p} - \boldsymbol{\pi}_0) = (\mathbf{A}'\mathbf{A})\sqrt{n}(\hat{\boldsymbol{\theta}} - \boldsymbol{\theta}_0) + o_p(1) \,.$$

Since the Jacobian has full rank at $\boldsymbol{\theta}_0$, $\mathbf{A}'\mathbf{A}$ is nonsingular. Thus, we can express $\hat{\boldsymbol{\theta}}$ in terms of $\mathbf{p}$ as

$$\sqrt{n}(\hat{\boldsymbol{\theta}} - \boldsymbol{\theta}_0) = (\mathbf{A}'\mathbf{A})^{-1}\mathbf{A}' \, \mathbf{Diag}(\boldsymbol{\pi}_0)^{-1/2}\sqrt{n}(\mathbf{p} - \boldsymbol{\pi}_0) + o_p(1) \,.$$
$$(12.15)$$

Now, the asymptotic distribution of $\mathbf{p}$ determines that of $\hat{\boldsymbol{\theta}}$. From (12.6), $\sqrt{n}(\mathbf{p} - \boldsymbol{\pi}_0)$ is asymptotically normal, with covariance matrix $[\mathbf{Diag}(\boldsymbol{\pi}_0) - \boldsymbol{\pi}_0\boldsymbol{\pi}_0']$. It follows that $\sqrt{n}(\hat{\boldsymbol{\theta}} - \boldsymbol{\theta}_0)$ is also asymptotically normal, with asymptotic covariance matrix

$$(\mathbf{A}'\mathbf{A})^{-1}\mathbf{A}' \, \mathbf{Diag}(\boldsymbol{\pi}_0)^{-1/2}$$
$$\times [\mathbf{Diag}(\boldsymbol{\pi}_0) - \boldsymbol{\pi}_0\boldsymbol{\pi}_0'] \times \mathbf{Diag}(\boldsymbol{\pi}_0)^{-1/2}\mathbf{A}(\mathbf{A}'\mathbf{A})^{-1} \,.$$

Using (12.11),

$$\boldsymbol{\pi}_0'\mathbf{Diag}(\boldsymbol{\pi}_0)^{-1/2}\mathbf{A}$$
$$= \boldsymbol{\pi}_0' \, \mathbf{Diag}(\boldsymbol{\pi}_0)^{-1/2}\mathbf{Diag}(\boldsymbol{\pi}_0)^{-1/2}(\partial\boldsymbol{\pi}/\partial\boldsymbol{\theta}_0)$$
$$= \mathbf{1}'(\partial\boldsymbol{\pi}/\partial\boldsymbol{\theta}_0) = \left(\sum_i \partial\pi_i/\partial\boldsymbol{\theta}_0\right)' = \mathbf{0}'$$

so the asymptotic covariance simplifies to $(\mathbf{A}'\mathbf{A})^{-1}$.

In summary, we have established the important result

$$\sqrt{n}(\hat{\boldsymbol{\theta}} - \boldsymbol{\theta}_0) \xrightarrow{d} N[\mathbf{0}, (\mathbf{A}'\mathbf{A})^{-1}] \,. \qquad (12.16)$$

The asymptotic covariance matrix of $\hat{\boldsymbol{\theta}}$ depends on the Jacobian $(\partial\boldsymbol{\pi}/\partial\boldsymbol{\theta}_0)$, and hence on the functional form for modeling $\boldsymbol{\pi}$ in terms of $\boldsymbol{\theta}$. Let $\hat{\mathbf{A}}$ denote $\mathbf{A}$ evaluated at the ML estimate $\hat{\boldsymbol{\theta}}$. We can estimate the covariance matrix by

$$\hat{\mathbf{Cov}}(\hat{\boldsymbol{\theta}}) = (\hat{\mathbf{A}}'\hat{\mathbf{A}})^{-1}/n \,. \qquad (12.17)$$

The asymptotic normality and covariance of $\hat{\boldsymbol{\theta}}$ follows more simply from general results for ML estimators. However, those results require stronger regularity conditions (Rao 1973, p. 364) than the ones assumed here. Suppose we take independent observations from $f(\mathbf{x}; \boldsymbol{\theta})$, some probability density or mass function. The ML estimator $\hat{\boldsymbol{\theta}}$ is efficient, in the sense that

$$\sqrt{n}(\hat{\boldsymbol{\theta}} - \boldsymbol{\theta}) \xrightarrow{d} N(\mathbf{0}, \mathbf{Inf}^{-1})$$

where **Inf** denotes the information matrix for a single observation. The (j, k) element of **Inf** is

$$-E\left(\frac{\partial^2 \log f(\mathbf{x}, \boldsymbol{\theta})}{\partial \theta_j \, \partial \theta_k}\right) = E\left[\frac{\partial \log f(\mathbf{x}, \boldsymbol{\theta})}{\partial \theta_j} \cdot \frac{\partial \log f(\mathbf{x}, \boldsymbol{\theta})}{\partial \theta_k}\right].$$

When f is the distribution of a single observation having multinomial probabilities $\{\pi_1(\boldsymbol{\theta}), \ldots, \pi_N(\boldsymbol{\theta})\}$, this element of **Inf** equals

$$\sum_{i=1}^{N} \left(\frac{\partial \log(\pi_i(\boldsymbol{\theta}))}{\partial \theta_j}\right)\left(\frac{\partial \log(\pi_i(\boldsymbol{\theta}))}{\partial \theta_k}\right)\pi_i(\boldsymbol{\theta}) = \sum_{i=1}^{N} \left(\frac{\partial \pi_i(\boldsymbol{\theta})}{\partial \theta_j}\right)\left(\frac{\partial \pi_i(\boldsymbol{\theta})}{\partial \theta_k}\right)\frac{1}{\pi_i(\boldsymbol{\theta})}.$$

This is the (j, k) element of $\mathbf{A'A}$. Thus the asymptotic covariance is $\mathbf{Inf}^{-1} = (\mathbf{A'A})^{-1}$.

For results of this section to be applicable, a ML estimator of $\boldsymbol{\theta}$ must exist and be a solution of the likelihood equations. This requires the following *strong identifiability* condition: For every $\epsilon > 0$, there exists a $\delta > 0$ such that if $\|\boldsymbol{\theta} - \boldsymbol{\theta}_0\| > \epsilon$, then $\|\boldsymbol{\pi}(\boldsymbol{\theta}) - \boldsymbol{\pi}_0\| > \delta$. This condition implies a weaker one that there cannot be two $\boldsymbol{\theta}$ values having the same $\boldsymbol{\pi}$ value. When there is strong identifiability and the other regularity conditions hold, the probability we can obtain an ML estimator as a root of the likelihood equations converges to 1 as $n \to \infty$. That estimator has the asymptotic properties given above of a solution of the likelihood equations. For proofs of these results, see Birch (1964a) and Rao (1973, pp. 360–362).

12.2.2 Asymptotic Distribution of Cell Probability Estimators

To obtain the asymptotic distribution of the model-based estimator $\hat{\boldsymbol{\pi}}$, we use the Taylor-series expansion

$$\hat{\boldsymbol{\pi}} = \boldsymbol{\pi}(\hat{\boldsymbol{\theta}}) = \boldsymbol{\pi}(\boldsymbol{\theta}_0) + \left(\frac{\partial \boldsymbol{\pi}}{\partial \boldsymbol{\theta}_0}\right)(\hat{\boldsymbol{\theta}} - \boldsymbol{\theta}_0) + o_p(n^{-1/2}). \qquad (12.18)$$

The size of the remainder term follows from $(\hat{\boldsymbol{\theta}} - \boldsymbol{\theta}_0) = O_p(n^{-1/2})$. Now $\boldsymbol{\pi}(\boldsymbol{\theta}_0) = \boldsymbol{\pi}_0$, and $\sqrt{n}(\hat{\boldsymbol{\theta}} - \boldsymbol{\theta}_0)$ is asymptotically normal with asymptotic covariance $(\mathbf{A'A})^{-1}$. By the delta method,

$$\sqrt{n}(\hat{\boldsymbol{\pi}} - \boldsymbol{\pi}_0) \xrightarrow{d} N\left[\mathbf{0}, \left(\frac{\partial \boldsymbol{\pi}}{\partial \boldsymbol{\theta}_0}\right)(\mathbf{A'A})^{-1}\left(\frac{\partial \boldsymbol{\pi}}{\partial \boldsymbol{\theta}_0}\right)'\right]. \qquad (12.19)$$

Suppose the model holds. For estimating $\boldsymbol{\pi}$, $\hat{\boldsymbol{\pi}} = \pi(\hat{\boldsymbol{\theta}})$ is more efficient than the sample proportion $\mathbf{p}$. More generally, for estimating a smooth function $g(\boldsymbol{\pi})$ of $\boldsymbol{\pi}$, $g(\hat{\boldsymbol{\pi}})$ has smaller asymptotic variance than $g(\mathbf{p})$. We next derive this result, discussed in Section 6.4.4. In the derivation, we delete the Nth component from $\mathbf{p}$ and $\hat{\boldsymbol{\pi}}$, so their covariance matrices are positive definite (Problem 12.16). The Nth proportion is linearly related to the first $N-1$ by the constraint that they sum to 1. Let $\boldsymbol{\Sigma} = \mathbf{Diag}(\boldsymbol{\pi}) - \boldsymbol{\pi}\boldsymbol{\pi}'$ denote the $(N-1) \times (N-1)$ covariance matrix of $\sqrt{n}\mathbf{p}$. Letting $\mathbf{1} = (1, 1, \ldots, 1)'$, the inverse of $\boldsymbol{\Sigma}$ is

$$\boldsymbol{\Sigma}^{-1} = \mathbf{Diag}(\boldsymbol{\pi})^{-1} + \mathbf{1}\mathbf{1}'/\pi_N. \tag{12.20}$$

This can be verified by noting that $\boldsymbol{\Sigma}\boldsymbol{\Sigma}^{-1}$ is the identity matrix. The parameter vector $\boldsymbol{\theta}$ has $t < N-1$ elements. We assume the $(N-1) \times t$ Jacobian $(\partial\boldsymbol{\pi}/\partial\boldsymbol{\theta}_0)$ has rank t.

Let $(\partial g/\partial\boldsymbol{\pi}_0) = (\partial g/\partial\pi_1, \ldots, \partial g/\partial\pi_{N-1})'$, evaluated at $\boldsymbol{\pi} = \boldsymbol{\pi}_0$. By the delta method,

$$\text{Asymp. Var}[\sqrt{n}g(\mathbf{p})] = \left(\frac{\partial g}{\partial\boldsymbol{\pi}_0}\right)'[\text{Cov}[(\sqrt{n}\mathbf{p})]\left(\frac{\partial g}{\partial\boldsymbol{\pi}_0}\right) = \left(\frac{\partial g}{\partial\boldsymbol{\pi}_0}\right)'\boldsymbol{\Sigma}\left(\frac{\partial g}{\partial\boldsymbol{\pi}_0}\right)$$

and

$$\text{Asymp. Var}[\sqrt{n}g(\hat{\boldsymbol{\pi}})] = \left(\frac{\partial g}{\partial\boldsymbol{\pi}_0}\right)'[\text{Asymp. Cov}(\sqrt{n}\hat{\boldsymbol{\pi}})]\left(\frac{\partial g}{\partial\boldsymbol{\pi}_0}\right)$$

$$= \left(\frac{\partial g}{\partial\boldsymbol{\pi}_0}\right)'\left(\frac{\partial\boldsymbol{\pi}}{\partial\boldsymbol{\theta}_0}\right)[\text{Asymp. Cov}(\sqrt{n}\hat{\boldsymbol{\theta}})]\left(\frac{\partial\boldsymbol{\pi}}{\partial\boldsymbol{\theta}_0}\right)'\left(\frac{\partial g}{\partial\boldsymbol{\pi}_0}\right).$$

Using (12.11) and (12.20),

$$\text{Asymp. Cov}(\sqrt{n}\hat{\boldsymbol{\theta}}) = (\mathbf{A}'\mathbf{A})^{-1} = [(\partial\boldsymbol{\pi}/\partial\boldsymbol{\theta}_0)'\mathbf{Diag}(\boldsymbol{\pi}_0)^{-1}(\partial\boldsymbol{\pi}/\partial\boldsymbol{\theta}_0)]^{-1}$$

$$= [(\partial\boldsymbol{\pi}/\partial\boldsymbol{\theta}_0)'\boldsymbol{\Sigma}^{-1}(\partial\boldsymbol{\pi}/\partial\boldsymbol{\theta}_0)]^{-1}.$$

Since $\boldsymbol{\Sigma}$ is positive definite and $(\partial\boldsymbol{\pi}/\partial\boldsymbol{\theta}_0)$ has rank t, $\boldsymbol{\Sigma}^{-1}$ and $[(\partial\boldsymbol{\pi}/\partial\boldsymbol{\theta}_0)'\boldsymbol{\Sigma}^{-1}(\partial\boldsymbol{\pi}/\partial\boldsymbol{\theta}_0)]^{-1}$ are also positive definite.

To show that Asymp. $\text{Var}[\sqrt{n}g(\mathbf{p})] \geqslant$ Asymp. $\text{Var}[\sqrt{n}g(\hat{\boldsymbol{\pi}})]$, we must show that

$$\left(\frac{\partial g}{\partial\boldsymbol{\pi}_0}\right)'\left\{\boldsymbol{\Sigma} - \left(\frac{\partial\boldsymbol{\pi}}{\partial\boldsymbol{\theta}_0}\right)\left[\left(\frac{\partial\boldsymbol{\pi}}{\partial\boldsymbol{\theta}_0}\right)'\boldsymbol{\Sigma}^{-1}\left(\frac{\partial\boldsymbol{\pi}}{\partial\boldsymbol{\theta}_0}\right)\right]^{-1}\left(\frac{\partial\boldsymbol{\pi}}{\partial\boldsymbol{\theta}_0}\right)'\right\}\left(\frac{\partial g}{\partial\boldsymbol{\pi}_0}\right) \geqslant 0.$$

But this quadratic form is identical to

$$(\mathbf{Y} - \mathbf{B}\boldsymbol{\zeta})'\boldsymbol{\Sigma}^{-1}(\mathbf{Y} - \mathbf{B}\boldsymbol{\zeta})$$

where $\mathbf{Y} = \boldsymbol{\Sigma}(\partial g / \partial \boldsymbol{\pi}_0)$, $\mathbf{B} = (\partial \boldsymbol{\pi} / \partial \boldsymbol{\theta}_0)$, and $\boldsymbol{\zeta} = (\mathbf{B}'\boldsymbol{\Sigma}^{-1}\mathbf{B})^{-1}\mathbf{B}'\boldsymbol{\Sigma}^{-1}\mathbf{Y}$. The result then follows from the positive definiteness of $\boldsymbol{\Sigma}^{-1}$.

This proof is based on one given by Altham (1984). Her proof uses standard properties of ML estimators, and applies whenever regularity conditions hold that guarantee those properties. Her proof applies not only to categorical data, but to any situation in which a model describes the dependence of a set of parameters $\boldsymbol{\pi}$ on some smaller set $\boldsymbol{\theta}$.

12.3 ASYMPTOTIC DISTRIBUTION OF RESIDUALS AND GOODNESS-OF-FIT STATISTICS

We next study the distributions of goodness-of-fit statistics for the model $\boldsymbol{\pi} = \boldsymbol{\pi}(\boldsymbol{\theta})$. We first derive the asymptotic joint distribution of standardized residuals. Deriving the large-sample chi-squared distribution for the Pearson statistic is then straightforward. We also show that the Pearson and likelihood-ratio statistics are asymptotically equivalent, when the model holds. Our presentation borrows from Rao (1973, Section 6b), Bishop et al. (1975, Chap. 14), and Cox (1984).

12.3.1 Joint Asymptotic Normality of p and $\hat{\boldsymbol{\pi}}$

We first express the joint dependence of $\mathbf{p}$ and $\hat{\boldsymbol{\pi}}$ on $\mathbf{p}$, in order to show the joint asymptotic normality of $\mathbf{p}$ and $\hat{\boldsymbol{\pi}}$. This limiting distribution determines large-sample distributions of statistics that depend on both $\mathbf{p}$ and $\hat{\boldsymbol{\pi}}$, such as cell residuals and summary goodness-of-fit statistics.

From (12.18) and (12.15),

$$\hat{\boldsymbol{\pi}} - \boldsymbol{\pi}_0 = \left(\frac{\partial \boldsymbol{\pi}}{\partial \boldsymbol{\theta}_0} \right) (\hat{\boldsymbol{\theta}} - \boldsymbol{\theta}_0) + o_p(n^{-1/2})$$

$$= \mathbf{D}(\mathbf{p} - \boldsymbol{\pi}_0) + o_p(n^{-1/2})$$

where $\mathbf{D} = \mathbf{Diag}(\boldsymbol{\pi}_0)^{1/2}\mathbf{A}(\mathbf{A}'\mathbf{A})^{-1}\mathbf{A}'\mathbf{Diag}(\boldsymbol{\pi}_0)^{-1/2}$. Therefore

$$\sqrt{n}\left(\begin{matrix} \mathbf{p} - \boldsymbol{\pi}_0 \\ \hat{\boldsymbol{\pi}} - \boldsymbol{\pi}_0 \end{matrix} \right) = \left(\begin{matrix} \mathbf{I} \\ \mathbf{D} \end{matrix} \right) \sqrt{n}(\mathbf{p} - \boldsymbol{\pi}_0) + o_p(1)$$

where $\mathbf{I}$ is a $N \times N$ identity matrix. By the delta method,

$$\sqrt{n}\begin{pmatrix} \mathbf{p} - \boldsymbol{\pi}_0 \\ \hat{\boldsymbol{\pi}} - \boldsymbol{\pi}_0 \end{pmatrix} \xrightarrow{d} N(\mathbf{0}, \boldsymbol{\Sigma}^*) \tag{12.21}$$

where

$$\boldsymbol{\Sigma}^* = \begin{pmatrix} \mathbf{Diag}(\boldsymbol{\pi}_0) - \boldsymbol{\pi}_0\boldsymbol{\pi}_0' & [\mathbf{Diag}(\boldsymbol{\pi}_0) - \boldsymbol{\pi}_0\boldsymbol{\pi}_0']\mathbf{D}' \\ \mathbf{D}[\mathbf{Diag}(\boldsymbol{\pi}_0) - \boldsymbol{\pi}_0\boldsymbol{\pi}_0'] & \mathbf{D}[\mathbf{Diag}(\boldsymbol{\pi}_0) - \boldsymbol{\pi}_0\boldsymbol{\pi}_0']\mathbf{D}' \end{pmatrix}. \tag{12.22}$$

The two matrix blocks on the main diagonal of $\boldsymbol{\Sigma}^*$ are $\mathrm{Cov}(\sqrt{n}\mathbf{p})$ and Asymp. $\mathrm{Cov}(\sqrt{n}\hat{\boldsymbol{\theta}})$, derived previously. The new information here is that Asymp. $\mathrm{Cov}(\sqrt{n}\mathbf{p}, \sqrt{n}\hat{\boldsymbol{\theta}}) = [\mathbf{Diag}(\boldsymbol{\pi}_0) - \boldsymbol{\pi}_0\boldsymbol{\pi}_0']\mathbf{D}'$.

12.3.2 Asymptotic Distribution of Residuals

The standardized cell residual, presented in Section 7.3, is

$$e_i = \frac{n_i - \hat{m}_i}{\hat{m}_i^{1/2}} = \frac{\sqrt{n}(p_i - \hat{\pi}_i)}{\hat{\pi}_i^{1/2}}.$$

We next derive the asymptotic distribution of $\mathbf{e} = (e_1, \ldots, e_N)'$. This is useful in itself, since $\mathbf{e}$ is a common diagnostic measure of lack of fit. It is also helpful in deriving the distribution of the Pearson X^2 statistic, since $X^2 = \Sigma\, e_i^2$.

The residuals are functions of $\mathbf{p}$ and $\hat{\boldsymbol{\pi}}$, which are jointly asymptotically normal. To use the delta method, we calculate

$$\partial e_i/\partial p_i = \sqrt{n}\hat{\pi}_i^{-1/2}, \qquad \partial e_i/\partial \hat{\pi}_i = -\sqrt{n}(p_i + \hat{\pi}_i)/2\hat{\pi}_i^{-3/2}$$

$$\partial e_i/\partial p_j = \partial e_i/\partial \hat{\pi}_j = 0 \quad \text{for } i \neq j.$$

That is,

$$\frac{\partial \mathbf{e}}{\partial \mathbf{p}} = \sqrt{n}\,\mathbf{Diag}(\hat{\boldsymbol{\pi}})^{-1/2} \quad \text{and}$$

$$\frac{\partial \mathbf{e}}{\partial \hat{\boldsymbol{\pi}}} = -(\tfrac{1}{2})\sqrt{n}[\mathbf{Diag}(\mathbf{p}) + \mathbf{Diag}(\hat{\boldsymbol{\pi}})]\,\mathbf{Diag}(\hat{\boldsymbol{\pi}})^{-3/2}. \tag{12.23}$$

Evaluated at $\mathbf{p} = \boldsymbol{\pi}_0$ and $\hat{\boldsymbol{\pi}} = \boldsymbol{\pi}_0$, these matrices equal $\sqrt{n}\,\mathbf{Diag}(\boldsymbol{\pi}_0)^{-1/2}$ and $-\sqrt{n}\,\mathbf{Diag}(\boldsymbol{\pi}_0)^{-1/2}$. Using (12.22), (12.23), and $\mathbf{A}'\boldsymbol{\pi}_0^{1/2} = \mathbf{0}$ (which follows from (12.11)), the delta method implies

$$\mathbf{e} \xrightarrow{d} N(\mathbf{0}, \mathbf{I} - \boldsymbol{\pi}_0^{1/2}\boldsymbol{\pi}_0^{1/2\prime} - \mathbf{A}(\mathbf{A}'\mathbf{A})^{-1}\mathbf{A}'). \tag{12.24}$$

Though asymptotically normal, standardized residuals behave less variably than standard normal random variables. The adjusted residual (Haberman 1973a) is the standardized residual divided by its estimated standard error. This statistic, which *is* asymptotically standard normal, equals

$$r_i = \frac{e_i}{\left[1 - \hat{\pi}_i - \sum_j \sum_k \frac{1}{\hat{\pi}_i} \left(\frac{\partial \pi_i}{\partial \hat{\theta}_j} \right) \left(\frac{\partial \pi_i}{\partial \hat{\theta}_k} \right) \hat{v}^{jk} \right]^{1/2}} \tag{12.25}$$

where $\hat{v}^{jk}$ denotes the element in row j and column k of $(\hat{\mathbf{A}}'\hat{\mathbf{A}})^{-1}$.

12.3.3 Asymptotic Distribution of Pearson Statistic

To prove the Pearson X^2 statistic has an asymptotic chi-squared distribution, we use the following relationship between normal and chi-squared distributions, given by Rao (1973, p. 188):

Let $\mathbf{Y}$ be multivariate normal with mean $\boldsymbol{\nu}$ and covariance matrix $\mathbf{B}$. A necessary and sufficient condition for $(\mathbf{Y} - \boldsymbol{\nu})'\mathbf{C}(\mathbf{Y} - \boldsymbol{\nu})$ to have a chi-squared distribution is $\mathbf{BCBCB} = \mathbf{BCB}$. The degrees of freedom equal the rank of $\mathbf{CB}$.

When $\mathbf{B}$ is non-singular, the condition simplifies to $\mathbf{CBC} = \mathbf{C}$.

The Pearson statistic is related to the standardized residuals by

$$X^2 = \mathbf{e}'\mathbf{e}$$

so we apply this result by identifying $\mathbf{Y}$ with $\mathbf{e}$, $\boldsymbol{\nu} = \mathbf{0}$, $\mathbf{C} = \mathbf{I}$, and $\mathbf{B} = \mathbf{I} - \boldsymbol{\pi}_0^{1/2}\boldsymbol{\pi}_0^{1/2}{}' - \mathbf{A}(\mathbf{A}'\mathbf{A})^{-1}\mathbf{A}'$. Since $\mathbf{C} = \mathbf{I}$, the condition for $(\mathbf{Y} - \boldsymbol{\nu})'\mathbf{C}(\mathbf{Y} - \boldsymbol{\nu}) = \mathbf{e}'\mathbf{e} = X^2$ to have a chi-squared distribution simplifies to $\mathbf{BBB} = \mathbf{BB}$. A direct computation using $\mathbf{A}'\boldsymbol{\pi}_0^{1/2} = \mathbf{0}$ shows $\mathbf{B}$ is idempotent, so the condition holds. Since $\mathbf{e}$ is asymptotically multivariate normal, X^2 is asymptotically chi-squared.

Rao (1973, pp. 73) noted that for symmetric idempotent matrices, the rank equals the trace. The trace of $\mathbf{I}$ is N; the trace of $\boldsymbol{\pi}_0^{1/2}\boldsymbol{\pi}_0^{1/2}{}'$ equals the trace of $\boldsymbol{\pi}_0^{1/2}{}'\boldsymbol{\pi}_0^{1/2} = \Sigma \pi_{i0} = 1$, which is 1; the trace of $\mathbf{A}(\mathbf{A}'\mathbf{A})^{-1}\mathbf{A}'$ equals the trace of $(\mathbf{A}'\mathbf{A})^{-1}(\mathbf{A}'\mathbf{A}) =$ identity matrix of size $t \times t$, which is t. Thus, the rank of $\mathbf{B} = \mathbf{CB}$ is $N - t - 1$, and the asymptotic chi-squared distribution has df $= N - t - 1$.

This is a remarkably simple result. When the sample size is large, the distribution of X^2 does not depend on $\boldsymbol{\pi}_0$ or on the form of the model. It depends only on the difference between the dimension of π (which is $N-1$) and the dimension of θ. When there are $t=0$ parameters, $\{\hat{m}_i = m_i = n\pi_i\}$ are given constants. Then X^2 is Pearson's (1900) statistic (3.7) for testing that multinomial probabilities equal certain specified values, and df $= N - 1$, as claimed by Pearson.

12.3.4 Asymptotic Distribution of Likelihood-Ratio Statistic

Suppose the number of cells N is fixed, and all $\pi_i > 0$. When the model holds, the likelihood-ratio statistic G^2 is asymptotically equivalent to X^2 as $n \to \infty$. To show this, we express

$$G^2 = 2 \sum_i n_i \log\left(\frac{n_i}{\hat{m}_i}\right) = 2n \sum_i p_i \log\left(1 + \frac{p_i - \hat{\pi}_i}{\hat{\pi}_i}\right)$$

and apply the expansion

$$\log(1 + x) = x - x^2/2 + x^3/3 - \cdots, \quad \text{for } |x| < 1 .$$

We identify x with $(p_i - \hat{\pi}_i)/\hat{\pi}_i$, which converges in probability to 0 when the model holds. For large n,

$$G^2 = 2n \sum_i [\hat{\pi}_i + (p_i - \hat{\pi}_i)]\left[\frac{p_i - \hat{\pi}_i}{\hat{\pi}_i} - \left(\frac{1}{2}\right)\frac{(p_i - \hat{\pi}_i)^2}{\hat{\pi}_i^2} + \cdots\right]$$

$$= 2n \sum_i \left[(p_i - \hat{\pi}_i) - \left(\frac{1}{2}\right)\frac{(p_i - \hat{\pi}_i)^2}{\hat{\pi}_i} + \frac{(p_i - \hat{\pi}_i)^2}{\hat{\pi}_i} + O_p(p_i - \hat{\pi}_i)^3\right]$$

$$= n \sum_i \frac{(p_i - \hat{\pi}_i)^2}{\hat{\pi}_i} + 2n\, O_p(n^{-3/2}) = X^2 + O_p(n^{-1/2}) = X^2 + o_p(1)$$

since $\sum (p_i - \hat{\pi}_i) = 0$ and $(p_i - \hat{\pi}_i) = (p_i - \pi_i) - (\hat{\pi}_i - \pi_i)$, both of which are $O_p(n^{-1/2})$. Thus, when the model holds, the difference between X^2 and G^2 converges in probability to 0. As a consequence G^2, like X^2, has an asymptotic chi-squared distribution with df $= N - t - 1$.

Other properties of G^2 are also quite simple to show. For instance, the parameter value that maximizes the likelihood is the one that minimizes G^2. To see this, let

$$G^2(\pi; p) = 2n \sum p_i \log(p_i/\pi_i)$$

and note that the kernel of the log likelihood is

$$L(\theta) = n \sum p_i \log \pi_i(\theta)$$

$$= -n \sum p_i \log \frac{p_i}{\pi_i(\theta)} + n \sum p_i \log p_i$$

$$= - (\tfrac{1}{2})G^2(\pi(\theta); \mathbf{p}) + n \sum p_i \log p_i \, .$$

The second term in the last expression does not depend on θ, so maximizing $L(\theta)$ is equivalent to minimizing G^2 with respect to θ.

A fundamental result for G^2 concerns comparisons of nested models. Suppose model M_2 is a special case of model M_1, and let t_2 and t_1 denote the numbers of parameters in the two models. Let $\{\hat{\pi}_{1i}\}$ and $\{\hat{\pi}_{2i}\}$ denote ML estimators of cell probabilities for the two models. Then

$$G^2(M_2) - G^2(M_1) = 2n \sum p_i \log(\hat{\pi}_{1i}/\hat{\pi}_{2i})$$

has the form of -2(log likelihood ratio) for testing the hypothesis that M_2 holds against the alternative that M_1 holds. Theory for likelihood-ratio tests suggests that when the simpler model holds, the asymptotic distribution of $G^2(M_2) - G^2(M_1)$ is chi-squared with $t_1 - t_2$ degrees of freedom. For details, see Bishop et al. (1975, pp. 525–526), Haberman (1974a, p. 108), and Rao (1973, pp. 418–419). The statistic $X^2(M_2 \mid M_1)$ given in (7.3) is simply a quadratic approximation for the G^2 difference. Haberman (1977a) noted that these tests can perform well even for large, sparse tables, as long as $t_1 - t_2$ is small compared to the sample size and no expected frequency has larger order of magnitude than the others.

12.3.5 Asymptotic Noncentral Distributions

Results in this chapter assume that a certain parametric model holds. Since any unsaturated model is unlikely to hold perfectly for a given application, one might question the scope of these results. This is not a problem if we remind ourselves that models are merely convenient approximations for reality, and we rarely completely believe them. For instance, the ML estimator $\hat{\theta}$ converges to a value θ_0 that we can regard as the best fit of the chosen model to reality. In this sense, inferences for θ give us information about a useful approximation for reality. Similarly, model-based inferences about cell probabilities are inconsistent for the true probabilities when the model does not hold; nevertheless, those inferences are consistent for describing a useful smoothing of reality.

For goodness-of-fit statistics, unlike model parameters, there is a relevant distinction between limiting behavior when the model holds and when it does not hold. When the chosen model holds, we have seen that X^2 and G^2 have a limiting chi-squared distribution, and the difference between them disappears as the sample size increases. When the model does not hold, X^2 and G^2 tend to grow unboundedly as n increases, and the difference between them does not go to zero. To obtain proper limiting distributions, we consider a sequence of situations for which the lack of fit diminishes as n increases. Specifically, suppose the model is $\pi = \mathbf{f}(\boldsymbol{\theta})$, but in reality the probabilities equal

$$\boldsymbol{\pi}_n = \mathbf{f}(\boldsymbol{\theta}) + \boldsymbol{\delta}/\sqrt{n} \ . \tag{12.26}$$

That is, the "best fit" of the model to the population indicates that the ith probability equals $f_i(\boldsymbol{\theta})$, but the true value differs from that by $\delta_i/\sqrt{n}$.

In (12.26), we index the true probability by the sample size n, and let the model lack of fit go to 0 at the rate $n^{-1/2}$ as $n \to \infty$. For this representation, Mitra (1958) showed that the Pearson statistic X^2 has a limiting noncentral chi-squared distribution, with df $= N - t - 1$ and noncentrality parameter

$$\lambda = \boldsymbol{\delta}' \, \mathbf{Diag}[\mathbf{f}(\boldsymbol{\theta})]^{-1}\boldsymbol{\delta} \ .$$

From (12.26), the noncentrality equals

$$\lambda = n \sum_{i=1}^{n} \frac{[\pi_{ni} - f_i(\theta)]^2}{f_i(\theta)} \ .$$

This has the form of X^2, with the sample values p_i and $\hat{\pi}_i$ replaced by population values π_{ni} and $f_i(\boldsymbol{\theta})$. Similarly the noncentrality of the likelihood-ratio statistic has the form of G^2, with the same substitution. Haberman (1974a, pp. 109–112) showed that, under certain conditions, G^2 and X^2 have the same limiting distribution; that is, their noncentrality values converge to a common value as $n \to \infty$.

Representation (12.26) means that for large samples, the noncentral chi-squared approximation is valid when the model is just barely incorrect. In practice, it is often reasonable to adopt (12.26) for fixed, finite n in order to approximate the distribution of X^2, even though we would not expect (12.26) to hold as we obtain more data. The alternative representation

$$\boldsymbol{\pi} = \mathbf{f}(\boldsymbol{\theta}) + \boldsymbol{\delta} \tag{12.27}$$

of a *fixed* amount by which π differs from $\mathbf{f}(\boldsymbol{\theta})$ as $n \to \infty$ may seem more natural. In fact, representation (12.27) is more appropriate than (12.26) for proving the test to be consistent (i.e., for convergence to 1 of the probability of rejecting the null hypothesis that the model holds). For representation (12.27), however, the noncentrality parameter λ grows unboundedly as $n \to \infty$, and there is no proper limiting distribution for X^2 and G^2. Drost et al. (1989) gave noncentral approximations using other sequences of alternatives.

When the model holds, $\delta = 0$ in either representation. That is, $f(\theta) = \pi(\theta)$, $\lambda = 0$, and the results in Sections 12.3.3 and 12.3.4 apply.

12.4 ASYMPTOTIC DISTRIBUTIONS FOR LOGLINEAR MODELS

For loglinear models, formulas in Section 6.4 for the asymptotic covariance matrices of $\hat{\theta}$ and $\hat{\pi}$ are special cases of ones derived in Section 12.2. To constrain probabilities to sum to 1, we express such models for multinomial sampling as

$$\pi = \exp(X\theta)/[1' \exp(X\theta)] \tag{12.28}$$

where X is a model matrix and $1' = (1, \ldots, 1)$. Letting x_i denote the ith row of X,

$$\pi_i = \pi_i(\theta) = \exp(x_i\theta)/\left[\sum_k \exp(x_k\theta)\right].$$

12.4.1 Asymptotic Covariance Matrices

A model affects covariance matrices through the Jacobian. Since

$$\frac{\partial \pi_i}{\partial \theta_j} = \frac{[1' \exp(X\theta)][\exp(x_i\theta)]x_{ij} - [\exp(x_i\theta)]\left[\sum_k x_{kj} \exp(x_k\theta)\right]}{[1' \exp(X\theta)]^2}$$

$$= \pi_i x_{ij} - \pi_i \sum_k x_{kj} \pi_k$$

we have

$$\partial \pi/\partial \theta = [\text{Diag}(\pi) - \pi\pi']X .$$

The information matrix at θ_0 is

$$\mathbf{A}'\mathbf{A} = (\partial \boldsymbol{\pi}/\partial \boldsymbol{\theta}_0)' \, \mathbf{Diag}(\boldsymbol{\pi}_0)^{-1}(\partial \boldsymbol{\pi}/\partial \boldsymbol{\theta}_0)$$

$$= \mathbf{X}'[\mathbf{Diag}(\boldsymbol{\pi}_0) - \boldsymbol{\pi}_0 \boldsymbol{\pi}_0']' \, \mathbf{Diag}(\boldsymbol{\pi}_0)^{-1}$$

$$\times \, [\mathbf{Diag}(\boldsymbol{\pi}_0) - \boldsymbol{\pi}_0 \boldsymbol{\pi}_0']\mathbf{X}$$

$$= \mathbf{X}'[\mathbf{Diag}(\boldsymbol{\pi}_0) - \boldsymbol{\pi}_0 \boldsymbol{\pi}_0']\mathbf{X} \, .$$

Thus, for multinomial loglinear models, $\hat{\boldsymbol{\theta}}$ is asymptotically normally distributed, with estimated covariance matrix

$$\hat{\mathrm{Cov}}(\hat{\boldsymbol{\theta}}) = \{\mathbf{X}'[\mathbf{Diag}(\hat{\boldsymbol{\pi}}) - \hat{\boldsymbol{\pi}}\hat{\boldsymbol{\pi}}']\mathbf{X}\}^{-1}/n \, . \tag{12.29}$$

Similarly, from (12.19), the estimated asymptotic covariance matrix of $\hat{\boldsymbol{\pi}}$ is

$$\hat{\mathrm{Cov}}(\hat{\boldsymbol{\pi}}) = [\mathbf{Diag}(\hat{\boldsymbol{\pi}}) - \hat{\boldsymbol{\pi}}\hat{\boldsymbol{\pi}}']\mathbf{X}\{\mathbf{X}'[\mathbf{Diag}(\hat{\boldsymbol{\pi}}) - \hat{\boldsymbol{\pi}}\hat{\boldsymbol{\pi}}']\mathbf{X}\}^{-1}$$

$$\times \, \mathbf{X}'[\mathbf{Diag}(\hat{\boldsymbol{\pi}}) - \hat{\boldsymbol{\pi}}\hat{\boldsymbol{\pi}}']/n \, . \tag{12.30}$$

From (12.24), the standardized residuals are asymptotically normal with

Asymp. Cov(**e**)

$$= \mathbf{I} - \boldsymbol{\pi}_0^{1/2}\boldsymbol{\pi}_0^{1/2\prime} - \mathbf{A}(\mathbf{A}'\mathbf{A})^{-1}\mathbf{A}'$$

$$= \mathbf{I} - \boldsymbol{\pi}_0^{1/2}\boldsymbol{\pi}_0^{1/2\prime} - \mathbf{Diag}(\boldsymbol{\pi}_0)^{-1/2}[\mathbf{Diag}(\boldsymbol{\pi}_0) - \boldsymbol{\pi}_0\boldsymbol{\pi}_0']\mathbf{X}$$

$$\times \, \{\mathbf{X}'[\mathbf{Diag}(\boldsymbol{\pi}_0) - \boldsymbol{\pi}_0\boldsymbol{\pi}_0']\mathbf{X}\}^{-1}\mathbf{X}'[\mathbf{Diag}(\boldsymbol{\pi}_0) - \boldsymbol{\pi}_0\boldsymbol{\pi}_0']\mathbf{Diag}(\boldsymbol{\pi}_0)^{-1/2} \, .$$

12.4.2 Connection with Poisson Loglinear Models

In this book, we represented loglinear models in terms of expected cell frequencies $\mathbf{m} = (m_1, \ldots, m_N)'$, where $m_i = n\pi_i$, using formulas of the form

$$\log \mathbf{m} = \mathbf{X}_a \boldsymbol{\theta}_a \, . \tag{12.31}$$

The model matrix $\mathbf{X}_a$ and parameter vector $\boldsymbol{\theta}_a$ in this formula are slightly different from $\mathbf{X}$ and $\boldsymbol{\theta}$ in multinomial model (12.28). The simpler expression (12.31) does not incorporate constraints on $\mathbf{m}$, and is more natural for the Poisson sampling model. For model (12.28), $\mathbf{m} = n\boldsymbol{\pi}$ satisfies

$$\log \mathbf{m} = \log n\boldsymbol{\pi} = \mathbf{X}\boldsymbol{\theta} + [\log n - \log(\mathbf{1}' \exp(\mathbf{X}\boldsymbol{\theta}))]\mathbf{1}$$

$$= \mathbf{X}\boldsymbol{\theta} + \mathbf{1}\mu$$

where $\mu = \log n - \log(1' \exp(\mathbf{X\theta}))]$. In other words, multinomial model (12.28) implies Poisson model (12.31) with

$$\mathbf{X}_a = [\mathbf{1} : \mathbf{X}] \quad \text{and} \quad \mathbf{\theta}_a = (\mu, \mathbf{\theta}')' . \tag{12.32}$$

From (12.32), the columns of $\mathbf{X}$ in the multinomial representation must be linearly independent of $\mathbf{1}$; that is, the parameter μ, which relates to the total sample size, does not appear in $\mathbf{\theta}$. The dimension of $\mathbf{\theta}$ is one less than the number of parameters reported in this text for Poisson loglinear models. For instance, for the saturated model, $\mathbf{\theta}$ has $N - 1$ elements for the multinomial representation, reflecting the sole constraint on $\mathbf{\pi}$ of $\Sigma \, \pi_i = 1$.

CHAPTER NOTES

Section 12.1: Delta Method

12.1 The book by Bishop et al. (1975, Chap. 14) is a good source for a thorough discussion of o_p and O_p notation and the delta method. Benichou and Gail (1989) discussed the delta method for statistics that are implicit functions of asymptotically normal random variables.

In applying the delta method to a function g of an asymptotically normal random vector $\mathbf{T}_n$, suppose the 1st-order, ... , $(a - 1)$st-order differentials of the function are zero at $\mathbf{\theta}$, but the ath-order differential is nonzero. A generalization of the delta method implies that $n^{a/2}[g(\mathbf{T}_n) - g(\mathbf{\theta})]$ has limiting distribution involving products of order a of components of a normal random vector. When $a = 2$, the limiting distribution is a quadratic form in a multivariate normal vector, which often has a chi-squared distribution. Serfling (1980, p. 124) gave details.

The jackknife and the bootstrap are alternative tools for estimating standard errors. They are particularly helpful when use of the delta method is questionable—for instance, for small samples, highly sparse data, or complex sampling designs. For details, see Fay (1985), Parr and Tolley (1982), and Simonoff (1986).

Section 12.3: Asymptotic Distributions of Residuals and Goodness-of-Fit Statistics

12.2 The square root of a Poisson random variable with expectation m is, for large m, approximately normal with standard deviation $\frac{1}{2}$ (Problem 12.7). This suggests an alternative statistic for testing goodness of fit, the *Freeman–Tukey* statistic, $4 \, \Sigma \, (\sqrt{n}_i - \sqrt{\hat{m}}_i)^2$. When the model holds, the difference between this statistic and X^2 is also $o_p(1)$ as $n \to \infty$. See Bishop et al. (1975, p. 514) for details.

The asymptotic results of this chapter do not apply when the number of cells N grows as the sample size $n \to \infty$, or when different expected frequencies grow at different rates. Haberman (1988) gave examples of how the consistency of X^2 breaks down when the asymptotics are not of standard type. Zelterman (1987) gave an adaptation of X^2 that Haberman noted had better behavior for some alternative asymptotic approaches.

PROBLEMS

12.1 Show:
  a. If $c > 0$, $n^{-c} = o(1)$ as $n \to \infty$.
  b. If c is a nonzero constant, $c z_n$ has the same order as z_n; that is, $o(c z_n)$ is equivalent to $o(z_n)$ and $O(c z_n)$ is equivalent to $O(z_n)$.
  c. $o(y_n) o(z_n) = o(y_n z_n)$, $O(y_n) O(z_n) = O(y_n z_n)$, $o(y_n) O(z_n) = o(y_n z_n)$.

12.2 Show that if X^2 has an asymptotic chi-squared distribution with fixed df as $n \to \infty$, then $X^2/n = o_p(1)$.

12.3

  a. Use Tchebychev's inequality to prove that if $E(X_n) = \mu_n$ and $\text{Var}(X_n) = \sigma_n^2 < \infty$, then $(X_n - \mu_n) = O_p(\sigma_n)$.
  b. Suppose $Y_1, \ldots, Y_n$ are independent observations with $E(Y_i) = \mu$ and $\text{Var}(Y_i) = \sigma^2$ for $i = 1, \ldots, n$. Let $\bar{Y}_n$ denote their sample mean. Apply (a) to show that $(\bar{Y}_n - \mu) = O_p(n^{-1/2})$.

12.4 Suppose Y is a Poisson random variable, with mean μ.
  a. For a constant c, show that

$$E[\log(Y + c)] = \log(\mu) + (c - \tfrac{1}{2})/\mu + O(\mu^{-2})$$

  (Hint: Note $\log(Y + c) = \log(\mu) + \log[1 + (Y + c - \mu)/\mu]$.)
  b. Suppose cell counts in a 2×2 table are independent Poisson random variables. To reduce bias in estimating the log odds ratio, argue that it is sensible to use the sample log odds ratio after adding $\tfrac{1}{2}$ to each cell.

12.5 Let p denote the sample proportion for n independent Bernoulli trials. Find the asymptotic distribution of the estimator $[p(1 - p)]^{1/2}$ of the Bernoulli standard deviation. What happens when $\pi = 0.5$?

12.6 Suppose T_n has a Poisson distribution with mean $m = n\mu$, for fixed $\mu > 0$. For large n, show the distribution of $\log(T_n)$ is approximately normal with mean $\log(m)$ and variance m^{-1}. (Hint: By the Central Limit Theorem, T_n/n is approximately $N(\mu, \mu/n)$ for large n.)

12.7 Refer to the previous problem.

    a. Show $\sqrt{T_n}$ has asymptotic variance $1/4$.

    b. For binomial data, show the asymptotic variance of $\sin^{-1}(\sqrt{p})$ is $1/4n$.

    c. The transformations in (a) and (b) are *variance-stabilizing*, producing random variables having asymptotic variances that are the same for all values of the parameter. Explain how these transformations could be useful for applying ordinary least squares to count data.

12.8 For a multinomial $(n, \{\pi_i\})$ distribution, show the correlation between p_i and p_j is $-[\pi_i \pi_j/(1 - \pi_i)(1 - \pi_j)]^{1/2}$. What does this equal when $\pi_i = 1 - \pi_j$ and $\pi_k = 0$ for $k \neq i, j$?

12.9 An animal population has N species, with population proportion π_i of the ith species. *Simpson's index of ecological diversity* (Simpson 1949) is $I(\pi) = 1 - \Sigma \pi_i^2$.

    a. Suppose two animals are randomly chosen from the population, with replacement. Show that $I(\pi)$ is the probability they are different species.

    b. Let p_i denote the proportion of a large random sample classified in species i, $i = 1, \ldots, N$. Show that the estimated asymptotic standard error of $I(\mathbf{p})$ is

$$2\left\{\left[\sum p_i^3 - \left(\sum p_i^2\right)^2\right]/n\right\}^{1/2}.$$

(For further information on diversity measures, see Rao 1982).

12.10 Suppose $\{n_i\}$ are independent Poisson random variables. If $\hat{\beta} = \Sigma a_i \log(n_i)$, show by the delta method that its estimated asymptotic variance is

$$\hat{\sigma}^2(\hat{\beta}) = \sum a_i^2/n_i.$$

(This formula applies to ML estimation of parameters for the saturated loglinear model, for which estimates are the same contrasts of $\{\log(n_i)\}$ that the parameters are of $\{\log(m_i)\}$. Formula (12.9) yields the asymptotic covariance structure of such estimators. Lee (1977) gave details.)

12.11 Assuming independent binomial samples, derive the asymptotic standard error of the log of the relative risk.

12.12 The sample size may need to be relatively large before the sampling distribution of sample gamma is approximately normal, especially if $|\gamma|$ is large (Rosenthal 1966). A transformation of gamma often converges more quickly to normality. Consider the Fisher-type transform $\hat{\xi} = \tanh^{-1}(\hat{\gamma}) = (\frac{1}{2})\log[(1 + \hat{\gamma})/(1 - \hat{\gamma})]$ (Agresti 1980, Carr et al. 1989, O'Gorman and Woolson 1988).

    a. Show that the asymptotic variance of $\hat{\xi}$ can be estimated by the asymptotic variance of $\hat{\gamma}$ multiplied by $(1 - \hat{\gamma}^2)^{-2}$.

    b. Explain how to construct a confidence interval for ξ and use it to obtain one for γ.

    c. Show that $\hat{\xi} = (\frac{1}{2})\log(C/D)$, which is half the log odds ratio in the 2×2 case.

12.13 Let $\phi^2(\mathbf{T}) = \Sigma(T_i - \pi_{i0})^2/\pi_{i0}$. Then $\phi^2(\mathbf{p}) = X^2/n$, where X^2 is the Pearson statistic for testing $H_0: \pi_i = \pi_{i0}$, $i = 1, \ldots, N$ (Section 3.2), and $n\phi^2(\boldsymbol{\pi})$ is the noncentrality for that test when $\boldsymbol{\pi}$ is the true value. Under H_0, why does the delta method not yield an asymptotic normal distribution for $\phi^2(\mathbf{p})$? (See Note 12.1.)

12.14 In an $I \times J$ contingency table, let θ_{ij} denote local odds ratio (2.7), and let $\hat{\theta}_{ij}$ denote the same function of sample proportions.

    a. Show Asymp. $\text{Cov}(\sqrt{n} \log \hat{\theta}_{ij}, \sqrt{n} \log \hat{\theta}_{i+1,j}) = -[1/\pi_{i+1,j} + 1/\pi_{i+1,j+1}]$.

    b. Show Asymp. $\text{Cov}(\sqrt{n} \log \hat{\theta}_{ij}, \sqrt{n} \log \hat{\theta}_{i+1,j+1}) = 1/\pi_{i+1,j+1}$.

    c. When $\hat{\theta}_{ij}$ and $\hat{\theta}_{hk}$ are based on mutually exclusive sets of cells, show Asymp. $\text{Cov}(\sqrt{n} \log \hat{\theta}_{ij}, \sqrt{n} \log \hat{\theta}_{hk}) = 0$.

    d. State the joint asymptotic distribution of the $(I-1)(J-1)$ estimators $\{\log \hat{\theta}_{ij}\}$.

12.15 ML estimates of $\{m_{ijk}\}$, and hence the X^2 and G^2 statistics, are not direct for model (XY, XZ, YZ). Alternative approaches may yield direct analyses. For $2 \times 2 \times 2$ tables, find a statistic for testing the hypothesis of no three-factor interaction, using the delta method with the asymptotic normality of $\log \hat{\theta}_{111}$, where

$$\hat{\theta}_{111} = \frac{p_{111}p_{221}/p_{121}p_{211}}{p_{112}p_{222}/p_{122}p_{212}}.$$

12.16 Refer to Section 12.2.2, with $\Sigma = \mathbf{Diag}(\boldsymbol{\pi}) - \boldsymbol{\pi}\boldsymbol{\pi}'$ the covariance matrix of $\mathbf{p} = (p_1, \ldots, p_{N-1})'$. Let

$$Z = c_i \quad \text{with probability } \pi_i, \quad i = 1, \dots, N - 1$$

$$= 0 \quad \text{with probability } \pi_N$$

and let $\mathbf{c} = (c_1, \dots, c_{N-1})'$.

a. Show $E(Z) = \mathbf{c}'\boldsymbol{\pi}$, $E(Z^2) = \mathbf{c}' \mathbf{Diag}(\boldsymbol{\pi})\mathbf{c}$, and $\mathrm{Var}(Z) = \mathbf{c}'\boldsymbol{\Sigma}\mathbf{c}$.

b. Suppose at least one $c_i \neq 0$, and all $\pi_i > 0$. Show $\mathrm{Var}(Z) > 0$, and deduce that $\boldsymbol{\Sigma}$ is positive definite.

c. If $\boldsymbol{\pi} = (\pi_1, \dots, \pi_N)'$ (so $\boldsymbol{\Sigma}$ is $N \times N$), prove $\boldsymbol{\Sigma}$ is not positive definite.

12.17 Consider the model for a 2×2 table, $\pi_{11} = \theta^2$, $\pi_{12} = \pi_{21} = \theta(1 - \theta)$, $\pi_{22} = (1 - \theta)^2$, where θ is unknown (see Problem 3.21).

a. Find the matrix $\mathbf{A}$ in (12.14) for this model.

b. Use $\mathbf{A}$ to obtain the asymptotic variance of the ML estimator $\hat{\theta}$. (As a check, it is simple to find it directly using the inverse of $-E\partial^2 L/\partial\theta^2$, where L is the log likelihood.) For which θ value is the variance maximized? What is the distribution of $\hat{\theta}$ if $\theta = 0$ or $\theta = 1$?

c. Find the asymptotic covariance matrix of $\sqrt{n}\,\hat{\boldsymbol{\pi}}$.

d. Find the degrees of freedom for the asymptotic distribution of X^2.

12.18 Refer to the model in Section 3.2.4. Obtain the asymptotic variance of $\hat{\pi}$.

12.19 Justify the use of *estimated* asymptotic covariance matrices. For instance, how do we know that $\hat{\mathbf{A}}'\hat{\mathbf{A}}$ is close to $\mathbf{A}'\mathbf{A}$, for large samples?

12.20 Suppose cell counts $\{n_i\}$ are independent Poisson random variables, where $m_i = E(n_i)$. Consider the Poisson loglinear model

$$\log \mathbf{m} = \mathbf{X}_a \boldsymbol{\theta}_a, \quad \text{where } \mathbf{m} = (m_1, \dots, m_N) .$$

Using arguments similar to those in Section 12.2, show that the large sample covariance matrix of $\hat{\boldsymbol{\theta}}_a$ can be estimated by $[\mathbf{X}_a' \mathbf{Diag}(\hat{\mathbf{m}})\mathbf{X}_a]^{-1}$, where $\hat{\mathbf{m}}$ is the ML estimator of $\mathbf{m}$.

12.21 For a given set of constraints on parameters, show that weak identifiability conditions hold for the independence loglinear model for a two-way table; that is, when two values for the parameter vector give the same probabilities, those parameter vectors must be identical.

12.22 Use the delta method, with derivatives (12.23), to derive the asymptotic covariance matrix in (12.24) for standardized residuals. Show that this matrix is idempotent.

12.23 In some situations, X^2 and G^2 take very similar values. Explain the joint influence on this event of (1) whether the model holds, (2) whether n is large, and (3) whether N is large.

12.24 Give $\mathbf{X}$ and $\boldsymbol{\theta}$ in multinomial representation (12.28) for the independence model for an $I \times J$ table. By contrast, give $\mathbf{X}_a$ for the corresponding Poisson loglinear model (12.31).

12.25 Using (12.19) and (12.29), derive the asymptotic covariance matrix (12.30).

12.26 The estimator $\hat{\pi}_{ij} = p_{i+} p_{+j}$ is the ML estimator of π_{ij} for the independence model. Suppose the independence model does not hold.

a. Show $E(p_{i+} p_{+j}) = \pi_{i+} \pi_{+j}(n-1)/n + \pi_{ij}/n$.

b. Show that

$$\text{Var}(p_{i+} p_{+j})$$
$$= [c_1 \pi_{i+}^2 \pi_{+j}^2 + c_2 \pi_{ij} \pi_{i+} \pi_{+j} + (n-1)(n-2)$$
$$\times (\pi_{i+} \pi_{+j}^2 + \pi_{i+}^2 \pi_{+j}) + c_3 \pi_{ij}^2$$
$$+ (n-1)(2\pi_{ij} \pi_{+j} + 2\pi_{ij} \pi_{i+} + \pi_{i+} \pi_{+j}) + \pi_{ij}]/n^3 ,$$

where $c_1 = (n-1)(6-4n)$, $c_2 = 2(n-1)(n-4)$, and $c_3 = (n-2)$.
(Hint: Let $\mathbf{X}_i = (X_{i1}, \ldots, X_{in})$, where $X_{ia} = 1$ if the ath observation falls in row i, and 0 otherwise. Note $p_{i+} = \Sigma_a X_{ia}/n$. Similarly, let $\mathbf{Y}_j = (Y_{j1}, \ldots, Y_{jn})$ indicate whether each observation is in column j. Evaluate $E(p_{i+} p_{+j})^2 = E[\Sigma_a \Sigma_b \Sigma_c \Sigma_d X_{ia} X_{ib} Y_{jc} Y_{jd}]/n^4$.)

c. Show $\text{MSE}(p_{i+} p_{+j})$ has the same formula given in (b), but with $c_1 = (n-1)(6-4n) + n(n-1)^2$, $c_2 = 2(n-1)(n-4) - 2n(n-1)^2$, and $c_3 = (n-2) + n(n-1)^2$. (This result was used in Section 6.4.4.)

CHAPTER 13

Estimation Theory for Parametric Models

Chapter 12 presented asymptotic theory for models for categorical data. This final chapter discusses fixed-sample-size estimation theory.

The models presented in this text are generalized linear models, or multivariate extensions of such models. Section 13.1 outlines ML theory for generalized linear models. Section 13.2 illustrates the theory for Poisson and multinomial loglinear models.

Other methods of estimation have similar asymptotic properties as maximum likelihood and may be more appropriate for certain purposes. Section 13.2 introduces quasi-likelihood estimation and its role in the treatment of overdispersion. Section 13.3 discussed the weighted least squares approach to fitting models for categorical data. Section 13.4 uses Bayesian methods to estimate cell probabilities in a contingency table. The final section outlines other methods of estimation for categorical data analysis.

13.1 MAXIMUM LIKELIHOOD FOR GENERALIZED LINEAR MODELS

Section 4.1 introduced generalized linear models (GLMs). Models studied in this text are GLMs with binomial or Poisson random component, or multivariate extensions of GLMs with multinomial random component. In this section we take a closer look at GLMs and their properties, outlining results from Nelder and Wedderburn (1972) and McCullagh and Nelder (1989).

13.1.1 Components of a Generalized Linear Model

The first component of a GLM, the *random component*, refers to the response variable, Y. Suppose the N observations on Y are independent, and denote their values by $(y_1, \ldots, y_N)$. We assume the probability function for y_i has form

$$f(y_i; \theta_i, \phi) = \exp\{[y_i\theta_i - b(\theta_i)]/a(\phi) + c(y_i, \phi)\}. \tag{13.1}$$

The parameter θ_i is called the *natural parameter*. The function $a(\phi)$ often has form $a(\phi) = \phi/\omega_i$ for known weight ω_i, and ϕ is called the *dispersion parameter*. For instance, when y_i is a mean of n_i independent readings, usually $\omega_i = n_i$.

When ϕ is a known constant, (13.1) simplifies to form (4.1) for the natural exponential family,

$$f(y_i; \theta_i) = a(\theta_i)b(y_i)\exp[y_iQ(\theta_i)].$$

We identify $Q(\theta)$ in (4.1) with $\theta/a(\phi)$ in (13.1), $a(\theta)$ in (4.1) with $\exp[-b(\theta)/a(\phi)]$ in (13.1), and $b(y)$ in (4.1) with $\exp[c(y, \phi)]$ in (13.1). The more general formula (13.1) is useful for two-parameter families, such as the normal or gamma, in which ϕ is a nuisance parameter. It is not needed for one-parameter families such as the binomial and Poisson.

General expressions for the first two moments of Y_i use terms in (13.1). Let $\ell(\theta_i, \phi; y_i) = \log f(y_i; \theta_i, \phi)$ denote the contribution of the ith observation to the log likelihood. Then

$$\ell(\theta_i, \phi; y_i) = [y_i\theta_i - b(\theta_i)]/a(\phi) + c(y_i, \phi)$$

and

$$\partial\ell/\partial\theta_i = [y_i - b'(\theta_i)]/a(\phi), \qquad \partial^2\ell/\partial\theta_i^2 = -b''(\theta_i)/a(\phi)$$

where $b'(\theta_i)$ and $b''(\theta_i)$ denote the first two derivatives of b evaluated at θ_i. We apply the likelihood results

$$E\left(\frac{\partial\ell}{\partial\theta}\right) = 0 \quad \text{and} \quad -E\left(\frac{\partial^2\ell}{\partial\theta^2}\right) = E\left(\frac{\partial\ell}{\partial\theta}\right)^2$$

which hold under regularity conditions satisfied by the exponential family (Cox and Hinkley 1974, Section 4.8). From the first formula,

$$\mu_i = E(Y_i) = b'(\theta_i). \tag{13.2}$$

The second formula implies

$$b''(\theta_i)/a(\phi) = E[(Y - b'(\theta_i))/a(\phi)]^2 = \text{Var}(Y_i)/[a(\phi)]^2$$

so that

$$\text{Var}(Y_i) = b''(\theta_i)a(\phi). \tag{13.3}$$

Let $x_{i1}, \ldots, x_{it}$ denote values of t explanatory variables for the ith observation. The *systematic component*, the second component of a GLM, relates parameters $\{\eta_i\}$ to the explanatory variables using a *linear predictor*

$$\eta_i = \sum_j \beta_j x_{ij}, \quad i = 1, \ldots, N.$$

In matrix form,

$$\boldsymbol{\eta} = \mathbf{X}\boldsymbol{\beta}$$

where $\boldsymbol{\eta} = (\eta_1, \ldots, \eta_N)'$, $\boldsymbol{\beta} = (\beta_1, \ldots, \beta_t)'$ are model parameters, and $\mathbf{X}$ is the $N \times t$ model matrix.

The *link function*, the third component of a GLM, connects the expectation μ_i of Y_i to the linear predictor by

$$\eta_i = g(\mu_i)$$

where g is a monotone, differentiable function. Thus, a GLM links the expected value of the response to the explanatory variables through the equation

$$g(\mu_i) = \sum_j \beta_j x_{ij}.$$

The function g for which $g(\mu_i) = \theta_i$ in (13.1) is called the *canonical link*. For it, there is the direct relationship

$$\theta_i = \sum_j \beta_j x_{ij}$$

between the natural parameter and the linear predictor. Since $\mu_i = b'(\theta_i)$, the canonical link is the inverse of the function b'.

13.1.2 Likelihood Equations

For N independent observations, the log likelihood is

$$L(\boldsymbol{\beta}) = \sum_i \log f(y_i; \theta_i, \phi) = \sum_i \ell_i$$

where $\ell_i = \ell(\theta_i, \phi; y_i)$, $i = 1, \ldots, N$. We use notation $L(\beta)$ to recognize that θ depends on the model parameter β.

To obtain the likelihood equations, we calculate

$$\frac{\partial \ell_i}{\partial \beta_j} = \frac{\partial \ell_i}{\partial \theta_i} \frac{\partial \theta_i}{\partial \mu_i} \frac{\partial \mu_i}{\partial \eta_i} \frac{\partial \eta_i}{\partial \beta_j} .$$

Since $\partial \ell_i / \partial \theta_i = [y_i - b'(\theta_i)]/a(\phi)$, and since $\mu_i = b'(\theta_i)$ and $\mathrm{Var}(Y_i) = b''(\theta_i)a(\phi)$,

$$\partial \ell_i / \partial \theta_i = (y_i - \mu_i)/a(\phi) ,$$

$$\partial \mu_i / \partial \theta_i = b''(\theta_i) = \mathrm{Var}(Y_i)/a(\phi) .$$

Also, since $\eta_i = \Sigma_j \beta_j x_{ij}$,

$$\partial \eta_i / \partial \beta_j = x_{ij} .$$

Finally, since $\eta_i = g(\mu_i)$, $\partial \mu_i / \partial \eta_i$ depends on the link function for the model. In summary,

$$\frac{\partial \ell_i}{\partial \beta_j} = \frac{(y_i - \mu_i)}{a(\phi)} \frac{a(\phi)}{\mathrm{Var}(Y_i)} \frac{\partial \mu_i}{\partial \eta_i} x_{ij} \tag{13.4}$$

and the likelihood equations are

$$\sum_{i=1}^{N} \frac{(y_i - \mu_i)x_{ij}}{\mathrm{Var}(Y_i)} \frac{\partial \mu_i}{\partial \eta_i} = 0 , \quad j = 1, \ldots, t . \tag{13.5}$$

The likelihood equations are nonlinear functions of β. Solving them for $\hat{\beta}$ requires iterative methods, discussed in the next subsection. The rate of convergence of $\hat{\beta}$ to β depends on the information matrix. Now

$$E\left(\frac{\partial^2 \ell_i}{\partial \beta_h \partial \beta_j}\right) = -E\left(\frac{\partial \ell_i}{\partial \beta_h}\right)\left(\frac{\partial \ell_i}{\partial \beta_j}\right)$$

$$= -E\left[\frac{(Y_i - \mu_i)x_{ih}}{\mathrm{Var}(Y_i)} \frac{\partial \mu_i}{\partial \eta_i} \frac{(Y_i - \mu_i)x_{ij}}{\mathrm{Var}(Y_i)} \frac{\partial \mu_i}{\partial \eta_i}\right]$$

$$= \frac{-x_{ih}x_{ij}}{\mathrm{Var}(Y_i)} \left(\frac{\partial \mu_i}{\partial \eta_i}\right)^2$$

so that

$$E\left(\frac{\partial^2 L(\boldsymbol{\beta})}{\partial\beta_h\,\partial\beta_j}\right) = -\sum_{i=1}^{N}\frac{x_{ih}x_{ij}}{\text{Var}(Y_i)}\left(\frac{\partial\mu_i}{\partial\eta_i}\right)^2.$$

The information matrix, which has elements $E[-\partial^2 L(\boldsymbol{\beta})/\partial\beta_h\,\partial\beta_j]$, equals

$$\mathbf{Inf} = \mathbf{X'WX}$$

where $\mathbf{W}$ is the diagonal matrix with elements

$$w_i = (\partial\mu_i/\partial\eta_i)^2/\text{Var}(Y_i) \tag{13.6}$$

on the main diagonal.

13.1.3 Fisher Scoring and Newton–Raphson

The iterative method used to fit GLMs is called *Fisher scoring*. It resembles the Newton–Raphson method, the distinction being that Fisher scoring uses the *expected value* of the second derivative matrix.

Let $\boldsymbol{\beta}^{(k)}$ denote the kth approximation for the ML estimate $\hat{\boldsymbol{\beta}}$. For the Newton–Raphson method, from (4.30),

$$\boldsymbol{\beta}^{(k+1)} = \boldsymbol{\beta}^{(k)} - (\mathbf{H}^{(k)})^{-1}\mathbf{q}^{(k)}$$

where $\mathbf{H}$ is the matrix having elements $\partial^2 L(\boldsymbol{\beta})/\partial\beta_h\,\partial\beta_j$, $\mathbf{q}$ is the vector having elements $\partial L(\boldsymbol{\beta})/\partial\beta_j$, and $\mathbf{H}^{(k)}$ and $\mathbf{q}^{(k)}$ are $\mathbf{H}$ and $\mathbf{q}$ evaluated at $\boldsymbol{\beta} = \boldsymbol{\beta}^{(k)}$. The formula for Fisher scoring is

$$\boldsymbol{\beta}^{(k+1)} = \boldsymbol{\beta}^{(k)} + (\mathbf{Inf}^{(k)})^{-1}\mathbf{q}^{(k)}$$

or

$$\mathbf{Inf}^{(k)}\boldsymbol{\beta}^{(k+1)} = \mathbf{Inf}^{(k)}\boldsymbol{\beta}^{(k)} + \mathbf{q}^{(k)} \tag{13.7}$$

where $\mathbf{Inf}^{(k)}$ is the kth approximation for the estimated information matrix; that is, $\mathbf{Inf}^{(k)}$ has elements $-E(\partial^2 L(\boldsymbol{\beta})/\partial\beta_h\,\partial\beta_j)$, evaluated at $\boldsymbol{\beta}^{(k)}$.

We next show a relation between ML estimation using Fisher scoring and *weighted least squares* estimation. The right-hand side of (13.7) is the vector having elements

$$\sum_j\left[\sum_i\frac{x_{ih}x_{ij}}{\text{Var}(Y_i)}\left(\frac{\partial\mu_i}{\partial\eta_i}\right)^2\beta_j^{(k)}\right] + \sum_i\frac{(y_i - \mu_i^{(k)})x_{ih}}{\text{Var}(Y_i)}\left(\frac{\partial\mu_i}{\partial\eta_i}\right)$$

where μ_i and $(\partial\mu_i/\partial\eta_i)$ are evaluated at $\boldsymbol{\beta}^{(k)}$. Thus,

$$\mathbf{Inf}^{(k)}\boldsymbol{\beta}^{(k)} + \mathbf{q}^{(k)} = \mathbf{X}'\mathbf{W}^{(k)}\mathbf{z}^{(k)} \ ,$$

where $\mathbf{W}^{(k)}$ is $\mathbf{W}$ (see (13.6)) evaluated at $\boldsymbol{\beta}^{(k)}$ and where $\mathbf{z}^{(k)}$ has elements

$$z_i^{(k)} = \sum_j x_{ij}\beta_j^{(k)} + (y_i - \mu_i^{(k)})\left(\frac{\partial\eta_i^{(k)}}{\partial\mu_i^{(k)}}\right)$$

$$= \eta_i^{(k)} + (y_i - \mu_i^{(k)})\left(\frac{\partial\eta_i^{(k)}}{\partial\mu_i^{(k)}}\right).$$

Equations (13.7) for Fisher scoring have form

$$(\mathbf{X}'\mathbf{W}^{(k)}\mathbf{X})\boldsymbol{\beta}^{(k+1)} = \mathbf{X}'\mathbf{W}^{(k)}\mathbf{z}^{(k)} \ .$$

These are the normal equations for using weighted least squares to fit a linear model for a dependent variable $\mathbf{z}^{(k)}$, when the model matrix is $\mathbf{X}$ and the weight matrix is $\mathbf{W}^{(k)}$. The equations have solution

$$\boldsymbol{\beta}^{(k+1)} = (\mathbf{X}'\mathbf{W}^{(k)}\mathbf{X})^{-1}\mathbf{X}'\mathbf{W}^{(k)}\mathbf{z}^{(k)} \ .$$

The vector $\mathbf{z}$ in this formulation is a linearized form of the link function at $\boldsymbol{\mu}$, evaluated at $\mathbf{y}$,

$$g(y_i) \approx g(\mu_i) + (y_i - \mu_i)g'(\mu_i) = \eta_i + (y_i - \mu_i)(\partial\eta_i/\partial\mu_i) = z_i \ .$$

This "adjusted" or "working" dependent variable $\mathbf{z}$ has ith element approximated by $z_i^{(k)}$ for the kth cycle of the iterative scheme. In that cycle, we regress $\mathbf{z}^{(k)}$ on $\mathbf{X}$ with weight $\mathbf{W}^{(k)}$ to obtain a new estimate $\boldsymbol{\beta}^{(k+1)}$. This estimate yields a new linear predictor value $\boldsymbol{\eta}^{(k+1)} = \mathbf{X}\boldsymbol{\beta}^{(k+1)}$ and a new adjusted-dependent-variable value $\mathbf{z}^{(k+1)}$ for the next cycle. The ML estimator is the limit of $\boldsymbol{\beta}^{(k)}$ as $k \to \infty$. In summary, the ML estimator results from iterative use of weighted least squares, in which the weight matrix changes at each cycle. The process is called *iterative reweighted least squares*.

A simple way to begin the iterative process uses the data $\mathbf{y}$ as the first estimate of $\boldsymbol{\mu}$. This determines the first estimate of the weight matrix $\mathbf{W}$ and hence the initial estimate of $\boldsymbol{\beta}$. Iterations proceed until changes in estimates between successive cycles are sufficiently small. At the initial step, it may be necessary to adjust some observations slightly so that $g(\mathbf{y})$, the initial value of $\mathbf{z}$, is finite. For instance, when g is the log link, we might replace $y_i = 0$ by $y_i = 10^{-6}$.

The asymptotic covariance matrix of $\hat{\beta}$ is the inverse of the information matrix, estimated by

$$\hat{\text{Cov}}(\hat{\beta}) = (\mathbf{X}'\hat{\mathbf{W}}\mathbf{X})^{-1}$$

where $\hat{\mathbf{W}}$ is $\mathbf{W}$ evaluated at $\hat{\beta}$. From (13.6), the form of $\mathbf{W}$ depends on the link chosen for the model.

13.1.4 Simplifications for Canonical Links

Certain simplifications occur when a GLM uses the canonical link. For that link,

$$\eta_i = \theta_i = \sum_j \beta_j x_{ij} \;.$$

When $a(\phi)$ in (13.1) is identical for all observations, the kernel of the log likelihood is $\sum y_i \theta_i$, which simplifies to

$$\sum_i y_i \left(\sum_j \beta_j x_{ij} \right) = \sum_j \beta_j \left(\sum_i y_i x_{ij} \right) \;.$$

Sufficient statistics for estimating β in the GLM are then

$$\sum_i y_i x_{ij} \;, \quad j = 1, \ldots, t \;.$$

For the canonical link,

$$\partial \mu_i / \partial \eta_i = \partial \mu_i / \partial \theta_i = \partial b'(\theta_i) / \partial \theta_i = b''(\theta_i)$$

so (13.4) simplifies to

$$\frac{\partial \ell_i}{\partial \beta_j} = \frac{(y_i - \mu_i)}{\text{Var}(Y_i)} \, b''(\theta_i) x_{ij} = \frac{(y_i - \mu_i) x_{ij}}{a(\phi)} \;.$$

The second derivatives of the log likelihood have components

$$\frac{\partial^2 \ell_i}{\partial \beta_h \partial \beta_j} = -\frac{x_{ij}}{a(\phi)} \left(\frac{\partial \mu_i}{\partial \beta_h} \right) \;.$$

These do not depend on the observations $\{y_i\}$, so

$$\partial^2 L(\beta) / \partial \beta_h \partial \beta_j = E[\partial^2 L(\beta) / \partial \beta_h \partial \beta_j] \;.$$

That is, $\mathbf{H} = -\mathbf{Inf}$, and the Newton–Raphson and Fisher scoring algorithms are identical. When $a(\phi)$ is identical for all observations, the likelihood equations are

$$\sum_i x_{ij} y_i = \sum_i x_{ij} \mu_i .$$

13.1.5 Goodness-of-Fit

A *saturated* GLM has as many parameters as observations, giving a perfect fit. A saturated model consigns all variation to the systematic component of the model. Let $\tilde{\theta}$ denote the estimate of θ for the saturated model, corresponding to estimated means $\tilde{\mu}_i = y_i$ for all i. For a given unsaturated model, the ratio

$$-2 \log\left(\frac{\text{maximum likelihood under model}}{\text{maximum likelihood under saturated model}} \right)$$

describes lack of fit. When the random component has $a(\phi) = \phi/\omega_i$, this measure equals

$$2 \sum \omega_i [y_i(\tilde{\theta}_i - \hat{\theta}_i) - b(\tilde{\theta}_i) + b(\hat{\theta}_i)] / \phi \tag{13.8}$$

$$= D(\mathbf{y}; \hat{\boldsymbol{\mu}})/\phi .$$

This is called the *scaled deviance*, and $D(\mathbf{y}; \hat{\boldsymbol{\mu}})$ is called the *deviance*. The greater the scaled deviance, the poorer the fit.

For two models, the second a special case of the first, the difference

$$D(\mathbf{y}; \hat{\boldsymbol{\mu}}_2) - D(\mathbf{y}; \hat{\boldsymbol{\mu}}_1) = 2 \sum \omega_i [y_i(\hat{\theta}_{1i} - \hat{\theta}_{2i}) - b(\hat{\theta}_{1i}) + b(\hat{\theta}_{2i})]$$

also has the form of the deviance. Under regularity conditions, the difference in scaled deviances has approximately a chi-squared distribution, with degrees of freedom equal to the difference between the number of parameters in the two models (McCullagh and Nelder 1983, pp. 233–236). For some GLMs (such as Poisson loglinear models for fixed N and large expected cell counts), the scaled deviance itself has an approximate chi-squared distribution. Methods for analyzing the deviance are generalizations of analysis of variance methods for normal linear models.

Examination of residuals helps us evaluate the fit for each observation. One type of residual for GLMs uses components of the deviance,

$$\sqrt{|d_i|} \times \text{sign}(y_i - \hat{\mu}_i)$$

where

$$d_i = 2\omega_i[y_i(\tilde{\theta}_i - \hat{\theta}_i) - b(\tilde{\theta}_i) + b(\hat{\theta}_i)]$$

and $D(\mathbf{y}; \hat{\boldsymbol{\mu}}) = \Sigma \, d_i$. An alternative is the *Pearson residual*,

$$e_i = \frac{y_i - \hat{\mu}_i}{[\hat{\text{Var}}(Y_i)]^{1/2}} \, .$$

Pierce and Schafer (1986) discussed various residuals for GLMs. They showed that deviance residuals are nearly normally distributed, after allowance for discreteness. See Green (1984), Jørgensen (1984), and McCullagh and Nelder (1989) for other definitions of residuals.

13.2 MAXIMUM LIKELIHOOD FOR LOGLINEAR MODELS

To illustrate ML theory for GLMs, we now apply the GLM framework to loglinear modeling of categorical data. We use the notation of the previous section, identifying cell counts as the random variables $\{Y_i, i = 1, \ldots, N\}$ and their expected values as $\{\mu_i\}$. In addition, we give Birch's arguments for the fundamental results quoted in Section 6.1.3 about ML estimation for loglinear modeling. We first consider Poisson sampling.

13.2.1 Poisson Loglinear Models

Suppose Y_i has Poisson distribution with mean μ_i. Then

$$
\begin{aligned}
f(y_i; \mu_i) &= \frac{e^{-\mu_i}\mu_i^{y_i}}{y_i!} \\
&= \exp[y_i \log(\mu_i) - \mu_i - \log(y_i!)] \\
&= \exp[y_i\theta_i - \exp(\theta_i) - \log(y_i!)]
\end{aligned}
$$

where $\theta_i = \log(\mu_i)$. This has form (13.1) with $b(\theta_i) = \exp(\theta_i)$, $a(\phi) = 1$, and $c(y_i, \phi) = -\log(y_i!)$. The natural parameter is $\theta_i = \log(\mu_i)$, and the mean and variance are $E(Y_i) = b'(\theta_i) = \exp(\theta_i) = \mu_i$ and $\text{Var}(Y_i) = b''(\theta_i) = \exp(\theta_i) = \mu_i$.

Since $g(\mu_i)$ equals the natural parameter θ_i when g is the log function, the canonical link is the log link, $\eta_i = \log(\mu_i)$. That link gives the loglinear model

$$\log \boldsymbol{\mu} = \mathbf{X}\boldsymbol{\beta} \, .$$

Since $\mu_i = \exp(\eta_i)$, $\partial\mu_i/\partial\eta_i = \exp(\eta_i) = \mu_i$, and likelihood equations (13.5) simplify to

$$\sum_i (y_i - \mu_i)x_{ij} = 0 .$$

For the loglinear models of Chapters 5–7 for nominal variables, for instance, the sufficient statistics $\{\Sigma_i \, y_i x_{ij}, \, j = 1, \dots, t\}$ are certain marginal tables. The likelihood equations equate the sufficient statistics to the corresponding marginal tables of fitted counts (Birch, 1963). Since

$$w_i = (\partial\mu_i/\partial\eta_i)^2/\mathrm{Var}(Y_i) = \mu_i$$

the estimated covariance matrix of $\hat{\beta}$ is $(\mathbf{X}'\hat{\mathbf{W}}\mathbf{X})^{-1}$, where $\hat{\mathbf{W}}$ is the diagonal matrix with elements of $\hat{\mu}$ on the main diagonal.

Suppose all cell counts are positive, and the model is parameterized so that $\mathbf{X}$ has full rank. Birch (1963) showed the likelihood equations are soluble, by noting that (for fixed $\{y_i > 0\}$) the kernel of the Poisson log likelihood

$$L = \sum [y_i \log(\mu_i) - \mu_i]$$

has individual terms converging to $-\infty$ as $\log(\mu_i) \to \pm\infty$; thus, the log likelihood is bounded above and attains its maximum at finite values of the model parameters. It is stationary at this maximum, since it has continuous first partial derivatives. Birch showed the likelihood equations have a unique solution, and the likelihood is maximized at that point. He proved this by showing that the matrix of values $\{-\partial^2 L/\partial\beta_h \partial\beta_j\}$ (i.e., the information matrix $\mathbf{X}'\mathbf{W}\mathbf{X}$) is nonsingular and nonnegative definite, and hence positive definite. Nonsingularity follows from $\mathbf{X}$ having full rank and the diagonal matrix $\mathbf{W}$ having positive elements $\{\mu_i\}$. Any quadratic form $\mathbf{c}'\mathbf{X}'\mathbf{W}\mathbf{X}\mathbf{c}$ equals $\Sigma_i[\sqrt{\mu_i}(\Sigma_j \, x_{ij}c_j)]^2 \geqslant 0$, so the matrix is also nonnegative definite.

For Poisson GLMs, $\hat{\theta}_i = \log \hat{\mu}_i$ and $b(\hat{\theta}_i) = \exp(\hat{\theta}_i) = \hat{\mu}_i$, whereas $\tilde{\theta}_i = \log y_i$ and $b(\tilde{\theta}_i) = y_i$ for the saturated model. Also $\phi = 1$, so the deviance and scaled deviance (13.8) equal

$$D(\mathbf{y}; \hat{\mu}) = 2 \sum [y_i \log(y_i/\hat{\mu}_i) - y_i + \hat{\mu}_i] .$$

This simplifies to the G^2 statistic when the fit satisfies $\Sigma \, y_i = \Sigma \, \hat{\mu}_i$, as is the case for loglinear models containing an intercept (grand mean) term. For a fixed number of cells N, it follows from Section 12.3.4 that the deviance has an approximate chi-squared distribution as $\{\mu_i\}$ grow unboundedly. The Pearson residual

$$e_i = (y_i - \hat{\mu}_i)/\sqrt{\hat{\mu}_i}$$

is the standardized residual introduced in Section 7.3, and $\Sigma\, e_i^2$ is the Pearson X^2 statistic.

13.2.2 Multinomial Loglinear Models

Similar results hold when we use ML estimation after conditioning on the total $n = \Sigma\, Y_i$ of the cell counts. When $\{Y_i, i = i, \ldots, N\}$ are independent Poisson random variables, the conditional distribution of $\{Y_i\}$ given n is multinomial with parameters $\{\pi_i = \mu_i/(\Sigma\, \mu_j)\}$. It is unnecessary to develop separate inferential theory for multinomial loglinear models, as shown by the following argument from Birch (1963), McCullagh and Nelder (1989, p. 211) and Palmgren (1981).

Express the Poisson loglinear model for expected cell counts $\{\mu_i\}$ as

$$\log \mu_i = \alpha + \mathbf{x}_i\boldsymbol{\beta}$$

where $(1, \mathbf{x}_i)$ is the ith row of the model matrix $\mathbf{X}$ and $(\alpha, \boldsymbol{\beta}')'$ is the model parameter vector. The Poisson log likelihood is

$$L = L(\alpha, \boldsymbol{\beta}) = \sum y_i \log(\mu_i) - \sum m_i$$
$$= \sum y_i(\alpha + \mathbf{x}_i\boldsymbol{\beta}) - \sum \exp(\alpha + \mathbf{x}_i\boldsymbol{\beta}) = n\alpha + \sum y_i\mathbf{x}_i\boldsymbol{\beta} - \tau$$

where $\tau = \Sigma\, \mu_i = \Sigma \exp(\alpha + \mathbf{x}_i\boldsymbol{\beta})$. Since $\log(\tau) = \alpha + \log[\Sigma \exp(\mathbf{x}_i\boldsymbol{\beta})]$, this log likelihood has the form

$$L = L(\tau, \boldsymbol{\beta}) = \left\{\sum y_i\mathbf{x}_i\boldsymbol{\beta} - n \log\left[\sum \exp(\mathbf{x}_i\boldsymbol{\beta})\right]\right\}$$
$$+ [n \log(\tau) - \tau]. \tag{13.9}$$

Now $\pi_i = \mu_i/(\Sigma\, \mu_j) = \exp(\alpha + \mathbf{x}_i\boldsymbol{\beta})/[\Sigma \exp(\alpha + \mathbf{x}_j\boldsymbol{\beta})]$, and $\exp(\alpha)$ cancels in the numerator and denominator. Thus, the first bracketed term in (13.9) is $\Sigma\, y_i \log(\pi_i)$, which is the multinomial log likelihood, conditional on the total cell count n. Unconditionally, $n = \Sigma\, Y_i$ has a Poisson distribution with expectation $\Sigma\, \mu_i = \tau$, so the second term in (13.9) is the Poisson log likelihood for n. Since $\boldsymbol{\beta}$ enters only in the first term, the ML estimator $\hat{\boldsymbol{\beta}}$ and its covariance matrix for the Poisson likelihood $L(\alpha, \boldsymbol{\beta})$

are identical to those for the multinomial likelihood. As noted in Section 12.4.2, the Poisson loglinear model requires one more parameter than the multinomial loglinear model.

A similar argument applies when there are several independent multinomial samples, rather than a single one. Each log likelihood term is a sum of components from different samples, but the Poisson log likelihood again decomposes into two parts. One part is a Poisson log likelihood for the independent sample sizes, and the other part is the sum of the independent multinomial log likelihoods.

13.2.3 Overdispersion and Quasi-Likelihood

The random component of a GLM assumes a certain distribution for the data. An alternative approach, *quasi-likelihood* estimation, assumes only a form for the functional relationship between the mean and the variance. That is, it assumes the variance has form

$$\text{Var}(Y) = V(\mu)$$

for some chosen function V. In the quasi-likelihood approach, proposed by Wedderburn (1974a), parameter estimates are the solutions of

$$\sum_{i=1}^{N} \frac{(y_i - \mu_i)x_{ij}}{V(\mu_i)} \left(\frac{\partial \mu_i}{\partial \eta_i} \right) = 0 , \quad j = 1, \ldots, t . \tag{13.10}$$

These *estimating equations* have the same form as the likelihood equations (13.5) for GLMs. They are not likelihood equations, however, without the additional assumption that $\{Y_i\}$ has distribution in the natural exponential family. Under that assumption, the variance function characterizes the distribution within the class of natural exponential families (Jørgensen, 1989).

Suppose we assume $\{Y_i\}$ are independent with

$$V(\mu) = \mu .$$

The quasi-likelihood estimates are the solution of (13.10) with $V(\mu_i)$ replaced by μ_i. Under the additional assumption that $\{Y_i\}$ satisfy (13.1), these estimates are also ML estimates. The exponential family (13.1) with $b(\theta) = \exp(\theta)$ and $a(\phi) = 1$ satisfies $V(\mu) = \mu$, since $E(Y) = b'(\theta)$ and $\text{Var}(Y) = b''(\theta)a(\phi)$. That case is simply the Poisson distribution. Thus, for $V(\mu) = \mu$, quasi-likelihood estimates are also ML estimates when the random component has a Poisson distribution.

Even when we expect data to be Poisson distributed, it is often sensible to permit the variance to *exceed* the mean. One cause for this is heterogeneity among subjects. Suppose at a fixed setting $\mathbf{x}$ of explanatory variables, $(Y \mid \mu)$ has mean μ and variance μ. If μ varies among subjects having that $\mathbf{x}$ value, with $E(\mu) = \lambda$, then

$$E(Y) = E[E(Y \mid \mu)] = E[\mu] = \lambda$$

$$\text{Var}(Y) = E[\text{Var}(Y \mid \mu)] + \text{Var}[E(Y \mid \mu)] = \lambda + \text{Var}(\mu).$$

For instance, suppose $(Y \mid \mu)$ has a Poisson distribution with mean μ, but μ itself has a gamma distribution with parameters (α, β). Then $\lambda = E(\mu) = \alpha/\beta$ and $\text{Var}(\mu) = \alpha/\beta^2 = \lambda/\beta$, so that the unconditional distribution of Y has $\text{Var}(Y) = [E(Y)](1 + \beta)/\beta$; in fact, Y has negative binomial distribution, for which the variance exceeds the mean (refer to Problems 3.16 and 3.17).

These considerations suggest an alternative to a Poisson GLM, in which the relationship between mean and variance has form

$$V(\mu) = \phi\mu$$

for some constant ϕ. Family (13.1) with $b(\theta) = \exp(\theta)$ and $a(\phi) = \phi$ satisfies this relationship, with $\mu = \exp(\theta)$ and $\text{Var}(Y) = \phi \exp(\theta)$. The canonical link is again the log link $\eta = \log(\mu)$. When the random component has form (13.1), the likelihood equations (13.5) for the GLM with log link are identical to those for Poisson loglinear models, so model parameter estimates are also identical. From (13.6),

$$w_i = (\partial\mu_i/\partial\eta_i)^2/\text{Var}(Y_i) = \mu_i^2/\phi\mu_i = \mu_i/\phi$$

so the estimated covariance matrix is ϕ times that for the Poisson loglinear model. The case $\phi > 1$ represents *overdispersion* for the Poisson model. Wedderburn (1974a) suggested estimating the dispersion parameter ϕ by $X^2/(N - t)$, where $X^2 = \Sigma(y_i - \hat{\mu}_i)^2/\hat{\mu}_i$.

Quasi-likelihood estimators make only second-moment assumptions about the random component, rather than full distributional assumptions. Nevertheless, McCullagh (1983) showed they have similar properties as ML estimators. Under quite general conditions, they are consistent and asymptotically normal. They are asymptotically efficient among estimators that are linear in $\{y_i\}$. When quasi-likelihood estimators are not ML, Cox (1983) and Firth (1987) suggested they still retain relatively high efficiency as long as the degree of overdispersion is moderate.

13.2.4   Overdispersion and Binomial GLMs

Results of Sections 13.1 and 13.2.3 also apply to models with a binomial random component, such as logistic and probit regression models. The individual binary observations have mean π and variance $\pi(1 - \pi)$. To permit overdispersion, we might instead assume the quadratic variance function

$$V(\pi) = \phi\pi(1 - \pi) .$$

For details, see McCullagh and Nelder (1989, Section 4.5) and Problems 3.18 and 13.2.

13.3   WEIGHTED LEAST SQUARES FOR CATEGORICAL DATA

For nearly all models in this text, we used the maximum likelihood (ML) method of parameter estimation. Weighted least squares (WLS) is an alternative method of estimation. Familiarity with the WLS method is useful for several reasons:

1. WLS computations have a standard form that is simple to apply for a wide variety of models. For certain analyses (e.g., fitting models for marginal distributions of multi-way tables), one can apply existing computer software to obtain WLS estimates, but not ML estimates.

2. Algorithms for calculating ML estimates often consist of iterative use of WLS. An example is the Fisher scoring method for generalized linear models discussed in Section 13.1.3.

3. When the model holds, WLS and ML estimators are asymptotically equivalent, both falling in the class of best asymptotically normal (BAN) estimators. For large samples, the estimators are approximately normally distributed around the parameter value, and the ratio of their variances converges to 1.

The classic article by Grizzle, Starmer, and Koch (1969) popularized WLS for categorical data analyses. In honor of those authors, WLS is often referred to as the *GSK method* when used for such analyses. This section summarizes the ingredients of the WLS approach.

13.3.1   Notation for WLS Approach

For a response variable Y having J categories, suppose there are independent multinomial samples of sizes $n_1, \ldots, n_I$ at I levels of an explanatory variable, or at I combinations of levels of several explanatory variables. Let $\boldsymbol{\pi} = (\boldsymbol{\pi}_1', \ldots, \boldsymbol{\pi}_I')'$, where

$$\boldsymbol{\pi}_i = (\pi_{1|i}, \pi_{2|i}, \ldots, \pi_{J|i})', \quad \text{with } \sum_j \pi_{j|i} = 1$$

denotes the conditional distribution of Y at level i of the explanatory variables. Let $\mathbf{p}$ denote corresponding sample proportions, with $\mathbf{V}$ their $IJ \times IJ$ covariance matrix. Since the I samples are independent,

$$\mathbf{V} = \begin{bmatrix} \mathbf{V}_1 & & & \mathbf{0} \\ & \mathbf{V}_2 & & \\ & & \ddots & \\ \mathbf{0} & & & \mathbf{V}_I \end{bmatrix}.$$

The covariance matrix of $\sqrt{n_i}\mathbf{p}_i$ is

$$n_i\mathbf{V}_i = \begin{bmatrix} \pi_{1|i}(1 - \pi_{1|i}) & -\pi_{1|i}\pi_{2|i} & \cdots & -\pi_{1|i}\pi_{J|i} \\ -\pi_{2|i}\pi_{1|i} & \pi_{2|i}(1 - \pi_{2|i}) & \cdots & -\pi_{2|i}\pi_{J|i} \\ \vdots & \vdots & & \vdots \\ -\pi_{J|i}\pi_{1|i} & -\pi_{J|i}\pi_{2|i} & \cdots & \pi_{J|i}(1 - \pi_{J|i}) \end{bmatrix}$$

(Section 12.1.5). Each set of proportions has $(J - 1)$ linearly independent elements.

The model has form

$$\mathbf{F}(\boldsymbol{\pi}) = \mathbf{X}\boldsymbol{\beta} \tag{13.11}$$

where $\mathbf{F}$ is a vector of $u \leq I(J - 1)$ response functions

$$\mathbf{F}(\boldsymbol{\pi}) = [F_1(\boldsymbol{\pi}), \ldots, F_u(\boldsymbol{\pi})]'$$

$\boldsymbol{\beta}$ is a $t \times 1$ vector of parameters, and $\mathbf{X}$ is a $u \times t$ model matrix of known constants having rank t. Let $\mathbf{F}(\mathbf{p})$ denote the sample response functions. We assume $\mathbf{F}$ has continuous second-order partial derivatives in an open region containing $\boldsymbol{\pi}$. This assumption enables us to use the delta method to obtain a large-sample normal distribution for $\mathbf{F}(\mathbf{p})$.

The asymptotic covariance matrix of $\mathbf{F}(\mathbf{p})$ depends on the $u \times IJ$ matrix

$$\mathbf{Q} = \left[\frac{\partial F_k(\boldsymbol{\pi})}{\partial \pi_{j|i}} \right]$$

for $k = 1, \ldots, u$ and all IJ combinations (i, j). Linear response models have response functions of form $\mathbf{F}(\boldsymbol{\pi}) = \mathbf{A}\boldsymbol{\pi}$ for a matrix of known constants $\mathbf{A}$, in which case $\mathbf{Q} = \mathbf{A}$. Loglinear and logit response functions have form $\mathbf{F}(\boldsymbol{\pi}) = \mathbf{K}\log(\mathbf{A}\boldsymbol{\pi})$ for certain matrices $\mathbf{K}$ and $\mathbf{A}$. In that case $\mathbf{Q} = \mathbf{K}[\mathrm{Diag}(\mathbf{A}\boldsymbol{\pi})]^{-1}\mathbf{A}$.

By the multivariate delta method (Section 12.1.6), the approximate large-sample covariance matrix of $\mathbf{F}(\mathbf{p})$ is

$$\mathbf{V}_F = \mathbf{Q}\mathbf{V}\mathbf{Q}' .$$

Let $\hat{\mathbf{V}}_F$ denote the sample version of $\mathbf{V}_F$, in which sample proportions are substituted in $\mathbf{Q}$ and $\mathbf{V}$. For subsequent formulas, this matrix must be nonsingular.

13.3.2 Inference Using the WLS Approach to Model Fitting

The WLS estimator is a generalization of the ordinary least squares estimator that permits sample responses to be correlated and have nonconstant variance. The WLS estimate of $\boldsymbol{\beta}$ in model (13.11) is

$$\mathbf{b} = (\mathbf{X}'\hat{\mathbf{V}}_F^{-1}\mathbf{X})^{-1}\mathbf{X}'\hat{\mathbf{V}}_F^{-1}\mathbf{F}(\mathbf{p}) .$$

This is the $\boldsymbol{\beta}$ value that minimizes the quadratic form

$$[\mathbf{F}(\mathbf{p}) - \mathbf{X}\boldsymbol{\beta}]'\hat{\mathbf{V}}_F^{-1}[\mathbf{F}(\mathbf{p}) - \mathbf{X}\boldsymbol{\beta}] .$$

The ordinary least squares estimate results when $\hat{\mathbf{V}}_F$ is a constant multiple of the identity matrix; that is, when the sample responses are uncorrelated with constant variance. The WLS estimator has an asymptotic multivariate normal distribution, with estimated covariance matrix

$$\hat{\mathrm{Cov}}(\mathbf{b}) = (\mathbf{X}'\hat{\mathbf{V}}_F^{-1}\mathbf{X})^{-1} .$$

The normal distribution applies when the sample size is sufficiently large that $\mathbf{F}(\mathbf{p})$ is close to normally distributed.

The residual term

$$W = [\mathbf{F}(\mathbf{p}) - \mathbf{X}\mathbf{b}]'\hat{\mathbf{V}}_F^{-1}[\mathbf{F}(\mathbf{p}) - \mathbf{X}\mathbf{b}]$$

$$= \mathbf{F}(\mathbf{p})'\hat{\mathbf{V}}_F^{-1}\mathbf{F}(\mathbf{p}) - \mathbf{b}'(\mathbf{X}'\hat{\mathbf{V}}_F^{-1}\mathbf{X})\mathbf{b}$$

describes model goodness of fit. Under the hypothesis $H_0: \mathbf{F}(\boldsymbol{\pi}) - \mathbf{X}\boldsymbol{\beta} = \mathbf{0}$, W has an asymptotic chi-squared distribution with df $= u - t$, the difference between the number of response functions and the number of model parameters.

Hypotheses about effects of explanatory variables have the form H_0: $\mathbf{C\beta = 0}$, where $\mathbf{C}$ is a known $c \times t$ matrix with $c \leq t$, having rank c. The estimator $\mathbf{Cb}$ of $\mathbf{C\beta}$ has an asymptotic normal distribution, with mean $\mathbf{0}$ under the null hypothesis and with covariance matrix estimated by $\mathbf{C(X'\hat{V}_F^{-1}X)^{-1}C'}$. Thus, the Wald statistic

$$W_C = \mathbf{b'C'[C(X'\hat{V}_F^{-1}X)^{-1}C']^{-1}Cb} \tag{13.12}$$

has an approximate chi-squared null distribution with df $= c$. This statistic also equals the difference between residual chi-squared statistics for the reduced model implied by H_0 and the full model.

For the special case H_0: $\beta_i = 0$, the $\mathbf{C}$ matrix for the Wald statistic is $\mathbf{C} = (0, \ldots, 0, 1, 0, \ldots, 0)$, where the 1 is in the ith position. Then $\mathbf{Cb} = b_i$, the term in brackets in (13.12) simplifies to the element in row i and column i of $\mathbf{(X'\hat{V}_F^{-1}X)^{-1}}$, and

$$W_C = b_i^2/\hat{V}\text{ar}(b_i)$$

has a single degree of freedom.

The estimate $\mathbf{b}$ yields predicted values $\hat{\mathbf{F}} = \mathbf{Xb}$ for the response functions. These predicted values are smoother than the sample response functions $\mathbf{F(p)}$. When the model holds, $\hat{\mathbf{F}}$ is asymptotically better than $\mathbf{F(p)}$ as an estimator of $\mathbf{F(\pi)}$ (Section 12.2.2). The estimated covariance matrix of the predicted values is

$$\hat{\mathbf{V}}_{\hat{F}} = \mathbf{X(X'\hat{V}_F^{-1}X)^{-1}X'} .$$

We can analyze the model fit by studying the residuals $\mathbf{F(p)} - \hat{\mathbf{F}}$. The residuals are orthogonal to the fit $\hat{\mathbf{F}}$, so

$$\text{Cov}[\mathbf{F(p)}] = \text{Cov}\{[\mathbf{F(p)} - \hat{\mathbf{F}}] + \hat{\mathbf{F}}\} = \text{Cov}[\mathbf{F(p)} - \hat{\mathbf{F}}] + \text{Cov}(\hat{\mathbf{F}})$$

and the estimated covariance matrix of the residuals equals

$$\hat{\mathbf{V}}_F - \hat{\mathbf{V}}_{\hat{F}} = \hat{\mathbf{V}}_F - \mathbf{X(X'\hat{V}_F^{-1}X)^{-1}X'} .$$

13.3.3 Scope of WLS vs ML Estimation

The WLS approach requires estimating the multinomial covariance structure of sample responses at each setting of the explanatory variables. It is inapplicable when explanatory variables are continuous, since there may be only one observation at each such setting. WLS also becomes less appropriate as the number of categorical explanatory variables increases, since there may be few observations at each of the many combinations of settings. Continuous explanatory variables or a large number of explanatory settings do not pose as severe a problem to the ML approach. When necessary, we can replace zero counts by very small constants so that ML estimates exist. Even for the ML approach, however, these factors adversely affect computing time and the quality of asymptotic approximations for goodness-of-fit statistics.

Suppose a certain model holds for categorical variables. For large cell expected frequencies, ML and WLS give similar results. Both estimators are in the class of best asymptotically normal estimators. For small samples, practical considerations often give an edge to ML estimation. For example, zero cell counts often adversely affec the WLS approach. The sample response functions may then be ill-defined or have a singular estimated covariance matrix. ML estimators, on the other hand, depend on the data only through the sufficient statistics. These statistics consist of terms such as marginal totals, which are less sparse than individual cell counts.

Because of potential problems with zero cell counts, some computer software for WLS estimation requires adding constants to empty cells, at least for fitting certain models. Results can depend strongly on the chosen constant, however. Adding a large constant (such as $\frac{1}{2}$) can have a conservative effect, with serious attenuation of parameter estimates and reduction in power. Adding a small constant may give too great an emphasis to the cell, since its estimated variance is then close to zero and it may have a strong influence in weighted analyses. A sensitivity analysis, comparing effects on the results of adding constants of various sizes, can be helpful.

An advantage of the WLS approach is simplicity of computation. However, as computer software becomes more widely available for ML analyses, this advantage loses its importance.

13.4 BAYESIAN INFERENCE FOR CATEGORICAL DATA

Bayesian inference is not as fully developed for categorical data analysis as in many other areas of statistics. This section presents an application where it does apply quite naturally—to smoothing cell counts in contingency tables.

Suppose our primary goal is to obtain good estimates of cell probabilities. We could use the sample proportions, which are ML estimates for the saturated model. These may have undesirable features, especially when the data are sparse. For instance, large tables may contain several empty cells. If these zero counts are sampling zeroes, rather than structural zeroes, it is not sensible to use 0.0 as the "best" estimate of a probability. In addition, Stein's results for estimating multivariate means (see Efron and Morris, 1977) suggest we can lower total mean squared error using biased estimators that shrink the sample proportions toward some average value.

Another option for estimating cell probabilities is to fit some model. Sometimes, though, there is no reason to expect any standard unsaturated model to describe the table well. For $I \times J$ cross-classifications of nominal variables, for instance, the independence model rarely fits well. When unsaturated models approximate the true relationship poorly, model-based estimators also have undesirable properties. Though they smooth the data, the smoothing is too severe for large samples. The model-based estimators are inconsistent, not converging to the true cell probabilities as the sample size increases.

A Bayesian approach to estimating cell probabilities can provide a compromise between the use of sample proportions and the use of model-based estimators. A model still provides part of the smoothing mechanism, with the Bayes estimators shrinking the sample proportions toward a set of proportions satisfying the model.

13.4.1 Bayesian Estimation of Binomial Proportion

Suppose cell counts $(n_1, \ldots, n_N)$ in a contingency table have a multinomial distribution with $n = \Sigma \, n_i$ and with parameters $\boldsymbol{\pi} = (\pi_1, \ldots, \pi_N)'$. We begin by illustrating basic ideas of Bayesian inference for the simple case $N = 2$. Then, the distribution of $(n_1, n_2) = (n_1, n - n_1)$ is binomial with parameter $\boldsymbol{\pi} = (\pi_1, \pi_2) = (\pi_1, 1 - \pi_1)$. The likelihood is proportional to

$$\pi_1^{n_1}(1 - \pi_1)^{n - n_1}$$

corresponding to n_1 "successes" and $n - n_1$ "failures" in n independent Bernoulli trials.

Since π_1 falls between 0 and 1, a natural prior density for π_1 is the beta,

$$g(\pi_1) = \frac{\Gamma(\beta_1 + \beta_2)}{\Gamma(\beta_1)\Gamma(\beta_2)} \; \pi_1^{\beta_1 - 1}(1 - \pi_1)^{\beta_2 - 1}, \quad 0 \leq \pi_1 \leq 1$$

for some choice of prior parameters $\beta_1 > 0$ and $\beta_2 > 0$. The mean of this prior distribution is $E(\pi_1) = \beta_1/(\beta_1 + \beta_2)$. When β_1 and β_2 exceed 1.0, the distribution is unimodal, with skew to the right when $\beta_1 < \beta_2$, skew to the left when $\beta_1 > \beta_2$, and symmetry when $\beta_1 = \beta_2$. The posterior density $h(\pi_1 \mid n_1)$ of π_1 given the data is proportional to the product of the prior density with the binomial likelihood, or

$$h(\pi_1 \mid n_1) \propto \pi_1^{n_1 + \beta_1 - 1}(1 - \pi_1)^{n - n_1 + \beta_2 - 1} \quad \text{for } 0 \leqslant \pi_1 \leqslant 1 .$$

The beta is the *conjugate* prior distribution. That is, the posterior is also beta, with parameters $n_1 + \beta_1$ and $n - n_1 + \beta_2$.

The mean of the posterior distribution is the most popular Bayesian estimator of a parameter. In a decision-theoretic framework (Ferguson 1967, p. 46), this is optimal when a squared error loss function $(T - \theta)^2$ describes the consequence of estimating a parameter θ by an estimator T. The mean of the beta posterior distribution for π_1 is

$$E(\pi_1 \mid n_1) = (n_1 + \beta_1)/(n + \beta_1 + \beta_2)$$
$$= wn_1/n + (1 - w)\beta_1/(\beta_1 + \beta_2)$$

where $w = n/(n + \beta_1 + \beta_2)$. The Bayes estimator is a weighted average of the sample proportion $p_1 = n_1/n$ and the mean $\beta_1/(\beta_1 + \beta_2)$ of the prior distribution. For a fixed choice of (β_1, β_2), the weight given to the sample increases as the sample size increases.

The Bayes estimator requires selecting prior parameters (β_1, β_2). When we feel completely ignorant about the value of π_1 (before seeing the data), we might choose a uniform prior distribution over the interval $[0, 1]$, which is the beta distribution with $\beta_1 = \beta_2 = 1$. The Bayes estimator of π_1 is then

$$E(\pi_1 \mid n_1) = (n_1 + 1)/(n + 2) .$$

This shrinks the sample proportion slightly toward $\frac{1}{2}$. When we observe only a single observation, a failure, the sample proportion is $0/1 = 0$. The Bayes estimator is $1/3$, considerably smoothed. If we observe 100 observations, all failures, the Bayes estimator is $1/102$—only a slight massaging of the sample proportion.

13.4.2 Dirichlet Prior and Posterior Distributions

Next we consider the case of $N > 2$ cells. The multinomial likelihood is proportional to

$$\prod_{i=1}^{N} \pi_i^{n_i} .$$

The Bayesian approach requires a multivariate prior distribution over potential π values. The multivariate generalization of the beta density is the *Dirichlet* density

$$g(\pi) = \frac{\Gamma\left(\sum \beta_i\right)}{\left[\prod_i \Gamma(\beta_i)\right]} \prod_{i=1}^{N} \pi_i^{\beta_i - 1} \quad \text{for } 0 \le \pi_i \le 1 \text{ all } i, \quad \sum_i \pi_i = 1$$

where the parameters $\{\beta_i > 0\}$. The expected value of π_i in this distribution is $\beta_i/(\sum_j \beta_j)$.

The posterior density is proportional to the product of the multinomial likelihood with the Dirichlet prior density, and is Dirichlet with parameters $\{n_i + \beta_i\}$. The Bayes estimator of π_i is

$$E(\pi_i \mid n_1, \ldots, n_N) = (n_i + \beta_i) / \left(n + \sum_j \beta_j\right) . \tag{13.13}$$

This formula suggests interpreting the Bayes estimator as a sample proportion in which the prior information corresponds to $\sum_j \beta_j$ trials with β_i outcomes of type i, $i = 1, \ldots, N$.

13.4.3 Data-Dependent Choice of Prior Distribution

Many scientists dislike the subjectivity of the Bayesian approach inherent in selecting a prior distribution. Influenced by the empirical Bayes approach of Robbins (1955) and others, statisticians have devised ways of letting the data suggest parameter values for use in the prior distribution. Fienberg and Holland (1970, 1973) proposed such analyses for categorical data. This subsection presents ideas taken primarily from their articles.

For the Dirichlet prior distribution, let $K = \sum \beta_i$ and $\gamma_i = E(\pi_i) = \beta_i/K$. The $\{\gamma_i\}$ are prior guesses for the cell probabilities. Bayes estimator (13.13) equals the weighted average

$$[n/(n + K)] p_i + [K/(n + K)] \gamma_i . \tag{13.14}$$

For a particular choice of $\{\gamma_i\}$, Fienberg and Holland (1970) showed that (13.14) has minimum total mean squared error when

$$K = \left(1 - \sum_i \pi_i^2\right) / \left[\sum_i (\gamma_i - \pi_i)^2\right] . \tag{13.15}$$

The optimal $K = K(\gamma, \pi)$ depends on the unknown probabilities π, so Fienberg and Holland used the estimate $K(\gamma, \mathbf{p})$ of K, in which the sample proportion $\mathbf{p}$ replaces π. As $\mathbf{p}$ falls closer to the prior guess γ, $K(\gamma, \mathbf{p})$ increases and the prior guess receives more weight in the posterior estimate.

This approach requires selecting a prior pattern $\{\gamma_i\}$ for the cell probabilities. Unless the nature of the data suggests certain values, we could select values that satisfy a simple model. For instance, for cross-classifications of two ordinal variables, we often expect a monotone association. It is then sensible to let $\{\gamma_{ij}\}$ satisfy the linear-by-linear association model (8.1), for which local log odds ratios are uniformly of one sign. For equal-interval scores (uniform association), the choices of the common local odds ratio and the row and column marginal probabilities determine $\{\gamma_{ij}\}$ (see Problem 6.39).

To avoid having to choose $\{\gamma_i\}$, we could let their values also be data-dependent. For a two-way table, Fienberg and Holland suggested $\{\hat{\gamma}_{ij} = p_{i+} p_{+j}\}$, for which the Bayes estimator shrinks sample proportions towards the fit of the independence model.

Data-based "pseudo-Bayes" estimators combine good characteristics of sample proportions and of model-based estimators. Like sample proportions, and unlike model-based estimators, they are consistent even when the model does not hold. Unless the model holds, the weight given the sample proportion increases to 1.0 as the sample size increases. Like model-based estimators, and unlike sample proportions, the Bayes estimators can incorporate certain special features of the data, such as ordered categories. This results in smoother estimates that can have much smaller total mean squared error than the sample proportions.

13.4.4 Salary and Experience Example

We illustrate this pseudo-Bayesian approach by estimating cell probabilities for Table 13.1, a cross-classification of monthly salary by years since degree. The table uses data reported by Simonoff (1987), taken from a study conducted by the Department of Energy. The sample consists of 147 nonsupervisory female employees having the Bachelors (but no higher) degree, who were practicing mathematics or statistics in 1981. Since both classifications are ordinal, we use the linear-by-linear association model (8.1) to impose a pattern on the parameters of the Dirichlet prior distribution. Let $\{\hat{\gamma}_{ij}\}$ denote ML estimates of $\{\pi_{ij}\}$ for that model. These are the component means of the Dirichlet prior distribution for $\{\pi_{ij}\}$. From Section 8.1.4, $\{\hat{\gamma}_{ij}\}$ match $\{p_{ij}\}$ in the marginal distributions and in the correlation.

Table 13.1 Cell Counts, and Pseudo-Bayes Estimates Using Uniform Association Model

Salary	Years Since Degree								
	0–2	3–5	6–8	9–11	12–14	15–17	18–23	24–29	30+
950–1350	7 (6.1)	1 (1.8)	1 (1.5)	0 (0.2)	0 (0.1)	0 (0.1)	0 (0.1)	2 (1.0)	0 (0.0)
1351–1750	10 (9.8)	6 (6.1)	5 (5.4)	3 (2.3)	0 (0.5)	1 (1.1)	1 (1.1)	1 (0.6)	0 (0.2)
1751–2150	12 (11.3)	14 (11.1)	7 (8.3)	1 (2.1)	4 (3.3)	2 (2.8)	2 (3.1)	1 (1.0)	2 (1.9)
2151–2550	0 (1.7)	1 (2.1)	8 (6.3)	3 (2.5)	3 (2.4)	3 (3.1)	5 (4.8)	0 (0.7)	4 (3.5)
2551–2950	0 (0.6)	0 (0.7)	3 (2.8)	2 (1.7)	0 (0.8)	6 (4.7)	5 (5.4)	2 (2.0)	7 (6.3)
2951–3750	1 (0.6)	0 (0.1)	1 (0.8)	0 (0.2)	1 (0.8)	1 (1.2)	6 (4.5)	0 (0.7)	2 (3.2)

Now let $\hat{K} = K(\hat{\boldsymbol{\gamma}}, \mathbf{p})$. The resulting pseudo-Bayes estimator (13.14) of π_{ij} is a weighted average of the sample proportion and the ML fit for the model. For fixed n, the weight given the sample proportion decreases as the fit of the model improves.

We used equal-interval scores in model (8.1). Mid-point scores give similar results. The data are sparse, but this model fits much better than the independence model, the G^2 values being 69.3 and 122.1 with degrees of freedom 39 and 40. For these data, $\hat{K} = 147.03$. The pseudo-Bayes estimates are

$$0.50p_{ij} + 0.50\hat{\gamma}_{ij} \, .$$

They are shown in Table 13.1. They are much smoother than the original sample counts, none of them equaling exactly zero. The pseudo-Bayes estimator that smooths towards the fit of the independence model is

$$0.71p_{ij} + 0.29(p_{i+}p_{+j})$$

reflecting the poorer fit of that model.

13.4.5 Improved Estimation by Smoothing Data

The smoothing induced by Bayes estimation can produce estimators that are substantially better than sample proportions. To illustrate, we show a numerical study comparing four estimators: (1) the sample proportion; (2) the ML estimator for the uniform association model; (3) the pseudo-Bayes estimator based on the independence fit $\hat{\gamma}_{ij} = p_{i+}p_{+j}$; and (4) the pseudo-Bayes estimator with $\{\hat{\gamma}_{ij}\}$ equaling the ML estimator for the uniform association model. We compared the estimators for sixteen different sets of probabilities, corresponding to

1. Two table sizes (3×3 and 6×6),
2. Two sample sizes ($n = 50$ and 200),
3. Two models (uniform association, with scores $1, 2, 3$ for 3×3 case and $1, 2, 3, 4, 5, 6$ for 6×6 case; nonuniform association—(8.1) with scores $1.0, 2.5, 3.0$ for 3×3 case and $1.0, 2.8, 4.2, 5.2, 5.8, 6.0$ for 6×6 case),
4. Two levels of association ($\beta = 0.1$ and 0.4 in (8.1)).

All cases used uniform marginal probabilities. We estimated total mean squared errors by simulating 5000 tables for each of the 16 conditions.

For each estimator $\hat{\pi}_{ij}$ of π_{ij}, Table 13.2 contains the values of

$$\frac{n \sum\limits_{k=1}^{5000} \sum\limits_{i} \sum\limits_{j} (\hat{\pi}_{ij,k} - \pi_{ij})^2}{(5000)IJ} \tag{13.16}$$

where $\hat{\pi}_{ij,k}$ denotes the value of $\hat{\pi}_{ij}$ in the kth randomly generated table. The standard errors of the estimates (13.16) are all less than 0.001. For the sample proportion estimator, we report the exact expected value of (13.16), which is $(1 - \Sigma \Sigma \pi_{ij}^2)/IJ$.

When the uniform association model holds, the amount of smoothing is substantial for the pseudo-Bayes estimator using that model. Its total mean squared error is much smaller than that for the sample proportion, and is nearly as small as that for the estimator based completely on the model. Compared to the independence-smoothed Bayes estimator, this Bayes estimator's performance improves as the strength of association increases and as the sample size increases.

The results for the eight cases where the uniform association model does *not* hold have greater relevance, since normally we would use Bayes smoothing when a model gives, at best, a rough approximation for the true structure. The pseudo-Bayes estimator smoothed by the uniform association model still does well. Its total mean squared error is quite a bit smaller than that for the sample proportion even when the sample size is as large as 200. When a model does not hold, as n increases it becomes more advantageous to use a Bayes estimator instead of the ML estimator for the model. However, it is only for the largest values of $I \times J$, β and n considered (the last row in Table 13.2) that the inconsistency of the ML model estimator starts to affect its performance seriously. This agrees with results in Section 6.4.4 showing that unless n is quite large, model-based estimators are often better than sample proportions even when the model does not hold.

13.4.6 Other Bayesian Estimators for Categorical Data

The Bayesian approach presented in this section focused directly on cell probabilities by using a Dirichlet prior distribution for them. Leonard (1975) and Laird (1978) gave an alternative Bayesian approach, focusing on parameters of the saturated loglinear model. For two-way tables, Laird let the association parameters have independent normal $N(0, \sigma^2)$ distributions. She also suggested empirical Bayesian analyses, estimating σ^2 by finding the value that maximizes an approximation for the marginal distribution of the cell counts, evaluated at the observed data. The use of

Table 13.2 Mean Squared Errors for Estimators of Cell Probabilities

				Estimator			
$I \times J$	β	n	Sample Proportion	Uniform-smoothed	Independence-smoothed	Uniform Association model
			Uniform Association Model Holds			
3×3	0.1	50	0.099	0.067	0.060	0.062
		200	0.099	0.068	0.064	0.062
	0.4	50	0.098	0.069	0.076	0.064
		200	0.098	0.067	0.095	0.062
6×6	0.1	50	0.027	0.012	0.013	0.009
		200	0.027	0.012	0.016	0.009
	0.4	50	0.027	0.014	0.018	0.012
		200	0.027	0.014	0.024	0.012
			Uniform Association Model Does Not Hold			
3×3	0.1	50	0.099	0.067	0.060	0.063
		200	0.099	0.068	0.065	0.064
	0.4	50	0.098	0.072	0.078	0.071
		200	0.098	0.084	0.098	0.097
6×6	0.1	50	0.027	0.013	0.013	0.010
		200	0.027	0.013	0.017	0.012
	0.4	50	0.027	0.016	0.019	0.016
		200	0.027	0.020	0.024	0.032

identical prior means for the association parameters is another way of inducing shrinkage towards the independence model.

13.5 OTHER METHODS OF ESTIMATION

This book has emphasized ML estimation, with brief expositions in the previous two sections on WLS and Bayes methods. This final section describes some alternative methods of estimation for categorical data.

13.5.1 Minimum Chi-Squared Estimators

Suppose we want to estimate π or θ, and we assume a model $\pi = \pi(\theta)$. Let $\tilde{\theta}$ denote a generic estimator of θ, for which $\tilde{\pi} = \pi(\tilde{\theta})$ is an estimator of π. The ML estimator $\hat{\theta}$ maximizes the likelihood. Section 12.3.4 showed that the ML estimator also minimizes the G^2 statistic for comparing observed and fitted proportions.

We can define other estimators by minimizing other measures of distance between $\pi(\theta)$ and $\mathbf{p}$. For instance, the value $\tilde{\theta}$ that minimizes the Pearson statistic

$$X^2[\pi(\theta), \mathbf{p}] = n \sum \frac{[p_i - \pi_i(\theta)]^2}{\pi_i(\theta)}$$

is called the *minimum chi-squared* estimate. A simpler estimate to calculate is the *minimum modified chi-squared* estimate, which minimizes

$$X^2_{\text{mod}}[\pi(\theta), \mathbf{p}] = n \sum \frac{[p_i - \pi_i(\theta)]^2}{p_i} . \tag{13.17}$$

This estimate is the solution for θ to the equations

$$\sum_i \frac{\pi_i(\theta)}{p_i} \left(\frac{\partial \pi_i(\theta)}{\partial \theta_j} \right) = 0 , \quad j = 1, \ldots, t \tag{13.18}$$

where t is the dimension of θ.

It is possible to specify a model using a set of *constraint equations* for π,

$$\{ g_j(\pi_1, \ldots, \pi_N) = 0 \} .$$

For instance, for an $I \times J$ table, the $(I-1)(J-1)$ constraint equations

$$\log \pi_{ij} - \log \pi_{i,j+1} - \log \pi_{i+1,j} + \log \pi_{i+1,j+1} = 0$$

specify the model of independence. The number of constraint equations is the same as the residual df for the model. Neyman (1949) noted that one can obtain minimum modified chi-squared estimates by minimizing the expression

$$\sum_{i=1}^{N} \frac{(p_i - \pi_i)^2}{p_i} + \sum_{j=1}^{N-t} \lambda_j g_j(\pi_1, \ldots, \pi_N)$$

with respect to π, where $\{\lambda_j\}$ are Lagrange multipliers.

When the constraint equations are linear in π, the resulting estimating equations are linear. In that case, Bhapkar (1966) showed that minimum modified chi-squared estimators are identical to WLS estimators. The minimum modified chi-squared statistic is then identical to the WLS residual chi-squared statistic for testing the fit of the model. When the constraint equations are nonlinear (e.g., the independence model), the WLS estimator is identical to the minimum modified chi-squared estimator based on a linearized version of the constraints; that is, when $g_j(\pi)$ is replaced by

$$g_j(\mathbf{p}) + (\pi - \mathbf{p})' \partial g_j(\pi) / \partial \pi$$

where the differential vector is evaluated at $\mathbf{p}$.

Neyman (1949) introduced minimum modified chi-squared estimators and showed that they and minimum chi-squared estimators are best asymptotically normal estimators. When the model holds, these estimators are asymptotically (as $n \to \infty$) equivalent to ML estimators. The asymptotic results of Chapter 12 for ML estimators also hold for these estimators.

When the model holds, different estimation methods (ML, WLS, minimum chi-squared, etc.) yield nearly identical estimates of parameters when the sample size is large. This happens partly because the estimators of θ are consistent, converging in probability to the true values as the sample size increases. When the model does not hold, estimates obtained by different methods can be quite different, even when n is large. The estimators converge to values for which the model gives the best approximation to reality, and this approximation is different when "best" is defined in terms of minimizing G^2 rather than minimizing X^2 or some other distance measure.

13.5.2 Minimum Discrimination Information

Kullback (1959) formulated the method of estimation by minimum discrimination information (MDI). The discrimination information for two probability vectors $\boldsymbol{\pi}$ and $\boldsymbol{\gamma}$ is

$$I(\boldsymbol{\pi}; \boldsymbol{\gamma}) = \sum \pi_i \log(\pi_i/\gamma_i) \,. \tag{13.19}$$

This measure of distance between $\boldsymbol{\pi}$ and $\boldsymbol{\gamma}$ is nonnegative, equaling 0 only when $\boldsymbol{\pi} = \boldsymbol{\gamma}$. Suppose the constraint equations for a model are linear,

$$\mathbf{A}\boldsymbol{\pi} = \boldsymbol{\beta} \,. \tag{13.20}$$

Then the MDI estimate is defined as the value of $\boldsymbol{\pi}$ that minimizes $I(\boldsymbol{\pi}; \mathbf{p})$, subject to these constraints.

Though some models (e.g., marginal homogeneity) have form (13.20), most models do not. Instead they satisfy constraints

$$\mathbf{A}\boldsymbol{\pi} = \mathbf{A}\mathbf{p} \tag{13.21}$$

such as given by likelihood equations for loglinear models (see (6.11)). In that case Gokhale and Kullback (1978) defined the MDI estimate as the one minimizing $I(\boldsymbol{\pi}; \boldsymbol{\gamma})$, subject to constraints (13.21), where $\boldsymbol{\gamma}$ is the vector with $\gamma_1 = \gamma_2 = \cdots$. In this sense, the MDI estimate is the simplest one to satisfy the constraints.

Gokhale and Kullback called constraints (13.21) *internal constraints*— they refer to the data inside the contingency table. They called constraints (13.20) *external constraints*—they are separate from the data. For the internal constraints problem, the MDI estimator is identical to the ML estimator (see Berkson 1972, Simon 1973). For the external constraints problem, the MDI estimator is not ML, but it has similar asymptotic properties. For instance, it is best asymptotically normal (BAN).

For internal constraints problems, Gokhale and Kullback recommended using the G^2 statistic to test goodness-of-fit. For the external constraints problem, though, they recommended using twice the minimized value of $I(\boldsymbol{\pi}; \mathbf{p})$,

$$2 \sum \tilde{\pi}_i \log(\tilde{\pi}_i/p_i)$$

where $\tilde{\boldsymbol{\pi}}$ is the MDI estimate. This statistic reverses the roles of $\mathbf{p}$ and $\boldsymbol{\pi}$ relative to G^2, much as X^2_{mod} in (13.17) reverses their roles relative to X^2.

Both statistics fall in the class of power divergence statistics defined by Cressie and Read (1984) (Problem 3.26), and have similar asymptotic properties.

13.5.3 Kernel Smoothing

Kernel estimation is a popular smoothing method for density estimation. Kernel estimates of cell probabilities in a contingency table have form

$$\tilde{\pi} = \mathbf{K}\mathbf{p} \tag{13.22}$$

for some matrix $\mathbf{K}$ containing nonnegative elements and having column sums equal to one.

For unordered multinomials with N categories, Aitchison and Aitken (1976) suggested the kernel function

$$k_{ij} = \lambda, \quad i = j$$
$$= (1 - \lambda)/(N - 1), \quad i \neq j$$

for $(1/N) \leqslant \lambda \leqslant 1$. The resulting kernel estimator of π has form

$$(1 - \alpha)\mathbf{p} + \alpha\mathbf{1}/N \tag{13.23}$$

where $\alpha = N(1 - \lambda)/(N - 1)$ and $\mathbf{1} = (1, 1, \ldots, 1)'$. The kernel estimator shrinks the sample proportion toward $(1/N, \ldots, 1/N)$. As λ decreases from 1 to $1/N$, the smoothing parameter α increases from 0 to 1.

Brown and Rundell (1985) proved that when no $\pi_i = 1$, there exists $\lambda < 1$ such that the total mean squared error is smaller for this kernel estimator than for the sample proportions. Results for other types of shrinkage estimators applied to multivariate means suggest that the improvement for the kernel estimator may be large when the sample size is small and the true cell probabilities are roughly equal.

Brown and Rundell formulated generalized kernels for multi-way contingency tables that may contain both nominal and ordinal variables. For a T-way table, let $\mathbf{L}_k$ be a stochastic matrix (i.e., row and column sums equal to 1) with elements

$$\ell_{k,ij} = \lambda_k, \quad i = j$$
$$= d_k(i, j)(1 - \lambda_k), \quad i \neq j,$$

$k = 1, \ldots, T$. They let **K** in (13.22) be the Kronecker product

$$\mathbf{K} = \mathbf{L}_1 \otimes \cdots \otimes \mathbf{L}_T .$$

When variable k is ordinal, $d_k(i, j)$ is chosen to be smaller for greater distances between categories i and j. If variable k is nominal, the natural choice is $d_k(i, j) = 1/(I_k - 1)$, where I_k is the number of categories for variable k. For fixed $\{\lambda_k\}$, collapsing the smoothed table gives the same result as smoothing the corresponding collapsing of the original table. For the case $\{\lambda_k = \lambda, k = 1, \ldots, T\}$, Brown and Rundell described ways of finding λ to minimize an unbiased estimate of the total mean squared error.

Titterington and Bowman (1985) described other approaches for classifications having ordered categories. For ordered categories, the Brown–Rundell and Titterington–Bowman kernels give probability estimates of form

$$\tilde{\pi}_i = (1 - \alpha)p_i + \alpha(\text{smoother}_i)$$

where the smoothing is designed to work well when true probabilities in nearby cells are similar.

13.5.4 Penalized Likelihood

The *penalized likelihood* method of estimation was introduced for density estimation by Good and Gaskins (1971). Simonoff (1983) and Titterington and Bowman (1985) applied it to estimating cell probabilities. The estimator is obtained by maximizing

$$L(\boldsymbol{\pi}) = \log \text{likelihood} - \phi(\boldsymbol{\pi})$$

where ϕ is a roughness penalty; that is, ϕ is a function that decreases as the probabilities $\boldsymbol{\pi}$ are smoother, in some sense. For two-way tables, Simonoff suggested the penalty function $\phi(\boldsymbol{\pi}) = \epsilon \, \Sigma \, \Sigma (\log \theta_{ij})^2$, which has the effect of shrinkage towards the independence estimator. Simonoff (1987) adapted his method for ordered classifications.

Like Bayesian and kernel estimators, penalized likelihood estimators are useful for giving cell probability estimates that are smoother than the sample proportions. Burman (1987), Titterington (1980) and Titterington and Bowman (1985) discussed many other ways to smooth categorical data. Much research remains to be done to compare the various methods.

CHAPTER NOTES

Section 13.1: Maximum Likelihood for Generalized Linear Models

13.1 For further discussion of GLMs and their generalizations, see Albert (1988), Andersen (1974, 1980), Hastie and Tibshirani (1987a,b), Jørgensen (1983, 1984, 1986, 1987, 1989), Nelder and Wedderburn (1972), Pregibon (1980, 1982b), and Wedderburn (1974a, 1974b, 1976), and books by Aitkin et al. (1989), Dobson (1983), and McCullagh and Nelder (1989). For further discussion of the relationship between iterative reweighted least squares and ML estimation, see Green (1984), Jennrich and Moore (1975), and Jørgensen (1984). Green, Jøorgensen, and Palmgren and Ekholm (1987) also discussed this relation for exponential family nonlinear models.

Standard errors from the information matrix are based on a quadratic approximation for the log likelihood. Confidence intervals using such standard errors can be misleading when this approximation is poor. Alternatively, we could form the confidence interval from the set of values that are not rejected when used as null hypothesis values in likelihood-ratio tests. See Aitken et al. (1989, Section 2.15) for an example.

Section 13.2: Maximum Likelihood for Loglinear Models

13.2 For further details on quasi-likelihood as well as its usage in modeling overdispersion, see Cox (1983), Firth (1987), McCullagh (1983, 1984, 1985), McCullagh and Nelder (1989), Nelder and Pregibon (1987), and Wedderburn (1974a, 1976).

Section 13.3: Weighted Least Squares for Categorical Data

13.3 When there is a *single* multinomial sample, the WLS approach in Section 13.3 for I independent multinomial samples follows by conditioning on the cell counts $(n_1, \ldots, n_I)$ constituting the joint marginal distribution of the explanatory variables. Applications of WLS include fitting mean response models (Section 9.6) and models for marginal distributions (Sections 11.3, 11.4). WLS has been used for a wide variety of applications in several articles by Gary Koch and his colleagues and graduate students in the Department of Biostatistics at the University of North Carolina. For further discussion of the WLS approach, see Bhapkar and Koch (1968), Forthofer and Lehnen (1981), Grizzle et al. (1969), Imrey et al. (1981), Koch et al. (1977), and Koch et al. (1985).

Section 13.4: Bayesian Inference for Categorical Data

13.4 When a true cell probability $\pi_i = 0$, the sample proportion $p_i = 0$ with probability one, and the sample proportion is better than any other estimator. Because there are parameter values for which the sample proportion is the optimal estimator, no other estimator (such as a Bayes estimator) is uniformly better than the sample proportion over the whole parameter space. Here, the criterion of comparison is the expected value of some measure of distance between the estimator and the parameter. Thus, the sample proportion is an *admissible* estimator, for standard loss functions (Johnson

1971). In this sense, the sample mean for multinomial data differs from the sample mean for a multivariate normal distribution, which is inadmissible (dominated by shrinkage estimators) when the dimension of the mean vector is at least three.

Bishop et al. (1975, pp. 430–432) and Ighadaro and Santner (1982) discussed alternative estimators for K in the pseudo-Bayes estimator. For other examples of Bayesian analyses of categorical data, see Agresti and Chuang (1989), Albert (1987, 1988), Altham (1969, 1971), Chuang (1982), Good (1965, 1976), Kadane (1985), Knuiman and Speed (1988), Laird (1978), Leonard (1975), Lindley (1964), Lindley and Smith (1972), Morris (1983), Spiegelhalter and Smith (1982), and Weisberg (1972). Efron and Morris (1977) presented an interesting application of empirical Bayesian estimation to proportion data. In estimating baseball batting averages for several major league players, they illustrated how empirical Bayes estimators can be far superior to sample proportions when the sample size is small, there are many proportions to estimate, or the true probabilities are roughly equal.

Section 13.5: Other Methods of Estimation

13.5 Berkson (1944, 1955, 1980) was a primary advocate of minimum chi-squared methods. For logistic regression, his *minimum logit chi-squared* estimators minimized a weighted sum of squares between sample logits and linear predictions. Mantel (1985) criticized minimum chi-squared methods, noting that their consistency requires group sizes to grow large, whereas ML is consistent however information goes to the limit. For instance, the Cochran–Mantel–Haenszel test (a ML score test) shows proper asymptotic behavior whether separate contingency tables become arbitrarily large or small tables arbitrarily numerous. See also Problem 13.24. For further discussion of estimation based on minimizing chi-squared, see Neyman (1949), Rao (1955, 1963), Bhapkar (1966), and Koch et al. (1985).

The method of minimum discrimination imformation was used for analyses of categorical data in a series of articles by Kullback and various co-authors. See, for instance, Ireland and Kullback (1968a, 1968b), Ireland et al. (1969), Ku et al. (1971), Ku and Kullback (1974), and the text by Gokhale and Kullback (1978).

Titterington (1980) surveyed the literature on kernel estimation for categorical data, including ways of choosing λ. Kernel methods are also useful for discrete regression modeling. For binary response data, Copas (1983) used kernel smoothing to display in a nonparametric manner the dependence on x of the probability $\pi(x)$ of success.

PROBLEMS

13.1 Suppose the response variable is binary. For n_i observations at the ith setting of the explanatory variables, let p_i denote the proportion of successes. Assume $\{n_i p_i, i = 1, \ldots, N\}$ are independent binomial (n_i, π_i) random variables.

a. Show $f(p_i; \theta_i, \phi)$ has form (13.1), with $b(\theta_i) = \log(1 + \exp(\theta_i))$ and $a(\phi) = \phi/\omega_i$ with $\phi = 1$ and $\omega_i = n_i$. Thus use (13.2) and (13.3) to obtain $E(p_i) = \pi_i$ and $\text{Var}(p_i) = \pi_i(1 - \pi_i)/n_i$.

b. Show the logit link $g(\pi_i) = \log[\pi_i/(1 - \pi_i)]$ is the canonical link.

c. Show that the deviance equals

$$2 \sum_{i=1}^{N} n_i p_i \log\left(\frac{n_i p_i}{n_i \hat{\pi}_i}\right) + 2 \sum_{i=1}^{N} (n_i - n_i p_i) \log\left(\frac{n_i - n_i p_i}{n_i - n_i \hat{\pi}_i}\right)$$

that is, $2 \sum$ observed[log(observed/fitted)], where the sum is taken over all $2N$ cells corresponding to successes and failures.

d. For GLMs with logit link, show that the linearized form of the link function is (4.34).

e. For the logit link, show that the asymptotic covariance matrix of $\hat{\beta}$ is $(\mathbf{X'WX})^{-1}$, where $\mathbf{W}$ is the diagonal matrix with elements $n_i \pi_i (1 - \pi_i)$ on the main diagonal.

13.2 Refer to the previous problem. Suppose we drop the binomial assumption, but assume

$$E(p_i) = \pi_i \quad \text{and} \quad \text{Var}(p_i) = \phi \pi_i (1 - \pi_i)/n_i$$

for some unknown constant ϕ.

a. If the random component has form (13.1), identify the functions $b(\theta)$ and $a(\phi)$.

b. For the canonical link, give equations that are likelihood equations if the random component has form (13.1), and obtain an expression for the estimated covariance matrix.

c. For a GLM with link $\log[\pi_i/(1 - \pi_i)]$, show that model parameter estimates are the same as for the binomial logit model, but the covariance matrix is ϕ times as great. (The case $\phi > 1$ is overdispersion for the binomial logit model.)

d. When we make the moment assumptions but do not assume the distribution of the random component is (13.1), are the estimates ML? Explain.

13.3 In a GLM, suppose $\text{Var}(Y_i) = V(\mu_i)$, for some function V. Show that for the link satisfying $g'(\mu) = V^{-1/2}(\mu)$, the weight matrix is the same at each cycle. Find this link for a Poisson random component.

13.4 For noncanonical links in a GLM, explain why the matrix of second partial derivatives may depend on the data. Illustrate using the probit model for a binary response.

13.5 Consider a GLM having $a(\phi) = \phi/n_i$ in (13.1). For the canonical link, find the sufficient statistics and likelihood equations.

13.6 Let Y_{ij} represent the jth value of a variable for group i, $i = 1, \ldots, I$, $j = 1, \ldots, N_i$, where $\{Y_{ij}\}$ can assume nonnegative integer values. Suppose $\{Y_{ij}\}$ are independent Poisson random variables, with $E(Y_{ij}) = \mu_i$.

    a. For this GLM, show the ML estimate of μ_i is $\hat{\mu}_i = \bar{y}_i = \Sigma_j y_{ij}/N_i$. (This model, the Poisson loglinear model, and the survival models of Section 6.6 are examples of *Poisson regression* models. For other examples, see Koch et al. 1986.)

    b. Simplify the expression for the deviance for this model. (For testing this model, it follows from Fisher (1970 p. 58, originally published 1925) that the deviance and the Pearson statistic $\Sigma\Sigma (y_{ij} - \bar{y}_i)^2/\bar{y}_i$ have approximate chi-squared distributions with $\mathrm{df} = \Sigma (N_i - 1)$. For a single group, Cochran (1954) referred to $\Sigma (y_j - \bar{y})^2/\bar{y}$ as the *variance test* for the fit of a Poisson distribution, since it compares the observed variance to the predicted variance $\bar{y}$ of the Poisson distribution.)

13.7 Cell counts in an $I \times J$ table have a multinomial distribution.

    a. Show that the probability mass function can be expressed as

$$d^n n! \prod \prod (n_{ij}!)^{-1} \exp\left[\sum_{i=1}^{I-1} \sum_{j=1}^{J-1} n_{ij} \log(\alpha_{ij}) \right.$$
$$\left. + \sum_{i=1}^{I-1} n_{i+} \log(\pi_{iJ}/\pi_{IJ}) + \sum_{j=1}^{J-1} n_{+j} \log(\pi_{Ij}/\pi_{IJ}) \right]$$

    where $\alpha_{ij} = \pi_{ij} \pi_{IJ}/\pi_{iJ} \pi_{Ij}$ and d is independent of the data.

    b. Give an alternative expression using local odds ratios $\{\theta_{ij}\}$, by showing that

$$\Sigma\Sigma n_{ij} \log \alpha_{ij} = \Sigma\Sigma s_{ij} \log \theta_{ij}$$

    where

$$s_{ij} = \sum_{a \leqslant i} \sum_{b \leqslant j} n_{ab}.$$

13.8 Consider the family of distributions

$$f(y; \theta, \phi) = \exp\{[y\theta - b(\theta)]/a(\phi) + c(y,\phi)\}.$$

    a. Show that the normal distribution has this form.

    b. Show that the negative binomial distribution does not have this form.

    c. Jørgensen (1986) showed that a more appropriate form for two-parameter discrete distributions is

$$f(y; \theta, \phi) = \exp\{y\theta - b(\theta)/a(\phi) + c(y,\phi)\} .$$

    Show that the negative binomial distribution has this form.

13.9 Refer to Birch's results in Section 13.2.

    a. Show that L has individual terms converging to $-\infty$ as $\log(\mu_i) \to \pm\infty$.

    b. Explain why positive definiteness of the information matrix implies the solution of the likelihood equations is unique, with likelihood maximized at that point.

13.10 Show that ML estimates for Poisson loglinear models are identical to those obtained by splitting the sample into several independent multinomial samples. Specifically, suppose a set of Poisson means $\{m_{ij}\}$ satisfy

$$\log m_{ij} = \alpha_j + \mathbf{x}_{ij}\boldsymbol{\beta} .$$

Decompose the Poisson log likelihood so that part refers to the column totals and part refers to the effect of conditioning on those totals.

13.11 Consider the model of marginal homogeneity for an $I \times I$ table.

    a. Express this model in the form $\mathbf{F}(\boldsymbol{\pi}) = \mathbf{0}$, where $\mathbf{F}$ has $I - 1$ elements.

    b. Show how to use WLS to test the hypothesis of marginal homogeneity. (This is simply Bhapkar's test (10.20).)

13.12 For the WLS approach with response functions $\mathbf{F}(\boldsymbol{\pi}) = \mathbf{K}[\log(\mathbf{A}\boldsymbol{\pi})]$, show that $\mathbf{Q} = \mathbf{K}[\mathbf{Diag}(\mathbf{A}\boldsymbol{\pi})]^{-1}\mathbf{A}$.

13.13 In the WLS approach, show that the value for $\boldsymbol{\beta}$ that minimizes the quadratic form $[\mathbf{F}(\mathbf{p}) - \mathbf{X}\boldsymbol{\beta}]'\hat{\mathbf{V}}_F^{-1}[\mathbf{F}(\mathbf{p}) - \mathbf{X}\boldsymbol{\beta}]$ is $(\mathbf{X}'\hat{\mathbf{V}}_F^{-1}\mathbf{X})^{-1}\mathbf{X}'\hat{\mathbf{V}}_F^{-1}\mathbf{F}(\mathbf{p})$.

13.14 The response functions $\mathbf{F}(\mathbf{p})$ have asymptotic covariance matrix $\mathbf{V}_F$. Derive the asymptotic covariance matrix of the WLS model parameter estimator $\mathbf{b}$ and the predicted values $\hat{\mathbf{F}} = \mathbf{X}\mathbf{b}$.

13.15 Consider the mean response model fitted in Section 9.6.2. Show how to use WLS for this analysis, by identifying the number of multinomial samples I, the number of responses J, the response functions $\mathbf{F}$, the model matrix $\mathbf{X}$, the parameter vector $\boldsymbol{\beta}$, and the estimated covariance matrix $\hat{\mathbf{V}}_F$.

13.16 Explain how WLS can be used to conduct the longitudinal growth curve analysis in Section 11.3.2.

13.17 Consider the Bayes estimator of a binomial parameter, using a beta prior distribution.

    a. Does any beta prior distribution produce a Bayes estimator that coincides with the ML estimator?

    b. Show that the ML estimator is a limit of Bayes estimators, for a certain sequence of parameter values for the beta prior distribution.

    c. Find an improper prior density (one for which its integral is not finite) such that the Bayes estimator coincides with the ML estimator. (In this sense, the ML estimator is a *generalized* Bayes estimator.)

13.18 For the loss function $w(\theta)(T - \theta)^2$, the Bayes estimator is the posterior expected value of $\theta w(\theta)$, divided by the posterior expected value of $w(\theta)$ (Ferguson 1967, p. 47).

    a. For estimating a binomial parameter π with loss function $(T - \pi)^2/[\pi(1 - \pi)]$, show the ML estimator of π is a Bayes estimator for the uniform prior distribution over $[0, 1]$.

    b. For the loss function in (a), show that the risk function (the expected loss, viewed as a function of π) is constant over the parameter space. (Bayes estimators with constant risk are *minimax*; that is, their maximum risk is no greater than the maximum risk for any other estimator.)

13.19 For the Dirichlet prior distribution for multinomial probabilities, show the posterior expected value of π_i is (13.13). Show this Bayes estimator is a weighted average of the sample proportion and the prior expected value of π_i.

13.20   For Bayes estimator (13.14), show that the total mean squared error is

$$[K/(n + K)]^2 \left[\sum (\pi_i - \gamma_i)^2 \right] + [n/(n + K)^2] \left[1 - \sum \pi_i^2 \right].$$

Show (13.15) is the value of K that minimizes this.

13.21   Use a pseudo-Bayes analysis to estimate cell probabilities in Table 8.2. Explain how you chose the prior distribution.

13.22   Give $I - 1$ constraint equations that specify the model of marginal homogeneity in an $I \times I$ table. For this model, explain why minimum modified chi-squared estimates are identical to WLS estimates.

13.23   Show that the loglinear model of no three-factor interaction is specified by $(I - 1)(J - 1)(K - 1)$ constraint equations, such as

$$\log[(\pi_{ijk} \pi_{i+1,j+1,k})/(\pi_{i+1,jk} \pi_{i,j+1,k})]$$
$$- \log[(\pi_{ij,k+1} \pi_{i+1,j+1,k+1})/(\pi_{i+1,j,k+1} \pi_{i,j+1,k+1})] = 0.$$

For this model, are WLS estimates the same as minimum modified chi-squared estimates?

13.24   Let y_i be the number of successes in n_i independent identical Bernoulli trials for group i, $i = 1, \ldots, N$. Suppose observations in different groups are independent. Consider the model that the success probability is the same in each group, and denote that common value by π.
   a. Show that the ML estimator of π is $p = (\Sigma \, y_i)/(\Sigma \, n_i)$.
   b. The minimum chi-squared estimator $\tilde{\pi}$ is the value of π minimizing

$$\sum_{i=1}^{N} \frac{((y_i/n_i) - \pi)^2}{\pi} + \sum_{i=1}^{N} \frac{((y_i/n_i) - \pi)^2}{1 - \pi}.$$

The second term results from comparing $(1 - y_i/n_i)$ to $(1 - \pi)$, the proportions in the second category for the ith group. If $n_1 = \cdots = n_N = 1$, show that $\tilde{\pi}$ minimizes $Np(1 - \pi)/\pi + N(1 - p)\pi/(1 - \pi)$. Hence show that

$$\tilde{\pi} = p^{1/2}/[p^{1/2} + (1-p)^{1/2}].$$

Note the bias toward $\frac{1}{2}$ in this estimator.

    c. Argue that as $N \to \infty$ with all $n_i = 1$, the ML estimator is consistent, but the minimum chi-squared estimator is not (Mantel 1985).

    d. For the ML estimate when all $n_i = 1$, show that $X^2 = n$, illustrating that goodness-of-fit statistics can be completely uninformative for sparse data.

13.25 Refer to the previous problem. For $N = 2$ groups with n_1 and n_2 independent observations, find the minimum modified chi-squared estimator of the common value, π. Compare it to the ML estimator.

13.26 Give constraint equations that specify the symmetry model. Indicate whether they are external or internal.

13.27 Show that the kernel estimator (13.23) is the same as the Bayes estimator (13.13) for the Dirichlet prior in which each prior parameter is $\beta_i = \alpha n/(1 - \alpha)N$. Using this result, suggest a way of letting the data determine the value of α in the kernel estimator.

13.28 Refer to Table 7.11. Consider the model that simultaneously assumes (7.10) as well as linear logit relationships for the marginal effects of age on breathlessness and wheeze.

    a. Specify $\mathbf{K}$, $\mathbf{A}$, and $\mathbf{X}$ for which this model has form $\mathbf{K}\log(\mathbf{A\pi}) = \mathbf{X\beta}$, and explain how to fit it using WLS.

    b. Using a computer routine (such as CATMOD in SAS), fit the model and interpret estimates.

APPENDIX A

Using Computer Software to Analyze Categorical Data

This appendix is a guide to statistical computer software for categorical data analyses. The guide is approximate, at best, because of the continual introduction of new programs and updating of existing ones.

We begin by listing several statistical computer packages that can perform categorical data analyses. Then we outline, by chapter, specific procedures for analyses described in this book, and we give several examples of appropriate code.

A.1 SOFTWARE PACKAGES

A.1.1 SAS

SAS (SAS Institute 1987) is a general-purpose software package that has procedures for a wide variety of statistical analyses. The main procedures for categorical data analyses are CATMOD, FREQ, and LOGISTIC.

Procedure CATMOD, developed by W. Stanish, is useful for building a wide variety of models for categorical data. At present, the main mode of analysis is weighted least squares, though ML is an option for logit and generalized logit models. The user can write a formula to construct a particular response for a model, or request one of the standard options, such as cumulative logit. CATMOD assumes independent multinomial sampling at each combination of levels of variables specified in the POPULATION statement. When that statement is deleted, it assumes independent multinomial samples at all combinations of levels of explanatory variables specified in the MODEL statement. At the end of the model statement, the user can request options such as the covariance matrix of the parameter estimates and the predicted values of the response. CATMOD specifies quantitative explanatory variables using a DIRECT statement. For further details on CATMOD, see Imrey (1985) and Stanish (1986).

Procedure FREQ in SAS computes measures of association and their estimated standard errors. It also can perform generalized Cochran–Mantel–Haenszel tests of partial association, and exact tests of independence in $I \times J$ tables.

Procedure LOGISTIC in SAS gives ML fitting of logit, probit, and extreme-value models for binary responses, and the cumulative logit model for ordinal responses. Explanatory variables can be categorical or continuous. The procedure incorporates model selection procedures and regression diagnostic options.

Other programs run on SAS that are not specifically supported by SAS (see SAS Institute 1986). The procedure LOGIST, developed by F. Harrell, is one of these. It uses ML estimation to fit logistic regression and cumulative logit models. Two other supplemental programs, MLOGIT and MPROBIT, perform fitting of multinomial logit and probit models.

A.1.2 GLIM

The interactive program GLIM (Numerical Algorithms Group 1986) is a low-cost package for the main frame or PC, sponsored by the Royal Statistical Society. GLIM is an acronym for "generalized linear interactive modeling." It is designed for ML fitting of generalized linear models.

GLIM gives the user several choices for each component of a generalized linear model. The $ERROR directive specifies the distribution of the random component. Options include normal, Poisson, binomial, and gamma. The $LINK directive specifies the link. Options include identity, log, logit, reciprocal, probit, complementary log-log, square root, and exponent. When the user does not specify a link, the default is the canonical link. The $FIT directive specifies the systematic component of the model. The $FACTOR directive specifies variables that are qualitative factors, the $UNITS directive specifies the length of data vectors (the number of observations for each variable), the $DATA directive lists the names for the variables, and the data follow the $READ statement. The $CALCULATE directive defines new variables using data or results of fitting models.

Several auxiliary statistics can be exhibited using the $DISPLAY directive. For instance, the statement

$DISPLAY E R V C S D

requests model parameter estimates, fitted values and residuals, the covariance matrix of the estimates, the correlation matrix of the estimates, the standard errors of differences of estimates, and the deviance and degrees of freedom. The $LOOK directive displays requested variables or statistics describing the fit of the model. For instance, $LOOK %X2 requests the Pearson goodness-of-fit statistic. GLIM codes parameter estimates for factors so the first level has estimate equal to zero. See Sections 5.1.3 and 5.3.5.

It is simple to use GLIM to fit logit models, probit models, extreme-value models, loglinear models, and models for survival and rates. Macros can be constructed to fit more complex models. The *GLIM Newsletter* regularly prints such macros. For further discussion of the use of GLIM in fitting generalized linear models, see Aitkin et al. (1989), Healy (1988), and Lindsey (1989).

A.1.3 BMDP

Like SAS, BMDP is a general-purpose statistical software package. The control language is a sequence of paragraphs separated by "/". The BMDP-LR program fits logistic regression models. Though it entails more work for the user, the nonlinear regression program BMDP-3R can fit many models for categorical data. The BMDP-4F program performs analyses for contingency tables. It can compute several measures of association and their estimated standard errors. It also gives ML fitting of the loglinear models for nominal variables discussed in Chapters 5–7. See Brown (1986) and Dixon (1981) for details.

A.1.4 SPSS^X

SPSS^X is another general-purpose statistical package, oriented towards needs of social scientists. The LOGLINEAR program, described by Clogg and Becker (1986) and Norušis (1988), gives ML fitting of a wide variety of loglinear models and corresponding logit models. In this program, the DESIGN statement specifies the form of the model. Optional output includes fitted values, standardized and adjusted residuals, and model parameter estimates and their correlations. This program can fit the standard loglinear and logit models described in Chapters 4–7. It can also fit the models for rates presented in Section 6.6 and the ordinal loglinear models of Chapter 8 and the equivalent logit models. It fits models having structural zeroes by attaching zero weights to certain cells.

The HILOGLINEAR program in SPSS[X] is designed to fit standard hierarchical loglinear models for nominal variables. The CROSSTABS program calculates several measures of association and gives significance tests for two-way tables. The PROBIT procedure does probit regression and logistic regression. The newest release also has a regular logistic regression program and a procedure for correspondence analysis.

LOGLINEAR and HILOGLINEAR are also available on their PC software, SPSS/PC+.

A.1.5 StatXact

StatXact is a statistical package giving exact analysis for categorical data methods and some nonparametric methods. Among its statistical procedures are Fisher's exact test and its generalizations for $I \times J$ tables, both for nominal and ordinal variables. It also can conduct exact tests of conditional independence and of equality of odds ratios in $2 \times 2 \times K$ tables, and exact confidence intervals for the common odds ratio in several 2×2 tables. The exact analyses were presented in a series of articles by C. Mehta, N. Patel, and various co-authors. StatXact uses Monte Carlo methods to approximate exact P-values and confidence intervals when a data set is too large for exact inference to be computationally feasible. StatXact runs on an IBM-PC compatible microcomputer having at least 512 K of RAM. It is distributed by Cytel Software Corporation, 137 Erie Street, Cambridge, MA 02139.

A.1.6 Other Programs

ANOAS is a FORTRAN program for ML estimation of ordinal loglinear models and more general association models for two-way tables. It can fit the linear-by-linear association model, row and column effects models, the RC model, the quasi-uniform association model, the quasi-independence model, and many other models. To obtain a copy, contact Dr. Clifford Clogg, Department of Statistics, Pennsylvania State University, University Park, PA 16802. Dr. Clogg also has programs for multi-group versions of the above models, latent class analysis, and analysis of rate data.

SYSTAT is a general-purpose program. It contains FORTRAN supplemental programs for logit, probit, and multinomial logit modeling. These programs work in PC or minicomputer environments. For further information, contact Systat Inc., 1800 Sherman Avenue, Evanston, IL 60201.

GAUSS is another general-purpose program. The loglinear analysis module, written by Dr. J. Scott Long, gives ML fitting using the Newton–Raphson method. The user can enter a model matrix, which permits fitting a wide variety of loglinear models, including models for ordinal variables or models that incorporate cell weights.

EGRET is a program that focuses on methods for epidemiological problems. It incorporates many survival models, as well as specialized methods for categorical data, such as logistic regression and exact methods. For further information, contact Statistics and Epidemiology Research Corporation, 909 Northeast 43rd Street, Suite 310, Seattle, WA 98105.

FREQ, a FORTRAN program listed in the text by Haberman (1979), provides ML fitting of loglinear models, using Newton–Raphson. ECTA (Fay and Goodman 1975) provides ML fitting of loglinear models by iterative proportional fitting. MULTIQUAL (Bock and Yates 1973) is another program for ML fitting of loglinear and logit models. GENCAT (Landis et al. 1976) fits models using WLS estimation.

A.2 LISTINGS OF COMPUTER ROUTINES BY CHAPTER

Chapter 2: Describing Two-Way Tables
The odds ratio, relative risk, and ordinal measures such as gamma and Kendall's tau-b are computed by SAS (PROC FREQ), BMDP (4F), and SPSS^X (CROSSTABS). The concentration and uncertainty measures are provided by SPSS^X (LOGLINEAR), and the tetrachoric correlation by BMDP (4F).

Chapter 3: Inference for Two-Way Tables
Chi-squared statistics for testing independence are reported by BMDP (4F), SAS (PROC FREQ), and SPSS^X (CROSSTABS, LOGLINEAR, HILOGLINEAR). BMDP (4F) and SAS (PROC FREQ) report estimated standard errors for several measures of association. Table A.1 uses SAS and Table A.2 uses BMDP to calculate chi-squared statistics and several measures of association (including gamma and Kendall's tau-b) and their estimated standard errors, for Table 3.4.

Several packages perform Fisher's exact test, including SAS (PROC FREQ), BMDP (4F), SPSS^X, and StatXact. SAS (PROC FREQ) and StatXact can perform exact tests when I and/or J exceeds 2. StatXact also performs exact conditional estimation for 2×2 tables. It computes the conditional ML estimate of θ, and exact confidence intervals for θ.

Table A.1 SAS Used to Calculate Chi-Squared Statistics, Measures of Association and Estimated Standard Errors

```
INPUT INCOME $ SATISF $ COUNT @@;
CARDS;
1 VD 20    1 LD 24    1 MS 80     1 VS  82
2 VD 22    2 LD 38    2 MS 104    2 VS 125
3 VD 13    3 LD 28    3 MS  81    3 VS 113
4 VD  7    4 LD 18    4 MS  54    4 VS  92
PROC FREQ ORDER = DATA; WEIGHT COUNT;
TABLE SATISF*INCOME / ALL;
```

Table A.2 BMDP Used to Calculate Pearson Chi-Squared and Ordinal Measures of Association and Estimated Standard Errors

```
/INPUT          VARIABLES ARE 2.
                FORMAT IS '(16F4.0)'
                TABLE IS 4, 4.
/VARIABLE       NAMES ARE SATISF, INCOME.
/CATEGORY       NAMES(1) ARE 'VERY DISSAT', 'LIT DISSAT',
                  'MOD SAT', 'VERY SAT'.
                CODES(1) ARE 1, 2, 3, 4.
                NAMES(2) ARE '<6', '6-15', '15-25', '>25'.
                CODES(2) ARE 1, 2, 3, 4.
/TABLES         COLUMNS ARE SATISF.
                ROWS ARE INCOME.
/STATISTICS     CHISQUARE. GAMMA.
/END
  20 24 80 82 22 38 104 125 13 28 81 113 7 18 54 92
```

Chapter 4: Models for Binary Response Variables

GLIM is designed for ML fitting of generalized linear models. Other packages having nonlinear regression programs can be adapted to fit generalized linear models. See Suissa and Adam (1987) for SAS.

Programs that can fit logistic regression models include GLIM, SAS (PROC LOGISTIC or PROBIT), and BMDP (LR). When all explanatory variables are categorical, SAS (PROC CATMOD) and SPSS[x] (LOGLINEAR) are also appropriate. Table A.3 uses SAS (LOGISTIC) to fit a logistic regression model to Table 4.1, and Table A.4 uses SAS (CATMOD) to fit a linear logit model to Table 4.4. The logit is the default response for CATMOD. The WEIGHT statement in CATMOD weights each combination of variables by the number of observations in that cell. The DIRECT statement specifies quantitative variables. Op-

Table A.3 SAS (LOGISTIC) Used to Fit Logistic Regression Model

```
INPUT LI REMISS @@;
CARDS;
  8 0      8 0    10 0    10 0    12 0    12 0    12 0    14 0    14 0
 14 0     16 0    16 0    16 0    18 1    20 0    20 1    20 1    22 0
 22 1     24 0    26 1    28 1    32 0    34 1    38 0    38 1    38 1
PROC LOGISTIC;
MODEL REMISS = LI;
```

Table A.4 SAS (CATMOD) Used to Fit Linear Logit Model

```
INPUT PRESSURE DISEASE COUNT @@;
CARDS;
111.5  1   3      111.5  0  153
121.5  1  17      121.5  0  235
  .
  .
  .
191.5  1   8      191.5  0   35
PROC CATMOD; WEIGHT COUNT;
DIRECT PRESSURE;
MODEL DISEASE = PRESSURE / ML NOGLS PRED = FREQ COVB;
```

tions in the MODEL statement include ML, requesting the maximum likelihood fit. PROC LOGISTIC has options for forward and backward variable selection procedures, and alternative link functions. It can report the covariance matrix of model parameters, Wald and likelihood-ratio tests for model parameters, and confidence intervals for probabilities.

The Cochran–Armitage trend test is one of many special cases available with PROC FREQ in SAS. It is also available in BMDP (4F), using the LIN option in the STATISTICS command. StatXact computes an exact P-value for this test, conditional on the observed margins.

Probit models can be fitted in SPSS[X] with the PROBIT procedure. Probit and extreme value models can be fitted in GLIM, and in SAS with PROC PROBIT, or with PROC LOGISTIC (using LINK = NORMIT or LINK = CLOGLOG). Table A.5 uses GLIM to fit logistic regression, probit and extreme-value models to Table 4.7. The $YVAR directive specifies that KILL (the number of beetles killed) is the response variable. In the $ERROR directive, BIN (or just B) specifies a binomial distribution for the random component, and NUMBER is the name of the variable giving the number of trials for each binomial observation. The $UNITS specifies that there are eight observations (binomial counts). The $FIT directive specifies that DOSE is the explanatory variable in the

Table A.5   GLIM Used to Fit Logit, Probit, and Extreme-Value Models

```
$UNITS 8
$DATA DOSE KILL NUMBER
$READ
1.691   6   59    1.724  13  60    1.755  18  62    1.784  28  56
1.811  52  63    1.837  53  59    1.861  61  62    1.884  60  60
$YVAR KILL
$ERROR BIN NUMBER
$FIT DOSE $
$DISPLAY E R V $

$LINK P
$FIT DOSE $

$LINK C
$FIT DOSE $

$CALCULATE SURVIVE = NUMBER – KILL $
$YVAR SURVIVE
$FIT DOSE $
```

model. Later $LINK directives change the link from the logit, which is the default for binomial random component, to probit (P) and complementary log-log (C). Using the C link with SURVIVE as the binomial response is equivalent to using the log-log link with KILL as the response.

PROC LOGISTIC in SAS presents Pearson and deviance residuals and several diagnostic measures for logistic regression proposed by Pregibon (1981). Appendices of the book by Breslow and Day (1980) contain FORTRAN programs for conditional logistic regression.

Chapter 6: Fitting Loglinear and Logit Models
Programs that can calculate $\{\hat{m}_i\}$, G^2 and associated statistics for loglinear models include BMDP (4F), GLIM, SPSS^X (HILOGLINEAR, LOGLINEAR), and SAS (CATMOD). The Newton–Raphson method is used in ML fitting with SAS, GLIM, and SPSS^X (LOGLINEAR), whereas BMDP (4F) and SPSS^X (HILOGLINEAR) use iterative proportional fitting (IPF). Stanish (1986) showed how to use PROC IML of SAS to implement IPF.

Loglinear models for Table 5.1 are fitted using SAS in Table A.6, BMDP in Table A.7, SPSS^X in Table A.8, and GLIM in Table A.9. In SAS, the LOGLIN statement defines loglinear model effects. The first

Table A.6   SAS Used to Fit Loglinear and Logit Models

```
INPUT D $ V $ P $ COUNT @@;
  IF COUNT = 0 THEN COUNT = 1E − 20;
CARDS;
WW Y 19      WW N 132
.
.
.
B B Y 6     B B N 97
PROC CATMOD; WEIGHT COUNT;
MODEL D *V* P = _RESPONSE_ / ML COVB PRED = FREQ;
LOGLIN D | V   D | P   V | P;

PROC CATMOD; WEIGHT COUNT;
MODEL P = D V / ML FREQ PRED = PROB;

PROC CATMOD; WEIGHT COUNT;
POPULATION D V;
MODEL P = V / ML FREQ PRED = PROB;
```

Table A.7   BMDP-4F Used to Fit Loglinear Models

```
/PROBLEM        TITLE IS 'DEATH PENALTY'.
/INPUT          VARIABLES ARE 3.
                FORMAT IS '(8F4.0)'.
                TABLE IS 2,2,2.
/VARIABLE       NAMES ARE PENALTY, VICTIM, DEFENDANT.
/TABLE          INDICES ARE PENALTY, VICTIM, DEFENDANT.
                SYMBOLS ARE P,V,D.
/CATEGORY       CODES(1) ARE 1,2.
                NAMES(1) ARE YES, NO.
                CODES(2) ARE 1,2.
                NAMES(2) ARE WHITE, BLACK.
                CODES(3) ARE 1,2.
                NAMES(3) ARE WHITE, BLACK.
/PRINT          MARGINAL IS 2.
                EXPECTED.
                STANDARDIZED.
                LAMBDA.
/FIT            ALL.
/FIT            MODEL IS VD,P.
/FIT            MODEL IS VD,VP.
/FIT            MODEL IS VD,VP,DP.
/FIT            MODEL IS PVD.
/END
  19 132  0  9 11 52  6 97
```

Table A.8 SPSS[x] Used to Fit Loglinear Models

```
DATA LIST LIST / P D V COUNT *
VALUE LABELS P 1 'YES' 2 'NO' / D 1 'WHITE' 2 'BLACK' /
  V 1 'WHITE' 2 'BLACK'
WEIGHT BY COUNT
HILOGLINEAR P(1, 2) D(1, 2) V(1, 2)
  /DESIGN D*V  D*P  V*P
BEGIN DATA
1 1 1  19
2 1 1 132
  .
  .
  .
2 2 2  97
END DATA
```

Table A.9 GLIM Used to Fit Loglinear Models

```
$UNITS 8
$FACTOR  D  2  V  2  P  2
$DATA D V P COUNT
$READ
1 1 1 19    1 1 2 132
  .
  .
2 2 1  6    2 2 2  97
$YVAR COUNT
$ERROR POIS
$FIT D+V+P: +D.V: +V.P: +D.P $
```

MODEL statement in Table A.6 fits model (DV, PD, PV). Alternatively, all two-factor terms can be requested using

$$\text{LOGLIN D} \mid \text{V} \mid \text{P @ 2;}$$

The ML option specifies a maximum likelihood fit, COVB requests the covariance matrix of parameter estimates, and PRED = FREQ gives the fitted values (PRED = PROB gives predicted cell probabilities and their standard errors). The second and third MODEL statements in Table A.6 fit logit models, the default in CATMOD. These logit models for death penalty verdict are equivalent to loglinear models (DV, PD, PV) and (DV, PV). If we omitted the POPULATION statement before the last model statement, SAS would fit the model for the marginal $P–V$ table (collapsed over D). PROC CATMOD in SAS treats zero counts as structural zeroes, so they must be replaced by small constants when they are actually sampling zeroes. Imrey (1985) and Stanish (1986) discussed the use of SAS for loglinear and logit models.

In Table A.7, the options EXPECTED, STANDARDIZED, and LAMBDA in BMDP-4F produce fitted values, standardized residuals, and parameter estimates and estimated standard errors. In Table A.8, the DESIGN statement in the HILOGLINEAR procedure for SPSS[X] requests model (DV, PD, PV). In Table A.9, the $ERROR POIS directive in GLIM specifies a Poisson random component, and the $FIT directive fits the sequence of models (D, V, P), (DV, P), (DV, PV), and (DV, PV, PD).

Loglinear models for rates and survival can be fitted in GLIM using the offset directive for the log exposure. Table A.10 uses GLIM to fit model (6.17) to Table 6.12. Models for rates can be fitted in SPSS[X] by identifying exposures as cell weights in the CWEIGHT option. They can be fitted in BMDP (4F) by specifying the exposures as initial values to be fitted to the marginals, in an INITIAL statement. The same procedures can be used for table standardization methods of Section 6.7, except that sample cell counts are used in the offset or as cell weights or initial values, and entries satisfying the model that have the target marginal distributions are entered as data.

Chapter 7: Selecting and Applying Loglinear Models

Forward selection and backward elimination procedures for loglinear models can be implemented using BMDP (4F) and SPSS[X] (HILOG-LINEAR).

Standardized residuals are reported by GLIM, BMDP (4F), and SPSS[X] (HILOGLINEAR). Adjusted residuals are calculated by SPSS[X] (LOGLINEAR), and can be obtained using GLIM (Defize 1980, Pregibon 1986).

Table A.10   GLIM Used for Survival Analysis

```
$UNITS 4
$FACTOR   AGE 2   VALUE 2
$DATA AGE VALUE COUNT EXPOSE
$READ
1 1 4 1259    1 2 1 2082    2 1 7 1417    2 2 9 1647
$CALCULATE LOGEXP = %LOG(EXPOSE) $
$OFFSET LOGEXP $
$ERROR POIS
$YVAR COUNT
$FIT AGE + VALVE $
```

Cochran–Mantel–Haenszel statistics for $2 \times 2 \times K$ tables are given by SAS (PROC FREQ) and BMDP (4F). SAS also reports generalized tests for $I \times J \times K$ tables. Table A.11 uses SAS for Table 7.9. PROC FREQ in SAS reports a large-sample test proposed by Breslow and Day (1980) for homogeneity of odds ratios. BMDP (4F) also gives such a test. StatXact gives Birch's exact test of conditional independence for $2 \times 2 \times K$ tables, Zelen's exact test of homogeneity of odds ratios, and an exact confidence interval for a common odds ratio.

O'Brien (1986) gave a SAS macro for computing powers using the noncentral chi-squared distribution.

Chapter 8: Loglinear-Logit Models for Ordinal Variables

GLIM, SPSS[X] (LOGLINEAR) and SAS (CATMOD, ML option) can fit all loglinear models in this chapter, using the Newton–Raphson method. The uniform association model and row or column effects models are fitted to Table 8.1 using GLIM in Table A.12, SPSS[X] in Table A.13, and SAS in Table A.14. In GLIM, the $CALCULATE directives calculate the scores for use in the uniform association and row effects models. Alternatively, we could enter data vectors containing the scores, or use the $ASSIGN directive to create score vectors. In SPSS[X], the statement COMPUTE UV = INCOME*SATISF sets up the cross-product scores for the uniform association model, and the statement DESIGN = SATISF INCOME UV fits that model. The statement COMPUTE U = INCOME sets up scores for the income classification, and the statement DESIGN = SATISF, INCOME, SATISF BY U fits the column effects model. See Norušis (1988) for further details.

In SAS (CATMOD), the default responses are the baseline-category logits (Section 9.1). The loglinear models are fitted by specifying the model matrix for a corresponding generalized logit model. For instance, the first model matrix in Table A.14 has size 12×4, since there are three logits in each of the four rows of Table 8.1 ($3 \times 4 = 12$), and since there

Table A.11 SAS Used for Cochran–Mantel–Haenszel Test

```
INPUT LEVEL $ DELAY $ CURED $ COUNT @@;
CARDS;
1 1 1 0    1 1 2 6    1 2 1 0    1 2 2 5
  :
  :
5 1 1 2    5 1 2 0    5 2 1 5    5 2 2 0
PROC FREQ; WEIGHT COUNT;
TABLES LEVEL * DELAY * CURED / CMH;
```

Table A.12   GLIM Used to Fit Uniform Association and Row Effects Models

```
$UNITS 16
$FACTOR INCOME 4   SATISF 4
$DATA INCOME SATISF COUNT
$READ
1 1  20    1 2  24    1 3  80    1 4  82
.
.
.
4 1  7    4 2  18    4 3  54    4 4  92
$CALCULATE UV = INCOME * SATISF $
$YVAR COUNT
$ERROR POIS
$FIT INCOME + SATISF + UV $
$CALCULATE V = SATISF $
$FIT INCOME + SATISF + INCOME.V $
```

Table A.13   SPSS^X Used to Fit Uniform Association and Column Effects Models

```
DATA LIST LIST /SATISF INCOME COUNT *
VALUE LABELS SATISF 1 'VERY DISSAT' 2 'LITTLE DISSAT' 3 'MOD
  SAT' 4 'VERY SAT'
  / INCOME 1 '<6' 2 '6–15' 3 '15–25' 4 '>25'
WEIGHT BY COUNT
COMPUTE UV = INCOME * SATISF
COMPUTE U = INCOME
LOGLINEAR SATISF(1, 4) INCOME(1, 4) WITH UV/
  DESIGN = SATISF INCOME UV/
  PRINT = ESTIM/
LOGLINEAR SATISF(1, 4) INCOME(1, 4) WITH U/
  DESIGN = SATISF, INCOME, SATISF BY U/
  PRINT = ESTIM
BEGIN DATA
1 1  20
.
.
.
4 4  92
END DATA
```

Table A.14 SAS Used to Fit Uniform Association and Row Effects Models

```
INPUT INCOME SATISF COUNT @@;
CARDS;
1 1 20    1 2 24    1 3 80    1 4 82
. . .
4 1 7     4 2 18    4 3 54    4 4 92
PROC CATMOD ORDER = DATA; WEIGHT COUNT;
POPULATION INCOME;
MODEL SATISF = (1 0 0    -3,   0 1 0 -2,  0 0 1 -1,
               1 0 0    -6,   0 1 0 -4,  0 0 1 -2,
               1 0 0    -9,   0 1 0 -6,  0 0 1 -3,
               1 0 0   -12,   0 1 0 -8,  0 0 1 -4)/ ML NOGLS PRED = FREQ;
PROC CATMOD ORDER = DATA; WEIGHT COUNT;
POPULATION INCOME;
MODEL SATISF = (1 0 0 -3   0   0,   0 1 0 -2 0   0,   0 0 1 -1 0   0,
               1 0 0  0  -3   0,   0 1 0  0 -2 0,   0 0 1  0 -1 0,
               1 0 0  0   0  -3,   0 1 0  0  0 -2,   0 0 1  0  0 -1,
               1 0 0  3   3   3,   0 1 0  2  2 2,   0 0 1  1  1  1)/ ML NOGLS;
```

497

are four parameters in the model for $\log(m_{ij}/m_{i4})$ implied by the linear-by-linear association model. The first three elements in each row of the model matrix pertain to the intercept parameters for the three logits. Using scores $\{u_i = i\}$ and $\{v_j = j\}$, the fourth element is $u_i(v_j - v_4)$, the coefficient of β in the logit model. The second model matrix, for the row effects model, uses the scaling $\mu_4 = -(\mu_1 + \mu_2 + \mu_3)$. Elements 4, 5, and 6 in each row of the model matrix are the coefficients $(v_j - v_4)$ of μ_1, μ_2, and μ_3 in the model for $\log(m_{ij}/m_{i4})$ implied by the row effects model. In SAS it is also possible to specify a logit model for adjacent-categories logits, using the keyword ALOGITS in a RESPONSE statement.

Generalized Mantel tests can be obtained using SAS (PROC FREQ). The user can select scores to obtain the correlation-type statistic (8.17) of Mantel, the ANOVA-type statistic of Birch and Landis et al., or versions of these statistics using rank-type scores.

Davis (1988) presented a SAS macro for fitting the RC model. Breen (1984) and Lindsey (1989) used GLIM to fit this and other association models. Becker (1989c) gave a program that fits the $RC(M)$ model.

Correspondence analysis is available with program CA in BMDP, procedure CORRESP in SAS, and procedure ANACOR in SPSS[X].

Chapter 9: Multinomial Response Models
Generalized logit models can be fitted directly using SAS (CATMOD). This program selects the final response category as the baseline. When all variables are categorical, generalized logit models can be fitted using programs that fit corresponding loglinear models, such as BMDP (4F), SPSS[X] (LOGLINEAR), or GLIM.

The LOGIT module in SYSTAT fits multinomial logit models, as does the supplemental program MLOGIT for SAS (for details, contact Dan Steinberg, Salford Systems, 3672 Caminito Cielo del Mar, San Diego, CA 92130).

The logit model (9.2) for a nominal response and ordinal explanatory variable can be fitted using programs for generalized logit models (e.g., CATMOD in SAS) or using a program (such as GLIM or SPSS[X]– LOGLINEAR) that can fit the corresponding column effects loglinear model.

Procedure LOGISTIC and supplementary procedure LOGIST in SAS provide ML fitting of cumulative logit models, and CATMOD in SAS provides WLS fitting. Table A.15 fits a cumulative logit model with an interaction term to Table 9.8, using LOGIST. In the PROC statement for LOGIST, K is one less than the number of response categories. Table A.16 fits cumulative logit model (9.16) to Table 8.6, using CATMOD. The RESPONSE CLOGITS statement specifies cumulative logits for the

Table A.15 SAS (LOGIST) Used to Fit Cumulative Logit Model

```
INPUT MENTAL SES LIFE;
LS = LIFE * SES;
CARDS;
4 1 1
 .
 .
 .
1 0 9
PROC LOGIST K = 3;
MODEL MENTAL = LIFE SES LS;
```

Table A.16 SAS (CATMOD) Used to Fit Cumulative Logit Model

```
INPUT HOSP $ OPER DUMP $ COUNT @@;
CARDS;
A 1 N 23  A 1 S 7  A 1 M 2  A 2 N 23  A 2 S 10  A 2 M 5
 .
 .
 .
D 3 N 14  D 3 S 8  D 3 M 3  D 4 N 13  D 4 S  6  D 4 M 4
PROC CATMOD ORDER = DATA; WEIGHT COUNT;
   POPULATION HOSP OPER;
   RESPONSE CLOGITS;
   DIRECT OPER;
MODEL DUMP = _RESPONSE_ OPER;
```

response functions. Cumulative logit and probit models and complementary log–log models can also be fitted with procedures PROBIT and LOGISTIC in SAS. The latter procedure also gives a score test of the proportional odds assumption that the effect parameters are identical for each cutpoint.

Hutchison (1984) showed how GLIM can fit cumulative link functions, with or without the assumption that effects are independent of the cutpoint. See Lindsey (1989) for a GLIM macro for the proportional odds case.

Table A.17 uses CATMOD in SAS to fit the mean response model to the olive preference data in Table 9.11.

Chapter 10: Models for Dependent Samples
Bhapkar's test of marginal homogeneity can be obtained using PROC CATMOD in SAS. Table A.18 gives code for the example in Section 10.3. Firth (1989), Firth and Treat (1988) and Lipsitz (1988) showed how GLIM can give an ML fit of the model of marginal homogeneity and

TABLE A.17 SAS Used to Fit Mean Response Model

INPUT URBAN $ LOCATION $ PREF $ COUNT @@;
CARDS;
1 M 1 20 1 M 2 15 1 M 3 12 1 M 4 17 1 M 5 16 1 M 6 28
.
.
O S 1 11 O S 2 9 O S 3 26 O S 4 19 O S 5 17 O S 6 24
PROC CATMOD ORDER = DATA; WEIGHT COUNT;
POPULATION URBAN LOCATION;
RESPONSE 1 2.5 4.5 6 7 8.5;
MODEL PREF = URBAN LOCATION / COVB;

Table A.18 SAS Used to Test Marginal Homogeneity

INPUT THEN $ NOW $ COUNT @@;
CARDS;
NE NE 11607 NE MW 100 NE S 366 NE W   124
.
.
W NE   63 W MW 176 W S 286   W W 10192
PROC CATMOD; WEIGHT COUNT;
  RESPONSE MARGINALS;
MODEL THEN * NOW = _RESPONSE_ / FREQ;
REPEATED TIME 2;

related models. Table A.19 shows Lipsitz's approach, whereby one expresses the I^2 expected frequencies in terms of parameters for the $(I-1)^2$ cells in the first $I-1$ rows and $I-1$ columns, the cell in the last row and last column, and $I-1$ marginal totals (which are the same for rows and columns). In Table A.19, M11 denotes expected frequency m_{11}, M1 denotes $m_{1+} = m_{+1}$, and so forth. This parameterization uses formulas such as $m_{14} = m_{1+} - m_{11} - m_{12} - m_{13}$, and $m_{41} = m_{+1} - m_{11} - m_{21} - m_{31} = m_{1+} - m_{11} - m_{21} - m_{31}$ for terms in the last column or last row. The marginal homogeneity fit is obtained using the Poisson random component and identity link. Lipsitz also gave a SAS IML routine for fitting this model.

Models of symmetry and quasi-symmetry can be fitted using standard loglinear models (Problem 10.27). The quasi-independence model can be fitted using BMDP (4F or 3R), SAS (CATMOD), GLIM, or SPSS[x] (LOGLINEAR, see Norušis 1988). SAS fits it by fitting the independence model to the amended table having main-diagonal counts replaced by zeroes (which CATMOD treats as structural zeroes). GLIM can fit it by adding to the independence model a parameter for each main-diagonal cell, or by fitting the independence model with zero weight attached to the main diagonal. Table A.20 shows both of these approaches in GLIM. We first generated levels for X and Y. For instance, %GL(4, 4) generates levels 1 through 4 in blocks of 4, resulting in level 1 for the first four units, 2 for the next four units, and so forth. We then defined four variables that are indicators for the main diagonal cells. Adding these to the independence model forces a perfect fit on the main diagonal. The second approach constructs a variable (WT) that equals 1 except on the main diagonal, where it is 0. Fitting the independence model with a $WEIGHT directive for this variable then gives the result. This is equivalent to deleting the main-diagonal counts from the data set (i.e.,

Table A.19 GLIM Used to Fit Marginal Homogeneity by ML

```
$UNITS 16
$DATA COUNT M11 M12 M13 M21 M22 M23 M31 M32 M33 M44 M1 M2 M3
$READ
```

COUNT	M11	M12	M13	M21	M22	M23	M31	M32	M33	M44	M1	M2	M3
11607	1	0	0	0	0	0	0	0	0	0	0	0	0
100	0	1	0	0	0	0	0	0	0	0	0	0	0
366	0	0	1	0	0	0	0	0	0	0	0	0	0
124	−1	−1	−1	0	0	0	0	0	0	0	1	0	0
87	0	0	0	1	0	0	0	0	0	0	0	0	0
13677	0	0	0	0	1	0	0	0	0	0	0	0	0
515	0	0	0	0	0	1	0	0	0	0	0	0	0
302	0	0	0	−1	−1	−1	0	0	0	0	0	1	0
172	0	0	0	0	0	0	1	0	0	0	0	0	0
225	0	0	0	0	0	0	0	1	0	0	0	0	0
17819	0	0	0	0	0	0	0	0	1	0	0	0	0
270	0	0	0	0	0	0	−1	−1	−1	0	0	0	1
63	−1	0	0	−1	0	0	−1	0	0	0	1	0	0
176	0	−1	0	0	−1	0	0	−1	0	0	0	1	0
286	0	0	−1	0	0	−1	0	0	−1	0	0	0	1
10192	0	0	0	0	0	0	0	0	0	1	0	0	0

```
$YVAR COUNT
$ERROR POIS $
$LINK IDENTITY $
$FIT M11 + M12 + M13 + M21 + M22 + M23 + M31 + M32 + M33 + M44
   + M1 + M2 + M3 $
$DIS R $
```

Table A.20   GLIM Used to Fit Quasi Independence

```
$UNITS 16
$DATA COUNT
$READ
11607  100   366      124
 .
 .
 .
    63  176  286   10192
$YVAR COUNT
$ERROR POIS
$CALC X = %GL(4,4) $
$CALC Y = %GL(4,1) $
$FAC X 4   Y 4 $
$CALC D1 = %EQ(X,1)&%EQ(Y,1) $
$CALC D2 = %EQ(X,2)&%EQ(Y,2) $
$CALC D3 = %EQ(X,3)&%EQ(Y,3) $
$CALC D4 = %EQ(X,4)&%EQ(Y,4) $
$FIT X + Y: + D1 + D2 + D3 +D4 $

$CALC WT = 1 - D1 - D2 - D3 - D4 $
$WEIGHT WT $
$FIT X + Y $
```

we could simply have fitted the independence model after entering the nine other observations as the data set).

McNemar's test, the test of symmetry, and kappa and its standard error are computed by BMDP (4F), using the MCN option in the STATISTICS statement. SPSS[X] also computes McNemar's test, using the MCNEMAR subcommand in the NPAR TESTS procedure.

The Bradley–Terry model can be fitted by fitting the quasi-symmetry model, or by fitting a quasi-independence model discussed by Imrey et al. (1976) and Fienberg and Larntz (1976). It can also be fitted using corresponding logit models. For instance, for the baseball data, one could set up seven artificial explanatory variables, one for each team. For a given observation, the variable for team i is 1 if it wins, -1 if it loses, and 0 if it is not one of the paired teams for that observation. One then fits the logit model having these artificial variables (one of which is redundant) as explanatory variables. One can delete the intercept term in SAS (CATMOD) using the NOINT option, and in GLIM by entering -1 in the list of terms in the $FIT directive.

Chapter 11: Analyzing Repeated Categorical Response Data
SPSS[X] reports Cochran's Q test, using the COCHRAN subcommand in

the NPAR TESTS procedure. The test can also be conducted using procedures for the Cochran–Mantel–Haenszel test.

The generalized Bhapkar test of marginal homogeneity can be computed using PROC CATMOD in SAS. Lipsitz (1988) used SAS and GLIM to obtain the ML fit. Table A.21 uses GLIM to test marginal homogeneity for Table 11.2, where for instance $M11P$ denotes m_{11+}. Under the assumption of marginal homogeneity, the eight cell expected frequencies are expressed in terms of m_{111}, m_{11+}, m_{1+1}, m_{+11}, m_{1++}, and m_{222} (since $m_{+1+} = m_{++1} = m_{1++}$). Note, for instance, that $m_{112} = m_{11+} - m_{111}$ and $m_{122} = m_{111} + m_{1++} - m_{11+} - m_{1+1}$.

The SAS manual (SAS Institute, 1987) gives several examples of the use of CATMOD for repeated measurement data. CATMOD provides predicted logits and their standard errors. The fitted probabilities are given by the transformation $\exp(\text{logit})/[1 + \exp(\text{logit})]$. The standard errors of the fitted probabilities equal the standard errors of the predicted logits multiplied by $\hat{\pi}(1 - \hat{\pi})$ for that predicted probability. This formula is based on applying the delta method to the transformation used to obtain predicted probabilities from predicted logits. Table A.22 uses CATMOD to fit the model described in Section 11.3.2. The second and third columns of the model matrix pertain to the diagnosis and treatment effects. The last two columns pertain to separate linear effects of time for the two treatments. The CONTRAST statement tests whether those slopes are the same. Table A.23 uses CATMOD to fit a cumulative logit model to the marginal distributions of Table 11.7.

Table A.21   GLIM Used to Fit Marginal Homogeneity by ML to Table 11.2

```
$ UNITS 8
$DATA COUNT M111 M11P M1P1 MP11 M1PP M222
$READ
 6   1   0   0   0 0 0
16  -1   1   0   0 0 0
 2  -1   0   1   0 0 0
 4   1  -1  -1   0 1 0
 2  -1   0   0   1 0 0
 4   1  -1   0  -1 1 0
 6   1   0  -1  -1 1 0
 6   0   0   0   0 0 1
$YVAR COUNT
$ERROR POIS
$LINK IDENTITY
$FIT M111 + M11P + M1P1 + MP11 + M1PP + M222 $
$LOOK %X2 $
$DIS R $
```

Table A.22 SAS Used to Fit Model for Marginal Logits

```
INPUT DIAGNOSE $ TREAT $ WEEK1 $ WEEK2 $ WEEK4 $ COUNT @@;
CARDS;
MILD      STAND  N  N  N  16
.
.
.
SEVERE  NEW     A  A  A  6
PROC CATMOD ORDER = DATA; WEIGHT COUNT;
  POPULATION DIAGNOSE TREAT;
  RESPONSE LOGIT;
MODEL WEEK1 * WEEK2 * WEEK4 = (1  1  1  0  0,
                              1  1  1  1  0,
                              1  1  1  2  0,
                              1  1  0  0  0,
                              1  1  0  0  1,
                              1  1  0  0  2,
                              1  0  1  0  0,
                              1  0  1  1  0,
                              1  0  1  2  0,
                              1  0  0  0  0,
                              1  0  0  0  1,
                              1  0  0  0  2)/ PROB COVB PRED;
     CONTRAST 'EQUAL SLOPES'
       ALL_PARMS 0  0  0  1  −1;
```

Table A.23 SAS Used to Fit Marginal Cumulative Logit Model

```
INPUT TREAT $ INITIAL $ FOLLOW $ COUNT @@;
  IF COUNT = 0 THEN COUNT = 1E − 8; CARDS;
ACTIVE    <20   <20    7      ACTIVE    <20  20–30   4
.
.
.
PLACEBO  <60  30–60  14      PLACEBO  >60   >60   22
PROC CATMOD  ORDER = DATA;  WEIGHT COUNT;
POPULATION TREAT;
RESPONSE CLOGIT;
MODEL INITIAL * FOLLOW = (1  0  0  1  1  1,
                          0  1  0  1  1  1,
                          0  0  1  1  1  1,
                          1  0  0  1  0  0,
                          0  1  0  1  0  0,
                          0  0  1  1  0  0,
                          1  0  0  0  1  0,
                          0  1  0  0  1  0,
                          0  0  1  0  1  0,
                          1  0  0  0  0  0,
                          0  1  0  0  0  0,
                          0  0  1  0  0  0)
(1 2 3 = 'CUTPOINTS', 4 = 'TREAT', 5 = 'OCCAS', 6 = 'TREAT * OCCAS');
```

A Twentieth-Century Tour of Categorical Data Analysis

To readers who have made it this far, I offer my congratulations for your hard work and patience. To develop a more complete understanding of methods for categorical data analysis, you may want to see how these methods have developed in this century. I suggest the following chronological list of 25 sources. These do not reflect my opinion about the "most important" articles, but I chose them because they convey a sense of how methodology has evolved. The bibliography contains complete references for these sources.

Pearson (1900)
Yule (1912)
Fisher (1922)
Bartlett (1935)
Berkson (1944)
Neyman (1949)
Cochran (1954)
Goodman and Kruskal (1954)
Roy and Mitra (1956)
Cox (1958a)
Mantel and Haenszel (1959)
Birch (1963)
Birch (1964b)

Caussinus (1965)
Goodman (1968)
Mosteller (1968)
Grizzle et al. (1969)
Goodman (1970)
Haberman (1970)
McFadden (1974)
Nelder and Wedderburn (1972)
Bishop et al. (1975)
Goodman (1979a)
McCullagh (1980)
Goodman (1986)

APPENDIX C

Chi-Squared Distribution Values for Various Right-Hand Tail Probabilities

				α			
df	0.250	0.100	0.050	0.025	0.010	0.005	0.001
1	1.32330	2.70554	3.84146	5.02389	6.63490	7.87944	10.828
2	2.77259	4.60517	5.99147	7.37776	9.21034	10.5966	13.816
3	4.10835	6.25139	7.81473	9.34840	11.3449	12.8381	16.266
4	5.38572	7.77944	9.48773	11.1433	13.2767	14.8602	18.467
5	6.62568	9.23635	11.0705	12.8325	15.0863	16.7496	20.515
6	7.84080	10.6446	12.5916	14.4494	16.8119	18.5476	22.458
7	9.03715	12.0170	14.0671	16.0128	18.4753	20.2777	24.322
8	10.2188	13.3616	15.5073	17.5346	20.0902	21.9550	26.125
9	11.3887	14.6837	16.9190	19.0228	21.6660	23.5893	27.877
10	12.5489	15.9871	18.3070	20.4831	23.2093	25.1882	29.588
11	13.7007	17.2750	19.6751	21.9200	24.7250	26.7569	31.264
12	14.8454	18.5494	21.0261	23.3367	26.2170	28.2995	32.909
13	15.9839	19.8119	22.3621	24.7356	27.6883	29.8194	34.528
14	17.1170	21.0642	23.6848	26.1190	29.1413	31.3193	36.123
15	18.2451	22.3072	24.9958	27.4884	30.5779	32.8013	37.697
16	19.3688	23.5418	26.2962	28.8454	31.9999	34.2672	39.252
17	20.4887	24.7690	27.5871	30.1910	33.4087	35.7185	40.790
18	21.6049	25.9894	28.8693	31.5264	34.8053	37.1564	42.312
19	22.7178	27.2036	30.1435	32.8523	36.1908	38.5822	43.820
20	23.8277	28.4120	31.4104	34.1696	37.5662	39.9968	45.315

df	α						
	0.250	0.100	0.050	0.025	0.010	0.005	0.001
21	24.9348	29.6151	32.6705	35.4789	38.9321	41.4010	46.797
22	26.0393	30.8133	33.9244	36.7807	40.2894	42.7956	48.268
23	27.1413	32.0069	35.1725	38.0757	41.6384	44.1813	49.728
24	28.2412	33.1963	36.4151	39.3641	42.9798	45.5585	51.179
25	29.3389	34.3816	37.6525	40.6465	44.3141	46.9278	52.620
26	30.4345	35.5631	38.8852	41.9232	45.6417	48.2899	54.052
27	31.5284	36.7412	40.1133	43.1944	46.9630	49.6449	55.476
28	32.6205	37.9159	41.3372	44.4607	48.2782	50.9933	56.892
29	33.7109	39.0875	42.5569	45.7222	49.5879	52.3356	58.302
30	34.7998	40.2560	43.7729	46.9792	50.8922	53.6720	59.703
40	45.6160	51.8050	55.7585	59.3417	63.6907	66.7659	73.402
50	56.3336	63.1671	67.5048	71.4204	76.1539	79.4900	86.661
60	66.9814	74.3970	79.0819	83.2976	88.3794	91.9517	99.607
70	77.5766	85.5271	90.5312	95.0231	100.425	104.215	112.317
80	88.1303	96.5782	101.879	106.629	112.329	116.321	124.839
90	98.6499	107.565	113.145	118.136	124.116	128.299	137.208
100	109.141	118.498	124.342	129.561	135.807	140.169	149.449

Bibliography

Adelbasit, K. M., and R. L. Plackett. 1983. Experimental design for binary data. *J. Amer. Statist. Assoc.* **78**: 90–98.

Agresti, A. 1976. The effect of category choice on some ordinal measures of association. *J. Amer. Statist. Assoc.* **71**: 49–55.

Agresti, A. 1980. Generalized odds ratios for ordinal data. *Biometrics* **36**: 59–67.

Agresti, A. 1983a. A survey of strategies for modeling cross-classifications having ordinal variables. *J. Amer. Statist. Assoc.* **78**: 184–198.

Agresti, A. 1983b. Testing marginal homogeneity for ordinal categorical variables. *Biometrics* **39**: 505–510.

Agresti, A. 1984. *Analysis of Ordinal Categorical Data.* New York: Wiley.

Agresti, A. 1988. A model for agreement between ratings on an ordinal scale. *Biometrics* **44**: 539–548.

Agresti, A., and C. Chuang. 1989. Model-based Bayesian methods for estimating cell proportions in cross-classification tables having ordered categories. *Comput. Statist. Data Anal.* **7**: 245–258.

Agresti, A., and A. Kezouh. 1983. Association models for multidimensional cross-classifications of ordinal variables. *Comm. Statist.* **A 12**: 1261–1276.

Agresti, A., and D. Wackerly. 1977. Some exact conditional tests of independence for R × C cross-classification tables. *Psychometrika* **42**: 111–125.

Agresti, A., and M. Yang. 1987. An empirical investigation of some effects of sparseness in contingency tables. *Comput. Statist. Data Anal.* **5**: 9–21.

Agresti, A., C. Chuang, and A. Kezouh. 1987. Order-restricted score parameters in association models for contingency tables. *J. Amer. Statist. Assoc.* **82**: 619–623.

Agresti, A., C. R. Mehta, and N. R. Patel. 1990. Exact inference for contingency tables with ordered categories. *J. Amer. Statist. Assoc.*, **85**: to appear.

Aitchison, J., and C. G. G. Aitken. 1976. Multivariate binary discrimination by the kernel method. *Biometrika* **63**: 413–420.

Aitchison, J., and S. D. Silvey. 1957. The generalization of probit analysis to the case of multiple responses. *Biometrika* **44**: 131–140.

Aitchison, J., and S. D. Silvey. 1958. Maximum likelihood estimation of parameters subjects to restraints. *Ann. Math. Statist.* **29**: 813–828.

508

Aitkin, M. 1979. A simultaneous test procedure for contingency table models. *Appl. Statist.* **28:** 233–242.

Aitkin, M. 1980. A note on the selection of log-linear models. *Biometrics* **36:** 173–178.

Aitkin, M., and D. Clayton. 1980. The fittinf of exponential, Weibull, and extreme value distributions to complex censored survival data using GLIM. *Appl. Statist.* **29:** 156–163.

Aitkin, M., B. Francis, and N. Raynal. 1987. Une étude comparative d'analyses des correspondances ou de classifications et des modèles de variables latentes ou de classes latentes. *Rev. Statist. Appl.* **35:** 53–82.

Aitkin, M., D. Anderson, B. Francis, and J. Hinde. 1989. *Statistical Modelling in GLIM.* Oxford: Clarendon Press.

Alam, K., and K. T. Wallenius. 1976. Positive dependence and monotonicity in conditional distributions. *Comm. Statist.* **A1:** 525–534.

Albert, A., and J. A. Anderson. 1984. On the existence of maximum likelihood estimates in logistic models. *Biometrika* **71:** 1–10.

Albert, J. H. 1987. Empirical Bayes estimation in contingency tables. *Comm. Statist.* **A 16:** 2459–2485.

Albert, J. H. 1988. Computational methods using a Bayesian hierarchical generalized linear model. *J. Amer. Statist. Assoc.* **83:** 1037–1044.

Aldrich, J. H., and F. D. Nelson. 1984. *Linear Probability, Logit, and Probit Models.* Sage University Paper series on Quantitative Applications in the Social Sciences 07–045. Beverly Hills and London: Sage Publications.

Allison, P. D. 1984. *Event History Analysis.* Beverly Hills, CA: Sage Publications.

Altham, P. M. E. 1969. Exact Bayesian analysis of a 2×2 contingency table and Fisher's 'exact' significance test. *J. Roy. Statist. Soc.* **B31:** 261–269.

Altham, P. M. E. 1970a. The measurement of association of rows and columns for an $r \times s$ contingency table. *J. Roy. Statist. Soc.* **B32:** 63–73.

Altham, P. M. E. 1970b. The measurement of association in a contingency table: three extensions of the cross-ratios and metric methods. *J. Roy. Statist, Soc.* **B32:** 395–407.

Altham, P. M. E. 1971. The analysis of matched proportions. *Biometrika* **58:** 561–576.

Altham, P. M. E. 1975. Quasi-independent triangular contingency tables. *Biometrics* **31:** 233–238.

Altham, P. M. E. 1976. Discrete variable analysis for individuals grouped into families. *Biometrika* **63:** 263–269.

Altham, P. M. E. 1978. Two generalizations of the binomial distribution. *Appl. Statist.* **27:** 162–167.

Altham, P. M. E. 1984. Improving the precision of estimation by fitting a model. *J. Roy. Statist. Soc.* **B46:** 118–119.

Amemiya, T. 1981. Qualitative response models: A survey. *J. Econom. Literature* **19:** 1483–1536.

Amemiya, T., and J. L. Powell. 1980. A comparison of the logit model and normal discriminant analysis when the independent variables are binary. Technical Report no. 320, Inst. Math. Studies in Social Sciences, Stanford University.

Andersen, A. H. 1974. Multidimensional contingency tables. *Scan. J. Statist.* **1:** 115–127.

Andersen, E. B. 1980. *Discrete Statistical Models with Social Science Applications.* Amsterdam: North-Holland.

Anderson, J. A. 1972. Separate sample logistic discrimination. *Biometrika* **59**: 19–35.

Anderson, J. A. 1984. Regression and ordered categorical variables (with discussion). *J. Roy. Statist. Soc.* **B46**: 1–30.

Anderson, J. A., and P. R. Philips. 1981. Regression, discrimination, and measurement models for ordered categorical variables. *Appl. Statist.* **30**: 22–31.

Anderson, T. W., and L. A. Goodman. 1957. Statistical inference about Markov chains. *Ann. Math. Statist.* **28**: 89–110.

Aranda-Ordaz, F. J. 1981. On two families of transformations to additivity for binary response data. *Biometrics* **68**: 357–363.

Aranda-Ordaz, F. J. 1983. An extension of the proportional hazards model for grouped data. *Biometrics* **39**: 109–117.

Armitage, P. 1955. Tests for linear trends in proportions and frequencies. *Biometrics* **11**: 375–386.

Ashford, J. R., and R. D. Sowden. 1970. Multivariate probit analysis. *Biometrics* **26**: 535–546.

Asmussen, S. and D. Edwards. 1983. Collapsibility and response variables in contingency tables. *Biometrika* **70**: 567–578.

Assakul, K. and C. H. Proctor. 1967. Testing independence in two way contingency tables with data subject to misclassification. *Psychometrika* **32**: 67–76.

Baglivo, J., D. Olivier, and M. Pagano. 1988. Methods for the Analysis of Contingency Tables with Large and Small Cell Counts. *J. Amer. Statist. Assoc.* **83**: 1006–1013.

Baker, R. J., M. R. B. Clarke, and P. W. Lane. 1985. Zero entries in contingency tables. *Comput. Statist. Data Anal.* **3**: 33–45.

Baptista, J., and M. C. Pike. 1977. Algorithm AS115: Exact two-sided confidence limits for the odds ratio in a 2×2 table. *Appl. Statist.* **26**: 214–220.

Bard, Y. 1974. *Nonlinear Parameter Estimates.* New York: Academic Press.

Barnard, G. A. 1945. A new test for 2×2 tables. *Nature* **156**: 177.

Barnard, G. A. 1947. Significance tests for 2×2 tables. *Biometrika* **34**: 123–138.

Barnard, G. A. 1949. Statistical inference. *J. Roy. Statist. Soc.* **B11**: 115–139.

Barnard, G. A. 1979. In contradiction to J. Berkson's dispraise: Conditional tests can be more efficient. *J. Statist. Plann. Inference* **3**: 181–188.

Bartlett, M. S. 1935. Contingency table interactions. *J. Roy. Statist. Soc.*, Supplement **2**, 248–252.

Beal, S. L. 1987. Asymptotic confidence intervals for the difference between two binomial parameters. *Biometrics* **43**: 941–950.

Becker, M. 1989a. Models for the analysis of association in multivariate contingency tables. *J. Amer. Statist. Assoc.* **84**: 1014–1019.

Becker, M. 1989b. On the bivariate normal distribution and association models for ordinal categorical data. *Statist. Probab. Lett.* **8**: 435–440.

Becker, M. 1989c. Maximum likelihood estimation of the RC(M) association model. *Appl. Statist.* **38**: to appear.

Becker, M., and C. C. Clogg. 1989. Analysis of sets of two-way contingency tables using association models. *J. Amer. Statist. Assoc.* **84**: 142–151.

Bedrick, E. J. 1987. A family of confidence intervals for the ratio of two binomial proportions. *Biometrics* **43**: 993–998.

Begg, C. B., and R. Gray. 1984. Calculation of polytomous logistic regression parameters using individualized regressions. *Biometrika* **71**: 11–18.

Ben-Akiva, M., and S. R. Lerman. 1985. *Discrete Choice Analysis: Theory and Application to Travel Demand*. Cambridge, MA: MIT Press.

Benedetti, J. K., and M. B. Brown. 1978. Strategies for the selection of loglinear models. *Biometrics* **34**: 680–686.

Benichou, J., and M. H. Gail. 1989. A delta method for implicitly defined random variables. *Amer. Statist.* **43**: 41–44.

Bennett, S. 1983. Analysis of survival data by the proportional odds model. *Statist. Medic.* **2**: 279–285.

Benzécri, J.-P. 1973. *L'Analyse des Données, Vol, 1: La Taxonomie, Vol. 2: L'Analyse des Correspondances*. Paris: Dunod.

Berkson, J. 1938. Some difficulties of interpretation encountered in the application of the chi-square test. *J. Amer. Statist. Assoc.* **33**: 526–536.

Berkson, J. 1944. Application of the logistic function to bio-assay. *J. Amer. Statist. Assoc.* **39**: 357–365.

Berkson, J. 1951. Why I prefer logits to probits. *Biometrics* **7**: 327–339.

Berkson, J. 1953. A statistically precise and relatively simple method of estimating the bioassay with quantal response, based on the logistic function. *J. Amer. Statist. Assoc.* **48**: 565–599.

Berkson, J. 1955. Maximum likelihood and minimum logit χ^2 estimation of the logistic function. *J. Amer. Statist. Assoc.* **50**: 130–162.

Berkson, J. 1972. Minimum discrimination information, the 'no interaction' problem, and the logistic function. *Biometrics* **28**: 443–468.

Berkson, J. 1978. In dispraise of the exact test. *J. Statist. Plann. Inference* **2**: 27–42.

Berkson, J. 1980. Minimum chi-square, not maximum likelihood! (plus discussion). *Ann. Statist.* **8**: 457–487.

Berry, K. J., and P. W. Mielke, Jr. 1988. Monte Carlo comparisons of the asymptotic chi-square and likelihood-ratio tests with the nonasymptotic chi-square test for sparse r × c tables. *Psychol. Bull.* **103**: 256–264.

Bhapkar, V. P. 1966. A note on the equivalence of two test criteria for hypotheses in categorical data. *J. Amer. Statist. Assoc.* **61**: 228–235.

Bhapkar, V. P. 1968. On the analysis of contingency tables with a quantitative response. *Biometrics* **24**: 329–338.

Bhapkar, V. P. 1970a. Categorical data analogs of some multivariate tests. pp. 85–110 in *Essays in Probability and Statistics*, ed. by R. C. Bose et al. Chapel Hill, NC: University of North Carolina Press.

Bhapkar, V. P. 1970b. On Cochran's Q test and its modification. pp. 255–267 in *Random Counts in Scientific Work*, Vol. 2, ed. G. P. Patil. University Park, PA: Pennsylvania State University Press.

Bhapkar, V. P. 1973. On the comparison of proportions in matched samples. *Sankhya*, **A35**: 341–356.

Bhapkar, V. P. 1979. On tests of marginal symmetry and quasi-symmetry in two and three-dimensional contingency tables. *Biometrics* **35**: 417–426.

Bhapkar, V. P. 1986. On conditionality and likelihood with nuisance parameters in models for contingency tables. Department of Statistics, Technical Report No. 253, University of Kentucky.

Bhapkar, V. P., and G. G. Koch. 1968. On the hypotheses of 'no interaction' in multidimensional contingency tables. *Biometrics* **24:** 567–594.

Bhapkar, V. P. and G. W. Somes. 1977. Distribution of Q when testing equality of matched proportions. *J. Amer. Statist. Assoc.* **72:** 658–661.

Bickel, P. J., E. A. Hammel, and J. W. O'Connell. 1975. Sex bias in graduate admissions: data from Berkeley. *Science* **187:** 398–403.

Billingsley, P. 1961. Statistical methods in Markov Chains. *Ann. Math. Statist.* **32:** 12–40.

Birch, M. W. 1963. Maximum likelihood in three-way contingency tables. *J. Roy. Statist. Soc.* **B25:** 220–233.

Birch, M. W. 1964a. A new proof of the Pearson-Fisher theorem. *Ann. Math. Statist.* **35:** 817–824.

Birch, M. W. 1964b. The detection of partial association I: The 2×2 case. *J. Roy. Statist. Soc.* **B26:** 313–324.

Birch, M. W. 1965. The detection of partial association II: The general case. *J. Roy. Statist. Soc.* **B27:** 111–124.

Bishop, Y. V. V. 1967. Multidimensional contingency tables: cell estimates. Ph. D. dissertation, Department of Statistics, Harvard University.

Bishop, Y. V. V. 1969. Full contingency tables, logits, and split contingency tables. *Biometrics* **25:** 119–128.

Bishop, Y. V. V. 1971. Effects of collapsing multidimensional contingency tables. *Biometrics* **27:** 545–562.

Bishop, Y. V. V., and F. Mosteller. 1969. Smoothed contingency table analysis. Chap. IV-3 of *The National Halothane Study*. Washington, DC: U.S. Government Printing Office.

Bishop, Y. V. V., S. E. Fienberg, and P. W. Holland. 1975. *Discrete Multivariate Analysis*. Cambridge, MA: MIT Press.

Bliss, C. I. 1934. The method of probits. *Science* **79:** 38–39 (Correction, 409–410).

Bliss, C. I. 1935. The calculation of the dosage-mortality curve. *Ann. Appl. Biol.* **22:** 134–167.

Blyth, C. R. 1972. On Simpson's paradox and the sure-thing principle. *J. Amer. Statist. Assoc.* **67:** 364–366.

Bock, R. D. 1970. Estimating multinomial response relations. pp. 453–479 in *Contributions to Statistics and Probability*, ed. by R. C. Bose. Chapel Hill, NC: University of North Carolina Press.

Bock, R. D., and L. V. Jones. 1968. *The Measurement and Prediction of Judgement and Choice*. San Francisco: Holden-Day.

Bock, R. D., and G. Yates. 1973. MULTIQUAL: Log-linear analysis of nominal or ordinal qualitative data by the method of maximum likelihood. Chicago: International Educational Services.

Bohrnstedt, G. W., and D. Knoke. 1982. *Statistics for Social Data Analysis*. Itasca, IL: Peacock.

Bonney, G. E. 1987. Logistic regression for dependent binary observations. *Biometrics* **43:** 951–973.

Borsch-Supan, A. 1987. *Econometric Analysis of Discrete Choice*. Lecture Notes in Economics and Mathematical Systems. Berlin: Springer-Verlag.

Bowker, A. H. 1948. A test for symmetry in contingency tables. *J. Amer. Statist. Assoc.* **43:** 572–574.

Box, J. F. 1978. *R. A. Fisher, The Life of a Scientist.* New York: John Wiley.

Bradley, R. A. 1976. Science, statistics and paired comparisons. *Biometrics* **32:** 213–240.

Bradley, R. A., and M. E. Terry. 1952. Rank analysis of incomplete block designs I. The method of paired comparisons. *Biometrika* **39:** 324–345.

Breen, R. 1984. Fitting non-hierarchical and association log linear models using GLIM. *Sociological Methods and Res.* **13:** 77–107.

Brennan, J. F. 1949. Evaluation of parameters in the Gompertz and Makeham equations. *J. Amer. Statist. Assoc.* **44:** 116–121.

Breslow, N. 1976. Regression analysis of the log odds ratio: a method for retrospective studies. *Biometrics* **32:** 409–416.

Breslow, N. 1981. Odds ratio estimators when the data are sparse. *Biometrika* **68:** 73–84.

Breslow, N. 1982. Covariance adjustment of relative-risk estimates in matched studies. *Biometrics* **38:** 661–672.

Breslow, N. 1984. Extra-Poisson variation in log-linear models. *Appl. Statist.* **33:** 38–44.

Breslow, N., and J. Cologne. 1986. Methods of estimation in log odds ratio regression models. *Biometrics* **42:** 949–954.

Breslow, N., and N. E. Day. 1975. Indirect standardization and multivariate models for rates, with reference to the age adjustment of cancer incidence and relative frequency data. *J. Chronic Diseases* **28:** 289–303.

Breslow, N., and N. E. Day. 1980, 1987. *Statistical Methods in Cancer Research*, vol. I: The Analysis of Case-Control Studies; vol. II: The Design and Analysis of Cohort Studies. Lyon: IARC.

Breslow, N., and K. Y. Liang. 1982. The variance of the Mantel–Haenszel estimator. *Biometrics* **38:** 943–952.

Breslow, N., and W. Powers. 1978. Are there two logistic regressions for retrospective studies? *Biometrics* **34:** 100–105.

Brier, S. S. 1980. Analysis of contingency tables under cluster sampling. *Biometrika* **67:** 591–596.

Bross, I. D. J. 1958. How to use ridit analysis. *Biometrics* **14:** 18–38.

Brown, C. C. 1975. On the use of indicator variables for studying the time-dependence of parameters in a response-time model. *Biometrics* **31:** 863–872.

Brown, M. B. 1974. Identification of the sources of significance in two-way contingency tables. *Appl. Statist.* **23:** 405–413.

Brown, M. B. 1976. Screening effects in multidimensional contingency tables. *Appl. Statist.* **25:** 37–46.

Brown, M. B. 1986. Categorical data analysis in BMDP: present and future. pp. 257–262 in *Computer Science and Statistics: Proc. 17th Symp. on the Interface.* Amsterdam: North-Holland.

Brown, M. B., and J. K. Benedetti. 1977. Sampling behavior of tests for correlation in two-way contingency tables. *J. Amer. Statist. Assoc.* **72:** 309–315.

Brown, M. B., and C. Fuchs. 1983. On maximum likelihood estimation in sparse contingency tables. *Comput. Statist. Data Anal.* **1:** 3–15.

Brown, P. J., and P. W. K. Rundell. 1985. Kernel estimates for categorical data. *Technometrics* **27:** 293–299.

Brownstone, D., and K. A. Small. Efficient estimation of nested logit models. *J. Bus. Econ. Statist.* **7**: 67–74.

Buckley, W. E. 1988. Concussions in football: A multivariate analysis. *Amer. J. Sports Med.* **16**: 51–56.

Burman, P. 1987. Smoothing sparse contingency tables. *Sankhya* **A49**: 24–36.

Burridge, J. 1981. A note on maximum likelihood estimation for regression models using grouped data. *J. Roy. Statist. Soc.* **B43**: 41–45.

Carr, G. J., K. B. Hafner, and G. G. Koch. 1989. Analysis of rank measures of association for ordinal data from longitudinal studies. *J. Amer. Statist. Assoc.* **84**: 797–804.

Caussinus, H. 1965. Contribution à l'analyse statistique des tableaux de correlation. *Ann. Fac. Sci. Univ. Toulouse* **29**: 77–182.

Chaloner, K., and K. Larntz. 1988. Optimal Bayesian design applied to logistic regression experiments. *J. Statist. Plann. Inference* **21**: 191–208.

Chambers, E. A., and D. R. Cox. 1967. Discrimination between alternative binary response models. *Biometrika* **54**: 573–578.

Chapman, D. G., and R. C. Meng. 1966. The power of chi-square tests for contingency tables. *J. Amer. Statist. Assoc.* **61**: 965–975.

Chapman, D. G., and J. Nam. 1968. Asymptotic power of chi-square tests for linear trends in proportions. *Biometrics* **24**: 315–327.

Chen, T. 1979. Log-linear models for categorical data with misclassification and double sampling. *J. Amer. Statist. Assoc.* **74**: 481–488.

Chen, T., and S. E. Fienberg. 1976. The analysis of contingency tables with incompletely classified data. *Biometrics* **32**: 133–144.

Chin, T., W. Marine, E. Hall, C. Gravelle, and J. Speers. 1961. The influence of Salk vaccination on the epidemic pattern and the spread of the virus in the community. *Amer. J. Hyg.* **73**: 67–94.

Chuang, C. 1982. Empirical Bayes methods for a two-way multiplicative-interaction model. *Comm. Statist.* **A11**: 2977–2989.

Chuang, C., D. Gheva, and C. Odoroff. 1985. Methods for diagnosing multiplicative-interaction models for two-way contingency tables. *Comm. Statist.* **A14**: 2057–2080.

Clayton, D. G. 1974. Some odds ratio statistics for the analysis of ordered categorical data. *Biometrika* **61**: 525–531.

Clogg, C. C. 1978. Adjustment of rates using multiplicative models. *Demography* **15**: 523–539.

Clogg, C. C. 1982a. Using association models in sociological research: Some examples. *Amer. J. Sociol.* **88**: 114–134.

Clogg, C. C. 1982b. Some models for the analysis of association in multiway cross-classifications having ordered categories. *J. Amer. Statist. Assoc.* **77**: 803–815.

Clogg, C. C., and M. Becker. 1986. Log-linear modeling with SPSS^X. pp. 263–269 in *Computer Science and Statistics: Proc. 17th Symp. on the Interface*, ed. by D. M. Allen. Amsterdam: North Holland.

Clogg, C. C., and S. R. Eliason. 1987. Some common problems in log-linear analysis. *Sociol. Methods Res.* **15**: 4–44.

Clogg, C. C., and S. R. Eliason. 1988. A flexible procedure for adjusting rates and proportions, including statistical methods for group comparisons. *Amer. Sociol. Rev.* **53**: 267–283.

Clogg, C. C., and L. A. Goodman. 1984. Latent structure analysis of a set of multidimensional contingency tables. *J. Amer. Statist. Assoc.* **79:** 762–771.

Clogg, C. C., and L. A. Goodman. 1985. Simultaneous latent structure analysis in several groups. *Sociol. Method.* **15:** 81–110.

Clogg, C. C., and J. W. Shockey. 1988. Multivariate analysis of discrete data. in *Handbook of Multivariate Experimental Psychology*, ed. by J. R. Nesselroade and R. B. Cattell. New York: Plenum Press.

Cochran, W. G. 1950. The comparison of percentages in matched samples. *Biometrika* **37:** 256–266.

Cochran, W. G. 1952. The χ^2 test of goodness-of-fit. *Ann. Math. Statist.* **23:** 315–345.

Cochran, W. G. 1954. Some methods of strengthening the common χ^2 tests. *Biometrics* **10:** 417–451.

Cochran, W. G. 1955. A test of a linear function of the deviations between observed and expected numbers. *J. Amer. Statist. Assoc.* **50:** 377–397.

Cohen, A., and H. B. Sackrowitz. 1988. Tests for independence in contingency tables with ordered alternatives. Unpublished manuscript.

Cohen, J. 1960. A coefficient of agreement for nominal scales. *Educ. Psychol. Meas.* **20:** 37–46.

Cohen, J. 1968. Weighted kappa: nominal scale agreement with provision for scaled disagreement or partial credit. *Psychol. Bull.* **70:** 213–220.

Cohen, J. E. 1976. The distribution of the chi-squared statistic under clustered sampling from contingency tables. *J. Amer. Statist. Assoc.* **71:** 665–670.

Conaway, M. R. 1989. Analysis of repeated categorical measurements with conditional likelihood methods. *J. Amer. Statist. Assoc.* **84:** 53–62.

Conover, W. J. 1974. Some reasons for not using the Yates continuity correction on 2×2 contingency tables (with comments). *J. Amer. Statist. Assoc.* **69:** 374–382.

Copas, J. B. 1973. Randomization models for the matched and unmatched 2×2 tables. *Biometrika* **60:** 467–476.

Copas, J. B. 1983. Plotting p against x. *Appl. Statist.* **32:** 25–31.

Copas, J. B. 1988. Binary regression models for contaminated data (with discussion). *J. Roy. Statist. Soc.* **B50:** 225–265.

Cormack, R. M. 1989. Log-linear models for capture-recapture. *Biometrics* **45:** 395–413.

Cornfield, J. 1956. A statistical problem arising from retrospective studies. in *Proc. Third Berkeley Symposium on Math. Statist. and Probab.*, ed. by J. Neyman, **4:** 135–148.

Cornfield, J. 1962. Joint dependence of risk of coronary heart disease on serum cholesterol and systolic blood pressure: a discriminant function analysis. *Fed. Proc.* **21**, Supplement No. 11: 58–61.

Cox, C. 1984. An elementary introduction to maximum likelihood estimation for multinomial models: Birch's theorem and the delta method. *Amer. Statist.* **38:** 283–287.

Cox, D. R. 1958a. The regression analysis of binary sequences. *J. Roy. Statist. Soc.* **B20:** 215–242.

Cox, D. R. 1958b. Two further applications of a model for binary regression. *Biometrika* **45:** 562–565.

Cox, D. R. 1970. *The Analysis of Binary Data.* (2nd edn. 1989. by D.R. Cox and E.J. Snell). London: Chapman and Hall.

Cox, D. R. 1972. Regression models and life tables (with discussion). *J. Roy. Statist. Soc.* **B74:** 187–220.

Cox, D. R. 1983. Some remarks on overdispersion. *Biometrika* **70:** 269–274.

Cox, D. R., and D. V. Hinkley. 1974. *Theoretical Statistics.* London: Chapman and Hall.

Cox, D. R., and D. Oakes. 1984. *Analysis of Survival Data.* London: Chapman and Hall.

Cox, M. A. A., and R. L. Plackett. 1980. Small samples in contingency tables. *Biometrika* **67:** 1–13.

Cramér, H. 1946. *Mathematical Methods of Statistics.* Princeton: Princeton University Press.

Cressie, N., and T. R. C. Read. 1984. Multinomial goodness-of-fit tests. *J. Roy. Statist. Soc.* **46:** 440–464.

Cressie, N., and T. R. C. Read. 1989. Pearson X^2 and the loglikelihood ratio statistic G^2: A comparative review. *Internat. Statist. Rev.* **57:** 19–43.

Crowder, M. J. 1978. Beta-binomial ANOVA for proportions. *Appl. Statist.* **27:** 34–37.

D'Agostino, R. B., W. Chase, and A. Belanger. 1988. The appropriateness of some common procedures for testing the equality of two independent binomial populations. *Amer. Statist.* **42:** 198–202.

Darroch, J. N. 1962. Interactions in multi-factor contingency tables. *J. Roy. Statist. Soc.* **B24:** 251–263.

Darroch, J. N. 1974. Multiplicative and additive interaction in contingency tables. *Biometrika* **61:** 207–214.

Darroch, J. N. 1976. No-interaction in contingency tables. In *Proc. 9th Int. Biometrics Conf.* **1:** 296–316.

Darroch, J. N. 1981. The Mantel-Haenszel test and tests of marginal symmetry; fixed-effects and mixed models for a categorical response. *Internat. Statist. Rev.* **49:** 285–307.

Darroch, J. N. 1986. Quasi-symmetry. pp. 469–473 in *Encyclopedia of Statistical Sciences,* Vol. 7. New York: John Wiley.

Darroch, J. N., and P. I. McCloud. 1986. Category distinguishability and observer agreement. *Austral. J. Statist.* **28:** 371–388.

Darroch, J. N., and D. Ratcliff. 1972. Generalized iterative scaling for log-linear models. *Ann. Math. Statist.* **43:** 1470–1480.

Darroch, J. N., S. L. Lauritzen, and T. P. Speed. 1980. Markov fields and log-linear interaction models for contingency tables. *Ann. Statist.* **8:** 522–539.

Das Gupta, S., and M. D. Perlman. 1974. Power of the noncentral F-test: Effect of additional variates on Hotelling's T^2-test. *J. Amer. Statist. Assoc.* **69:** 174–180.

David, H. A., 1988. *The Method of Paired Comparisons.* Oxford: Oxford University Press.

Davis, C. S. 1988. Estimation of row and column scores in the linear-by-linear association model for two-way contingency tables. pp. 946–951 in *Proc. 13th Annual SAS Users Group International Conf.* Cary, NC: SAS Institute, Inc.

Davis, L. J. 1986a. Exact tests for 2 by 2 contingency tables. *Amer. Statist.* **40:** 139–141.

Davis, L. J. 1986b. Relationship between strictly collapsible and perfect tables. *Statist. Probab. Lett.* **4:** 119–122.

Davis, L. J. 1990. Collapsibility of likelihood ratio tests in multidimensional contingency tables. *Comm. Statist.* **A19:** to appear.

Day, N. E., and D. P. Byar. 1979. Testing hypotheses in case-control studies—Equivalence of Mantel-Haenszel statistics and logit score tests. *Biometrics* **35:** 623–630.

Defize, P. R. 1980. The calculation of adjusted residuals for log-linear models in GLIM. *GLIM Newsletter* **1:** 41.

Delany, M. F., and C. T. Moore. 1987. American alligator food habits in Florida. Unpublished manuscript.

Deming, W. E. and F. F. Stephan. 1940. On a least squares adjustment of a sampled frequency table when the expected marginal totals are known. *Ann. Math. Statist.* **11:** 427–444.

Demo, D. H., and K. D. Parker. 1987. Academic achievement and self-esteem among black and white college students. *J. Social Psychol.* **127:** 345–355.

Dempster, A. P., N. M. Laird, and D. B. Rubin. 1977. Maximum likelihood from incomplete data via the EM algorithm (with discussion). *J. Roy. Statist. Soc.* **B39:** 1–38.

DiFrancisco, W., and Z. Critelman. 1984. Soviet political culture and covert participation in policy implementation. *Amer. Polit. Sci. Rev.* **78:** 603–621.

Dixon, W. J. 1983. *BMDP Statistical Software* 1981. Berkeley: Univ. of Calif. Press.

Dobson, A. J. 1983. *An Introduction to Statistical Modelling.* London: Chapman and Hall.

Doll, R., and A. B. Hill. 1952. A study of the aetiology of carcinoma of the lung. *British. Med. J.* **2:** 1271–1286.

Doll, R., and R. Peto. 1976. Mortality in relation to smoking: Twenty years' observations on male British doctors. *British Med. J.* **2:** 1525–1536.

Domencich, T. A., and D. McFadden. 1975. *Urban Travel Demand: A Behavioral Analysis.* Amsterdam: North-Holland.

Donner, A., and W. W. Hauck. 1986. The large-sample efficiency of the Mantel-Haenszel estimator in the fixed-strata case. *Biometrics* **42:** 537–545.

Doolittle, M. H. 1888. Association ratios. *Bull. Philos. Soc. Washington* **10:** 83–87 and 94–96.

Draper, N. R., and H. Smith. 1981. *Applied Regression Analysis*, 2nd ed. New York: John Wiley.

Drost, F. C., W. C. M. Kallenberg, D. S. Moore, and J. Oosterhoff. 1989. Power approximations to multinomial tests of fit. *J. Amer. Statist. Assoc.* **84:** 130–141.

Ducharme, G. R., and Y. Lepage. 1986. Testing collapsibility in contingency tables. *J. Roy. Statist. Soc.* **B48:** 197–205.

Dyke, G. V., and H. D. Patterson. 1952. Analysis of factorial arrangements when the data are proportions. *Biometrics* **8:** 1–12.

Edwardes, M. 1988. A new theory for the analysis of ordered categorical data, and an application with the Mann-Whitney test. Unpublished manuscript.

Edwards, A. W. F. 1963. The measure of association in a 2×2 table. *J. Roy. Statist. Soc.* **A126:** 109–114.

Efron, B. 1975. The efficiency of logistic regression compared to normal discriminant analysis. *J. Amer. Statist. Assoc.* **70:** 892–898.

Efron, B. 1978. Regression and ANOVA with zero-one data: measures of residual variation. *J. Amer. Statist. Assoc.* **73:** 113–121.

Efron, B., and C. Morris. 1977. Stein's paradox in statistics. *Sci. Amer.* **236:** 119–127.

Ekholm, A., and J. Palmgren. 1987. Correction for misclassification using doubly sampled data. *J. Official Statist.* **3:** 419–429.

El-Khorazaty, M. N., P. B. Imrey, G. G. Koch, and H. B. Wells. 1977. Estimating the total number of events with data from multiple-record systems: a review of methodological strategies. *Internat. Statist. Rev.* **45:** 129–157.

Elliott, G. C. 1988. Interpreting higher order interactions in log-linear analysis. *Psychol. Bull.* **103:** 121–130.

Escoufier, Y. 1982. L'analyse des tableaux de contingence simples et multiples. in *Proc. Internat. Meeting on the Analysis of Multidimensional Contingency Tables* (Rome, 1981), ed. by R. Coppi. *Metron* **40:** 53–77.

Espeland, M. A., and C. L. Odoroff. 1985. Log-linear models for doubly sampled categorical data fitted by the EM algorithm. *J. Amer. Statist. Assoc.* **80:** 663–670.

Farewell, V. T. 1979. Some results on the estimation of logistic models based on retrospective data. *Biometrika* **66:** 27–32.

Farewell, V. T. 1982. A note on regression analysis of ordinal data with variability of classification. *Biometrika* **69:** 533–538.

Fay, R. 1985. A jackknifed chi-squared test for complex samples. *J. Amer. Statist. Assoc.* **80:** 148–157.

Fay, R. 1986. Causal models for patterns of nonresponse. *J. Amer. Statist. Assoc.* **81:** 354–365.

Fay, R., and L. A. Goodman. 1975. ECTA program: Description for users. Chicago: Department of Statistics, University of Chicago.

Ferguson, T. S. 1967. *Mathematical Statistics: A Decision Theoretic Approach.* New York: Academic Press.

Fienberg, S. E. 1970a. An iterative procedure for estimation in contingency tables. *Ann. Math. Statist.* **41:** 907–917.

Fienberg, S. E. 1970b. Quasi-independence and maximum likelihood estimation in incomplete contingency tables. *J. Amer. Statist. Assoc.* **65:** 1610–1616.

Fienberg, S. E. 1972a. The analysis of incomplete multi-way contingency tables. *Biometrics* **28:** 177–202 (correction **29:** 829).

Fienberg, S. E. 1972b. The multiple recapture census for closed populations and incomplete 2^k contingency tables. *Biometrika* **59:** 591–603.

Fienberg, S. E. 1979. The use of chi-squared statistics for categorical data problems. *J. Roy Statist. Soc.* **B41:** 54–64.

Fienberg, S. E. 1980a. *The Analysis of Cross-Classified Categorical Data*, 2nd edn. Cambridge, MA: MIT Press.

Fienberg, S. E. 1980b. Fisher's contributions to the analysis of categorical data. pp. 75–84 in *R. A. Fisher: An Appreciation*, ed. by S. E. Fienberg and D. V. Hinkley. Berlin: Springer-Verlag.

Fienberg, S. E. 1984. The contributions of William Cochran to categorical data analysis. pp. 103–118 in *W. G. Cochran's Impact on Statistics*, ed. by P. S. R. S. Rao and J. Sedransk. New York: John Wiley.

Fienberg, S. E., and P. W. Holland. 1970. Methods for eliminating zero counts in contingency tables. pp. 233–260 in *Random Counts on Models and Structures*, ed. by G. P. Patil. University Park, PA: Pennsylvania State University Press

Fienberg, S. E., and P. W. Holland. 1973. Simultaneous estimation of multinomial cell probabilities. *J. Amer. Statist. Assoc.* **68:** 683–690.

Fienberg, S. E., and K. Larntz. 1976. Loglinear representation for paired and multiple comparison models. *Biometrika* **63:** 245–254.

Fienberg, S. E., and W. M. Mason. 1979. Identification and estimation of age-period-cohort models in the analysis of discrete archival data. *Sociol. Method.* 1–67.

Fienberg, S. E., and M. Meyer. 1983. Iterative proportional fitting. pp. 275–279 in *Encyclopedia of Statistical Sciences*, Vol. 4. New York: John Wiley.

Finney, D. J. 1971. *Probit Analysis*, 3rd edn. Cambridge: Cambridge University Press.

Firth, D. 1987. On the efficiency of quasi-likelihood estimation. *Biometrika* **74**: 233–245.

Firth, D. 1989. Marginal homogeneity and the superposition of Latin squares. *Biometrika* **76**: 179–182.

Firth, D., and B. R. Treat. 1988. Square contingency tables and GLIM. *GLIM Newsletter* No. 16, 16–20.

Fisher, R. A. 1922. On the interpretation of chi-square from contingency tables, and the calculation of P. *J. Roy. Statist. Soc.* **85**: 87–94.

Fisher, R. A. 1924. The conditions under which chi-square measures the descrepancy between observation and hypothesis. *J. Roy. Statist. Soc.* **87**: 442–450.

Fisher, R. A. 1934, 1970. *Statistical Methods for Research Workers.* (originally published 1925, 14th edn. 1970) Edinburgh: Oliver and Boyd.

Fisher, R. A. 1935a. *The Design of Experiments.* (8th edn., 1966) Edinburgh: Oliver and Boyd.

Fisher, R. A. 1935b. Appendix to article by C. Bliss. *Ann. Appl. Biol.* **22**: 164–165.

Fisher, R. A. 1935c. The logic of inductive inference (with discussion). *J. Roy. Statist. Soc.* **98**: 39–82.

Fisher, R. A. 1956. *Statistical Methods for Scientific Inference.* Edinburgh: Oliver and Boyd.

Fisher, R. A. 1962. Confidence limits for a cross-product ratio. *Austral. J. Statist.* **4**: 41.

Fisher, R. A., and F. Yates. 1938. *Statistical Tables.* Edinburgh: Oliver and Boyd.

Fitzpatrick, S., and A. Scott. 1987. Quick simultaneous confidence intervals for multinomial proportions. *J. Amer. Statist. Assoc.* **82**: 875–878.

Fix, E., J. L. Hodges, and E. L. Lehmann. 1959. The restricted chi-square test. in *Probability and Statistics: The Harald Cramér Volume.* ed. by U. Grenander. New York: John Wiley.

Fleiss, J. L. 1975. Measuring agreement between two judges on the presence or absence of a trait. *Biometrics* **31**: 651–659.

Fleiss, J. L. 1981. *Statistical Methods for Rates and Proportions*, 2nd edn. New York: John Wiley.

Fleiss, J. L., and J. Cohen. 1973. The equivalence of weighted kappa and the intraclass correlation coefficient as measures of reliability. *Educ. Psychol. Meas.* **33**: 613–619.

Fleiss, J. L., and B. S. Everitt. 1971. Comparing the marginal totals of square contingency tables. *British J. Math. Statist. Psychol.* **24**: 117–123.

Fleiss, J. L., J. Cohen, and B. S. Everitt. 1969. Large-sample standard errors of kappa and weighted kappa. *Psychol. Bull.* **72**: 323–327.

Follman, D. A., and D. Lambert. 1989. Generalizing logistic regression by nonparametric mixing. *J. Amer. Statist. Assoc.* **84**: 295–300.

Forthofer, R. N., and R. G. Lehnen. 1981. *Public Program Analysis, A New Categorical Data Approach.* Belmont, CA: Lifetime Learning Publications.

Fowlkes, E. B. 1987. Some diagnostics for binary logistic regression via smoothing. *Biometrika* **74**: 503–515.

Fowlkes, E. B., A. E. Freeny, and J. Landwehr. 1988. Evaluating logistic models for large contingency tables. *J. Amer. Statist. Assoc.* **83:** 611– 622.

Francom, S. F., C. Chuang-Stein, and J. R. Landis. 1989. A log-linear model for ordinal data to characterize differential change among treatments. *Statist. Med.* **8:** 571–582.

Freedman, D., R. Pisani, and R. Purves. 1978. *Statistics.* New York: Norton.

Freeman, D. H., Jr. and T. R. Holford. 1980. Summary rates. *Biometrics* **36:** 195–205.

Freeman, G. H., and J. H. Halton. 1951. Note on an exact treatment of contingency, goodness of fit and other problems of significance. *Biometrika* **38:** 141–149.

Freeman, M. F., and J. W. Tukey. 1950. Transformations related to the angular and the square root. *Ann. Math. Statist.* **21:** 607–611.

Fuchs, C. 1982. Maximum likelihood estimation and model selection in contingency tables with missing data. *J. Amer. Statist. Assoc.* **77:** 270–278.

Gabriel, K. R. 1966. Simultaneous test procedures for multiple comparisons on categorical data. *J. Amer. Statist. Assoc.* **61:** 1081–1096.

Gabriel, K. R. 1971. The biplot graphic display of matrices with applications to principal component analysis. *Biometrika* **58:** 453–467.

Gaddum, J. H. 1933. Reports on biological standards. III. Methods of biological assay depending on a quantal response. *Spec. Rep. Ser. Med. Res. Counc., London,* no. 183.

Gail, M. H. 1978. The analysis of heterogeneity for indirect standardized mortality ratios. *J. Roy. Statist. Soc.* **A141:** 224–234.

Gail, M. H., and J. J. Gart. 1973. The determination of sample sizes for use with the exact conditional test in 2×2 comparative trials. *Biometrics* **29:** 441–448.

Gail, M. H., and N. Mantel. 1977. Counting the number of $r \times c$ contingency tables with fixed margins. *J. Amer. Statist. Assoc.* **72:** 859–862.

Gallagher III, B. J., B. J. Jones, and L. P. Barakat. 1987. The attitudes of psychiatrists toward etiological theories of schizophrenia. *J. Clin. Psychol.* **43:** 438–443.

Gart, J. J. 1962. Approximate confidence limits for the relative risk. *J. Roy. Statist. Soc.* **B24:** 454–463.

Gart, J. J. 1969. An exact test for comparing matched proportions in crossover designs. *Biometrika* **56:** 75–80.

Gart, J. J. 1970. Point and interval estimation of the common odds ratio in the combination of 2×2 tables with fixed margins. *Biometrika* **57:** 471–475.

Gart, J. J. 1971. The comparison of proportions: a review of significance tests, confidence intervals and adjustments for stratification. *Rev. Int. Statist. Rev.* **39:** 148–169.

Gart, J. J., and J. Nam. 1988. Approximate interval estimation of the ratio of binomial parameters: A review and corrections for skewness. *Biometrics* **44:** 323–338.

Gart, J. J., and D. G. Thomas. 1972. Numerical results on approximate confidence limits for the odds ratio. *J. Roy. Statist. Soc.* **B34:** 441–447.

Gart, J. J., and J. R. Zweifel. 1967. On the bias of various estimators of the logit and its variance with applications to quantal bioassay. *Biometrika* **54:** 181–187.

Gart, J. J., H. M. Pettigrew, and D. G. Thomas. 1985. The effect of bias, variance estimation, skewness and kurtosis of the empirical logit on weighted least squares analyses. *Biometrika* **72:** 179–190.

Genter, F. C., and V. T. Farewell. 1985. Goodness-of-link testing in ordinal regression models. *Canad. J. Statist.* **13:** 37–44.

Gilbert, G. N. 1981. *Modelling Society*. London: George Allen & Unwin.

Gilula, Z. 1983. Latent conditional independence in two-way contingency tables: a diagnostic approach. *British. J. Math. Statist. Psychol.* **36**: 114–122.

Gilula, Z. 1984. On some similarities between canonical correlation models and latent class models for two-way contingency tables. *Biometrika* **71**: 523–529.

Gilula, Z. 1986. Grouping and association in two-way contingency tables: A canonical correlation analytic approach. *J. Amer. Statist. Assoc.* **81**: 773–779.

Gilula, Z., and S. Haberman. 1986. Canonical analysis of contingency tables by maximum likelihood. *J. Amer. Statist. Assoc.* **81**: 780–788.

Gilula, Z. and S. Haberman. 1988. The analysis of multivariate contingency tables by restricted canonical and restricted association models. *J. Amer. Statist. Assoc.* **83**: 760–771.

Gilula, Z., A. Krieger, and Y. Ritov. 1988. Ordinal association in contingency tables: some interpretive aspects. *J. Amer. Statist. Assoc.* **83**: 540–545.

Gimotty, P., and M. B. Brown. 1987. The effect of imputed values on the distribution of the goodness-of-fit chi-square statistic. *Comput. Statist. Data Anal.* **5**: 201–213.

Glass, D. V., ed. 1954. *Social Mobility in Britain*. Glencoe, IL: Free Press.

Gleser, L. J., and D. S. Moore. 1985. The effect of positive dependence on chi-squared tests for categorical data. *J. Roy. Statist. Soc.* **47**: 459–465.

GLIM Newsletters (1979–) Oxford: Numerical Algorithms Group.

Glonek, G., J. N. Darroch, and T. P. Speed. 1988. On the existence of maximum likelihood estimators for hierarchical loglinear models. *Scand. J. Statist.* **15**: 187–193.

Godambe, V. P. 1980. On sufficiency and ancillarity in the presence of a nuisance parameter. *Biometrika* **67**: 155–162.

Gokhale, D. V., and S. Kullback. 1978. *The Information in Contingency Tables*. New York: Marcel Dekker.

Goldberg, J. D. 1975. The effects of misclassification on the bias in the difference between two proportions and the relative risk in the fourfold table. *J. Amer. Statist. Assoc.* **70**: 561–567.

Good, I. J. 1963. Maximum entropy for hypothesis formulation, especially for multidimensional contingency tables. *Ann. Math. Statist.* **34**: 911–934.

Good, I. J. 1965. *The Estimation of Probabilities: An Essay on Modern Bayesian Methods*. Cambridge, MA: MIT Press.

Good, I. J. 1976. On the application of symmetric Dirichlet distributions and their mixtures to contingency tables. *Ann. Statist.* **4**: 1159–1189.

Good, I. J., and R. A. Gaskins. 1971. Nonparametric roughness penalties for probability densities. *Biometrika* **58**: 255–277.

Goodman, L. A. 1962. Statistical methods for analyzing processes of change. *Amer. J. Sociol.* **68**: 57–78.

Goodman, L. A. 1964a. Simultaneous confidence intervals for contrasts among multinomial populations. *Ann. Math. Statist.* **35**: 716–725.

Goodman, L. A. 1964b. Simultaneous confidence intervals for cross-product ratios in contingency tables. *J. Roy. Statist. Soc.* **B26**: 86–102.

Goodman, L. A. 1964c. Interactions in multi-dimensional contingency tables. *Ann. Math. Statist.* **35**: 632–646.

Goodman, L. A. 1965. On simultaneous confidence intervals for multinomial proportions. *Technometrics* **7:** 247–254.

Goodman, L. A. 1968. The analysis of cross-classified data: independence, quasi-independence, and interactions in contingency tables with or without missing entries. *J. Amer. Statist. Assoc.* **63:** 1091–1131.

Goodman, L. A. 1969. On partitioning chi-square and detecting partial association in three-way contingency tables. *J. Roy. Statist. Soc.* **B31:** 486–498.

Goodman, L. A. 1970. The multivariate analysis of qualitative data: interaction among multiple classifications. *J. Amer. Statist. Assoc.* **65:** 226–256.

Goodman, L. A. 1971a. The analysis of multidimensional contingency tables: stepwise procedures and direct estimation methods for building models for multiple classifications. *Technometrics* **13:** 33–61.

Goodman, L. A. 1971b. The partitioning of chi-square, the analysis of marginal contingency tables, and the estimation of expected frequencies in multidimensional contingency tables. *J. Amer. Statist. Assoc.* **66:** 339–344.

Goodman, L. A. 1972. Some multiplicative models for the analysis of cross-classified data. pp. 649–696 in *Proceedings of the 6th Berkeley Symposium*, ed. by L. Le Cam et al., vol. 1. Berkeley: University of California Press.

Goodman, L.A. 1973a. The analysis of multidimensional contingency tables with some variables are posterior to others: A modified path analysis approach. *Biometrika* **60:** 179–192.

Goodman, L. A. 1973b. Causal analysis of data from panel studies and other kinds of surveys. *Amer. J. Sociol.* **78:** 1135–1191.

Goodman, L. A. 1974. Exploratory latent structure analysis using both identifiable and unidentifiable models. *Biometrika* **61:** 215–231.

Goodman, L. A. 1979a. Simple models for the analysis of association in cross-classifications having ordered categories. *J. Amer. Statist. Assoc.* **74:** 537–552.

Goodman, L. A. 1979b. Multiplicative models for square contingency tables with ordered categories. *Biometrika* **66:** 413–418.

Goodman, L. A. 1981a. Association models and canonical correlation in the analysis of cross-classifications having ordered categories. *J. Amer. Statist. Assoc.* **76:** 320–334.

Goodman, L. A. 1981b. Association models and the bivariate normal for contingency tables with ordered categories. *Biometrika* **68:** 347–355.

Goodman, L. A. 1983. The analysis of dependence in cross-classifications having ordered categories, using log-linear models for frequencies and log-linear models for odds. *Biometrics* **39:** 149–160.

Goodman, L. A. 1985. The analysis of cross-classified data having ordered and/or unordered categories: Association models, correlation models, and asymmetry models for contingency tables with or without missing entries. *Ann. Statist.* **13:** 10–69.

Goodman, L. A. 1986. Some useful extensions of the usual correspondence analysis approach and the usual log-linear models approach in the analysis of contingency tables. *Internat. Statist. Rev.* **54:** 243–309.

Goodman, L. A. 1987. New methods for analyzing the intrinsic character of qualitative variables using cross-classified data. *Amer. J. Sociol.* **93:** 529–583.

Goodman, L. A., and W. H. Kruskal. 1979. *Measures of Association for Cross Classifications*. New York: Springer-Verlag (contains articles appearing in *J. Amer. Statist. Assoc.* in 1954, 1959, 1963, 1972)

Graubard, B. I., and E. L. Korn. 1987. Choice of column scores for testing independence in ordered 2 × K contingency tables. *Biometrics* **43:** 471–476.

Green, P. J. 1984. Iteratively weighted least squares for maximum likelihood estimation and some robust and resistant alternatives (with discussion). *J. Roy. Statist. Soc.* **B46:** 149–192.

Greenacre, M. J. 1981. Practical correspondence analysis. pp. 119–146 in *Interpreting Multivariate Data*, ed. V. Barnett. Chichester: John Wiley.

Greenacre, M. J. 1984. *Theory and Applications of Correspondence Analysis.* New York: Academic Press.

Greenacre, M. J., and T. Hastie. 1987. The geometric interpretation of correspondence analysis. *J. Amer. Statist. Assoc.* **82:** 437–447.

Greenland, S. 1989. Generalized Mantel-Haenszel estimators for K 2 × J tables. *Biometrics* **45:** 183–191.

Grizzle, J. E. 1967. Continuity correction in the chi-square test for 2 × 2 tables. *Amer. Statist.* **21:** 28–32.

Grizzle, J. E., C. F. Starmer, and G. G. Koch. 1969. Analysis of categorical data by linear models. *Biometrics* **25:** 489–504.

Gross, S. T. 1981. On asymptotic power and efficiency of tests of independence in contingency tables with ordered classifications. *J. Amer. Statist. Assoc.* **76:** 935–941.

Gurland, J., I. Lee, and P. A. Dahm. 1960. Polychotomous quantal response in biological assay. *Biometrics* **16:** 382–398.

Haber, M. 1980. A comparison of some continuity corrections for the chi-squared test on 2 × 2 tables. *J. Amer. Statist. Assoc.* **75:** 510–515.

Haber, M. 1982. The continuity correction and statistical testing. *Internat. Statist. Rev.* **50:** 135–144.

Haber, M. 1985. Maximum likelihood methods for linear and log-linear models in categorical data. *Comput. Statist. Data Anal.* **3:** 1–10.

Haber, M. 1986. An exact unconditional test for the 2 × 2 comparative trial. *Psychol. Bull.* **99:** 129–132.

Haber, M. 1987. A comparison of some conditional and unconditional exact tests for 2 by 2 contingency tables. *Comm. Statist.* **B16:** 999–1013.

Haber, M. 1989. Do the marginal totals of a 2 × 2 contingency table contain information regarding the table proportions? *Comm. Statist.* **A18:** 147–156.

Haber, M., and M. B. Brown. 1986. Maximum likelihood methods for loglinear models when expected frequencies are subject to linear constraints. *J. Amer. Statist. Assoc.* **81:** 477–482.

Haberman, S. J. 1970. The General Log-linear Model. Ph. D. dissertation, University of Chicago.

Haberman, S. J. 1973a. The analysis of residuals in cross-classification tables. *Biometrics* **29:** 205–220.

Haberman, S. J. 1973b. Log-linear models for frequency data: Sufficient statistics and likelihood equations. *Ann. Statist.* **1:** 617–632.

Haberman, S. J. 1974a. *The Analysis of Frequency Data.* Chicago: University of Chicago Press.

Haberman, S. J. 1974b. Log-linear models for frequency tables with ordered classifications. *Biometrics* **36:** 589–600.

Haberman, S. J. 1977a. Log-linear models and frequency tables with small expected cell counts. *Ann. Statist.* **5**: 1148–1169.

Haberman, S. J. 1977b. Maximum likelihood estimation in exponential response models. *Ann. Statist.* **5**: 815–841.

Haberman, S. J. 1978, 1979. *Analysis of Qualitative Data*, vols. 1 and 2. New York: Academic Press.

Haberman, S. J. 1981. Tests for independence in two-way contingency tables based on canonical correlation and on linear-by-linear interaction. *Ann. Statist.* **9**: 1178–1186.

Haberman, S. J. 1982a. The analysis of dispersion of multinomial responses. *J. Amer. Statist. Assoc.* **77**: 568–580.

Haberman, S. J. 1982b. Measures of association. pp. 130–137 in *Encyclopedia of Statistical Sciences*, vol. 1. New York: John Wiley.

Haberman, S. J. 1988. A warning on the use of chi-squared statistics with frequency tables with small expected cell counts. *J. Amer. Statist. Assoc.* **83**: 555–560.

Haldane, J. B. S. 1940. The mean and variance of χ^2, when used as a test of homogeneity, when expectations are small. *Biometrika* **31**: 346–355.

Haldane, J. B. S. 1955. The estimation and significance of the logarithm of a ratio of frequencies. *Ann. Human Genet.* **20**: 309–311.

Hamada, M., and C. F. J. Wu. 1990. A critical look at accumulation analysis and related methods. *Technometrics* **32**: to appear.

Harkness, W. L., and L. Katz. 1964. Comparison of the power functions for the test of independence in 2×2 contingency tables. *Ann. Math. Statist.* **35**: 1115–1127.

Hastie, T., and R. Tibshirani. 1987a. Generalized additive models: Some applications. *J. Amer. Statist. Assoc.* **82**: 371–386.

Hastie, T., and R. Tibshirani. 1987b. Non-parametric logistic and proportional odds regression. *Appl. Statist.* **36**: 260–276.

Hauck, W. W. 1989. Odds ratio inference from stratified samples. *Comm. Statist.* **A18**: 767–800.

Hauck, W. W., and S. Anderson. 1986. A comparison of large-sample confidence interval methods for the difference of two binomial probabilities. *Amer. Statist.* **40**: 318–322.

Hauck, W. W., and A. Donner. 1977. Wald's test as applied to hypotheses in logit analysis. *J. Amer. Statist. Assoc.* **72**: 851–853.

Hauck, W. W., and A. Donner. 1988. The asymptotic relative efficiency of the Mantel-Haenszel estimator in the increasing-number-of-strata case. *Biometrics* **44**: 379–384.

Haynam, G. E., Z. Govindarajulu, and F. C. Leone. 1970. Tables of the cumulative non-central chi-square distribution, in *Selected Tables in Mathematical Statistics*, ed. by H. L. Harter and D. B. Owen. Chicago: Markham.

Healy, M. J. R. 1988. *GLIM: An Introduction*. Oxford: Clarendon Press

Hedlund, R. D. 1978. Cross-over voting in a 1976 presidential primary. *Public Opinion Quart.* **41**: 498–514.

Helmes, E., and G. C. Fekken. 1986. Effects of psychotropic drugs and psychiatric illness on vocational aptitude and interest assessment. *J. Clin. Psychol.* **42**: 569–576.

Hennekens, C. H., and J. E. During. 1987. *Epidemiology in Medicine*. Boston: Little, Brown.

Henry, N. W. 1983. Latent structure analysis. pp. 497–504 in *Encyclopedia of Statistical Sciences*, vol. 4. New York: John Wiley.

Hill, M. O. 1974. Correspondence analysis: a neglected multivariate method. *Appl. Statist.* 23: 340–354.

Hirji, K. F., C. R. Mehta, and N. R. Patel. 1987. Computing distributions for exact logistic regression. *J. Amer. Statist. Assoc.* 82: 1110–1117.

Hirotsu, C. 1982. Use of cumulative efficient scores for testing ordered alternatives in discrete models. *Biometrika* 69: 567–577.

Hirschfeld, H. O. 1935. A connection between correlation and contingency. *Cambridge Philos. Soc. Proc.* (*Math. Proc.*) 31: 520–524.

Hochberg, Y. 1977. On the use of double sampling schemes in analysing categorical data with misclassification error. *J. Amer. Statist. Assoc.* 72: 914–921.

Hocking, R. R., and H. H. Oxspring. 1971. Maximum likelihood estimation with incomplete multinomial data. *J. Amer. Statist. Assoc.* 66: 65–70.

Hocking, R. R., and H. H. Oxspring. 1974. The analysis of partially categorized contingency data. *Biometrics* 30: 469–483.

Hodges, J. L., Jr. 1958. Fitting the logistic by maximum likelihood. *Biometrics* 14: 453–461.

Hoem, J. M. 1987. Statistical analysis of a multiplicative model and its application to the standardization of vital rates: A review. *Internat. Statist. Rev.* 5: 119–152.

Holford, T. R. 1976. Life tables with concomitant information. *Biometrics* 32: 587–597.

Holford, T. R. 1978. The analysis of pair-matched case-control studies, a multivariate approach. *Biometrics* 34: 665–672.

Holford, T. R. 1980. The analysis of rates and of survivorship using log-linear models. *Biometrics* 36: 299–305.

Holmes, M. C. and R. E. O. Williams. 1954. The distribution of carriers of *Streptococcus pyogenes* among 2413 healthy children. *J. Hyg.* (*Cambridge*) 52: 165–179.

Holmquist, N. S., C. A. McMahon, and O. D. Williams. 1967. Variability in classification of carcinoma in situ of the uterine cervix. *Arch. Pathol.* 84: 334–345.

Holt, D., A. J. Scott, and P. D. Ewings. 1980. Chi-squared tests with survey data. *J. Roy. Statist. Soc.* A143: 303–320.

Hosmane, B. 1987. An empirical investigation of chi-square tests for the hypothesis of no three-factor interaction in $I \times J \times K$ contingency tables. *J. Statist. Comput. Simulation.* 28: 167–178.

Hosmer, D. W., and S. Lemeshow. 1980. A goodness-of-fit test for the multiple logistic regression model. *Comm. Statist.* A10: 1043–1069.

Hosmer, D. W., and S. Lemeshow. 1989. *Applied Logistic Regression.* New York: John Wiley.

Hout, M., O. D. Duncan, and M. E. Sobel. 1987. Association and heterogeneity: Structural models of similarities and differences. *Sociol. Method.* 17: 145–184.

Hutchison, D. 1984. Ordinal variable regression using the McCullagh (proportional odds) model. *GLIM Newsletter* 9: 9–17.

Hutchinson, T. P. 1979. The validity of the chi-square test when the expected frequencies are small: a list of recent research references. *Comm. Statist.* A8: 327–335.

Ighodaro, A., and T. Santner. 1982. Ridge type estimators of multinomial cell probabilities, in S. Gupta and J. O. Berger (Eds.) *Statistical Decision Theory and Related Topics III*, Vol. 2. New York: Academic Press.

Imrey, P. B. 1985. SAS software for log-linear models. in *Proc. 10th Annual SAS Users Group International Conference.* Cary, NC: SAS Institute Inc.

Imrey, P. B., W. D. Johnson, and G. G. Koch. 1976. An incomplete contingency table approach to paired-comparison experiments. *J. Amer. Statist. Assoc.* **71**: 614–623.

Imrey, P. B., G. G. Koch, and M. E. Stokes. 1981. Categorical data analysis: Some reflections on the log linear model and logistic regression. Part I: Historical and methodological overview. *Internat. Statist. Rev.* **49**: 265–283.

Imrey, P. B., G. G. Koch, and M. E. Stokes. 1982. Categorical data analysis: Some reflections on the log linear model and logistic regression. Part II: Data analysis. *Internat. Statist. Rev.* **50**: 35–63.

Ireland, C. T., and S. Kullback. 1968a. Minimum discrimination information estimation. *Biometrics* **24**: 707–713.

Ireland, C. T., and S. Kullback. 1968b. Contingency tables with given marginals. *Biometrika* **55**: 179–188.

Ireland, C. T., H. H. Ku, and S. Kullback. 1969. Symmetry and marginal homogeneity of an r × r contingency table. *J. Amer. Statist. Assoc.* **64**: 1323–1341.

Irwin, J. O. 1935. Tests of significance for differences between percentages based on small numbers. *Metron* **12**: 83–94.

Irwin, J. O. 1949. A note on the subdivision of χ^2 into components. *Biometrika* **36**: 130–134.

Iverson, G. R. 1979. Decomposing chi-square: A forgotten technique. *Sociol. Methods Res.* **8**: 143–157.

Jennings, D. E. 1986. Outliers and residual distributions in logistic linear regression. *J. Amer. Statist. Assoc.* **81**: 987–990.

Jennings, M. K. 1987. Residues of a movement: The aging of the American protest generation. *Amer. Political Sci. Rev.* **81**: 367–382.

Jennings, M. K., and R. G. Niemi. 1968. The transmission of political values from parent to child. *Amer. Political Sci. Rev.* **62**: 169–184.

Jennrich, R. I., and R. H. Moore. 1975. Maximum likelihood estimation by means of nonlinear least squares. *Proc. Statistical Computing Section, Amer. Statist. Assoc.* 57–65.

Johnson, B. M. 1971. On the admissible estimators for certain fixed sample binomial problems. *Ann. Math. Statist.* **42**: 1579–1587.

Johnson, W. 1985. Influence measures for logistic regression: another point of view. *Biometrika* **72**: 59–65.

Johnson, W. D., and G. G. Koch. 1978. Linear models analysis of competing risks for grouped survival times. *Internat. Statist. Rev.* **46**: 21–51.

Jolayemi, E. T. and M. B. Brown. 1984. The choice of a log-linear model using a C_p-type statistic. *Comput. Statist. Data Anal.* **2**: 159–165.

Jones, M. P., T. W. O'Gorman, J. H. Lemke, and R. F. Woolson. 1989. A Monte Carlo investigation of homogeneity tests of the odds ratio under various sample size considerations. *Biometrics* **45**: 171–181.

Jørgensen, B. 1983. Maximum likelihood estimation and large-sample inference for generalized linear and nonlinear regression models. *Biometrika* **70**: 19–28.

Jørgensen, B. 1984. The delta algorithm and GLIM. *Internat. Statist. Rev.* **52**: 283–300.

Jørgensen, B. 1986. Some properties of exponential dispersion models. *Scand. J. Statist.* **13**: 187–197.

Jørgensen, B. 1987. Exponential dispersion models (with discussion). *J. Roy. Statist. Soc.* **B49**: 127–162.

Jørgensen, B. 1989. The theory of exponential dispersion models and analysis of deviance. Preliminary version of book on generalized linear models.

Kadane, J. B. 1985. Is victimization chronic? A Bayesian analysis of multinomial missing data. *J. Econometrics* **29:** 47–67.

Kempthorne, O. 1979. In dispraise of the exact test: Reactions. *J. Statist. Plann. Inference* **3:** 199–213.

Kendall, M. G. 1938. A new measure of rank correlation. *Biometrika* **30:** 81–93.

Kendall, M. G. 1945. The treatment of ties in rank problems. *Biometrika* **33:** 239–251.

Kendall, M., and A. Stuart. 1979. *The Advanced Theory of Statistics*, vol 2: Inference and Relationship, 4th edn. New York: Macmillan.

Kenward, M. G., and B. Jones. 1987. A log linear model for binary and cross-over data. *Appl. Statist.* **36:** 192–204.

Killion, R. A., and D. A. Zahn. 1976. A bibliography of contingency table literature. *Internat. Statist. Rev.* **44:** 71–112.

Kleinman, J. C. 1975. Proportions with extraneous variance: two independent samples. *Biometrics* **31:** 737–743.

Knuiman, M. W., and T. P. Speed. 1988. Incorporating prior information into the analysis of contingency tables. *Biometrics* **44:** 1061–1071.

Koch, G. G., and V. P. Bhapkar. 1982. Chi-square tests. pp. 442–457 in *Encyclopedia of Statistical Sciences*, vol. 1. New York: John Wiley.

Koch, G. G., and S. Edwards. 1985. Logistic regression. pp. 128–133 in *Encyclopedia of Statistical Sciences*, vol. 5. New York: John Wiley.

Koch, G. G., and D. W. Reinfurt. 1971. The analysis of categorical data from mixed models. *Biometrics* **27:** 157–173.

Koch, G. G., S. S. Atkinson, and M. E. Stokes. 1986. Poisson regression. pp. 32–41 in *Encyclopedia of Statistical Sciences*, vol. 7. New York: John Wiley.

Koch, G. G., P. B. Imrey, and D. W. Reinfurt. 1972. Linear model analysis of categorical data with incomplete response vectors. *Biometrics* **28:** 663–692.

Koch, G. G., D. H. Freeman, and J. L. Freeman. 1975. Strategies in the multivariate analysis of data from complex surveys. *Internat. Statist. Rev.* **43:** 59–78.

Koch, G. G., J. R. Landis, J. L. Freeman, D. H. Freeman, and R. G. Lehnen. 1977. A general methodology for the analysis of experiments with repeated measurement of categorical data. *Biometrics* **33:** 133–158.

Koch, G. G., I. A. Amara, G. W. Davis, and D. B. Gillings. 1982. A review of some statistical methods for covariance analysis of categorical data. *Biometrics* **38:** 563–595.

Koch, G. G., P. B. Imrey, J. M. Singer. S. S. Atkinson, and M. E. Stokes. 1985. *Lecture Notes for Analysis of Categorical Data*. Montreal: Les Presses de L'Université de Montréal.

Koehler, K. 1986. Goodness-of-fit tests for log-linear models in sparse contingency tables. *J. Amer. Statist. Assoc.* **81:** 483–493.

Koehler, K., and K. Larntz. 1980. An empirical investigation of goodness-of-fit statistics for sparse multinomials. *J. Amer. Statist. Assoc.* **75:** 336–344.

Koehler, K., and J. Wilson. 1986. Chi-square tests for comparing vectors of proportions for several cluster samples. *Comm. Statist.* **A15:** 2977–2990.

Koopman, P. A. R. 1984. Confidence limits for the ratio of two binomial proportions. *Biometrics* **40:** 513–517.

Kraemer, H. C. 1979. Ramifications of a population model for κ as a coefficient of reliability. *Psychometrika*. **44:** 461–472.

Kraemer, H. C. 1983. Kappa coefficient. pp. 352–354 in *Encyclopedia of Statistical Sciences*, vol. 4. New York: John Wiley.

Kraemer, H. C., and S. Thiemann. 1987. *How Many Subjects*. Beverly Hills: Sage Publications.

Kruskal, W. H. 1958. Ordinal measures of association. *J. Amer. Statist. Assoc.* **53:** 814–861.

Ku, H. H., and S. Kullback. 1974. Loglinear models in contingency table analysis. *Amer. Statist.* **28:** 115–122.

Ku, H. H., R. N. Varner, and S. Kullback. 1971. Analysis of multidimensional contingency tables. *J. Amer. Statist. Assoc.* **66:** 55–64.

Kullback, S. 1959. *Information Theory and Statistics*. New York: John Wiley.

Kullback, S., M. Kupperman, and H. H. Ku. 1962. Tests for contingency tables and Markov chains. *Technometrics* **4:** 573–608.

Kupper, L. L., and J. K. Haseman. 1978. The use of a correlated binomial model for the analysis of certain toxicological experiments. *Biometrics* **34:** 69–76.

Kupper, L. L., C. Portier, M. D. Hogan, and E. Yamamoto. 1986. The impact of litter effects on dose-response modeling in teratology. *Biometrics* **42:** 85–98.

Kuritz, S. J., J. R. Landis, and G. G. Koch. 1988. A general overview of Mantel–Haenszel methods: Applications and recent developments. *Ann. Rev. Public Health* **9:** 123–160.

Läärä, E., and J. N. S. Matthews. 1985. The equivalence of two models for ordinal data. *Biometrika* **72:** 206–207.

Lagakos, S. W. 1979. General right censoring and its impact on the analysis of survival data. *Biometrics* **35:** 139–156.

Laird, N. M. 1978. Empirical Bayes methods for two-way contingency tables. *Biometrika* **65:** 581–590.

Laird, N. M., and D. Olivier. 1981. Covariance analysis of censored survival data using log-linear analysis techniques. *J. Amer. Statist. Assoc.* **76:** 231–240.

Lancaster, H. O. 1949. The derivation and partition of χ^2 in certain discrete distributions. *Biometrika* **36:** 117–129.

Lancaster, H. O. 1951. Complex contingency tables treated by partition of χ^2. *J. Roy. Statist. Soc.* **B13:** 242–249.

Lancaster, H. O. 1961. Significance tests in discrete distributions. *J. Amer. Statist. Assoc.* **56:** 223–234.

Lancaster, H. O., and M. A. Hamdan. 1964. Estimation of the correlation coefficient in contingency tables with possible nonmetrical characters. *Psychometrika* **29:** 383–391.

Lancaster, H. O. 1969. *The Chi-Squared Distribution*. New York: John Wiley.

Landis, J. R., and G. G. Koch. 1975. A review of statistical methods in the analysis of data arising from observer reliability studies, Parts I, II. *Statist. Neerlandica* **29:** 101–123, 151–161.

Landis, J. R., and G. G. Koch. 1977a. The measurement of observer agreement for categorical data. *Biometrics* **33:** 159–174.

Landis, J. R., and G. G. Koch. 1977b. An application of hierarchical kappa-type statistics in the assessment of majority agreement among multiple observers. *Biometrics* **33:** 363–374.

Landis, J. R., E. R. Heyman, and G. G. Koch. 1978. Average partial association in three-way contingency tables: A review and discussion of alternative tests. *Internat. Statist. Rev*, **46**: 237–254.

Landis, J. R., W. M. Stanish, J. L. Freeman, and G. G. Koch. 1976. A computer program for the generalized chi-square analysis of categorical data using least squares. *Comput. Prog. Biomed.* **6**: 196–231.

Landis, J. R., J. M. Lepkowski, C. S. Davis, and M. Miller. 1987. Cumulative logit models for weighted data from complex sample surveys. *Proc. Soc. Statist. Section, Amer. Statist. Assoc.*, 165–170.

Landwehr, J. M., D. Pregibon, and A. C. Shoemaker. 1984. Graphical methods for assessing logistic regression models. *J. Amer. Statist. Assoc.* **79**: 61–71.

Larntz, K. 1978. Small-sample comparison of exact levels for chi-squared goodness-of-fit statistics. *J. Amer. Statist. Assoc.* **73**: 253–263.

Larson, M. G. 1984. Covariate analysis of competing-risks data with log-linear models. *Biometrics* **40**: 459–469.

Lawal, H. B. 1984. Comparisons of the X^2, Y^2, Freeman-Tukey and Williams improved G^2 test statistics in small samples of one-way multinomials. *Biometrika* **71**: 415–418.

Lawal, H. B., and G. J. G. Upton. 1980. An approximation to the distribution of the X^2 goodness-of-fit statistic for use with small expectations. *Biometrika* **67**: 447–453.

Lawal, H. B., and G. J. G. Upton. 1984. On the use of X^2 as a test of independence in contingency tables with small cell expectations. *Austral. J. Statist.* **26**: 75–85.

Lazarsfeld, P. F., and N. W. Henry. 1968. *Latent Structure Analysis*. Boston: Houghton Mifflin.

Lebart, L., A. Morineau, and K. M. Warwick. 1984. *Multivariate Descriptive Statistics: Correspondence Analysis and Related Techniques for Large Matrices*, trans. E. M. Berry. New York: John Wiley.

Lee, E. T. 1974. A computer program for linear logistic regression analysis. *Computer Prog. Biomed.* **4**: 80–92.

Lee, S. K. 1977. On the asymptotic variances of $\hat{u}$ terms in loglinear models of multidimensional contingency tables. *J. Amer. Statist. Assoc.* **72**: 412–419.

Lefkopoulou, M., D. Moore, and L. Ryan. 1989. The analysis of multiple correlated binary outcomes: application to rodent teratology experiments. *J. Amer. Statist. Assoc.* **84**: 810–815.

Lehmann, E. L. 1966. Some concepts of dependence. *Ann. Math. Statist.* **37**: 1137–1153.

Leonard, T. 1975. Bayesian estimation methods for two-way contingency tables. *J. Roy. Statist. Soc.* **B37**: 23–37.

Lesaffre, E., and A. Albert. 1989. Multiple-group logistic regression diagnostics. *Appl. Statist.* **38**: 425–440.

Lewis, B. N. 1962. On the analysis of interaction in multidimensional contingency tables. *J. Roy. Statist. Soc.* **A125**: 88–117.

Liang, K. Y. 1984. The asymptotic efficiency of conditional likelihood methods. *Biometrika* **71**: 305–313.

Liang, K. Y., and S. G. Self. 1985. Tests for homogeneity of odds ratios when the data are sparse. *Biometrika* **72**: 353–358.

Liang, K. Y., and S. L. Zeger. 1986. Longitudinal data analysis using generalized linear models. *Biometrika* **73**: 13–22.

Liang, K. Y., and S. L. Zeger. 1988. On the use of concordant pairs in matched case-control studies. *Biometrics* **44**: 1145–1156.

Light, R. J., and B. H. Margolin. 1971. An analysis of variance for categorical data. *J. Amer. Statist. Assoc.* **66**: 534–544.

Lilienfeld, A. M., and D. E. Lilienfeld. 1980. *Foundations of Epidemiology*, 2nd edn. Oxford: Oxford University Press.

Lindley, D. V. 1964. The Bayesian analysis of contingency tables. *Ann. Math. Statist.* **35**: 1622–1643.

Lindley, D. V., and A. Smith. 1972. Bayesian estimates for the linear model (with discussion). *J. Roy. Statist. Soc.* **B34**: 1–42.

Lindsay, B., C. C. Clogg, and J. Grego. 1989. Semi-parametric estimation in the Rasch model and in some related exponential response models, including a simple latent class model for item analysis. Unpublished manuscript.

Lindsey, J. K. 1989. *The Analysis of Categorical Data Using GLIM*. New York: Springer-Verlag.

Lipsitz, S. 1988. Methods for analyzing repeated categorical outcomes. Unpublished PhD Dissertation, Department of Biostatistics, Harvard University.

Little, R. J. 1989. Testing the equality of two independent binomial proportions. *Amer. Statist.* **43**: 283–288.

Little, R. J., and D. B. Rubin. 1987. *Statistical Analysis with Missing Data*. New York: John Wiley.

Lloyd, C. J. 1988. Some issues arising from the analysis of 2×2 contingency tables. *Austral. J. Statist.* **30**: 35–46.

Luce, R. D. 1959. *Individual Choice Behavior*. New York: John Wiley.

Madansky, A. 1963. Tests of homogeneity for correlated samples. *J. Amer. Statist. Assoc.* **58**: 97–119.

Maddala, G. S. 1983. *Limited-Dependent and Qualitative Variables in Econometrics*. Cambridge: Cambridge University Press.

Madsen, M. 1976. Statistical analysis of multiple contingency tables: Two examples. *Scand. J. Statist.* **3**: 97–106.

Magidson, J. 1982. Qualitative variance, entropy, and correlation ratios for nominal dependent variables. *Social Sci. Res.* **10**: 177–194.

Mann, J. I., M. P. Vessey, M. Thorogood, and R. Doll. 1975. Myocardial infarction in young women with special reference to oral contraceptive practice. *British J. Med.* **2**: 241–245.

Mantel, N. 1963. Chi-square tests with one degree of freedom: extensions of the Mantel-Haenszel procedure. *J. Amer. Statist. Assoc.* **58**: 690–700.

Mantel, N. 1966. Models for complex contingency tables and polychotomous dosage response curves. *Biometrics* **22**: 83–95.

Mantel, N. 1985. Maximum likelihood vs. minimum chi-square. *Biometrics* **41**: 777–781.

Mantel, N., and D. P. Byar. 1978. Marginal homogeneity, symmetry and independence. *Comm. Statist.* **A7**: 953–976.

Mantel, N., and W. Haenszel. 1959. Statistical aspects of the analysis of data from retrospective studies of disease. *J. Natl. Cancer Inst.* **22**: 719–748.

Mantel, N., and B. F. Hankey. 1978. A logistic regression analysis of response time data where the hazard function is time dependent. *Comm. Statist.* **A7**: 333–347.

McCullagh, P. 1977. A logistic model for paired comparisons with ordered categorical data. *Biometrika* **64**: 449–453.

McCullagh, P. 1978. A class of parametric models for the analysis of square contingency tables with ordered categories. *Biometrika* **65**: 413–418.

McCullagh, P. 1980. Regression models for ordinal data (with discussion). *J. Roy. Statist. Soc.* **B42**: 109–142.

McCullagh, P. 1982. Some applications of quasisymmetry. *Biometrika* **69**: 303–308.

McCullagh, P. 1983. Quasi-likelihood functions. *Ann. Statist.* **11**: 59–67.

McCullagh, P. 1984. Generalized linear models. *Europ. J. Oper. Res.* **16**: 285–292.

McCullagh, P. 1985. Quasi-likelihood functions. pp. 464–467 in *Encyclopedia of Statistical Sciences*, vol. 7. New York: John Wiley.

McCullagh, P. 1986. The conditional distribution of goodness-of-fit statistics for discrete data. *J. Amer. Statist. Assoc.* **81**: 104–107.

McCullagh, P., and J. A. Nelder. 1983, 2nd edn. 1989. *Generalized Linear Models*. London: Chapman and Hall.

McCutcheon, A. L. 1987. *Latent Class Analysis*. Beverly Hills: Sage Publications.

McFadden, D. 1974. Conditional logit analysis of qualitative choice behavior. pp. 105–142 in *Frontiers in Econometrics*, ed. by P. Zarembka. New York: Academic Press.

McFadden, D. 1981. Econometric models of probabilistic choice. pp. 198–272 in *Structural Analysis of Discrete Data with Econometric Applications*, ed. by C. F. Manski and D. McFadden. Cambridge, MA: MIT Press.

McFadden, D. 1982. Qualitative response models. pp. 1–37 in *Advances in Econometrics*, ed. by W. Hildebrand. Cambridge: Cambridge University Press.

McFadden, D. 1984. Econometric analysis of qualitative response models. pp. 1395–1457 in *Handbook of Econometrics*, vol. II, ed. by Z. Griliches and M. D. Intriligator. Amsterdam: North-Holland.

McNemar, Q. 1947. Note on the sampling error of the difference between correlated proportions or percentages. *Psychometrika* **12**: 153–157.

Meeks, S. L., and R. B. D'Agostino. 1983. A model for comparisons with ordered categorical data. *Comm. Statist.* **A12**: 895–906.

Mehta, C. R., and N. R. Patel. 1983. A network algorithm for performing Fisher's exact test in r × c contingency tables. *J. Amer. Statist. Assoc.* **78**: 427–434.

Mehta, C. R., N. R. Patel, and A. A. Tsiatis. 1984. Exact significance testing to establish treatment equivalence with ordered categorical data. *Biometrics* **40**: 819–825.

Mehta, C. R., N. R. Patel, and R. Gray. 1985. Computing an exact confidence interval for the common odds ratio in several 2 by 2 contingency tables. *J. Amer. Statist. Assoc.* **80**: 969–973.

Mehta, C. R., N. R. Patel, and P. Senchaudhuri. 1988. Importance sampling for estimating exact probabilities in permutational inference. *J. Amer. Statist. Assoc.* **83**: 999–1005.

Mendenhall, W. M., R. R. Million, D. E. Sharkey, and N. J. Cassis. 1984. Stage T3 squamous cell carcinoma of the glottic larynx treated with surgery and/or radiation therapy. *Internat. J. Radiat. Oncol. Biol. Phys.* **10**: 357–363.

Miettinen, O. S., 1969. Individual matching with multiple controls in the case of all-or-none responses. *Biometrics* **25**: 339–355.

Miettinen, O. S., and M. Nurminen. 1985. Comparative analysis of two rates. *Statist. Med.* **4**: 213–226.

Miller, R. G. 1981. *Simultaneous Statistical Inference.* 2nd edn. New York: Springer-Verlag.

Minkin, S. 1987. On optimal design for binary data. *J. Amer. Statist. Assoc.* **82**: 1098–1103.

Mitchell, T. J., and B. W. Turnbull. 1979. Log-linear models in the analysis of disease prevalence data from survival/sacrifice experiments. *Biometrics* **35**: 221–234.

Mitra, S. K. 1958. On the limiting power function of the frequency chi-square test. *Ann. Statist.* **29**: 1221–1233.

Moore, D. F. 1986. Asymptotic properties of moment estimates for overdispersed counts and proportions. *Biometrika* **73**: 583–588.

Moore, D. S. 1977. Generalized inverses, Wald's method, and the construction of chi-squared tests of fit. *J. Amer. Statist. Assoc.* **72**: 131–137.

Moore, D. S. 1986. Tests of chi-squared type. pp. 63–95 in *Goodness-of-Fit Techniques*, ed. by R. D'Agostino and M. A. Stephens. New York: Marcel Dekker.

Morgan, S. P., and J. D. Teachman. 1988. Logistic regression: description, examples, and comparisons. *J. Marriage Family* **50**: 929–936.

Morris, C. N. 1975. Central limit theorems for multinomial sums. *Ann. Statist.* **3**: 165–188.

Morris, C. N. 1983. Parametric empirical Bayes inference: Theory and applications. *J. Amer. Statist. Assoc.* **78**: 47–55.

Morrison, D. 1972. Upper bounds for correlations between binary outcomes and probabilistic predictions. *J. Amer. Statist. Assoc.* **67**: 68–70.

Moses, L. E., J. D. Emerson, and H. Hosseini. 1984. Analyzing data from ordered categories. *N. Engl. J. Med.* **311**: 442–448.

Mosteller, F. 1952. Some statistical problems in measuring the subjective response to drugs. *Biometrics* **8**: 220–226.

Mosteller, F. 1968. Association and estimation in contingency tables. *J. Amer. Statist. Assoc.* **63**: 1–28.

Mote, N. L., and R. L. Anderson. 1965. An investigation of the effect of misclassification on the properties of χ^2-tests in the analysis of categorical data. *Biometrika* **52**: 95–109.

Mukerjee, R. 1987. On zero cells in log-linear models. *Sankhya*, Ser. B, **49**: 97–102.

Mullins, E. J., and P. Sites. 1984. The origins of contemporary eminent black Americans: A three-generation analysis of social origin. *Amer. Sociol. Rev*, **49**: 672–685.

Nair, V. N. 1986. Testing in industrial experiments with ordered categorical data (with discussion). *Technometrics* **28**: 283–311.

Nair, V. N. 1987. Chi-squared-type tests for ordered alternatives in contingency tables. *J. Amer. Statist. Assoc.* **82**: 283–291.

Nelder, J., and D. Pregibon. 1987. An extended quasi-likelihood function. *Biometrika* **74**: 221–232.

Nelder, J., and R. W. M. Wedderburn. 1972. Generalized linear models. *J. Roy. Statist. Soc.* **A135**: 370–384.

Nerlove, M., and S. J. Press. 1973. Univariate and multivariate log-linear and logistic models. Rand Corp. Technical Report R-1306–EDA/NIH, Santa Monica, CA.

Neyman, J. 1949. Contributions to the theory of the χ^2 test. pp. 239–273 in *Proceedings of the First Berkeley Symposium on Mathematical Statistics and Probability*, ed. by J. Neyman. Berkeley: University of California Press.

Norušis, M. J. 1988. *SPSS^X Advanced Statistics Guide*, 2nd edn. New York: McGraw-Hill.

Numerical Algorithms Group 1986. *The GLIM System Release 3.77 Manual.* Downers Grove, IL: Numerical Algorithms Group.

O'Brien, R. G. 1986. Using the SAS system to perform power analyses for log-linear models. pp. 778–784 in *Proc. Eleventh Annual SAS Users Group Conference*. Cary, NC: SAS Institute Inc.

O'Gorman, T. W., and R. F. Woolson. 1988. Analysis of ordered categorical data using the SAS system. pp. 957–963 in *Proc. Thirteenth Annual SAS Users Group Conference*. Cary, NC: SAS Institute Inc.

Pack, S. E. 1986. Hypothesis testing for proportions with over-dispersion. *Biometrics* **42**: 967–972.

Pagano, M., and Halvorsen, K. T. 1981. An algorithm for finding the exact significance levels of $r \times c$ contingency tables. *J. Amer. Statist. Assoc.* **76**: 931–934.

Paik, M. 1985. A graphic representation of a three-way contingency table: Simpson's paradox and correlation. *Amer. Statist.* **39**: 53–54.

Palmgren, J. 1981. The Fisher information matrix for log-linear models arguing conditionally in the observed explanatory variables. *Biometrika* **68**: 563–566.

Palmgren, J. 1987. Precision of double sampling estimators for comparing two probabilities. *Biometrika* **74**: 687–694.

Palmgren, J., and A. Ekholm. 1987. Exponential family non-linear models for categorical data with errors of observation. *Appl. Stochastic Models Data Anal.* **3**: 111–124.

Parr, W. C., and H. D. Tolley. 1982. Jackknifing in categorical data analysis. *Austral. J. Statist.* **24**: 67–79.

Patefield, W. M. 1982. Exact tests for trends in ordered contingency tables. *Appl. Statist.* **31**: 32–43.

Patnaik, P. B. 1949. The non-central χ^2 and F-distributions and their applications. *Biometrika* **36**: 202–232.

Paul, S. R. 1985. Estimation of the parameters of a clumped binomial model via the EM algorithm. *Amer. Statist.* **39**: 136–139.

Paul, S. R., and R. L. Plackett. 1978. Inference sensitivity for Poisson mixtures. *Biometrika* **65**: 591–602.

Paul, S. R., K. Y. Liang, and S. G. Self. 1989. On testing departure from the binomial and multinomial assumptions. *Biometrics* **45**: 231–236.

Pearson, E. S. 1947. The choice of a statistical test illustrated on the interpretation of data classified in 2×2 tables. *Biometrika* **34**: 139–167.

Pearson, K. 1900. On a criterion that a given system of deviations from the probable in the case of a correlated system of variables is such that it can be reasonably supposed to have arisen from random sampling. *Philos. Mag.*, Series 5, **50**: 157–175. (Reprinted 1948 in *Karl Pearson's Early Statistical Papers*, ed. by E. S. Pearson, Cambridge: Cambridge University Press.)

Pearson, K. 1904. Mathematical contributions to the theory of evolution XIII: On the theory of contingency and its relation to association and normal correlation. *Draper's Co. Research Memoirs. Biometric Series*, no. 1. (Reprinted 1948 in *Karl Pearson's Early Papers*, ed. by E. S. Pearson, Cambridge: Cambridge University Press.)

Pearson, K. 1913. On the probable error of a correlation coefficient as found from a fourfold table. *Biometrika* **9**: 22–27.

Pearson, K. 1922. On the χ^2 test of goodness of fit. *Biometrika* **14**: 186–191.

Pearson, K., and D. Heron. 1913. On theories of association. *Biometrika* **9**: 159–315.

Pettitt, A. N. 1984. Proportional odds models for survival data and estimates using ranks. *Appl. Statist.* **33**: 169–175.

Pierce, D. A., and D. W. Schafer. 1986. Residuals in generalized linear models. *J. Amer. Statist. Assoc.* **81:** 977–983.

Plackett, R. L. 1962. A note on interactions in contingency tables. *J. Roy. Statist. Soc.* **B24:** 162–166.

Plackett, R. L. 1964. The continuity correction in 2×2 tables. *Biometrika* **51:** 327–337.

Plackett, R. L. 1965. A class of bivariate distributions. *J. Amer. Statist. Assoc.* **60:** 516–522.

Plackett, R. L. 1981. *The Analysis of Categorical Data*, 2nd edn. London: Griffin.

Plackett, R. L., and S. R. Paul. 1978. Dirichlet models for square contingency tables. *Comm. Statist.* **A7:** 939–952.

Pratt, J. W. 1981. Concavity of the log likelihood. *J. Amer. Statist. Assoc.* **76:** 103–106.

Pregibon, D. 1980. Goodness of link tests for generalized linear models. *Appl. Statist.* **29:** 15–24.

Pregibon, D. 1981. Logistic regression diagnostics. *Ann. Statist.* **9:** 705–724.

Pregibon, D. 1982a. Resistant fits of some commonly used logistic models with medical applications. *Biometrics* **38:** 485–498.

Pregibon, D. 1982b. Score tests in GLIM with applications. pp. 87–97 in *Lecture Notes in Statistics, 14: GLIM 82, Proceedings of the International Conference on Generalised Linear Models*, ed. by R. Gilchrist. New York: Springer-Verlag.

Prentice, R. 1976. Use of the logistic model in retrospective studies. *Biometrics* **32:** 599–606.

Prentice, R. 1986. Binary regression using an extended beta-binomial distribution, with discussion of correlation induced by covariate measurement errors. *J. Amer. Statist. Assoc.* **81:** 321–327.

Prentice, R., and L. A. Gloeckler. 1978. Regression analysis of grouped survival data with application to breast cancer data. *Biometrics* **34:** 57–67.

Prentice, R., and R. Pyke. 1979. Logistic disease incidence models and case-control studies. *Biometrika* **66:** 403–412.

Press, S. J., and S. Wilson. 1978. Choosing between logistic regression and discriminant analysis. *J. Amer. Statist. Assoc.* **73:** 699–705.

Price, C. J., C. A. Kimmel, J. D. George, and M. C. Marr. 1987. The developmental toxicity of diethylene glycol dimethyl ether in mice. *Fund. Appl. Toxicol.* **8:** 115–126.

Quine, M. P., and E. Seneta. 1987. Bortkiewicz's data and the law of small numbers. *Internat. Statist. Rev.* **5:** 173–181.

Radelet, M. 1981. Racial characteristics and the imposition of the death penalty. *Amer. Sociol. Rev.* **46:** 918–927.

Raftery, A. E. 1986. Choosing models for cross-classifications. *Amer. Sociol. Rev.* **51:** 145–146.

Rao, C. R. 1955. Theory of the method of estimation by minimum chi-square. *Bull. Inst. Internat. Statist.* **35:** 25–32.

Rao, C. R. 1957. Maximum likelihood estimation for the multinomial distribution. *Sankhya* **18:** 139–148.

Rao, C. R. 1958. Maximum likelihood estimation for the multinomial distribution with an infinite number of cells. *Sankhya* **20:** 211–218.

Rao, C. R. 1963. Criteria of estimation in large samples. *Sankhya* **25:** 189–206.

Rao, C. R. 1973. *Linear Statistical Inference and its Applications*, 2nd edn. New York: Wiley.

Rao, C. R. 1982. Diversity: its measurement, decomposition, apportionment and analysis. *Sankhya* **A44:** 1–22.

Rao, J. N. K., and A. J. Scott. 1981. The analysis of categorical data from complex sample surveys: chi-squared tests for goodness-of-fit and independence in two-way tables. *J. Amer. Statist. Assoc.* **76:** 221–230.

Rao, J. N. K., and A. J. Scott. 1987. On simple adjustments to chi-square tests with sample survey data. *Ann. Statist.* **15:** 385–397.

Rao, J. N. K., and D. R. Thomas. 1988. The analysis of cross-classified categorical data from complex sample surveys. *Sociol. Method.* **18:** 213–270.

Rasch, G. 1961. On general laws and the meaning of measurement in psychology. pp. 321–333 in *Proc. 4th Berkeley Symp. Math. Statist. Probab.*, vol. 4, ed. J. Neyman. Berkeley: University of California Press.

Read, T. R. C., and N. A. C. Cressie. 1988. *Goodness-of-Fit Statistics for Discrete Multivariate Data.* New York: Springer-Verlag.

Rice, W. R. 1988. A new probability model for determining exact P-values for 2×2 contingency tables when comparing binomial proportions. *Biometrics* **44:** 1–22.

Robbins, H. 1955. An empirical Bayes approach to statistics. pp. 157–164 in *Proc. 3rd Berkeley Symp. Math. Statist. Probab.*, vol. 1, ed. by J. Neyman. Berkeley, CA: University of California Press.

Robins, J., N. Breslow, and S. Greenland. 1986. Estimators of the Mantel–Haenszel variance consistent in both sparse data and large-strata limiting models. *Biometrics* **42:** 311–323.

Rochon, J. 1989. The application of the GSK method to the determination of minimum sample sizes. *Biometrics* **45:** 193–205.

Rosenbaum, P. R., and D. R. Rubin. 1983. The central role of the propensity score in observational studies for causal effects. *Biometrika* **70:** 41–55.

Rosenthal, I. 1966. Distribution of the sample version of the measure of association, gamma. *J. Amer. Statist. Assoc.* **61:** 440–453.

Rosner, B. 1989. Multivariate methods for clustered binary data with more than one level of nesting. *J. Amer. Statist. Assoc.* **84:** 373–380.

Roy, S. N., and M. A. Kastenbaum. 1956. On the hypothesis of no "interaction" in a multiway contingency table. *Ann. Math. Statist.* **27:** 749–757.

Roy, S. N., and S. K. Mitra. 1956. An introduction to some nonparametric generalizations of analysis of variance and multivariate analysis. *Biometrika* **43,** 361–376.

Rubin, D. 1976. Inference and missing data. *Biometrika* **63:** 581–592.

Ryan, L. 1989. Dose response models for developmental toxicity. Unpublished manuscript.

Santner, T. J., and M. K. Snell. 1980. Small-sample confidence intervals for $p_1 - p_2$ and p_1/p_2 in 2×2 contingency tables. *J. Amer. Statist. Assoc.* **75:** 386–394.

SAS Institute Inc. 1986. *SUGI Supplemental Library User's Guide*, Version 5 ed. Cary, NC: SAS Institute, Inc.

SAS Institute Inc. 1987. *SAS/STAT Guide for Personal Computers*, Version 6. Cary, NC: SAS Institute, Inc.

Schluchter, M. D, and K. L. Jackson. 1989. Log-linear analysis of censored survival data with partially observed covariates. *J. Amer. Statist. Assoc.* **84:** 42–52.

Schmidt, P., and R. P. Strauss. 1975. The prediction of occupation using multiple logit models. *Internat. Econ. Rev.* **16:** 471–486.

Schwarz, G. 1978. Estimating the dimension of a model. *Ann. Statist.* **6:** 461–464.

Semenya, K., G. G. Koch, M. E. Stokes, and R. N. Forthofer. 1983. Linear models methods for some rank function analyses of ordinal categorical data. *Comm. Statist.* **A12:** 1277–1298.

Sekar, C. C., and W. E. Deming. 1949. On a method of estimating birth and death rates and the extent of registration. *J. Amer. Statist. Assoc.* **44:** 101–115.

Serfling, R. 1980. *Approximation Theorems in Statistics.* New York: John Wiley.

Sewell, W. H., and A. M. Orenstein. 1965. Community of residence and occupational choice. *Amer. J. Sociol.* **70:** 551–563.

Shapiro, S., D. Slone, L. Rosenberg, D. Kaufman, P. D. Stolley, and O. S. Miettinen. 1979. Oral contraceptive use in relation to myocardial infarction. *Lancet, i:* 743–746.

Shapiro, S. H. 1982. Collapsing contingency tables—a geometric approach. *Amer. Statist.* **36:** 43–46.

Shih, W. J. 1987. Maximum likelihood estimation and likelihood ratio test for square tables with missing data. *Statist. Med.* **6:** 91–97.

Shuster, J., and D. Downing. 1976. Two-way contingency tables for complex sampling schemes. *Biometrika* **63:** 271–276.

Simon, G. 1973. Additivity of information in exponential family probability laws. *J. Amer. Statist. Assoc.* **68:** 478–482.

Simon, G. 1974. Alternative analyses for the singly-ordered contingency table. *J. Amer. Statist. Assoc.* **69:** 971–976.

Simon, G. 1978. Efficacies of measures of association for ordinal contingency tables. *J. Amer. Statist. Assoc.* **73:** 545–551.

Simonoff, J. 1983. A penalty function approach to smoothing large sparse contingency tables. *Ann. Statist.* **11:** 208–218.

Simonoff, J. 1986. Jackknifing and bootstrapping goodness-of-fit statistics in sparse multinomials. *J. Amer. Statist. Assoc.* **81:** 1005–1111.

Simonoff, J. 1987. Probability estimation via smoothing in sparse contingency tables with ordered categories. *Statist. Probab. Lett.* **5:** 55–63.

Simonoff, J. 1988. Detecting outlying cells in two-way contingency tables via backwards-stepping. *Technometrics* **30:** 339–345.

Simpson, E. H. 1949. The measurement of diversity. *Nature* **163:** 688.

Simpson, E. H. 1951. The interpretation of interaction in contingency tables. *J. Roy. Statist. Soc.* **B13:** 238–241.

Small, K. A. 1987. A discrete choice model for ordered alternatives. *Econometrica* **55:** 409–424.

Small, K. A. 1988. Discrete choice econometrics (Book reviews). *J. Math. Psychol.* **32:** 80–87.

Smith, K. W. 1976. Table standardization and table shrinking: Aids in the traditional analysis of contingency tables. *Social Forces* **54:** 669–693.

Snapinn, S. M., and R. D. Small. 1986. Tests of significance using regression models for ordered categorical data. *Biometrics* **42:** 583–592.

Snedecor, G. W. 1937. *Statistical Methods.* Ames, Iowa: Iowa State Press.

Snell, E. J. 1964. A scaling procedure for ordered categorical data. *Biometrics* **20:** 592–607.

Somers, R. H. 1962. A new asymmetric measure of association for ordinal variables. *Amer. Sociol. Rev.* **27:** 799–811.

Somes, G. W. 1982. Cochran's Q statistic. pp. 24–26 in *Encyclopedia of the Statistical Sciences*, vol. 2. New York: John Wiley.

Somes, G. W. 1986. The generalized Mantel-Haenszel statistic. *Amer. Statist.* **40:** 106–108.

Spiegelhalter, D. J., and A. F. M. Smith. 1982. Bayes factors for linear and log-linear models with vague prior information. *J. Roy. Statist. Soc.* **B44**: 377–387.

Spitzer, R. L., J. Cohen, J. L. Fleiss, and J. Endicott. 1967. Quantification of agreement in psychiatric diagnosis. *Arch. Gen. Psychiatry* **17**: 83–87.

Srole, L., T. S. Langner, S. T. Michael, P. Kirkpatrick, M. K. Opler, and T. A. C. Rennie. 1978. *Mental Health in the Metropolis: The Midtown Manhattan Study*, rev. edn. New York: NYU Press.

Stanish, W. M. 1986. Categorical data analysis strategies using SAS software. pp. 239–256 in *Computer Science and Statistics: The Interface*, ed. by D. M. Allen. New York: Elsevier Science Publishers (North-Holland)

Stanish, W. M., D. B. Gillings, and G. G. Koch. 1978. An application of multivariate ratio methods for the analysis of a longitudinal clinical trial with missing data. *Biometrics* **34**: 305–317.

Stevens, S. S. 1951. Mathematics, measurement, and psychophysics. pp. 1–49 in *Handbook of Experimental Psychology*, ed. by S. S. Stevens. New York: John Wiley.

Stigler, S. 1986. *The History of Statistics: The Measurement of Uncertainty Before 1900*. Cambridge, MA: Harvard University Press.

Stiratelli, R., N. Laird, and J. H. Ware. 1984. Random-effects models for serial observations with binary response. *Biometrics* **40**: 1025–1035.

Stram, D. O., L. J. Wei, and J. H. Ware. 1988. Analysis of repeated ordered categorical outcomes with possibly missing observations and time-dependent covariates. *J. Amer. Statist. Assoc.* **83**: 631–637.

Stuart, A. 1955. A test for homogeneity of the marginal distributions in a two-way classification. *Biometrika* **42**: 412–416.

Stukel, T. A. 1988. Generalized logistic models. *J. Amer. Statist. Assoc.* **83**: 426–431.

Sugiura, N. and M. Otake. 1974. An extension of the Mantel-Haenszel procedure to K $2 \times C$ contingency tables and the relation to the logit model. *Comm. Statist.* **A3**: 829–842.

Suissa, S. and J. Adam. 1987. Generalized linear regression for discrete data using SAS. *Amer. Statist.* **41**: 241.

Suissa, S., and J. J. Shuster. 1984. Are uniformly most powerful unbiased tests really best? *Amer. Statist.* **38**: 204–206.

Suissa, S. and J. J. Shuster. 1985. Exact unconditional samples sizes for the 2 by 2 binomial trial. *J. Roy. Statist. Soc.* **A148**: 317–327.

Sundberg, R. 1975. Some results about decomposable (or Markov-type) models for multidimensional contingency tables: Distribution of marginals and partitioning of tests. *Scand. J. Statist.* **2**: 71–79.

Tanner, M. A., and M. A. Young. 1985. Modelling agreement among raters. *J. Amer. Statist. Assoc.* **80**: 175–180.

Tarone, R. E., and J. J. Gart. 1980. On the robustness of combined tests for trends in proportions. *J. Amer. Statist. Assoc.* **75**: 110–116.

Tarone, R. E., J. J. Gart, and W. W. Hauck. 1983. On the asymptotic relative efficiency of certain noniterative estimators of a common relative risk or odds ratio. *Biometrika* **70**: 519–522.

Tavaré, S. 1983. Serial dependence in contingency tables. *J. Roy. Statist. Soc.* **B45**: 100–106.

Tavaré, S., and P. M. E. Altham. 1983. Serial dependence of observations leading to contingency tables, and corrections to chi-squared statistics. *Biometrika* **70**: 139–144.

Tenenbein, A. 1970. A double sampling scheme for estimating from binomial data with misclassifications. *J. Amer. Statist. Assoc.* **65:** 1350–1361.

Theil, H. 1969. A multinomial extension of the linear logit model. *Internat. Econ. Rev.* **10:** 251–259.

Theil, H. 1970. On the estimation of relationships involving qualitative variables. *Amer. J. Sociol.* **76:** 103–154.

Thomas, D. G. 1971. Exact confidence limits for the odds ratio in a 2×2 table. *Appl. Statist.* **20:** 105–110.

Thomas, D. G. 1975 Exact and asymptotic methods for the combination of 2×2 tables. *Comput. Biomed. Res.* **8:** 423–446.

Thompson, R., and R. J. Baker. 1981. Composite link functions in generalised linear models. *Appl. Statist.* **30:** 125–131.

Thompson, W. A. 1977. On the treatment of grouped observations in life studies. *Biometrics* **33:** 463–470.

Thornes, B., and J. Collard. 1979. *Who Divorces?* London: Routledge & Kegan Paul.

Titterington, D. M. 1980. A comparative study of kernel-based density estimates for categorical data. *Technometrics* **22:** 259–268.

Titterington, D. M., and A. W. Bowman. 1985. A comparative study of smoothing procedures for ordered categorical data. *J. Statist. Comput. Simul.* **21:** 291–312.

Tocher, K. D. 1950. Extension of the Neyman-Pearson theory of tests to discontinuous variates. *Biometrika* **37:** 130–144.

Tostesen, A. N. A., and C. B. Begg. 1988. A general regression methodology for ROC curve estimation. *Medical Decision Making* **8:** 204–215.

Train, K. 1986. *Qualitative Choice Analysis: Theory, Econometrics, and an Application.* Cambridge, MA: MIT Press.

Tsiatis, A. A. 1980. A note on the goodness-of-fit test for the logistic regression model. *Biometrika* **67:** 250–251.

Tsutakawa, R. K. 1980. Selection of dose levels for estimating a percentage point of a logistic quantal response curve. *Appl. Statist.* **29:** 25–33.

Upton, G. J. G. 1982. A comparison of alternative tests for the 2×2 comparative trial. *J. Roy. Statist. Soc.* **A145:** 86–105.

U.S. Bureau of the Census. 1984. *Statistical Abstract of the United States: 1985*, 105th edn. Washington, DC.

van der Heijden, P. G. M., and J. de Leeuw. 1985. Correspondence analysis: a complement to log-linear analysis. *Psychometrika* **50:** 429–447.

van der Heijden, P. G. M., A. de Falguerolles, and J. de Leeuw. 1989. A combined approach to contingency table analysis using correspondence analysis and log-linear analysis (with discussion). *Applied Statistics* **38:** 249–292.

Wald, A. 1943. Tests of statistical hypotheses concerning several parameters when the number of observations is large. *Trans. Amer. Math. Soc.* **54:** 426–482.

Walker, S. H., and D. B. Duncan. 1967. Estimation of the probability of an event as a function of several independent variables. *Biometrika* **54:** 167–179.

Ware, J. H., S. Lipsitz, and F. E. Speizer. 1988. Issues in the analysis of repeated categorical outcomes. *Statist. Med.* **7:** 95–107.

Wedderburn, R. W. M. 1974a. Quasi-likelihood functions, generalised linear models, and the Gauss-Newton method. *Biometrika* **61:** 439–447.

Wedderburn, R. W. M. 1974b. Generalised linear models specified in terms of constraints. *J. R. Statist. Soc.* **B36:** 449–454.

Wedderburn, R. W. M. 1976. On the existence and uniqueness of the maximum likelihood estimates for certain generalized linear models. *Biometrika* **63:** 27–32.

Weisberg, H. I. 1972. Bayesian comparison of two ordered multinomial populations. *Biometrics* **28:** 859–867.

Wermuth, N. 1976a. Model search among multiplicative models. *Biometrics* **32:** 253–263.

Wermuth, N. 1976b. Exploratory analyses of multidimensional contingency tables. pp. 279–295 in *Proc. 9th Internat. Biometrics. Conf.*, vol. 1. The Biometric Society.

Wermuth, N. 1987. Parametric collapsibility and the lack of moderating effects in contingency tables with a dichotomous response variable. *J. Roy. Statist. Soc.* **B49:** 353–364.

White, A. A., J. R. Landis, and M. M. Cooper. 1982. A note on the equivalence of several marginal homogeneity test criteria for categorical data. *Internat. Statist. Rev.* **50:** 27–34.

Whitehead, J. 1980. Fitting Cox's regression model to survival data using GLIM. *Appl. Statist.* **29:** 268–275.

Whittaker, J., and M. Aitkin. 1978. A flexible strategy for fitting complex log-linear models. *Biometrics* **34:** 487–495.

Whittemore, A. S. 1978. Collapsibility of multidimensional contingency tables. *J. Roy. Statist. Soc.* **B40:** 328–340.

Wilks, S. S. 1935. The likelihood test of independence in contingency tables. *Ann. Math. Statist.* **6:** 190–196.

Wilks, S. S. 1938. The large-sample distribution of the likelihood ratio for testing composite hypotheses. *Ann. Math. Statist.* **9:** 60–62.

Williams, D. A. 1982. Extra-binomial variation in logistic linear models. *Appl. Statist.* **31:** 144–148.

Williams, D. A. 1984. Residuals in generalized linear models. pp. 59–68 in *Proc. 12th Internat. Biometrics Conf.* The Biometric Society.

Williams, D. A. 1987. Generalized linear model diagnostics using the deviance and single case deletions. *Appl. Statist.* **36:** 181–191.

Williams, E. J. 1952. Use of scores for the analysis of association in contingency tables. *Biometrika* **39:** 274–289.

Williams, O. D., and J. E. Grizzle. 1972. Analysis of contingency tables having ordered response categories. *J. Amer. Statist. Assoc.* **67:** 55–63.

Wilson, J. R. 1989. Chi-square tests for overdispersion with multiparameter estimates. *Appl. Statist.* **38:** 441–453.

Wood, C. L. 1978. Comparison of linear trends in binomial proportions. *Biometrics* **34:** 496–504.

Woolf, B. 1955. On estimating the relation between blood group and disease. *Ann. Human Genet. (London)* **19:** 251–253.

Woolson, R. F., and W. R. Clarke. 1984. Analysis of categorical incomplete longitudinal data. *J. Roy Statist. Soc.* **A147:** 87–99.

Worsley, K. J. 1987. Un example d'identification d'un modèle log-linéaire grace a une analyse des correspondances. *Rev. Statist. Appl.* **35:** 13–20.

Wrigley, N. 1985. *Categorical Data Analysis for Geographers and Environmental Scientists.* London: Longman.

Wu, C. F. J. 1985. Efficient sequential designs with binary data. *J. Amer. Statist. Assoc.* **80:** 974–984.

Yates, F. 1934. Contingency tables involving small numbers and the χ^2 test. *J. Roy. Statist. Soc. Supplement* **1:** 217–235.

Yates, F. 1948. The analysis of contingency tables with grouping based on quantitative characters. *Biometrika* **35:** 176–181.

Yates, F. 1984. Tests of significance for 2×2 contingency tables. *J. Royal Statist. Soc.* **A147:** 426–463.

Yule, G. U. 1900. On the association of attributes in statistics. *Philos. Trans. Roy. Soc. London* **A194:** 257–319.

Yule, G. U. 1903. Notes on the theory of association of attributes in statistics. *Biometrika* **2:** 121–134.

Yule, G. U. 1906. On a property which holds good for all groupings of a normal distribution of frequency for two variables, with applications to the study of contingency tables for the inheritance of unmeasured qualities. *Proc. Royal Soc.* **A77:** 324–336.

Yule, G. U. 1912. On the methods of measuring association between two attributes (with discussion). *J. Roy. Statist. Soc.* **75:** 579–642.

Zeger, S. L. and B. Qaqish. 1988. Markov regression models for time series: a quasi-likelihood approach. *Biometrics* **44:** 1019–1031.

Zeger, S. L., K. Liang, and S. G. Self. 1985. The analysis of binary longitudinal data with time-independent covariates. *Biometrika* **72:** 31–38.

Zeger, S. L., K.-Y. Liang, and P. S. Albert. 1988. Models for longitudinal data: a generalized estimating equation approach. *Biometrics* **44:** 1049–1060.

Zelen, M. 1971. The analysis of several 2×2 contingency tables. *Biometrika* **58:** 129–137.

Zelen, M. 1972. Exact significance tests for contingency tables embedded in a 2^n classification. pp. 737–757 in *Proc. 6th Berkeley Symp. Math. Statist. Probab.*, vol. 1, ed. by L. LeCam, J. Neyman, and E. L. Scott. Berkeley: University of California Press.

Zellner, A., and T. Lee. 1965. Joint estimation of relationships involving discrete random variables. *Econometrica* **33:** 383–394.

Zelterman, D. 1987. Goodness-of-fit tests for large sparse multinomial distributions. *J. Amer. Statist. Assoc.* **82:** 624–629.

Index of Examples

Index of Selected Notation

Note: Page number in parentheses indicates first appearance in text.

ASE, asymptotic standard error (54)

C, number of concordant pairs (21)

D, number of discordant pairs (21)
Diag, diagonal matrix (114)

e_i, standardized residual (108)

G^2, likelihood-ratio statistic (48)
$G^2(M)$, (95)
$G^2(I)$, (97)
$G^2(M_2|M_1)$, (96)

I, number of rows (8)

J, number of columns (8)

K, number of strata in 3-way table (138)

L, log-likelihood function (40)
$L \times L$, linear-by-linear association model (269)

m_{ij}, expected frequency (41)
ML, maximum likelihood (40)
MSE, mean squared error (183)
M_1, M_2, models, M_2 special case of M_1 (95)

n, total sample size ($= \Sigma\, n_i$ in contingency table) (11)
n_{ij}, n_{ijk}, cell count (11)

N, number of units in generalized linear model, (e.g., number of Poisson cell counts in a contingency table) (80)

p_{ij}, sample cell proportion (11)

u_i, row score (263)

v_j, column score (263)

X^2, Pearson statistic (43)
(X, Y, Z), loglinear model symbols (144)
(XY, XZ, YZ), loglinear model symbols (144)

$z_{\alpha/2}$, standard normal percentage point (55)

α_{ij}, odds ratio (18)

Φ, standard normal cdf (90)

λ, noncentrality (98)
λ_i^X, λ_{ij}^{XY}, λ_{ijk}^{XYZ}, terms in loglinear model (143)

θ, odds ratio (15)
θ_{ij}, local odds ratio (18)
$\theta_{ij(k)}$, conditional odds ratio (145)
θ_{ijk}, ratio of odds ratios (278)

π_{ij}, cell probability (9)
π_{i+}, row probability total (9)
$\pi_{j|i}$, conditional probability in row i (9)

$\chi^2(\alpha)$, chi-squared percentage point (98)

Author Index

Subject Index